The Ecology of Natural Resources

Second Edition

The Ecology of Natural Resources

Second Edition

I. G. Simmons
Professor of Geography, University of Bristol

Edward Arnold

First published 1974
by Edward Arnold (Publishers) Ltd.
41 Bedford Square, London WC1B 3DQ
Reprinted 1975, 1977
Second edition first published 1981

British Library Cataloguing in Publication Data
Simmons, Ian Gordon
The ecology of natural resources 2nd edition
1. Natural resources
I. Title
333 HC55

ISBN 0-7131-6328-3 Pbk

Typeset in Plantin Light by Asco Trade Typesetting Ltd., Hong Kong
Printed and Bound in Great Britain by
Butler & Tanner Ltd., Frome and London

Contents

Preface

There are now so many books dealing with man's relationships with his environment that yet another deserves some apologia for adding to them. My reasons for writing it spring from several years of teaching in this field, during which I found that many of the books I read and recommended to students were either too limited in scope or spatial coverage or too strident in their viewpoint for my taste. This book is the end-product of my reaction to that situation. I hope it will be primarily useful for students reading appropriate courses at second- and third-year levels in British universities, and for upper-division courses in North America; but I would stress that it is intended as an introduction and no more. Each component section has its own greater complexities, and I hope that some readers will use this book as a springboard for deeper studies. Beyond these people, other readers with an interest in the subjects discussed here may find in it some materials upon which to base their views as citizens.

To use the word 'ecology' in the title demands some further explanation, since it is a word whose meaning has latterly become not only elastic but stretched so far that it is unlikely to hold anything up. I take it to mean the study of the relationships of living organisms to each other and their inanimate environment, and I include man as one of those organisms. I have avoided the term 'human ecology' because I think that the focal concept of ecology is a holistic viewpoint which is broken up a little more every time a qualifying adjective is attached, and I prefer to conceive of an 'ecology' in which, from time to time and place to place, one or other of the components may be dominant. In cities it is man, at the North Pole it is nature. But the admission of man means not only a particular species of tool- and material-using beast but a cultural animal also. If, as I believe and as will become apparent as the book progresses, ecology teaches us about the limits imposed by the dynamics and structure of natural systems, then we must also realize that all adjustments within this envelope must be made through the medium of culture. Thus economics, ethology and ethics all have a role to play in bringing about more harmonious interactions between man and nature. The special contribution, if any, of the geographer lies in his interest in (would that I could say understanding of) the points at which both natural and social systems meet.

Trying to cover such a wide field of knowledge and opinion brings two major problems. The first is that important material may be overlooked and, equally important, wrong interpretations may be made of published data and views. Oliver Goldsmith suggested that 'a book may be amusing with numerous errors, or it may be very dull without a single absurdity', and I would rather be classified in the first category: I shall be surprised if those thus offended do not make my mistakes clear to me. A second difficulty is the ageing of material between final draft and publication, especially in many of the areas covered in this book. As a partial attack on this problem, every effort has been made to reduce the period between the submission of the manuscript and the publication of the book.

To acknowledge adequately all the help I have had would require a chapter in itself, but some debts are so great that they must be mentioned here. The basic idea of this type of book was conceived in that fertile womb, the Berkeley campus of the University of California. I am grateful to the American Council of Learned Societies for their award of an American Studies Program Fellowship, which made possible that year of study, fruitful also in fields other than resource studies, and to James Parsons, Chairman during 1964–5, who made me so welcome at Berkeley, then and since. During that year I first met Dan Luten, under whose genial but mind-stretching tutelage I have subsequently always regarded myself, even when the physical distance between us has been considerable. I have also benefited greatly from a Winston Churchill Memorial Trust Travelling Fellowship held in 1971 and 1972, which enabled me to visit North America and Japan to gather valuable information and meet interesting people. During the academic year 1972–3, I was Visiting Professor at York University, Toronto, and as well as paying a handsome salary, that lively institution provided time in which to prepare the final draft. Florence Davies, my secretary at York, typed furiously but accurately, checked references, packed parcels of manuscript and, most important of all, acted as intermediary between me and the Secretarial Services unit who actually produced the final typescript. The Chairman of the Department of Geography, Bill Found, who probably realized his Visiting Professor was more a visitor than a professor, bore this situation with a calm cheerfulness for which I am very grateful.

From my undergraduate days onwards, I have had the benefit of advice from and discussion with Palmer Newbould, currently of the New University of Ulster; much of this took place in contexts where that admirable material, ethyl alcohol, was consumed, and he can always claim that I thus misinterpreted totally what he told me. Since 1962 my membership of the Department of Geography at Durham has been a secure base from which to work and travel, and in particular its Head, Professor W. B. Fisher, has striven to provide conditions conducive to the production of academic work. Early drafts were typed by the grossly overladen secretarial staff of the Department, among whom Suzanne Eckford must be specially thanked. During 1969–70 a particularly lively group of students wrote essays along the themes of this book; none has been directly plagiarized, but some provided interest and stimulus, especially those by Cathy Goulder, Richard de Bastion, John Richardson, John Button, Alastair Steel, Jill Evered and Roger Weatherley.

My wife Carol deserves more than a paragraph to herself, for she has read, commented on, punctuated, corrected and re-read practically every word of this book, more than once while coping with domestic chores as well as this academic drudgery; only I know just how much she has helped.

Finally, this book is dedicated to my parents, Chris and Charles Simmons, as an inadequate, if sincere, expression of filial gratitude.

Preface to the Second Edition

I am firstly grateful to all those who bought copies of the first edition, or recommended it to their students, during the years 1974–80: I am glad that the book has been of use to so many people. I am also pleased to be able to thank collectively all those who made constructive comments in letters and in reviews. With their help some unspotted misprints were removed in the second printing, for example, and they have all helped prepare the ground for this new edition. Now that he has left Edward Arnold, the great encouragement and help that I received from John Davey can be properly acknowledged.

This edition has been brought up to date statistically as far as possible but more importantly it has changed conceptually as well. The most obvious difference is the greatly increased emphasis given to energy, not only in itself as a resource but as a linkeage with many other resource processes. But discerning readers will notice shifts in some of the material on population and on development, for example, and there is a new section on tourism; these aim to take account of new attitudes in these fields. The overhauling that I have given the book has resulted in about half of it being rewritten: I hope this will make it more readable.

I am hopeful too that in addition to retaining its original type of users the book will now appeal to those students whose texts, originally issued in the heady years of 1970–76, did not sell in large enough numbers for them to merit new editions and who now need a more up-to-date book. The number of enrollments in courses of a resources-environment nature has dropped since those years but the need for them is still there and I hope this book will continue to play a useful role.

Parts of this edition were typed by my secretary, Mrs Mary Southcott, and I would like to add her name to those who are to be thanked. The revision was prepared at the University of Bristol and I am grateful to Peter Haggett and my colleagues for not loading me with the quantity of work that would have made its completion unlikely.

I. G. Simmons
Bristol, March 1980

Names of organizations abbreviated in text and bibliography

AAAS	American Association for the Advancement of Science (Washington, DC)
AAG	Association of American Geographers (Washington, DC)
BTA	British Travel Association (London)
FAO	The Food and Agriculture Organization of the United Nations (Rome)
FDA	Food and Drug Administration (Washington, DC)
IBP	International Biological Programme (London)
IUCN	International Unions for the Conservation of Nature and Natural Resources (Gland, Switzerland)
NAS/NRC	National Academy of Sciences/National Research Council (Washington, DC)
OECD	Organization for Economic Co-operation and Development (Paris)
RFF	Resources for the Future Inc. (Washington, DC)
UNESCO	United Nations Educational, Social and Cultural Organization (Paris)
UNEP	United Nations Environmental Programme (Nairobi)

Acknowledgements

The author and publishers gratefully acknowledge permission to reprint or modify copyright material:

Blackwell Scientific Publications for Fig. 1.2*; McGraw-Hill Book Co. for Fig. 1.3; The Open University for Figs. 1.4 and 6.2; W. B. Saunders & Co. for Figs. 1.5, 1.6 and 7.7; Academic Press Inc. for Fig. 1.7; Heidelberg and Springer-Verlag for Fig. 1.8; Macmillan Publishing Co. Inc. for Figs. 1.9 and 1.11; Franz Steiner Verlag and Professor H. Leith for Fig. 1.10; Pion Ltd. for Figs. 1.14 and 11.1; Brookhaven National Library for Fig. 1.16; Wadsworth Publishing Co. for Fig. 1.17; John Wiley and Sons for Figs. 3.1, 4.7, 8.3, and 12.1; Oxford University Press (East Africa) for Fig. 3.2; Belknap Press for Fig. 3.3; US National Parks Service and US Government Printing Office, Washington DC for Fig. 3.4; Her Majesty's Stationery Office for Figs. 3.5 and 11.11; David and Charles Ltd. for Fig. 4.3; Department of Geography, University of Waterloo, Ontario for Fig. 4.6; Methuen and Co. Ltd. for Figs. 5.1 and 5.2; Institute of Civil Engineers for Fig. 5.3; Department of Agriculture, Alberta for Figs. 5.4 and 5.6; *New Scientist* for Fig. 5.5; *Ambio* for Fig. 5.7; Food and Agriculture Organization, Rome for Fig. 6.1; Chapman and Hall Ltd. for Fig. 6.3; W. H. Freeman and *Scientific American* for Figs. 7.1, 8.2, 8.4, 10.1, 11.6, 12.2 and 12.3; Chatto and Windus and Frederick Praeger Inc. for Figs. 7.2 and 7.3 A & B; North Holland Publishing Co. for Figs. 7.4 and 7.5; Centre for Agricultural Strategy, University of Reading for Fig. 7.6; National Academy of Sciences, Washington DC for Figs. 8.1, 10.5, 10.6 and 10.7; Sinauer Associates Inc. for Figs. 8.5 and 8.6; Elsevier, Amsterdam for Fig. 9.1; Ann Arbor Science Publishers Inc. for Fig. 9.2; Blackie & Co., Glasgow for Figs. 10.2 and 10.3; *Bulletin of the Atomic Scientists* for Fig. 10.4; IPC for Fig. 10.8; Publishing Sciences Group for Fig. 10.9; Reidel, Dordecht for Figs. 11.2 and 11.3; Plenum Press for Fig. 11.4; Kodansha and Academic Press for Fig. 11.5; Metcalfe and Eddy Inc. for Fig. 11.7; Organization for Economic Co-operation and Development for Fig. 11.8; US Government Printing Office, Washington DC for Fig. 11.9; Microfilms International Marketing Corporation for Fig. 11.10; International Planned Parenthood Federation for Fig. 12.4; Prentice-Hall Inc. and University of Chicago Press for Fig. 12.5; University of Hawaii Press for Fig. 13.1; Prentice-Hall Inc. for Fig. 1.4.

* Figure numbers refer to this book. Detailed citations may be found by consulting captions and bibliography.

Part I

Introduction

1

Nature and resources

Man is a material-using animal. Everything he uses, from the food needed to keep him alive to the objects he fabricates, whether tools or sculptures, comes from the substances of the planet on which he lives. Wastes are then returned to the biological and abiotic systems of the earth. And because of his acquisition of culture, man desires to use these systems for non-utilitarian purposes of a recreational or spiritual kind.

If we look more closely at how man utilizes and processes materials from his surroundings, we see that a first group consists of resources which are used in the processes of the metabolism of his body, such as food and water (Table 1.1). These allow the growth and renewal of tissue and provide energy for chemical processes such as movement. The energy is transformed from chemical energy to heat and given off to the atmosphere, from whence it cannot be reclaimed. As a result of energy consumption and metabolic processes, excretory matter is produced. This often contains mineral substances which could be of value, for instance as plant food, and so use does not here necessarily mean loss. The fact that in industrial societies these 'waste' materials are usually dumped in the nearest large body of water is not relevant to the basic chemistry of the processes.

The substances which go to make up our bodies are theoretically available for re-use upon our death—as T. S. Eliot put it:

> . . . and ashes to the earth
> Which is already flesh, faeces
> Bone of man and beast, cornstalk and leaf.

The pattern of material use within the body is genetically and biologically determined, but the disposal of waste products and dead individuals is subject to considerable variation because there are different cultural practices among biologically similar groups of men.

The use of materials outside the human body, whether raw or chemically or biologically processed, is likewise subject to variations in practice. Many of these materials are non-renewable resources, like metals or stone; but some, such as wood and wood products like paper, come from renewable resources. Re-use of the objects possessed after their normal life is usually theoretically possible except where a transformation process is used to get rid of it, as in the disposal of waste paper by burning. That the opportunities to re-use materials may not be taken is again irrelevant to the basic characteristics of the resource.

TABLE 1.1 Daily human metabolic turnover

Male of 154 lb (69·8 kg)						
Input					*Output*	
		Water	2,220 g		Water	2,542 g
Protein	80 g				Solids	61 g
Fats	150 g	Food	523 g	BECOMES		
Carbohydrates	270 g				Carbon dioxide	928 g
Solids and minerals	23 g	Oxygen	862 g			
					Other substances (CO, H_2, CH_4, H_2S, NH_3 plus organic compounds)	54 g

Source: McHale 1972

A third group comprises resources used outside the body, the gathering of which leaves them unaltered. Such features of our surroundings as scenery, wildlife (if observed rather than hunted) and water, for swimming or sailing, remain unchanged by our use of them for recreation and aesthetic satisfaction. Attrition of scenery may occur because there are too many users, or water may be fouled by the sailors and thus rendered unusable for the swimmers, but the potential exists for entirely non-transformational use.

The total flow of a material from its state in nature through its period of contact with man to its disposal can be termed a *resource process* (Firey 1960), since the human decision to use the material enables it to be labelled a resource. The same term can be used for environments used for example for recreation or wildlife conservation. Resource processes can be studied in a variety of ways: we can, for instance, see them as a set of interactions between living and non-living components of the biosphere in all their various solid, liquid and gaseous phases, and man may play a role of varying dominance in these systems; such a viewpoint is generally termed ecological.

A second approach is to explore the manner in which the distribution of resources to people is achieved in various societies in attempts to match the demand and supply, i.e. the viewpoint of economics. The relationship between man's culture and his surroundings can be inspected in a behavioural context, thus studying the psychological activity leading up to the use of the earth's substances and habitats: this is clearly part of the field of ethology. Another method of study judges how man ought to use the biosphere and its resources, both for particular cultures and for the species as a whole: this a particular branch of ethics within the general field of behavioural studies.

This book aims to emphasize the first of these four categories, the ecological point of view; but it recognizes that the objective, scientific study demanded by ecology does not portray the totality of man's interaction with the systems of this planet which comprises resource processes; it allows that many other factors, and in particular those usually designated as cultural, are of considerable importance. This includes politics, not dealt with much in this book, but clearly relevant and including some biological ideas in its thinking (Mackenzie 1978). Therefore the next section of Part I expands upon the themes of nature and culture, especially the outgrowths of the latter such as economics and ethology.

Nature

The natural world may be studied in many ways, and the classification and cataloguing of phenomena have occupied many workers during human history. For our present purposes, studies which emphasize the connections between the various components, and especially the interaction of living organisms with their abiotic environments, are most useful; hence the emphasis upon ecology, which can also encompass the impact of man upon natural systems. Since ecology is the study of living organisms and their relationship to each other and their surroundings, it it therefore mostly a study of the biosphere, which is influenced by the lithosphere, (as in soil parent material), and the atmosphere, (as in the incidence of climatic factors). The aggregate may appropriately be called the ecosphere. Since most of man's resources come from the ecosphere and since in garnering them he has usually changed the ecology, the relevance of ecological study to our present theme cannot be gainsaid.

Perhaps the basic lesson of ecology is that in every ecological system there are natural limits to the total amount of living matter that can be supported. These may be set by features as basic as the amount of solar radiation and moisture incident upon that part of the earth. The gradient from the lushness of equatorial forests to the sparse life of the high tundra is evidence of this. On the other hand it may be spatial: a rocky islet may be filled to capacity at the nesting time of a particular bird and thus will determine the population level of those creatures, even if the sea around is teeming with fish waiting to give themselves up. Another important limitation is often the supply of a nutrient element: nitrogen and phosphorus frequently play this role. Ecology hence provides an envelope criterion for resource processes by telling us whether in the long run a particular process is possible or impossible. If it tells us that the maximum number of people that can be fed on the incident solar radiation trapped by photosynthesis is x billion, then this is an absolute limit. If it tells us that the continued practice of shifting agriculture or grass-burning will result in devastating soil erosion, then the presence of an absolute limit has been demonstrated and cultural adjustment must seek to work within it.

Ecosystems

The ecology of an area can be considered as a set of objects together with the relationships between the objects and their attributes. The view of ecology encouraged by such a framework is obviously one which stresses interaction between parts and the mechanisms which control such connectivities and is called a systems approach. We can therefore designate a special class of systems which have ecological components, and call them *ecosystems*.

The term ecosystem was coined in 1935 by Tansley, but the concept has a much longer history, many attempts having been made to characterize the immense complexity and holistic character of the natural world. Thus the terms *microcosm*, *naturcomplex*, *holocoen* and *biosystem* have all been used from time to time for what is now generally called the ecosystem, with the major exception of the Soviet Union, where the term *biogeocoenose* is used with a roughly equivalent meaning. As Tansley (1935) stressed, the term ecosystem includes not only the organisms, but also the whole complex of physical factors forming what we call the environment.

A more rigorous definition has been given by E. P. Odum (1959):

> any area of nature that includes living organisms and non-living substances interacting to produce an exchange of materials between the living and non-living parts is an ecological system or ecosystem.

Two important characteristics stand out. Firstly the concept can be applied at any scale: a drop of water inhabited by protozoa is an ecosystem; so is the whole planet. This immediately introduces the problem of boundary definition: few ecosystems can satisfactorily be defined in space because one or more of their components overlaps with another system. A pond may appear to be a clearly bounded system, but the behaviour of wild ducks belies this simplicity.

A second feature is the reciprocity between living and non-living parts of the system. Not only does the 'environment' affect the organisms, but they in turn may change it. This has long been realized from studies of ecological succession, but in resource studies it becomes important to realize that use of the biotic member of an ecosystem as a resource may bring about changes in the non-living part too. The relationship between a forest and its soils is a very closely interconnected one, for example, and the replacement of deciduous trees by conifers in the temperate zone may bring about a change in soil type from a Brown Earth to a Podzol.

The scientific study of ecosystems is clearly difficult: mere inventory of all the components will not tell us about the connectivity which has been so strongly stressed, and research has indicated that two of the most important pegs upon which to hang functional studies of ecosystems are the flows of energy and matter within a given (and often imperfectly bounded) ecosystem.

Energy in ecosystems

In all ecosystems, as indeed in all studies of resource processes, the role of energy is crucial. Without the input of energy from the sun, life could not exist; the movements of air and water in the atmosphere are driven by this source, and even non-renewable resources of a geological nature have been formed as the result of solar energy. Coal and limestone are obvious instances and sandstones derive from weathering processes that are themselves impelled by inputs of solar energy. Nothing more fundamental to the nature of the planet can be imagined (Woodwell 1970a, Gates 1971).

The total amount of solar energy that reaches the earth's surface is probably in the order of 3,400 kcal/m^2/day. This is an average figure since the amount (flux density) is different from place to place. The maximum conversion of this energy by photosynthesis absorbs c. 170 kcal/m^2/day (5 per cent); the average is inevitably lower (H. T. Odum 1971). The whole of organic life depends upon this small fraction, for the only way of fixing solar energy for use by living creatures is by green plants; our stores of energy from coal and petroleum are merely fixed solar energy in a compressed and fossilized form.

Photosynthesis is the process by which green plants trap energy and incorporate it in complex organic molecules which then form food for themselves and other organisms. In a very complex series of reactions, water and carbon dioxide combine to synthesize sugars which may then be metabolized to starches, or which by the addition of mineral nutrients will form the very complex molecules (amino-acids and proteins, for example) which are the basis of living matter. But not all the energy gained by photosynthesis appears as plant tissue, for some is used by the metabolism of the plant in the process of respiration.

As an example, we may examine the data given by Whittaker (1975) for a deciduous forest in the eastern USA:

Insolation (visible spectrum)	56,000 cal/cm^2/yr
Rate of creation of organic material by plants (Net Primary Production or NPP)	510 cal/cm^2/yr

Of the original incident energy, therefore, only 0·91 per cent appears as plant material: not only does respiration take its toll but some of the light is the wrong wavelength for photosynthesis, and some is reflected from the surface of leaves. A marine example from the North Sea is parallel:

Insolation (visible spectrum)	473 cal/cm^2/day
NPP of plankton	0·314 cal/cm^2/day

These are daily figures for the summer only and here the efficiency of transformation is 0·066 per cent and suggests not only the scattering of light by the water but that other factors may be limiting productivity. But clearly the NPP is the basic step in an ecosystem since all other life will depend upon the plant material created then, and in the case of resources of a biological kind, production ecology is ideally part of the knowledge about the resource process.

In natural ecosystems many of the plants are eaten by herbivorous animals. These in turn may be consumed by carnivores and thus emerges the idea of a simple food chain. The important feature, so far as energy is concerned, is that at each stage of the food chain energy is lost to the system. This is a direct consequence of the operation of the second law of thermodynamics, which states that all energy which undergoes a change of form will tend to be transformed into heat energy. Thus the 'concentrated' potential energy present tend to be transformed into heat energy, and the 'concentrated' potential energy present in living tissue is 'dispersed' as heat by the metabolic processes of the organisms. In a simple food chain, even if the herbivores were to eat all the plant material, the use of energy and hence dispersal of heat would ensure that a much smaller amount of energy per unit area became visible animal tissue. This quantity of organic matter is called secondary production. It follows that carnivores have even less energy available for their consumption, and should there be yet another carnivore then the amounts of energy available to it will be very small indeed.

If we take an oakwood in southern England (Varley 1974) then:

Insolation	1×10^6 kJ/m²/yr
NPP of trees and shrubs	26×10^3 kJ/m²/yr
NPP of main herbivores (caterpillars, voles and mice)	51·2 kJ/m²/yr
NPP of main carnivores (spiders, shrews, tawny owls)	3·16 kJ/m²/yr

(Production studies are usually in either calories or Joules: 1 cal = 4·2 J)

These data are extracted from a more complex series of feeding relationships but show that the efficiency with which each stage (called a trophic level) converts energy is very low. The implications for resource use are clear: the further away from the primary production stage that man takes his crop, the less energy per unit area will be available, and so, to be efficient, he must act as a herbivore.

Rarely is all the tissue at one trophic level cropped by the next: a terrestrial herbivore usually eats only the aerial parts of a plant; similarly many herbivores die without being eaten by a carnivore. At death, therefore, another type of food chain forms in which non-living organic material is the energy source. This second chain is usually called the detritus or decomposer chain and the organisms in it are taxonomically very diverse, but an especially important group are the fauna and flora of soils which break down dead organic matter into its components: fungi and bacteria are notable components of this group (Fig. 1.1).

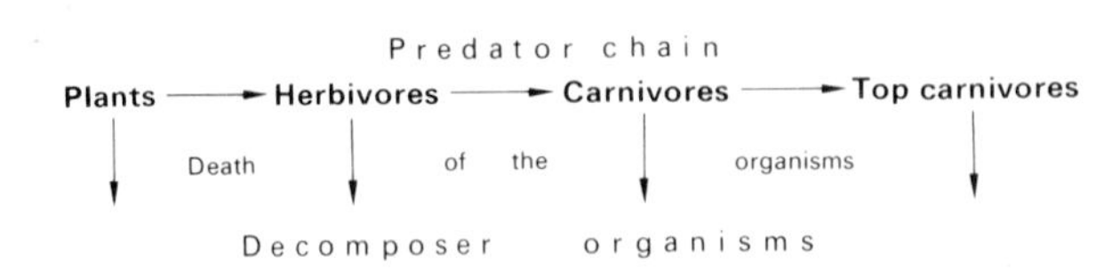

Fig. 1.1 A schematic diagram of trophic level structure of an ecosystem showing the basic energy paths through the organisms, beginning at the green plants. Not shown at each transfer stage is the loss of energy as heat

In addition, there are parasites on many organisms. These can be regarded either as a separate food chain or as components of the predator chain representing the next trophic level.

To summarize, energy enters the ecosystem as free, solar energy and leaves it as heat, having undergone changes from a 'concentrated' to a 'dispersed' state. Within the ecosystem is found energy-rich organic matter which upon the death of the organism, either plant, animal or fungus, undergoes decomposition. The complex organic ma-

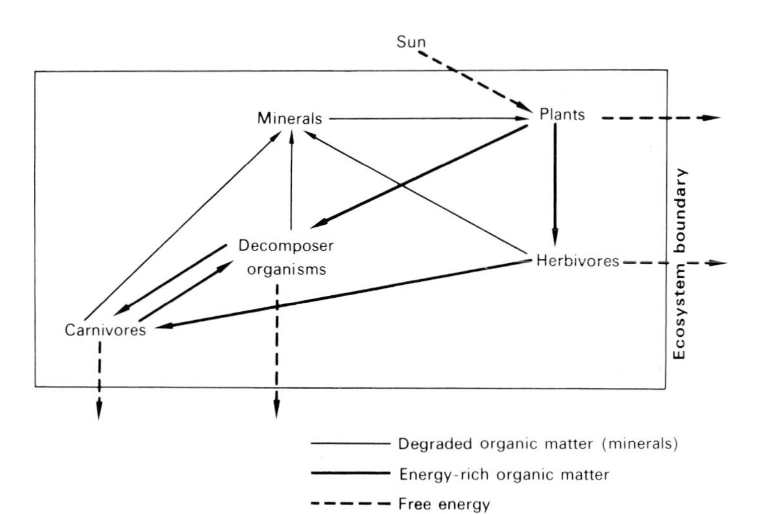

Fig. 1.2 The form in which energy and matter move through an ecosystem: the 'free energy' traversing the ecosystem boundaries outwards is the heat loss. Within the system, matter travels either as energy-rich organic matter or as mineral matter low in energy content.
Source: O'Connor 1964

Sun
Plants
Herbivores
Carnivores

1

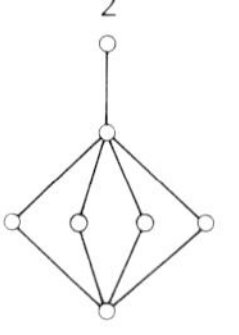

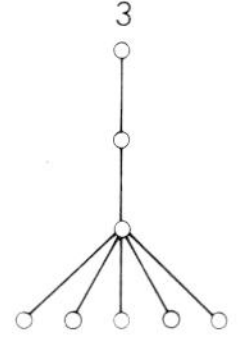

Sun
Plants
Herbivores
Carnivores

4

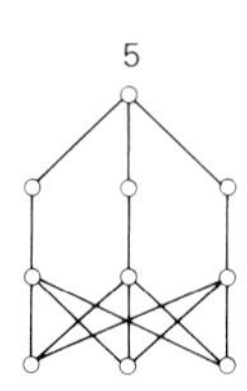

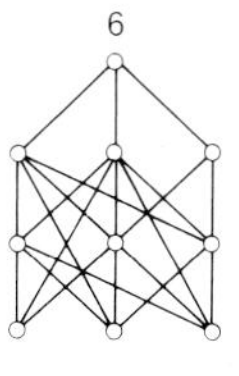

o = 1 species

Fig. 1.3 Representations of trophic level relationships. In (1) a herbivore grazes several species of plants but has only one predator; in (2) a single species of plant is eaten by four herbivores which form the prey of only one carnivore; in (3) the single herbivorous species is the food of five species of carnivores. In (4) three herbivores range across three species of plant but are eaten by a prey-specific predator; (5) and (6) depict various states of diversivory and show a relatively high number of energy pathways. They would suffer a lower loss of stability if a key species were to disappear.
Source: Watt 1968

terials are broken down to relatively simple inorganic compounds, with consequent dispersal of energy (Fig. 1.2).

Needless to say, the situation in nature and even in much modified ecosystems is more complicated. A herbivore may feed off many species of plants and in turn be eaten by several species of predators. A predator may have a preferred food source but shift to others in a time of scarcity. The diversivore habit is not uncommon, where an animal will eat plants, other animals and also be a scavenger on dead material: many bears are diversivorous and doubtless until affected by modern squeamishness our own species could rank thus. At any rate we are currently both herbivores and carnivores. Watt (1968) has diagrammed some of the different types of relationship that can exist at different trophic levels in ecosystems (Fig. 1.3). At even more complicated levels, it is probably more realistic to talk about *food webs* rather than chains, since the points of contact (at which energy is transferred) are so many (Fig. 1.4). Even this concept is perhaps insufficiently close to reality, and Elton (1966) has expanded it into the idea of a *species network*, where not only food but other relationships, such as competition for space and other forms of competitive interference, are considered. These latter factors, he notes, do not necessarily cause any immediate transfer of energy or materials from one species to another; relationships within such a network do not imply simultaneous activity or existence.

The implication of such models for resource processes are quite simple: the multiple interactions within the networks mean that when man crops a species as a resource, consequent shifts within the system will probably occur. Because of the complexity of

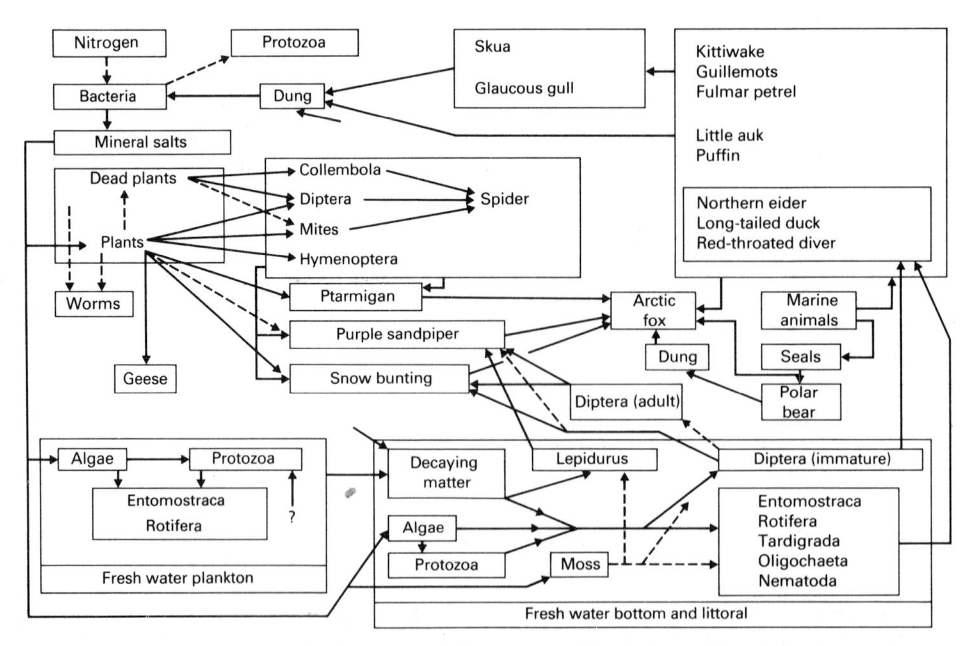

Fig. 1.4 A simplified food web for Bear Island, Spitsbergen. Originally published by Summerhayes and Elton in 1923, this is a classic early example of the concept of the web. Note the important role of the sea. Source: Collier et al 1973

most systems, these are unlikely to be predictable (although increased knowledge, and the use of computers is improving this situation), and inevitably some of the consequences have been deleterious to further human activity and the continued viability of the ecosystem.

The loss of energy as heat at each trophic level in the ecosystem ensures that the amount of potential energy decreases through the species network. Thus the numbers of organisms and the amount of living tissue (usually measured as dry weight per unit area and called *biomass*) normally diminish through the stages of secondary and tertiary productivity. A series of pyramids can be seen to exist which reflect this. The first is of numbers: in a measured area, the number of plants is many times the number of herbivores, which in turn greatly exceeds the quantity of carnivorous individuals. Carnivores and top carnivores are, therefore, relatively rare species (Fig. 1.5). Such a relationship is made even clearer when biomass or standing crop is inspected (this is not the same as productivity, since biomass is a static measurement at one point in time whereas productivity is a rate). Again, the biomass of plants is much higher than any of the dependent organisms. The calorific value of the tissues at various trophic levels is perhaps the best guide to energy relations, and here the pyramidal form of the numbers is confirmed (Fig. 1.6). There is a tendency for the size of the animals to get larger as the predator chain is followed: being bigger than your prey is an obvious advantage for a predacious species.

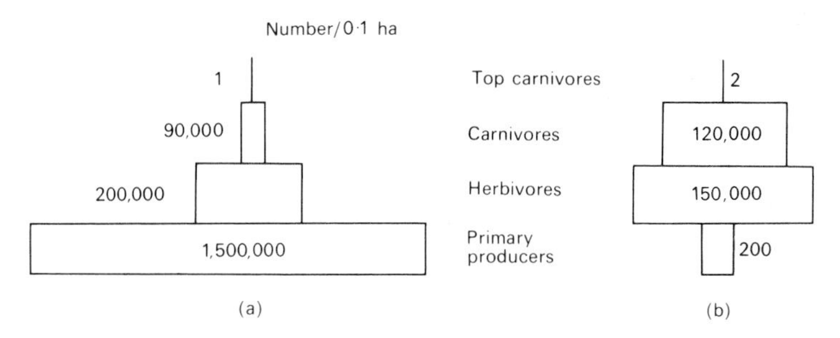

Fig. 1.5 Number pyramids of two ecosystems, exclusive of micro-organisms and soil animals. (a) is a grassland in summer, (b) a temperate forest in summer where the producers are trees which are large organisms and few in number.
Source: E. P. Odum 1971

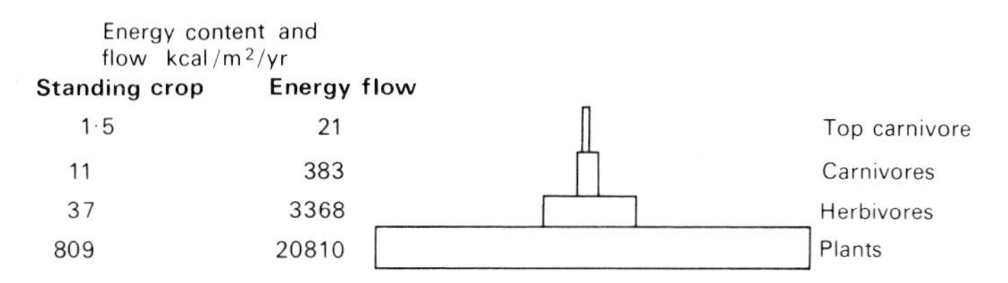

Fig. 1.6 An energy-content pyramid for Silver Springs, Florida. The pyramid represents the data for energy flow through the trophic levels in the course of a year. The standing crop biomass (left-hand column of figure) is considerably less than the yearly flow in this instance, but in, for example, a grassland it might be almost equivalent to the annual productivity.
Source: E. P. Odum 1971

Inorganic substances

Living things require between 30–40 of the 90 chemical elements which occur in nature. Their supplies come from many elements and compounds which undergo constant cycling at a variety of scales: some have gaseous phases and hence involve the atmosphere in their ecosystems; others are either in a solid or dissolved state and remain in the terrestrial and aqueous parts of the biosphere. In the first category, CO_2 is an obvious fundamental substance since it is required by plants for photosynthesis. If this were not continually produced by respiration and by breakdown of organic material, then the plant cover of the world would exhaust the atmospheric supplies in a year or so. Constant cycling of nitrogen, oxygen and water is also necessary for the support of life. Many mineral elements with sedimentary cycles, such as phosphorus, calcium and magnesium, are also essential for the growth of living tissue. These elements, along with many others, may impose checks upon the populations of a component of any ecosystem if they are in short supply.

In nature, an ecosystem receives its inputs of inorganic materials from a number of sources. As with energy, the plant is the initial point of incorporation. The atmosphere

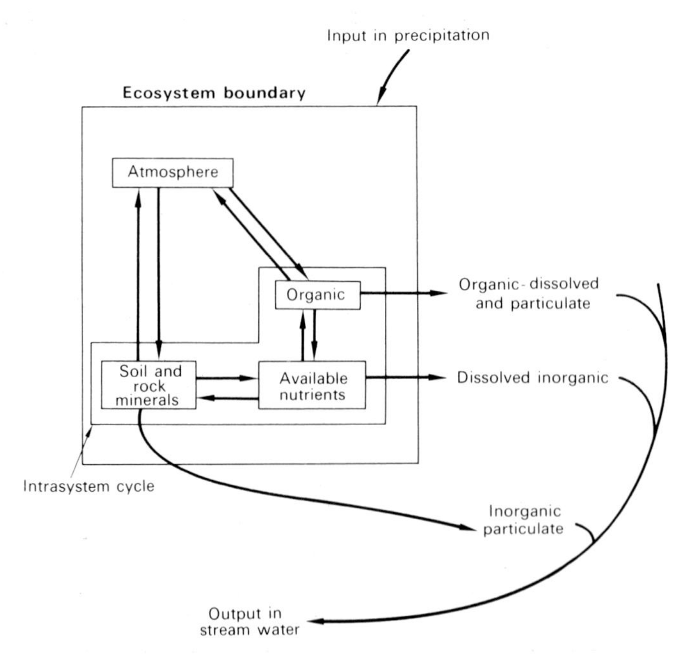

Fig. 1.7 A diagram of the flow and storage of essential elements through a terrestrial ecosystem. They cross the ecosystem boundary in an aqueous medium: in via precipitation and out via runoff. The atmosphere is regarded here as part of the system, i.e. that part of it performing gaseous exchange with the green plants. Within the system the lithosphere and the atmosphere also act as *de novo* sources of mineral elements in contrast to those which cycle between the organic matter and the pool of available nutrients, and replace those lost to output in stream water.
Source: Bormann and Likens 1969

contributes CO_2 to plants, also N_2 to species with nitrogen-fixing bacteria living symbiotically with them, and in turn the plants give off O_2 needed for the respiration of animals. The weathering of the rocks of the earth's crust further contributes basic minerals such as calcium, magnesium and phosphorus: they may also find their way into the atmosphere to be rained out (Fig. 1.7). Water makes its contribution both by assisting in weathering processes in the soil and sometimes, in the form of floods, by spreading nutrients around, as well as transporting them to the sea if they have been lost to the organic components of ecosystems. Its role in the transpiration stream is critical in the transfer of nutrients from the soil to the plant in terrestrial ecosystems.

In the natural state, the flow of nutrients is conserved within an ecosystem. Input and loss are usually small (in terrestrial systems especially) compared with the volume which circulates within the system. In a forest, for example, minerals originating in the rocks enter the soil, become part of the tree, descend at leaf-fall, are mineralized by the soil fauna and flora and then are again available for uptake by the tree (Fig. 1.8). Litter fungi are especially important in forests, where they act as holding sinks for nutrients such as Ca, Fe, Cu, Na, P and Zn. Rhizomorphs in tropical forests may hold mineral concentrations of up to 85 times that of the leaves and have a 99·9 per cent efficiency in retaining them against leaching (Stark 1972). The turnover times of nutrients in the various compartments of ecosystems seems to get longer with increasing latitude. Jordan and Kline (1972) quote 10·5 yr for the cycling time of a tropical rain forest and

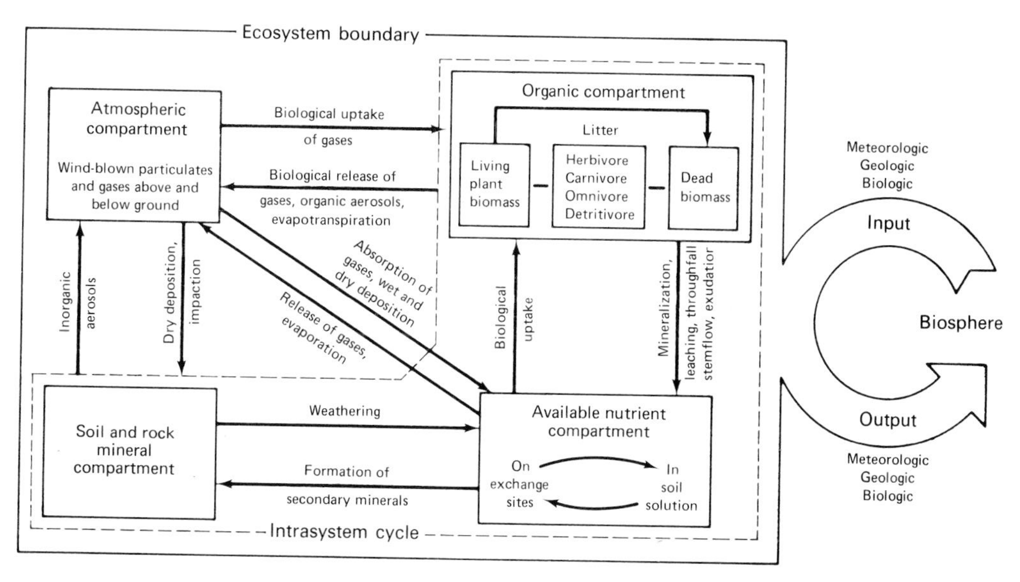

Fig. 1.8 A model depicting nutrient relationship in a terrestrial ecosystem. Major sites of accumulation and major exchange pathways within the ecosystem are shown. Nutrients that, because they have no prominent gaseous phase, continually cycle within the boundaries of the ecosystem between the available nutrient, organic matter, and mineral components tend to form an intrasystem cycle. Fluxes across the ecosystem's boundaries link individual ecosystems with the remainder of the biosphere.
Source: Likens et al 1977

42·7 yr for the taiga in the USSR. Animals have varying functions: in terrestrial decomposer chains they may perform a crucial role in physically commuting organic debris, and in the oceans the smallest zooplankton appear to be the key factor in the circulation of phosphorus and probably also of nitrogen; large terrestrial animals, however, appear to play a trivial role in nutrient circulation. Where climate seasonally inhibits many of the soil biota, as in boreal coniferous forests, unmineralized organic matter piles up on the forest floor and so fire may play a vital role as decomposers. In natural terrestrial systems, the living organisms are very important in the retention of essential elements, and about 50 per cent of the total reserve appears to be incorporated in both living and dead organic matter. Succession appears to be a process by which enough nutrients are accumulated to make possible the rise of succeeding populations, and a mature community perpetuates its stability by conserving its essential elements (Pomeroy 1970). A stable system such as a forest retains most of its nutrients by circulating them within the soil-vegetation subsystem, and losses to runoff are balanced by inputs into the system, as shown by Whittaker (1975) for calcium in a New York forest (Fig. 1.9). If the nutrient pathways are diverted by the destruction of one

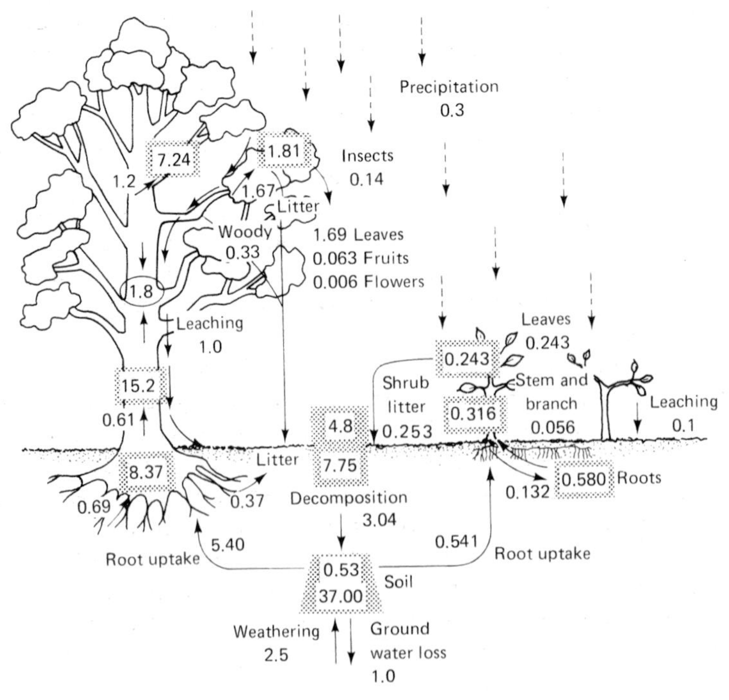

Fig. 1.9 The calcium cycle for an oak-pine forest in New York. Numbers in boxes are pools (g/m²); unboxed numbers are flows (g/m²/yr). The circled number (1·8 g/m²/yr) is net uptake into net accumulation of organic matter. In the soil boxes the upper compartment is exchangeable calcium; the lower, total calcium. Input from precipitation and weathering equals loss plus net accumulation.
Source: Whittaker 1975

component of the organic material, such as the vegetation (for example, by clear-cutting or catastrophic fire), then there is a rapid loss of mineral elements and particulate material, thus contributing to eutrophication and siltation downstream. After disturbances, successional species, such as shrubs and small trees, may play an important role in the rebuilding of the cycling and retention of nutrients which may take 60–80 years in northern hardwood forests (Likens et al 1978). In New Hampshire studies, the pin cherry (*Prunus pennsylvanica*) had an annual uptake of nitrogen 50 per cent greater than that of undisturbed forest, and the rapid growth of such pioneer trees clearly acts to minimize nutrient losses from the ecosystem by the channelling of water from runoff to evapo-transpiration and so reducing losses by erosion and in solution; by producing shade and hence reducing rates of decomposition of organic matter so that the supply of soluble ions available for loss via runoff is lessened; and by incorporating into the vegetation any nutrients that happen to become available. This is just one of a number of mechanisms which act to reduce nitrogen loss from disturbed ecosystems (Marks and Bormann 1972, Vitousek et al 1979).

Such cycles at the scale of local or regional ecosystems are part of global cycles. For example, the major flows in the world's nitrogen cycle have been modelled by Delwiche (1970) and the critical points can be identified. In the case of nitrogen, the atmosphere is the main reservoir and the loss of N_2 to sediments is apparently balanced by the gain from volcanic action, and indeed the N_2 content of the air may have increased during geological time. The fixation of atmospheric nitrogen is limited to a few, but abundant, organisms like the free living bacteria *Azotobacter* and *Clostridium*, symbiotic nodule bacteria on leguminous plants like *Rhizobium*, and some blue-green Algae; these are the keys to the movement of N_2 from the air reservoir into the productivity cycle since no higher plants are able to fix nitrogen alone: legumes only do so with the help of their symbiotic bacteria. Human effects include the role of the fertilizer industry in fixing atmospheric N_2 at a current rate of 26 per cent of the natural biological fixation and the emission of oxides of nitrogen from the combustion of fossil fuels at 11 per cent of the biological rate (Söderlund and Svensson 1976).

Another element essential for living tissue is phosphorus, whose reservoir is the crust and sediments of the earth. Thus the release of P into ecosystems is very slow, whereas its loss can be very rapid, especially where soil erosion occurs. Such phosphorus is washed into the deep sea sediments and is for all practical purposes lost to ecosystems although a little is returned to the land as guano and fish. Phosphorus mined and added to the land amounts to ca $12{\cdot}6 \times 10^6$ t/yr, which can be compared with 176×10^6 t/yr uptake by plants from soil and an estimated transfer of $2{\cdot}5–12{\cdot}3 \times 10^6$ t/yr from soils to freshwater. Human influence is not negligible at certain points of the cycle, therefore, although the supply of phosphate fertilizer will probably be determined mostly by the price of energy since the crustal reserve is so large at $1–10^{13} \times 10^6$ t of which $3{,}000–9{,}000 \times 10^6$ are probably mineable (Pierrou 1976).

Man requires all 40 essential elements for his metabolism and most of the rest for extrasomatic cultural activities. The overall tendency of his use of these materials has been to speed up the natural cycles so that pulses occur as the eutrophication of many water bodies because of the output of phosphorus-enriched water from agricultural and urban areas (Hutchinson 1970). One of the results of the acceleration of natural cycles has been to speed up the movements of crustal elements into the oceans, from which their recovery, even if technically possible, is likely to be very expensive and ecologically disruptive. Resource processes which 'short-circuit' such accelerated flows seem therefore to be highly desirable.

Biological productivity and cropping

It will be apparent that biological productivity (the rate of appearance of energy and matter as living tissue) is the basis of all our renewable biological resources, and it is involved with abiotic parts of the planet too. The metabolism of plants and animals is instrumental in maintaining the gaseous balance of the atmosphere, and the world's hydrological cycle interacts at several points with biological production. Again, 'fossil' matter sources appear in productivity considerations because of the limiting effects upon plants of inadequate mineral supplies, and in man-dominated ecosystems because of fuel-energy inputs into resource processes such as agriculture.

Biological productivity is therefore of considerable importance in human affairs, but its study is relatively new and subject still to wide margins of error. The measurement of rates of production, for example, is a difficult task involving sophisticated instrumentation in a field setting. Often the only practical way of proceeding is to measure standing crop biomass; this, however, may be very different from annual productivity since there are likely to be seasonal differences or, in the case of short-lived organisms, a turnover rate several times higher than the biomass at any one point in time. In order to reduce some of the errors, and to standardize measurements internationally, the International Biological Programme for the study of Biological Productivity and its relation to Human Welfare was carried out during the mid-1960s–1970s. Research programmes for each major biome have been aimed at providing comparable information on biological productivity.

Results of measurements of primary productivity from various ecosystems reveal, in general, the pattern expected from the visible biomass which is expressed in the physiognomy of the vegetation. The biomass and NPP (Table 1.2) clearly reflect the intuitive values suggested by the physiognomy of the vegetation type but certain features of the pattern (shown as a map in Fig 1.10) may be emphasized. The high values for the tropical forests are to be expected but less generally realized are the high values for estauaries and tropical reefs (commonly called coral reefs but more properly algal reefs) and for wetlands such as swamps and marshes. At the other end of the spectrum, although deserts and tundra have entirely expected values, the low productivity of the open oceans (which constitute 92 per cent of the oceans) places them in the same category as the very arid and very cold terrestrial biomes (Plates 1 and 2, pages 19 and 20). Thus production on land dominates the world pattern, partly because of the shallow depth of light penetration into the seas and the loss of nutrients from plankton in the deeper oceans. Mature terrestrial communities, too, tend to hold onto their nutrients in the 'tight' kinds of cycle described above.

Cultivated land occupies a relatively low place in the ranking of productivity. The

TABLE 1.2 Net primary production and related characteristics of the biosphere

1	*2*	*3*	*4*	*5*	*6*	*7*	*8*
		Net primary production (dry matter)			*Biomass (dry matter)*		
Ecosystem type	*Area (10^6 km^2)*	*Normal range ($g/m^2/year$)*	*Mean ($g/m^2/year$)*	*Total (10^9 t/year)*	*Normal range (kg/m^2)*	*Mean (kg/m^2)*	*Total (10^9 t)*
Tropical rain forest	17·0	1,000–3,500	2,200	37·4	6–80	45	765
Tropical seasonal forest	7·5	1,000–2,500	1,600	12·0	6–60	35	260
Temperate forest:							
evergreen	5·0	600–2,500	1,300	6·5	6–200	35	175
deciduous	7·0	600–2,500	1,200	8·4	6–60	30	210
Boreal forest	12·0	400–2,000	800	9·6	6–40	20	240
Woodland and shrubland	8·5	250–1,200	700	6·0	2–20	6	50
Savanna	15·0	200–2,000	900	13·5	0·2–15	4	60
Temperate grassland	9·0	200–1,500	600	5·4	0·2–5	1·6	14
Tundra and alpine	8·0	10–400	140	1·1	0·1–3	0·6	5
Desert and semidesert scrub	18·0	10–250	90	1·6	0·1–4	0·7	13
Extreme desert— rock, sand, ice	24·0	0–10	3	0·07	0–0·2	0·02	0·5
Cultivated land	14·0	100–4,000	650	9·1	0·4–12	1	14
Swamp and marsh	2·0	800–6,000	3,000	6·0	3–50	15	30
Lake and stream	2·0	100–1,500	400	0·8	0–0·1	0·02	0·05
Total continental:	*149*		*782*	*117·5*		*12·2*	*1837*
Open ocean	332·0	2–400	125	41·5	0–0·005	0·003	1·0
Upwelling zones	0·4	400–1,000	500	0·2	0·005–0·1	0·02	0·008
Continental shelf	26·6	200–600	360	9·6	0·001–0·04	0·001	0·27
Algal beds and reefs	0·6	500–4,000	2,500	1·6	0·04–4	2	1·2
Estuaries (excluding marsh)	1·4	200–4,000	1,500	2·1	0·01–4	1	1·4
Total marine:	*361*	—	*155*	*55·0*	—	*0·01*	*3·9*
Full total:	*510*	—	*336*	*172·5*	—	*3·6*	*1841*

Source: Whittaker and Likens 1975

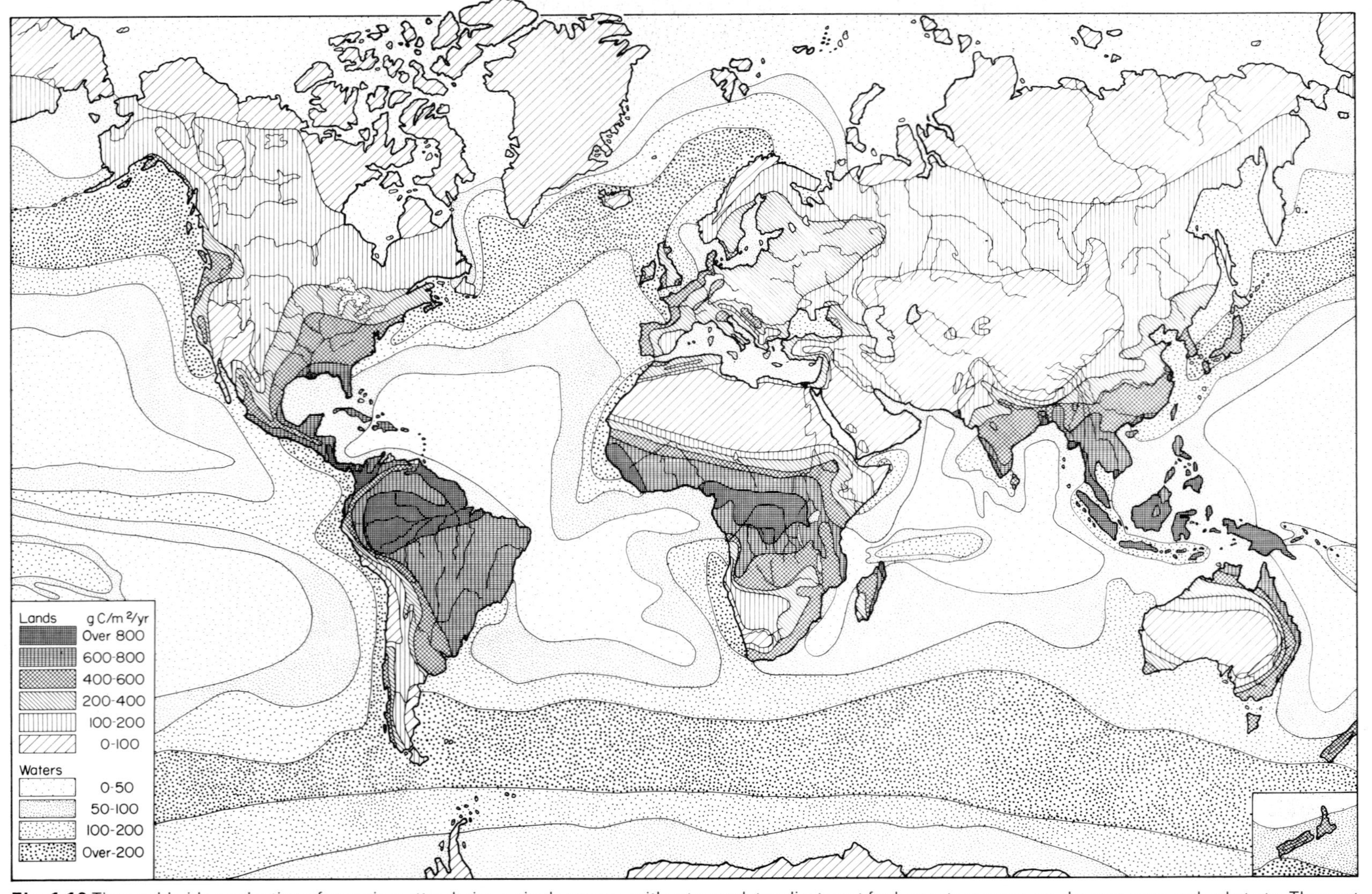

Fig. 1.10 The worldwide production of organic matter during a single season, without complete adjustment for losses to consumers, decomposers and substrate. Thus net primary production may exceed the values shown here, but economically usable products may be much less. Values are g carbon/m^2/yr, which should be multiplied by 2·2 to give approximate dry matter equivalents in g/m^2/yr.
Source: Lieth 1965

Plate 1 An estuary in west Wales. These habitats have a very high biological productivity and biotic diversity, and are often nurseries for offshore fisheries. In economic terms they are valued mostly as potentially reclaimable land for industry or for waste disposal. *(Aerofilms Ltd, London)*

seasonality and short growing season of crops, the fact that individual plants are often widely spaced with bare ground between them, and the consequently low quantity of chlorophyll per unit area all contribute to the apparently poor performance of crop plants (Table 1.3) even though modernized agriculture receives 'subsidies' of fossil-fuel derived energy in the form of machinery, pumped water, pesticides and fertilizers.

Nevertheless, some of them have very high productivity during their growing season, and Newbould (1971a) points out the control of seasonality of production which can be exercized over crops, and the differences in biochemical quality of the organic matter may make purely quantitative comparisons between wild and cultivated plants rather meaningless. Within the global figure of Table 1.2, certain variations can be detected: at lower latitudes, for example, Gramineae excel at dry matter production: sugar cane and Napier-grass (*Pennisetum purpureum*) have high standing crop values. In sub-tropical latitudes, the sorghums and Bermuda grass (*Cynodon dactylon*) do very well, and beyond them, sugar beet, rye grass and wheat are highly productive at temperate latitudes (Loomis and Gerakis 1975). But such data parallel those for natural biomes in the sense that dry matter production is not necessarily of an economically desirable and culturally acceptable crop. One of the lessons of NPP studies is that more work needs to be done on harvesting the yields of highly productive wild plants in a culturally acceptable form rather than replacing those biomes with plants or plant-animal systems of much lower productivity.

Even though not at present directly cropped, the NPP of a region or nation does

Plate 2 An arid steppe in California. Productivity and diversity are low, but owing to proximity to large urban centres, demands to convert such areas to recreational housing or to allow unrestricted access by cross-country vehicles are very high. The ecosystems easily break down under such pressures and regrowth is very slow. *(I. G. Simmons)*

TABLE 1.3 Yearlong NPP of some cultivated and natural ecosystems

System	*NPP ($g/m^2/yr$)*
Spartina saltmarsh, Georgia	3285
Desert, Nevada	40
20–35 yr-old pine plantation, England	2190
20–35 yr-old deciduous plantation, England	1095
Wheat, world average	343
Rice, world average	496
Potatoes, world average	400
Sugar cane, world average	1726
Mass algal culture, outdoors	4526

Source: E. P. Odum 1959
Note: No account is taken of fossil fuel subsidies to any system, nor of the biochemical quality of the organic production. The actual yearly productivity of the temperate crops is of course concentrated into a relatively short growing season.

represent its long-term renewable resources, if we accept that non-renewable resources such as metals cannot be infinitely re-used. The quantity of NPP per (human) capita thus gives an indication of the likelihood of self-sufficiency of a nation for organic production, of which food is the most important element. Such calculations as made by Eyre (1978) assume that trade is not important and also obscure the role of subsidies of fossil fuel except where this is deemed to represent food-purchasing power. But it can give us a new type of population-resource region (see chapter 12) on an ecological basis.

If there are difficulties in measuring primary productivity, then they are greatly compounded when secondary productivity is considered. Food webs are often so complex that an animal cannot be assigned to a single trophic level (see Fig 1.4) and at the simplest level of perplexity, animals move; and harvesting of them for biomass or calorific value measurements may produce problems, especially if the species is uncommon. A measure often adopted is the ratio

$$\frac{\text{(calories of growth)}}{\text{(calories consumed)}} = \text{net growth efficiency}$$

which for beef cattle on grassland is about 4 per cent. The equivalent measurement for pigs, young beef animals and young chickens is of the order of 20 per cent, which is clearly used to advantage in modern intensive farming methods for 'baby beef' and broiler chickens. In more natural systems, efficiencies are much lower. As examples we may quote a Tanzanian grassland with an above-ground NPP of 747 kcal/m/yr and a herbivore biomass of 14·9 kcal/m/yr with a productivity of 3·1 kcal/m/yr (Wiegert and Evans 1967). Meathop Wood, a deciduous woodland in northern England, had a total NPP of 6247 kcal/m/yr, of which only 14 kcal/m/yr was consumed by all herbivores. So secondary productivity was constrained not only by primary productivity but by the degree to which the NPP is utilized by the herbivores and the efficiency by which they convert the NPP to animal tissue.

The summarizing theme of all discussions of biological productivity is that of limits. These include the overall limit dictated by the quantity of solar radiation incident upon the earth and the efficiency of photosynthesis in fixing the energy since normally only 0·1–3·0 per cent of the incident radiation appears as NPP. Inorganic nutrients may also prescribe constraints by virtue of their short supply and the length of time they take to go through cycling processes. And since there is a loss of energy at each step through a food web, there are severe limitations imposed upon the productivity of the second and subsequent trophic levels. At the same time, we recognize that large quantities of organic matter flow through the decomposer level and that this has generally been undervalued as a resource.

Populations

In an ecosystem the flows of energy and matter in the form of organisms, and the adjustment of the individuals to the space dimension of their habitat, are expressed in terms of the dynamics of population of a given species. To understand the dynamics of the numbers of plants, animals and men is to be a good way down the road to the rational cropping of an ecosystem.

In the animal world as in man, the numbers of individuals of a species are determined by the relationship between natality and mortality. Usually these quantities are expressed in terms of time: thus a population increases when birth rate exceeds death rate. Conversely, if the death rate exceeds the birth rate then a population will dwindle and die out.

There is an incremental rate of recruitment to a breeding population and the general form of the equation for such an expansion is (Boughey 1968):

$$N_t = N_0 e^{rt} \qquad (1)$$

where N_t = numbers at time t
N_0 = number at time zero
$_e$ = base of natural logarithms
r = rate of population increase
t = time elapsed

which can be expressed logarithmically as

$$\log_e N_t = \log_e N_0 + rt \qquad (2)$$

A growth curve derived from these equations is shown in Fig. 1.11. Such a rate of growth is called exponential, and its capacity for effecting rapid increases in numbers is very high. For example, the time taken to double a population can be tabulated thus:

Rate of increase % p.a.	*Number of years to double population*
0·5	139
1	70
2	35
3	23
4	18

From this we see that quite low percentage rates will produce high absolute numbers quite quickly, a fact not unfamiliar to people repaying mortgage loans at 15 per cent, for instance. As an extreme instance of the potential of the exponential curve we may quote the example of a bacterium which divides into two every 20 minutes. This could produce a colony 1 ft deep over the surface of the earth in 15 days; 1 hour later the layer would be 6 ft deep. That this does not happen is due to a number of factors which are generally termed *environmental resistance*, a term which subsumes a great many influences which affect animal populations (Figs. 1.11 and 1.12).

Without such resistance, a population could fulfil its biotic potential to expand indefinitely. In reality, the exponential curve always levels off, at a value which is termed the carrying capacity of the habitat (K). This is represented by the expression:

$$N = \frac{K}{1 + e^{rt}} \qquad (3)$$

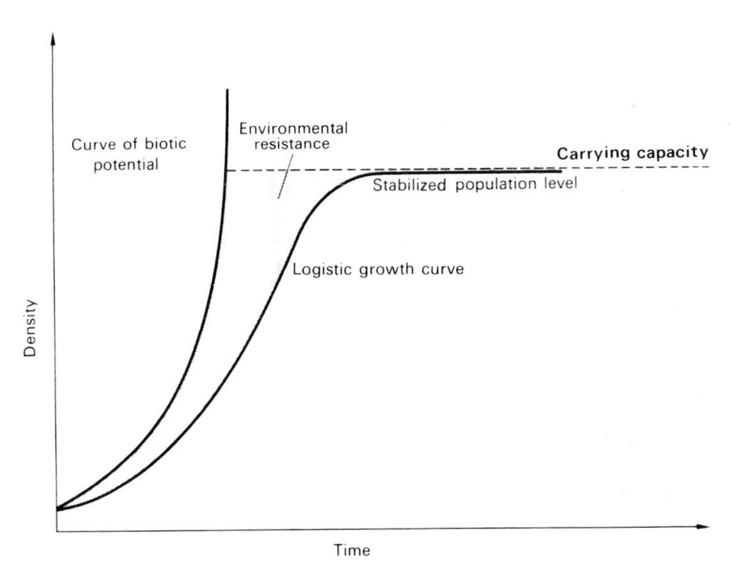

Fig. 1.11 The logistic growth curve flattens out at the carrying capacity, owing to increasing environmental resistance such as the factors in Fig. 1.12. If it were not for that factor, the curve of biotic potential would produce immense absolute numbers of any organism.
Source: Boughey 1968

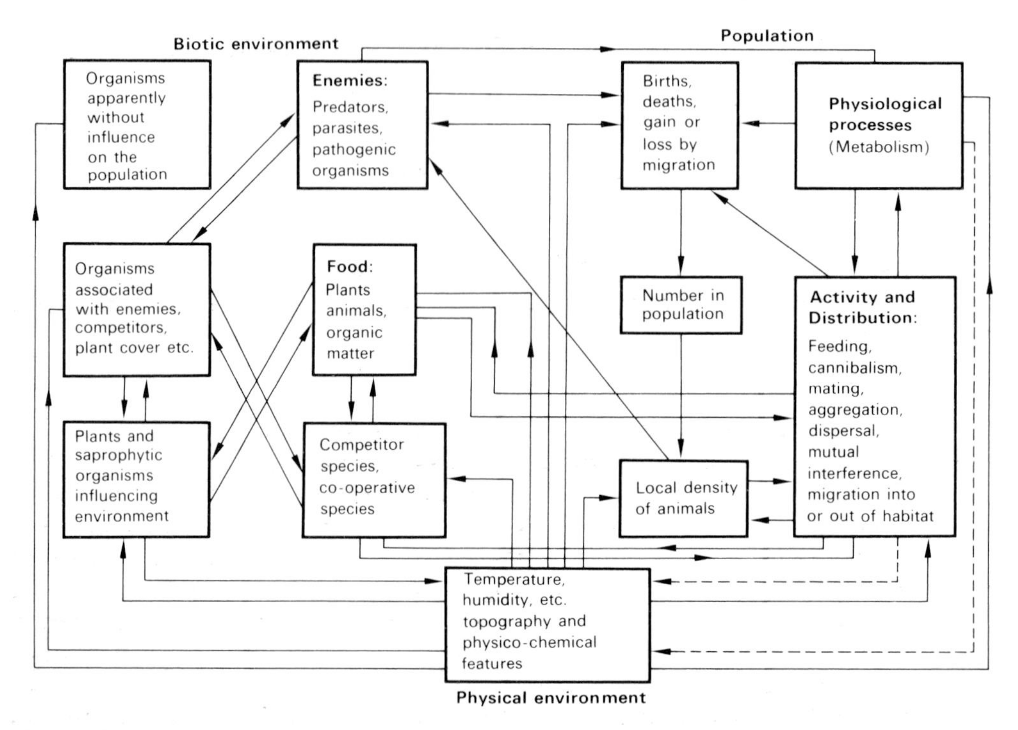

Fig. 1.12 A diagram of the major influences affecting animal populations. These influences produce the environmental resistance which prevents the population exceeding its carrying capacity except in the short term.
Source: Solomon 1969

for a population growing continuously and without catastrophes to an upper asymptote where $r = 0$. It may approach this value gradually from below or it may overshoot and, by virtue of high mortality or low fertility, fall back (with the possibility of oscillations) to the carrying capacity level (Fig. 1.13). The concept of carrying capacity has been elaborated by Dasmann (1964b), who suggests that there is firstly a survival capacity where there is enough food for survival but for neither vigour nor optimum growth, and where slight changes in ambient conditions can be disastrous. An optimum capacity appears superior since there is adequate nutrition and individual growth except perhaps for a few individuals. A third type is based largely on density, and is the tolerance capacity or the level at which territorial considerations at an intraspecific level force surplus individuals to migrate, or to be denied basic necessities like food, or the opportunity to reproduce. Populations at the survival level may also exhibit such characteristics. Some parallel with the human population may be seen if we equate the survival and tolerance capacity groups with the poorest nations and the optimum

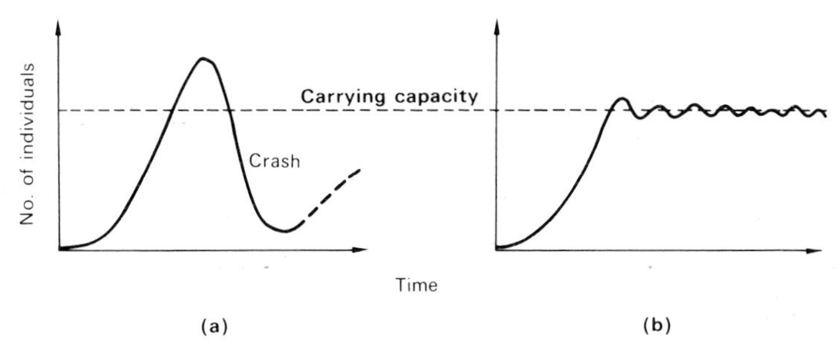

Fig. 1.13 Some populations rise rapidly, exceed the carrying capacity of their ecosystem and 'crash'; others level off near the carrying capacity, although oscillating around that level rather than maintaining it unvaryingly.

capacity with areas like North America and Western Europe; however, the density tolerance considerations might extend to the cities of the richer people too.

If we compare the dynamics of the human populations with that of animals, some similarities emerge. The curve for the numbers of the species, upon which there seems a wide measure of agreement between authorities, resembles a J-curve, but so far with only slight indications that an upper level may be reached in the foreseeable future; there is the added factor that for man K appears to be variable though presumably there is some upper absolute 'super-K' which cannot be exceeded (see p. 331). There are fears that the human population may breed itself beyond the carrying capacity of the earth, and this has renewed interest in the mechanisms by which animal populations regulate their numbers, especially since many of them stabilize below the numbers which would be expected if all the energy and space resources of their ecosystems were used.

There is some disagreement among zoologists as to the causes of the phenomenon, some of which may be explained by their investigations in particular kinds of environments, and it is not clear which if any of the mechanisms apply to man. Andrewartha and Birch (1954) stress the importance of factors external to the study population, such as food supply at critical periods and sudden shifts in weather, and note the lack of stability of the numbers of any one species even when the whole ecosystem is apparently not prone to large-scale fluctuations. Much of this kind of pattern operates in unfavourable environments or concerns organisms with a short life-span, and by contrast workers in more benign areas such as Lack (1966) stress the importance of density-dependent factors such as intraspecific competition and territoriality, and the role of biological controls such as predation. The homeostatic mechanism, according to Wynne-Edwards (1962), operates via the social behaviour of many animals. Features such as territoriality (in birds especially), the emergence of a hierarchy of dominance (usually called a peck-order and often a male phenomenon), and male sexual displays which are usually designed to secure status among other males, not to woo the female, are all ways in which breeding can be confined to a selected group in the population; in times of low rates of natality there are presumably eugenic consequences with survival value also. The need for the inception of the limitation mechanisms seems to be established at mass displays of either the whole population (the aerobatics of starling flocks are a possible example) or of the males, as in displays at the beginning of the breeding season which involve a number of individuals, like the virtual tournaments of sage grouse or ruffs.

Objections to such interpretations have been raised, e.g., by Hutchinson (1978): is it likely that a genetically determined mechanism can produce an altruistic population which in refraining from breeding will benefit all its descendants, when those individuals which do reproduce will leave more descendants than those practising the limitation?

It is tempting to seek analogies in human populations. Mass synchronous social displays take place every morning and evening in large cities (often more or less coincidentally with the starlings) and are called 'rush hours'; further human displays occur on Saturday afternoons except in high summer: these are largely composed of males and bright plumage is not uncommon. It would be worth seeing if any foundation would finance a project to see if commuters and football fans tend to have fewer children than the rest of the population. With more scientific logic, however, Wynne-Edwards (1962) concludes that *Homo sapiens* has lost its population regulation mechanisms. At the hunter-gatherer stage of culture this capacity was almost certainly present and contemporary 'primitive' groups provide evidence of population control, especially by infanticide. With the coming of the production of food surpluses by agriculture and the ability to store them against hard times, the control of reproduction passed from society to individuals. It has remained there from the agricultural revolution to the present day; current controversy over 'the right to reproduce' centres fundamentally on whether society should reclaim that control and enforce it through cultural measures such as contraception and induced abortion. Writers such as Stott (1962) produce evidence that stress in animals is often induced by overcrowding in advance of food shortages, and it triggers off mechanisms which reduce fertility or the survival rate of the young; he adduces similar evidence for the human species, based on gynaecological studies, and concludes that a world population 'crash' by starvation will be forestalled by increased rates of the incidence of sterility, perinatal death and unhealthy children, together with malformed and mentally impaired people. Avoidance of such unpleasant conditions must come via the cultural alternatives which are open to us. To quote and add an emphasis to G. E. Hutchinson (1978): 'It is very doubtful if any satisfactory mechanism can evolve by natural selection, *unaided by conscious foresight*, to limit growth of a whole population in such a way that resources are conserved for future generations'.

Culture

Man sees the world around him through the spectacles of culture, and so nature is thus transformed into resources. The elements of values, behaviour and technology which are fused together to make up culture are very varied, and the mix is different for diverse times and places: for some the spectacles are rose-tinted whereas others seem to see their world through distorting lenses. The sensory perception of the environment and the psychological translation and information of that knowledge into a decision to act, or not to act, upon the environment is a complex study, since between the perceptive input and the executive output lies a shadowy set of values conditioned by experience, imagination, fantasy, and other assorted intangibles derived from both rational and irrational sources. The ability to act is dependent largely upon the effective technology that a group or individual possesses, but behind the motivation lie the more elusive factors outlined above. There are, of course, cases where technology is used for its own sake: the alleged propensity of engineers to build dams whenever and wherever possible, irrespective of the need, is perhaps an example.

In general terms, the world exists not in absolute dimensions but in people's heads. The study of this state essentially consists of trying to understand the workings of the 'black box' of the human brain or group of brains, between the perceptive input and the output of action. The ways in which investigation may proceed vary according to the period of historical time involved. For the distant and preliterate past, there is only inference from the data furnished by archaeology. From such sources it is difficult to make firm statements about the relationships between man and nature in times long ago; it is not, however, impossible to speculate usefully, particularly if contemporary and near-recent ethnological parallels may be accepted. In the later chapter on the relations between unmanipulated ecosystems and hunting-gathering-fishing societies, some observations about this type of relationship will be found. In the case of the literate past, we have writings as a source of evidence for man's attitudes to nature. Some of these will deal explicitly with such a theme, others only implicitly, and there is in published work a great concentration upon the intellectual history of the West in this particular instance; other cultures have in general been little explored from this point of view. The use of written sources as a guide to themes in the relationship of man and nature carries over into the present, where many sources yield both clear-cut and covert clues about the ways in which men view their environment. But those who write about such phenomena may not be typical of those who act, and here the study of the environmental cognition or perception of resource managers of all kinds by means of the methods of psychological research has become of some interest and importance.

Man and nature in the past

As hinted above, the evidence derived from written sources can sometimes be criticized as unrepresentative of the reality of its age: a partial sample of a social stratum far removed, perhaps, in values and motivations from those actually engaged in dirtying their hands with real resources. Yet such writers are the formers of images of the world, no less than the recorders of other people's images. They deserve consideration, for many people act according to the image of the world that they have in their heads rather than in the light of objective 'scientific' information, and in any case the whole intellectual climate of opinion of a given time may determine whether or not 'scientific' knowledge is used, or indeed, misused. Keynes said, 'a study of the history of opinion is a necessary preliminary to the emancipation of the mind', and to discern trends in man's view of nature, where externalized in communicable forms such as writing and art, is a necessary preliminary to any study of man–nature relationships, and the ecology of resources is no exception.

From the times of classical antiquity to the eighteenth century, as Glacken (1967) points out, certain themes have dominated man's view of nature in the West. (The rest of the world had its traits also, but they are more diverse than those of the West, and less widely exported.) The idea of a designed earth, for example, is strong in the Judaic-Christian tradition. This essentially theological idea envisages the earth created for man or else for all life with man at the apex of a chain of being. Before the coming of evolutionary theory and ecology this was the West's great attempt to formulate a holistic concept of nature. In this way as many phenomena as possible were brought within the scope of the central theme, demonstrating a unity which was the achievement of a Creator. Although the seventeenth and eighteenth centuries saw criticism of this idea (not of the concept of design but of the relation of this order to the creative activity of the Deity), it could be extended to accommodate the theory of evolution when this began to emerge. At present, the study of ecology, whose message is essentially that of seeing systems as wholes and of perceiving the order within them, is a logical heir to this tradition of thought. Part of it also is the concept of 'man's place in nature': the postulation of a division of man from the rest of nature was recognized for instance in Genesis (I, 28–9) which says to men 'be fruitful and increase, fill the earth and subdue it, rule over the fish in the sea, the birds of heaven and every living thing that moves on the earth'. The persistence of this theme is symbolized by the papal encyclical of 1967, 'The development of peoples', which reasserts the idea of the creation of the earth by God for man. This book, while not aspiring to the authority of those two documents, also accepts that man and nature are different and must be reconciled.

A counter-argument to this theme finds it impossible to admit of a world being created especially for men when they themselves are often so wicked, and when the physical constitution of the world is obviously so imperfect as a habitation for them. In this tradition are the ideas of St Francis of Assisi, who asserted that, although man might be at the apex of creation, this did not mean that all life was for him and at his disposal. The same thoughts are exemplified in present-day protagonists of wildlife, who assert that plants and animals must exist in their own right rather than just by leave of, or for the use of, man.

A second major strand has been the study of environmental influence upon culture which carried with it the corollary of environmental limitations, and so we find Malthus's concern over the adequacy of food supplies falling within this theme. As is well known, he claimed that the growth of population would always outstrip the means of subsistence, and hence the potential perfectability of man being promulgated by some of his contemporaries was, for environmental reasons, an impossibility.

Thomas Malthus also dilated upon the modification of the earth to accommodate the growing populations of men, thus participating in the long-lived philosophical theme of man as a modifier of nature, a third major strand of thought. In some forms this was an optimistic tradition: man's skill was to put the final touches to God's unfinished work, or he was to be a bringer of order to, and custodian of, nature. If Ray and Buffon were optimists, then Malthus and, later, George Perkins Marsh were pessimists, for they saw chaos instead of order and profligacy instead of stewardship. Malthus in particular did not believe that nature would improve or that institutional reform could alter his basic principle of population. Nature was niggardly and man was slothful (except, presumably, where the passion between the sexes was concerned): optimism about the future of man had to be guarded and the possibility envisaged of only limited progress.

Marsh (1864) was mainly concerned with places rather than people. His writing about the changes wrought by man pointed out that not all these had been injurious to the environment, for soils had been improved and marshes drained so that civilized life might be enjoyed. He also pointed out that environmental change was a geological fact in many cases, but he is best known for his strictures on the misuse of nature that resulted in soil erosion, silting, fires and other downgrade processes. In many ways he is the intellectual forefather of the present all-pervasive concern with the disruptive effects of the technology made possible by our command of energy sources. His words might well adorn the portals of every resource agency in the world:

> man has too long forgotten that the earth was given to him for usufruct alone, not for consumption, still less for profligate waste.

Man and nature today

The themes of man's place in nature, of his subjugation of other components of the planet and his modification of its systems, cannot be said to be dead, although their context has been completely altered by the events of the nineteenth century. In practical terms the industrialization of the West was single-minded and inexorable; it produced the greatest alteration of nature, the greatest inroads on the world's resources and the greatest contamination then experienced. One effect seems to have been the heightening of confusion about man's place in nature. Omnipotence seemed to be manifest, yet the lot of many was so obviously unimproved.

One twentieth-century reaction to drifting man, cut off from the earth by the advent of urbanism, has been labelled alienation. It appears to be partly due to the low quality of contemporary man-nature relationships and partly a result of the sheer numbers in our society: the old and the new are vividly portrayed by J. B. Priestley and Jacquetta Hawkes (1955), in their contrast of the Indians of New Mexico and the 'anglos' of South Texas. One literary expression of alienation comes in, for example, the plays of Samuel Beckett, for whose characters the world is unreal and cannot be related to themselves. Man no longer belongs in any natural setting and his externalities (including the environment) are alien and almost certainly hostile. It can thus be treated aggressively or at the very least be regarded as a storehouse whose depletion is of concern only to the owners. The inventiveness and applicability of nineteenth-century technology meant also that the resources of the world appeared to be infinite, for new means of making them accessible and new markets for material products went together.

The generally cornucopian view of nature (provided that the proper social structure is present) promulgated by Marx and his later followers is an example of the attitudes to the relationship of man and nature which the nineteenth century produced. Most Marxists reject Malthusian views and so their materialism is usually regarded as being hostile to harmony and stability in man-nature relations (Fry 1976) although some attempts at reconciling the views of Marxians and Malthusians have been made, for example by H. E. Daly (1971). The general attitude that nature imposes no limits on human capabilities lives on. Colin Clark (1977), for example, values population growth because it forces innovation and social progress, and large populations enable economies of scale to be practised. He considers that it is possible to feed many more people than are at present alive but does not consider the environmental implications of doing so, and has been strongly criticized for being selective of only those aspects of population and economic growth which bring benefits. Deriving from a similar intellectual ancestry is much democratic socialist thinking, as of Crosland (1971) in the Labour Party of Britain. He contends that only continuing growth creates enough wealth to clear up the mess left by nineteenth-century industrialization, and that any diminution in the rate of economic growth results from the snobbery of the middle class in trying to prevent the workers from attaining what the wealthy already have. A slightly ameliorated version of the same document is taken by the committee of authors who wrote a British document (Verney 1972) in connection with the UN Stockholm Conference of 1972. The traditions of the mastery of man has become a central idea in, for example, contemporary Russian versions of Marxism and has led to various environmental disruptions (Goldman 1971, 1972). Traditional Chinese ideas about the harmony of man and nature have since 1949 been overthrown by the instilling of the necessity for man to dominate; but as Yi Fu Tuan (1968, 1970) has shown, the dichotomy between the philosophy and the actuality in ancient China regarding the treatment of nature was very wide, owing no doubt to the divorce between philosophers and peasants.

Alienation of a more intuitive character is expressed in the West in the 'dropout' or 'alternative society' concepts practised by some people, and it is significant that a return to older, land-based ways of self-sufficiency is advocated by many of them. More conventionally, a scientific alienation is taking place. As a result of scientists' abandonment of control of their discoveries, various catastrophes could occur and foreseers of doom have thus begun to emerge from science to postulate anew the 'dismal theorem' which Malthus first stated. The numbers of man and his treatment of the planet will, they aver, condemn *Homo sapiens* to speedy extinction: from starvation, from nuclear war, from poisonous compounds in all parts of the environment or from the sheer breakdown caused by the impossibility of controlling and keeping healthy so many millions of people. The examination of this position will form one of the concluding parts of this work.

The foregoing discussion has emphasized a dichotomy between man and nature. Whereas 'primitive' people seem to have enjoyed a close association with nature (probably too close for our present ideas of comfort), literate man has erected a dualism, in which 'progress' has been linked with control over nature. For many commentators it becomes an increasingly unsatisfactory idea and to them the future would scarcely seem to permit the indifference to the natural world, our environment and provider of our resources, which this attitude allows and may even encourage: it appears healthier for man to regard the planet less as a set of commodities for use and more as a community of which he forms a part. They concede that such an attitude will mean the abandonment of the central theme of our intellectual heritage and especially of its religious accompaniment, its anthropocentricity. To do that requires a revolution in thought, involving essentially the realization that our survival as a species is dependent upon non-human processes; an idea antihetical to most of the traditions of thought discussed above. But too often they advocate, or are interpreted to advocate, a return to some pastoral idyll. There are already too many people for such a relapse and we cannot reverse the time-trajectory of our man-directed systems. In managing the earth for our survival there is no way in which technology can be abandoned: what appears to be essential is a deeper knowledge of the relations between man-made machine-dominated systems and the bioenvironmental systems, and ways in which a stable co-existence can be procured.

Economics

The practice of economics is today the single most important feature governing the relations between man and nature. Based simply on the premise that no substance is in unlimited supply for all individuals, it aims at bringing together supply and demand by the mechanism of price. The concept appears to be identical whether the currency be conch shells, gold or Eurodollars. As far as resources are concerned, economists suggest that they are very largely created by price. If a material becomes scarce, then its raised price will make economic its extraction from poorer quality sources or will enforce a substitution, the invention and production of which is one of the proper roles of science and technology.

But the mechanism by which price is determined also has environmental relevance. In Marxian economics, for example, the value of a product is related solely to the amount of human labour that has gone into it and so both resources and environment have no value of their own. By contrast there is the consumer theory of value: if we want something badly enough we will pay heavily for it, as with oil where the price is 10–100 times the cost of production and transport. A kingdom, we may remember, was once offered in exchange for a horse. There is also a production theory of value which has gained impetus since 1973 which relates to the amount of energy used in the production of goods. If energy were very cheap then the Marxian view would probably be the closest to reality but at present the consumer theory seems to be the best fit to real conditions.

In some ways, economics is not always totally suitable for regulating human resource needs at present. At the most elementary level, price cannot always be equated with value, or to put it the other way round, some things of value have no price. They may be thought to be beyond such considerations or there may be no accurate method of fixing a price. Nobody knows the true value of a large area of unspoilt land of outstanding scenic beauty, for all the methods normally used to fix prices fail in some respect or other. In this case, large numbers of people attach value to such an area simply because it is there; they may have no intention at all of using it. When a controversy arose over the construction of a dam in Upper Teesdale in NE England, the 'conservationist' party could not assess the 'value' of the rare arctic-alpine plants, some of which would be destroyed, whereas the Water Board could easily quantify the 'value' of the extra water to be gained from the impoundment.

Economics has long been concerned with a rational man who optimizes his spending in accordance with carefully chosen criteria. But Galbraith (1967) has argued that the manipulation of the market by large industrial corporations backed particularly by demand manipulation through advertising has made nonsense of the myth of 'consumer sovereignty'. Economics therefore is less and less concerned with the individual's basic needs and spontaneous demands: rather is it more and more determined that he shall buy what industry thinks he ought to have. Keynes, too, pointed out that the interest of the individual is not necessarily the same as the general interest. Hardin's (1968) paper on the 'tragedy of the commons' put this in a resources perspective for he uses the example of an area of common grazing where it is in every individual's interest to pasture more animals but after a certain number is reached, then every new animal's presence is contrary to the common interest.

This relates to another generality, namely that some resources, such as food uncontaminated by residual pesticides, or clean air, or indeed the whole complex of values usually called 'environmental quality', are not in the market, and hence are not subject to the choices normally experienced there. Unless we are very rich we cannot go into the resource supermarket and buy an absence of air pollution, for example, as we could iron ore or wood pulp. In connection with the side-effects of particular processes, we should note that the market mechanism frequently overlooks what are called 'external diseconomies'; the full social costs are not taken into account because they are paid for, usually by society at large, elsewhere in the economy (Mishan 1967). The increasing use of 'disposable' articles, and the whole industry of packaging, is an example in point: in the price of elegantly packed shirts the packing does not include the costs of removing it from the household, converting it to ash and carbon dioxide, and taking the ash where it will not be deposited on one's other shirt which is hanging out to dry. The external diseconomies of widespread private ownership of motor cars in relation to air pollution, noise and premature death are not reflected in the bargain price which the dealers seize every opportunity to acquaint us with. Neither are market economics well fitted to respond to problems which force themselves suddenly upon resource managers or the general public, nor to cope with the long lead times which complex technology needs in order to develop substitutes.

The outgrowth of these and other trends in values to which economics appears to have given rise and which have been accepted by most politicians is a series of priorities which have been unquestioned at least since the industrial revolution but which now are beginning to be examined more critically. Initially, attention has focused upon the high place given to the production and consumption of commodities, where a rapid rate of throughput (achieved where necessary by 'built-in obsolescence') is sought. To sustain and enhance this rate, indices like the GNP are watched as barometers of the nation's health, and if the GNP should fail to rise at a pre-ordained rate then dire consequences are forecast and everybody is exhorted to work harder (Boulding 1970). The world view adopted appears to regard the population as being a system with an infinite pool of resources from which to draw and with a sink of infinite size down which to flush unwanted by-products and other discarded materials. The thrust of industrial states, both capitalist and socialist, is towards increased production and throughput of materials, without any concern for what biologists would call a dynamic equilibrium.

Although some economists still argue very strongly for economic growth as conventionally measured (e.g., Beckerman 1973, 1974), a more radical set of ideas has been put forward, trying to elaborate an economics of the steady state with no growth, or very little, in economic terms in the industrial nations (Boulding 1970, 1972). There has been as well a great quantity of writing tackling, for example, the economic problems of facing up to external diseconomies such as pollution. Can the market mechanism be made to cope with this problem, and who should pay for a cleaner environment? Or is it always essential to have legislation by central government? The investigation of what now governs the price and supply of materials and in particular the role of energy, the economic relations of rich and poor nations, and the construction of new measures of human welfare, all include considerations which involve the whole resource process rather than the portion of it which has for long been the concern of economists (Krutilla 1972, Lecomber 1975, 1979, Pearce and Rose 1975, Kneese 1977, Cottrell 1978).

Environmental perception

The role of the study of environmental perception or cognition has been briefly mentioned. The articulate expressions of attitudes may be seminal in determining long-term trends, but the majority of people make decisions at a more intuitive and less verbalized and rationalized level. Just as we have examined man—nature relationships viewed in terms of scientific and intellectual history, now we may turn to their study from within: how more ordinary individuals view the world about them and their use of it (Sewell and Burton 1971). Many aspects of the man—nature interaction result in patterns of resource use which are not rational in the eyes of Western science. The irrationalities which we detect in both industrial and non-industrial societies are plainly the result of conflict within the various elements of the resource process. In particular, the role of culture is very strong, and while we may study objectively the results of a particular cultural trait in terms of resource use, it is valuable to be able to penetrate the mind of the individual and see the way in which decisions come to be made (Saarinen 1969). The study of the environmentally orientated mental process of the individual 'resource manager' faced with the problem of what to do with his effluent may act as a key to the understanding of many of the irrationalities in resource management and use.

A number of processes are involved. There is firstly the sensory perception of the environment. At the 'primitive' level this is confined to the organs of the human body, but remote-sensing technology is becoming very important. Starting with conventional aerial photography, this has developed into a tool of immense precision using satellite photography, multi-spectral sensing, radar, and computer rectification of received images. In this way, photographs of infra-red reflection for example can be used to study variations in primary productivity, disease infestations of plants, or sources of thermal contamination of water. The United States currently has in orbit a satellite specifically designed for resources work and which transmits imagery from a number of optical-mechanical scanners and microwave bands. Radar imagery is of course independent of weather and diurnal conditions. Inventory and mapping of ecosystems and resources, quantification of ecosystem flows of matter and energy, and the monitoring of change are all facilitated by the application of this kind of technology (Holz 1973, Johnson 1971, Rudd 1974, Barrett and Curtis 1976, Peel, Curtis and Barrett 1977, Sabins 1978). It may be noted, however, that for some substances, especially those in the earth's crust, the age-old methods of trial borings or diggings have to be adopted in the end. But the perception of resources is becoming ever more keen and the inventory of potential supplies more complete, although clearly we are a long way from knowledge of the entire stock of the world's potential resources.

The next stage is also very important. It is here that culture, in its myriad facets, plays a leading part, for the transformation of the perception of the source of a material into the cognition of a resource occurs at this stage. Here resources 'become'; the 'cultural appraisal' takes place. The values informing the cultural appraisal are many and varied. Knowledge, especially of an objective scientific kind like knowing the uses of a particular forest tree or the edibility of a fungus, is central; but so are various types of prejudice, experience and imagination. An item of knowledge may become ritualized into a prejudice, as may have happened with pork in the Middle East; religion and magic are often powerful at this stage, hence the selection of a settlement site may in some cultures be determined less by the environmental suitability than by the interpretation of the numinous qualities of the place by the practitioner of spiritual affairs. Numerous examples of apparently irrational cognition of the qualities of the environment can be discerned, and incomplete appreciation of the properties of a resource is termed cognitive dissonance.

Even in the most sophisticated societies, the completeness of the resource manager's cognition is imperfect. The constraints are many: the costs of obtaining additional information, the lack of technological knowledge, human fallibility and chance are all possibilities. The complexity of the ecology of environment and its social veneer means that the capacity of the human mind for formulating and solving complex problems (even when aided by the computer) is very small compared with the size of the problems whose solution is required for objectively rational behaviour in the real world. Resource managers then become, according to Kates (1962),

> men bounded by inherent computational disabilities, products of their time and place (who) seek to wrest from their environment those elements that make a more satisfactory life for them and their fellows.

In deciding between alternative resource processes, managers may elect to choose consciously, or may adopt more reflex attitudes such as habitual choice with its recourse to traditional or repetitive behaviour, or may engage in unconscious or trivial choice. The choice must then be implemented. Apart from institutional considerations, the primary factor here is the technology available to the manager: the extension, as it were, of his arm. Although physically and temporally separate from the cognition process, it is in reality also part of the cognition process, for knowledge of technical capability is an important part of the decision-making process. The whole system can be approximately summed up in diagrammatic form (Fig. 1.14). The reality is much more complex, especially in the number and nature of feedback loops, but the diagram presents a useful summary model.

Whatever the complexity of the processes, the outcome in terms of cognition of, and adjustment to, the resources of the environment is of some importance. For sophisticated groups this result may mean the difference between comfort and discomfort, profit and loss; for subsistence cultures, life and death may be at stake. Nevertheless, before the advent of Western culture, some 'primitive' groups had been notably successful in adapting to extreme environments in which there appears to our eyes to be a paucity of resources. The Bushmen of the Kalahari Desert are an often-quoted example: to them the recognition of sources of moisture in their arid surroundings is the critical element. Thus all manner of plants and animals are used for food and some of these probably function largely as sources of moisture. The people become adept at spotting the traces of delicacies such as buried ostrich eggs, for example, which add a valuable element to their diet. The Eskimo groups of northern North America were also notably successful in adjusting to the food and material supplies of their unyielding environment. This success traditionally involved limitations of the population level of a group by means of infanticide and the exposure of non-productive old people. Contact

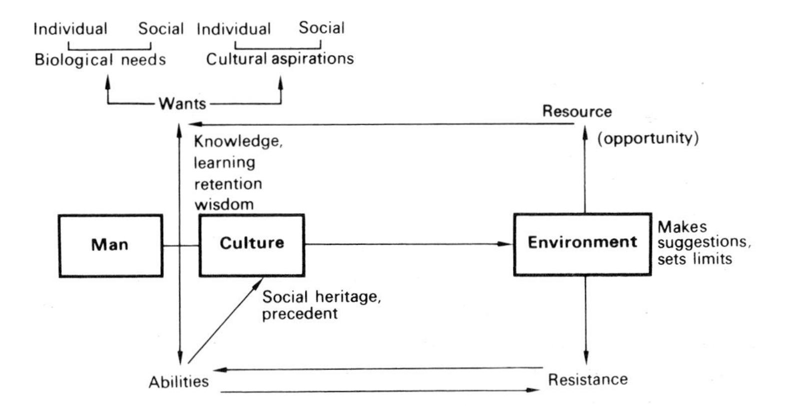

Fig. 1.14 A diagram of some of the factors involved in environmental perception. In reality there must be many more feedback loops than can be suggested in such a scheme, and culture appears to be very diverse.
Source: O'Riordan 1971b

with American culture has changed some of the Eskimo ways, and Sonnenfeld (1966) has suggested that landscape preferences among them may be related not only to their economy but to the degree of 'Westernization' which they have undergone.

Environmental cognition by more advanced people is often a very pragmatic consideration. It may often be important, for example, in the recognition of the probability of environmental hazards where these phenomena are not regular. This is particularly so in areas where such a hazard may make a great deal of economic difference to an individual or group. All kinds of hazards are mis-perceived or else call forth an inappropriate response: factories are sited on flood-plains, towns built along active faults and hurricane warnings ignored (White 1974, Burton and White 1978). Small wonder that some writers think that no matter how dire the warnings of resource shortages or environmental degradation, there will always be a significant number of people who will not believe it could happen.

The significance of environmental cognition and subsequent behaviour towards the local resources is not confined to instances of 'economic' activity such as agriculture, industry and settlement. Lucas (1965) has shown how the perception of the qualities of the Boundary Water Canoe Area (Minnesota-Ontario) varies with the type of user. Those who paddled their own canoe were willing to tolerate a lower level of obvious human use (i.e. logging) of the surrounding forest than those who used outboard motors. Clearly this has significance for the 'wilderness' type of recreationist for whom ecological virginity and the absence of fellow humans are paramount. Management of the resources of such an area can thus be directed at maintaining a certain mental condition through supplying the correct perceptive stimuli which assure the user that he is in a wilderness area. This may have considerable application in crowded countries where wild recreation areas are difficult to find.

Different backgrounds can lead to differing opinions as to the significance of a given body of facts. This is very noticeable in the gap between some social scientists and most natural scientists over the 'population explosion' and its effects on resources and environment. The former tend (with notable exceptions) to be optimistic and to regard sustained population growth as an acceptable and even desirable phenomenon; the latter are often Malthusian in outlook and make gloomy prognoses. These divergent views of the relations of population and resources are clearly germane to the theme of this book and will be examined in more detail in Part III.

Ecology and economics

Of all the attitudes to the relationships of man and nature that have been discussed, ecology and economics have become most important, and of these two the latter is dominant. Its purpose, as Caldwell (1971) points out, is simple, direct and obvious: if material wealth is seen as the natural and proper goal of man's activities, then the mastery of man over nature (and the pre-eminence of economic thought over ecological thought) becomes a fulfilment of human destiny. The perspectives of ecology are different from those of economics, for they stress limits rather than continued growth, stability rather than continuous 'development', and they operate on a different time-scale, for the amortization period of capital is replaced by that of the evolution of ecosystems and of organisms. So some reconciliation of the two systems of thought might be held to be desirable, in which the findings of one science might be translated with some precision into its impact upon the other, and the values suggested by ecology might become the operational dicta of economics and *vice versa* (Rapport and Turner 1977).

Attempts to find isomorphic concepts and to transfer the language of economics into the realities of environment have been made for example by Boulding (1966b, 1971) when he compares vital flows in both systems, such as energy in one and money in the other. Economic accountings of 'production' of goods and services such as GNP are in fact mostly measurements of decay, since cars and clothes wear out and food and gasoline are oxidized. The bigger the economic system, therefore, the more materials that have to be destroyed to maintain it. Holling (1969) also draws parallels which rely on common features between ecological and economic systems, such as high diversity and complex interactive pathways, distinctive historical character and important spatial attributes, and important structural properties such as thresholds and limits. These led Holling to simulate the process of recreational land acquisition as a comparative process to that of predation in ecological systems.

The possibility of energy as an intermediary between ecology and economics is fairly obvious, since it is important in ecosystems at all scales and can also be bought and sold; H. T. Odum (1971) attempts to provide energy flows for a set of marine bays in Texas which bring together all the flows of the ecosystem and man's industrial and recreational use of it, principally by allocating a dollar value to each kcal of energy expended in any activity, e.g. each visitor is assumed to expend 3,000 kcal/day in the area. An extension of such work has attempted to place monetary values on the 'work' done 'free' by natural ecosystems, e.g. in processing sewage in places where there are no treatment plants. Even more explicit is Deevey's (1971) statement that energy is the 'key to economic growth and the proximate cause of environmental pathologies', and he ranks the energy flows of the world in the following ratios:

Plant biosphere	*Industry*	*Husbandry*	*Personal*
300	25	2	1

noting that only the first-named is actually production, the others being consumption, and that the figure for industry represents a level of organization which he finds superior to pastoralism. He stresses, however, that if any of the 25 units is 'invested' in making the 300 non-renewable in a long-term perspective, then this is courting trouble; yet much of the use of energy by man in the cause of gaining economic wealth appears to be doing just that. One of the most stimulating contributions has come from Georgescu-Roegen (1971, 1975, 1976) who characterizes most economic activity as building up entropy (i.e. disorder in physical terms) in the bioenvironmental systems of the planet. Since the concept of entropy is linked with the nature of energy and with the relations of order and variety, links with ecological thinking and measurements may well emerge.

Energy appears to be a workable intermediary between ecology and economics in an empirical and descriptive context: we can say that so much energy has produced a particular degree of ecological change. But prediction is much more problematic, for some ecosystem energy-flows may be totally disrupted by the removal of one component, possibly by a very small input of man-manipulated energy, whereas others may require a gross perturbation before they are deflected from their stable state or their successional trajectory; without knowing most of the functional and structural characteristics of an ecosystem, the impact of economically directed energy flows is impossible to foretell except in the most general terms. The relations between order and diversity and the use of tools such as information theory may help to build a body of empirical and theoretical knowledge about this critical interaction, as may other scientific techniques; it is clearly a field in which the practical applications would be considerable.

Interactions

As has been stressed above, energy forms a medium of connection between man and his surroundings, no matter in what form it comes. The first and most obvious of the forms is manpower itself, which requires only normal survival conditions for its basic effectiveness although this is exhanced by social organization. Other forms of energy have to be harnessed and therefore at least to some extent understood before they can become useful ways of doing work. Fire was doubtless the first, followed by animal power made possible by the use of domesticated beasts in order to perform various tasks, particularly those associated with agriculture. Inanimate sources of power were for long dominated by wind and water, the latter being especially responsible for much industrial development before the nineteenth century; it persists chiefly in the form of hydro-electric power. But the great surge of industrialization dating from the nineteenth century is linked to the elaboration of ways in which to exploit fossil fuels, dominated at first by coal and more recently by oil and natural gas. In the mid-twentieth century atomic power has been added, where the energy of the nucleus of the atom is released in a controlled fashion to provide immense quantities of energy as electricity.

It is possible therefore to divide the history of man—environment relationships into periods characterized by cultural stages in which man's access to sources of energy plays a key role.

The first level we can distinguish is that of the hunter-fisher-gatherer. Once the dominant way of life of man, only a few groups employing this mode of subsistence have persisted to the present day. They have had access only to human energy for their environmental relations, which are dominated by the acquisition of food. The more advanced types have been able to channel this energy technologically: the blowpipe, bow and slingstick represent ways of concentrating the power conferred by human anatomy. Using these weapons, their access to environmental resources concentrated on the cropping of biotic organisms for food, with preferences for large mammals and fish being exhibited by many groups. Where a cutting edge or penetration power was required, then stone, especially in the form of flint and chert, was a common inorganic resource. Many other food sources such as small mammals, birds, berries and fruits were taken, but we may speculate that the expansion of human numbers took place up to the limits of a preferred food source rather than a total supply. Even so, such an economy required considerable space, and territoriality was probably well developed in the various biomes colonized by such people where the extremes of both desert and tundra have been occupied at this level of technology, as well as the less hostile life zones. In suitable environments access to a partly controllable energy source like fire was undoubtedly the most powerful way of altering the ecology of their territory. With it beasts might be run towards traps or over cliffs, and underbrush cleared for easier sighting of game. In forested terrain the regular use of fire would gradually produce a more open habitat, as is happening in regularly burnt savannas today.

The early farms of the Neolithic of south-west and south-east Asia and meso-America represent a tremendous change in the resource base of men. Now the food supply was under much more direct control and it was more concentrated; much less territory was required than for a purely hunting and gathering existence, although these activities persisted. Energy use at first was human-based only, via the hoe and digging stick, but the introduction of light ploughs made possible the use of draft animals. As metals like copper, bronze and iron came into use, so specialists in their fabrication emerged, and there is evidence from certain outcrops of rock, in Britain for example, that axe-factories with a far-flung trade network were established. The soil became a perceived resource, as did the forests, grasslands and savannas which provided the background for much of the early agriculture of the shifting cultivation type. In Bronze Age Britain, for example, evidence of the selective clearance of lime (*Tilia* sp.) trees has been interpreted by Turner (1962) as indicating a knowledge that good agricultural soils would be found underneath such trees. Even relatively primitive forms of agriculture, such as that described by Rappaport (1971), show favourable energy relationships for the societies involved. Tono-yam gardens in New Guinea, for example, had a calorie yield: input ratio of 20 : 1 and sweet potatoes of 18 : 1, energy expenditures in weeding, harvesting and carrying the crop all being counted. As might be predicted from ecosystem energetics, the introduction of pigs reduced the ratio to 2 : 1 at best, and it was often $< 1 : 1$, the pigs being valued then for their protein rather than their calorific value, and for non-dietary purposes.

The success of agriculture, coupled with increased knowledge about the storage of surpluses, may well have led to the removal of the decisions about family size from the group, as in hunting societies, to the family, where it has subsequently remained; one result was usually a considerable increase in population levels whenever agriculture was introduced or evolved.

The uncertainties of crop failure from blight or drought are combated by more advanced agriculturalists of the types found in ancient Mesopotamia or medieval Europe. Environmental resources already known assumed new degrees of importance: soil, for example, may be critical to success, and so the selection of the best soils for a particular crop, the draining of fields as by ridge and furrow ploughing, and the maintenance of soil fertility by replacing lost nutrients in the form of manure became subject to group decision and control, as in the manorial system of medieval Europe. Fossil sources of fertility like the calcareous Crag rock, which was used to add lime to the sandy fields of East Anglia, enter the resource nexus, and the uncultivated land or 'waste' beyond the fields was seen as a feeding ground for protein sources such as pigs as well as a source of raw materials and fuel. The overcoming of water shortages through irrigation is an aspect of environmental management and alteration. Its success in arid areas like the ancient Near East led to the accumulation of permanent surpluses which in turn permitted the development of urbanization. Ecological mistakes were sometimes made, as in Mesopotamia, where salinification of the irrigated fields and the choking of the irrigation canals with silt from overgrazed watersheds may have caused the downfall of great cities (Jacobsen and Adams 1958). By contrast, the agriculture of western Europe has been markedly stable in its environmental relations, although it is currently beginning to show some signs of breakdown in soil structure. Another effect of agricultural surpluses was the development of urban specialisms such as writing, which is crucial to centralized authority. In this manner, official attitudes to and uses of environment can be promulgated, and the archaeological evidence of the state religions of Egypt, for example, reveals a high degree of attunement to the hydrology and ecology of the Nile. In medieval Europe, authority might through the possession of literacy order an inventory of the resources of a kingdom: by no other means could Domesday Book have been compiled.

The age of the use of fossil fuels, which started in earnest in eighteenth- and nineteenth-century Europe and which flourishes unabated today, has marked a great difference of kind as well as degree in the environmental relations of man. Access to energy stored in coal and oil, with the later addition of hydro-electric power, has allowed the use of the earth's resources on a scale hitherto unimaginable. The steady improvement of technology has meant the recovery of progressively poorer deposits, so that scarcity, except on a regional basis, is practically unknown. Availability of relatively cheap energy encouraged moves towards energy-intensive economies (rather than labour-intensive ones) in industrial nations, a process only now looking as if it is coming to an end. A great environmental impact has been felt from the use of these energy sources in the mining, smelting and use of minerals and other materials of the earth's crust, so that every industrial region is marked with a million holes and heaps.

The environmental effects of this 'industrial revolution' have been massive and are well catalogued. Urbanization has been the most noticeable: in many ways quite unobjectionable but all the same providing concentrated sources of atmospheric pollutants, together with industrial and domestic effluents to affect the biota (and other men) far beyond the cities of their origin. There have been many other effects also, such as the devastation of forests to provide lumber for building and pulp for paper; more efficient hunting techniques have brought whales to the brink of extinction and some species of plants and animals have been exterminated due to habitat alteration or to increased accessibility allowing more thorough hunting: the passenger pigeon owes its demise to a combination of all these. Agriculture too has become industrialized, with energy from fossil fuels replacing the work formerly done by man. Planting and seeding have become mechanized and are preceded by commercial preparation of the materials;

industrial fertilizers have replaced the virtually closed cycle of cropping and manuring; weeding is carried out by mechanical and chemical means, as is the preparation of the soil; and insecticide use further reflects the binding of agriculture to the industrial world. H. T. Odum (1971) has calculated that an agricultural population living at 170 persons/km^2 can support 32 times that number in cities. A production yield of \$24·30/ha for grain in the USA, for example, is due to input worth \$21·87/ha for goods and services from the industrialized culture. The energy flows of an industrialized agriculture are summarized in Fig. 1.15.

Man has so far found he is able to remake the world according to his own plans, and his attitude to the natural world (at any rate the Western attitude, which at the height of European imperialism meant the attitude of much of the globe) has been by and large to maximize economic gain in the short run. Further expansion of activity has often been seen as a cure for present problems, an attitude given much credibility by the explosive growth and impressive achievements of science and technology. Of late there has been a swing from the unbridled optimism of the nineteenth century, and concern over the protection of nature and scenery, moves towards pollution control and anxiety over the effects of population growth on the biosphere have all become manifest in industrial nations; most less developed countries are as yet inclined to feel that these negative aspects of economic growth are bearable if they bring with them the enhanced prosperity enjoyed by industrialized countries.

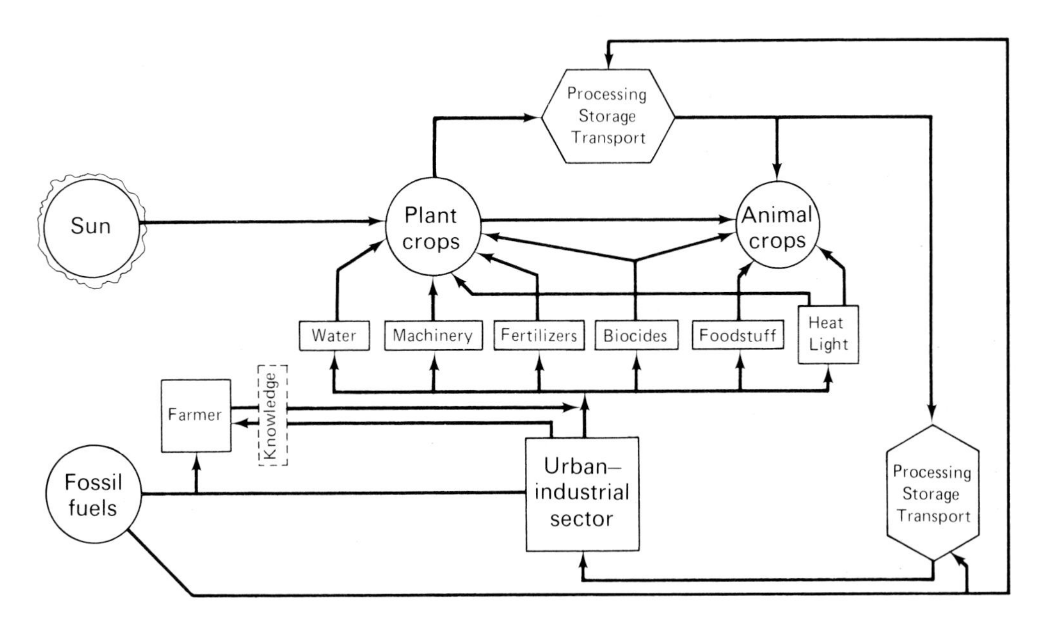

Fig. 1.15 Simplified network diagram of energy flows in modern agriculture. The addition of fossil fuels means that the agricultural population can support 32 times its own number in the urban-industrial sector. The quantity of energy applied to the crops via the machinery, fertilizer, etc. pathways is the energy density for the farming system and beyond certain levels produces spin-off effects not shown here. Wastes are not shown on this chart.

The year 1945 was the beginning of the atomic age, with the release from the Los Alamos bottle of a genie of quite ferocious power, and one whose destructive potential has been subject to considerable augmentation since then. Energy from nuclear fission is used for power generation too but is also available for boats, mostly submarines. Nuclear fusion is currently difficult to control, but if it becomes more tractable, and available in small doses, we shall then have a virtually unlimited source of energy. We need to be reminded, of course, that whatever comes from planetary material is eventually finite, and that energy made to perform work is dissipated as heat. The rate of increase of the energy consumption of the USA, for instance, if projected for a few hundred years would make its land surface as hot as the sun.

The introduction of large-scale power is dominated by its cost in relation to the rising expense of the recovery of fossil fuels; and the true cost of nuclear power is not yet known, since at present its use is subsidized by fossil fuel utilization in the process of prospecting for nuclear fuel, for example, and in many other ways. The potential environmental effects of nuclear energy are immense. In an unlimited nuclear war, a great part of the surface phenomena of the earth would be devastated and the legacy of the explosions would live on physically during the lives of the radioactive fallout particles and genetically for innumerable generations of whatever organisms were left: imaginative playwrights and SF writers seem to favour either rats or insects as the ecological dominants. Extraordinary effects could be deliberately wrought. Edward Teller has suggested blowing huge holes deep in the earth's crust in which to store rubbish, and the possibilities of thus 'digging' a second and larger Panama Canal at sea level are under active exploration (Rubinoff 1968). Tables 1.4 and 1.5 help to summarize the role of energy as a crucial element in man's environmental relations. Table 1.4 shows in particular the relative productivities, in both dry matter and energy content, of different food-producing systems, in which the dramatic increase between non-fuel-subsidized systems and those with fossil fuel input can be seen and extrapolated to its extreme condition in algal culture.

Table 1.5 puts food production into the wider context of a classification of ecosystems by their energy flow characteristics and intensities. At the top are the natural systems in

TABLE 1.4 Food yield for man

	Edible portion of net primary production	
Level of agriculture	*Dry matter (kg/ha/yr)*	*Energy (kcal/m²/yr)*
Food-gathering culture	0·4–20	0·2–10
Agriculture without fuel subsidy	50–2,000	25–1,000
Energy-subsidized grain agriculture	2,000–20,000	1,000–10,000
Theoretical energy-subsidized algal culture	20,000–80,000	10,000–40,000

Source: E. P. Odum 1971

TABLE 1.5 Ecosystem types classified by energy flow

Ecosystem type	Annual Energy Flow ($Kcal/m^2/yr$) Range		Average (Estimated)
1 Unsubsidized natural solar-powered ecosystems; e.g. open oceans, upland forests. Man's role: hunter-gatherer, shifting cultivation	1,000–10,000		2,000
2 Naturally subsidized solar-powered ecosystems. E.g. tidal estuary, lowland forests, coral reef. Natural processes aid solar energy input: e.g. tides, waves bring in organic matter or do recycling of nutrients so most energy from sun goes into production of organic matter. These are the most productive natural ecosystems on the earth. Man's role: fisherman, hunter-gatherer	10,000–50,000	The most productive natural ecosystems and the most productive agriculture seem to have upper limits of c. 50,000 $Kcal/m^2/yr$.	20,000
3 Man-subsidized solar-powered ecosystems. Food and fibre producing ecosystems subsidized by human energy as in simple farming systems or by fossil fuel energy as in advanced mechanized farming systems. E.g. Green Revolution crops are bred to use not only solar energy but fossil energy as fertilizers, pesticides and often pumped water. Applies to some forms of aquaculture also.	10,000–50,000		20,000
4 Fuel-powered urban-industrial systems. Fuel has replaced the sun as the most important source of immediate energy. These are the wealth-generating systems of the economy and also the generators of environmental contamination: in cities, suburbs and industrial areas. They are parasitic upon types 1–3 for life support (e.g. oxygen supply) and for food; possibly fuel also although this more likely comes from under the ground except in LDCs where wood is still an important domestic fuel.	100,000–3,000,000		2,000,000

Source: Odum 1975

which man functions only at a hunter-gatherer level, and in which neither nature nor man provided any energy 'subsidy'. Category 2 designates natural 'subsidies' where nature provides extra energy input: the case of the tides bringing energy and nutrients as detritus and carrying away wastes from estuaries and algal (coral) reefs is the best example; another is the case of riverine forests such as those of the Coast Redwood in California where there is a regular input of silt from floods. Systems 3 and 4 are affected by progressively heavier intensities of fossil fuel input to the point where the fuel-powered system gets most of its immediate energy from those sources rather than from the sun and is parasitic on the other systems for oxygen and food. Beyond even this system is the energy support needed for a lunar expedition, which is about $2{\cdot}7 \times 10^{12}$ kcal/cap/day, an astronomical quantity.

Man and ecosystems

Abstraction about the crucial role of energy flows in man's relations with the natural world are not the only generalizations that can be made about his interactions with ecosystems. His effect upon other features of their metabolism, such as limiting levels and their self-regulating mechanisms, must also be briefly considered, as must the outcomes of the magnitudes of the manipulations which he exerts, i.e. the transformation of natural ecological systems into resource processes, whose landscape expression is often identified as land use.

Human effects upon ecosystem processes

There are a number of facets of the ecosystem concept which relate to resource processes, among which the notion of limiting factors occupies an important place. First enunciated by Liebig as a 'law of the minimum' with respect to the supply of chemical materials essential for plant growth, it has widened to include more factors. The idea of tolerance is now used, since it is realized that the continuing presence of an organism depends on the completeness of a set of complex conditions. Tolerance, not only of 'environmental' factors but of intraspecific aggression and interspecific competition, varies with regard to each particular factor and may change synergistically in combination with other features of the organism's environment.

In any natural ecosystem the overall limiting factor must be the amount of solar energy incident upon it, but within this context many other boundaries may operate. The supply of a particular mineral nutrient may not only limit plant growth but because of its importance in animal metabolism may limit the number of animals too. The discovery of the importance of trace elements like boron in animal nutrition has highlighted the way in which a limiting factor may result from a quantitatively small component of an ecosystem. In his activities, man may alleviate a critical limit, as by using chemical fertilizers. Alternatively, he may introduce a new lower limit: the disposal of untreated sewage into coastal waters for instance, where its sheer physical presence may decrease the amount of light reaching the littoral and sublittoral vegetation and hence limit productivity. Such an action also puts a limit on the recreational use of the system.

Another feature of ecosystems which can be relatively easily measured is its diversity, usually expressed as a species : number ratio or a species : area ratio. These ratios increase during the early and middle stages of succession but appear to decline slightly once a steady state or climax condition is reached. Diversity is taken to indicate a state of complexity where energy flux and transformational efficiency are at their highest. A species may have a wide variety of food sources and an equally diverse array of predators so that any accidental disturbance will be damped down. Thus has arisen the hypothesis that stability of ecosystems is a function of their diversity (MacArthur 1955). Such an idea is probably too simple to apply generally, and other research has indicated, for example, that stability at herbivore and carnivore trophic levels increases with the number of competitor species (Watt 1965); and in one study, greater diversity at one trophic level was accompanied by lower stability at the next higher level (Hurd *et al.* 1972). Again, in some systems a key component can determine the stability of the whole of the system, as suggested by Paine (1969) in the case of the Crown of Thorns starfish irruptions. In the case of ecosystems cropped to provide resources or into which wastes are disposed, the important feature is the ability of the system to return to an original condition after a perturbation or even its attainment of a new level of stability under a permanent stress. Stability thresholds have been studied empirically, as in the case of trampling on alpine tundras (e.g., Willard and Marr 1970), and a good deal of research on computer modelling is designed to allow prediction of the changes that will follow a given natural or man-made input into the ecosystem (Woodwell and Smith 1969). The analysis and prediction of the resilience of ecosystems to human modification is nevertheless at an early stage and is a field in which considerable advances would be of great practical importance (Margalef 1968, Goodman 1975, Hill 1975, May 1975).

Ecosystems also have a multiplicity of homeostatic mechanisms, called feedback loops in cybernetic terms, which tend to maintain the system in a stable condition. By means of the growth, reproduction, mortality and immigration of the organisms, together with the process involving the abiotic components of the system, the quantities and rates of movement of matter and energy are controlled.

Instability in systems is a consequence of disorder and has been compared by Schultz (1967) to the thermodynamic concept of entropy: systems in which disorder is increasing are becoming highly entropic; those which are building up order are negentropic. Life itself would thus appear to be a massive and continuing accumulation of negative entropy. If it were possible to measure the rate of change of entropy or its real equivalent in ecosystems, then this parameter would possibly be a measure of the 'health' of systems from the human point of view. Its role in the manipulation of resource processes would be helpful, since man's main activities seem to be to remove the homeostatic mechanisms and promote instability, sometimes to the point of outright destruction: consider the example of an overgrazed range which first of all has rapidly declining plant productivity (and hence animal yields) and then undergoes soil erosion.

In ecological terms cropping means the removal of energy-rich matter from a particular ecosystem at a particular spatial scale. The products of the resource usage are then put back either into the same ecosystem or into another. Energy is irrecoverably being given off as heat, but the other products are not in fact lost to the world ecosystem, if we except astronauts' excreta. Man's cropping is at both herbivore and carnivore level but is less usual at top carnivore level because of the relative scarcity of the animals at this stage, unless they are gregarious organisms such as some species of fish. Solitary carnivores may often be utilized for culturally desirable products such as eagle's feathers and leopard skins. The detritus chain also supplies animal products: *moules marinières* are a crop from such a source.

The input of man consists of energy and matter in various forms. Much of the energy input into all resource processes is now from fossil fuels and some matter is also fossil, in the form of mined rock. The whole process of cropping is accomplished by the application of man-directed energy and matter. The mobility of man over the surface of the globe conferred by access to energy sources and the development of technology has also meant the transmission of species of plants and animals from their natural habitats to other places. Many of them are unable to survive in their new lands, but some have been spectacularly successful, as with the rapid spread of the starling into the urbanized eastern seaboard of the USA; the history of the rabbit in Australia is perhaps the best known. Success seems to occur either when there is a vacant niche that no native animal has succeeded in filling, or when the introduced species can outcompete the indigenous biota. Often too, the transplanted organism leaves behind its natural predators and may not have any in its new home. Highly manipulated ecosystems, such as cities, waste ground and agricultural land, frequently provide habitats for introduced species.

In the pursuit of crops, the removal of competitors becomes important because they represent energy and matter that man cannot or will not garner. Plants other than the chosen crop become weeds; animals that compete for forage at the herbivore level or predate upon the chosen crop species become pests. 'Weed' and 'pest', it must be emphasized, are cultural concepts since there are no such things in natural ecosystems. Such simplification reduces the energy flow through the system by reducing its diversity (Fig. 1.16). Inducing the food chains to converge upon man may well increase the net crop for man and make it more concentrated in protein, but the overall flux of energy in such systems is lower than in the natural state, since gathering and processing stages result in net losses of energy which are subsidized by fossil fuels in industrial countries (Fig. 1.17). Man very often puts the brake on succession (which represents the building up of diversity) and maintains systems at an early stage. Overcropping may well cause an ecosystem to revert to a very early stage of succession, one with much bare ground and hence susceptible to soil erosion for instance, and in some places this may mean a change to a more xeric kind of vegetation. Overgrazing on the semi-arid grasslands and the steppes of western North America or the Saharan fringes has enlarged the desert areas, and though this is not proven, the outbreaks of locusts could very well be a consequence of the years of overgrazing of the parts of the world in which they occur: removing any woodland or scrub that might harbour bird predators, for example, and providing suitable sites of oviposition in open sandy soils. The British grouse moor, reduced by management practices to a monoculture of heather (*Calluna vulgaris*) and grouse (*Lagopus scoticus*), with all other plant species burnt off and all

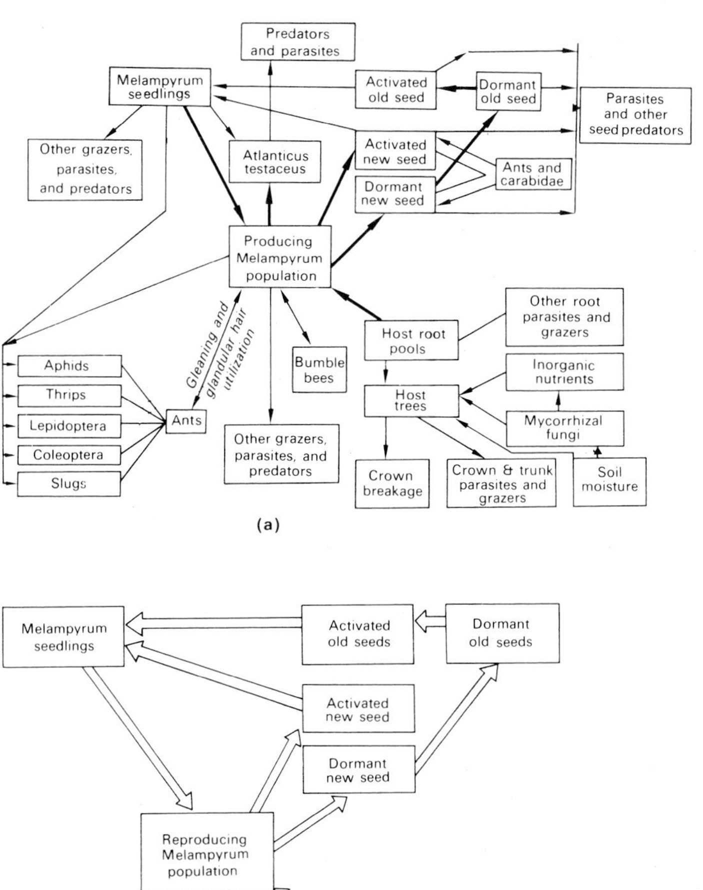

Fig. 1.16 An ecosystem centring around a plant parasitic on tree roots, *Melampyrum lineare*, in Michigan: (a) the natural system, showing biological relationships; (b) the system after the application of insecticides had eliminated certain components. The simplified nature of the ecosystem in (b) is apparent.
Source: Cantlon 1969

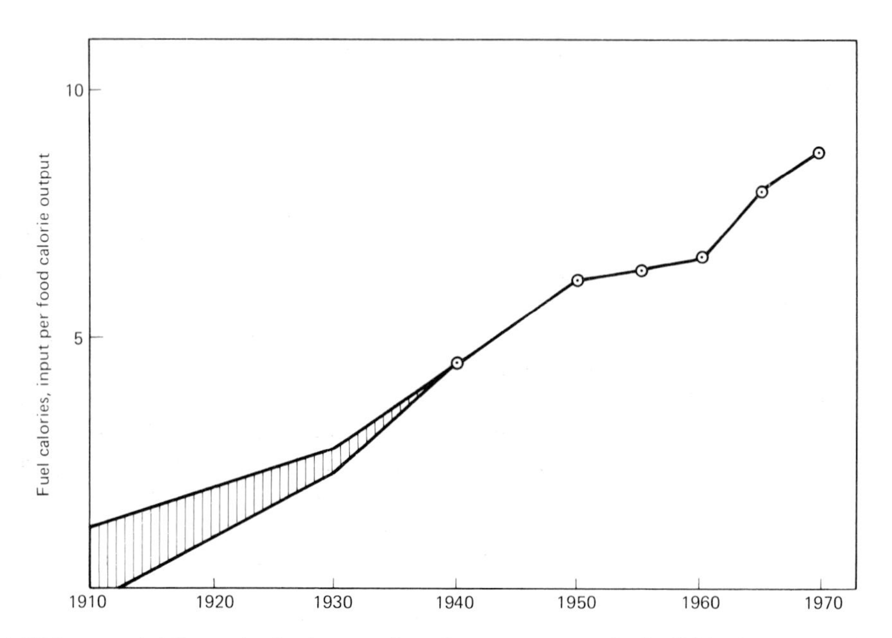

Fig. 1.17 Energy subsidies to the food system from farm to consumer in the USA: the number of calories put into the system to obtain one calorie of food. Values for 1910–1937 cannot be fully documented so a range of probable values is given.
Source: Steinhart & Steinhart 1974

other animal species shot off, is highly unstable since heather bark beetle can spread very rapidly, as can grouse ringworm. Man's response to instability is usually to increase his inputs of matter and energy, thus producing an ever-increasing spiral of manipulation in which the inputs needed to promote temporary stability themselves require higher inputs to achieve equilibrium at a later stage.

One aspect of simplification is the extinction of biota. By various methods, from outright killing off for food or pleasure to the alteration of habitat, a wide variety of plant and animal species have been eliminated at local or regional and world scales. Where these groups were rare anyway and possibly towards the end of their evolutionary term, then man has perhaps merely accelerated natural process. But it is difficult to imagine this to be true of the passenger pigeon, whose flocks were said to have darkened the sun at noon in the Mid-West of the nineteenth century but the last specimen of which died in a Cincinnati zoo in the 1920s. Neither is it likely to be so of the blue whale or the Javan rhinoceros or the oryx, to name but a few of the animals on IUCN's list of endangered species. Such devastation must be placed in perspective: the phenomenon of 'Pleistocene overkill' indicates a long history of man's effect upon animal population. Extinction of biota appears therefore to have been a long-term concomitant of man's status as an ecological dominant, although there are now powerful cultural currents, in the Western world at any rate, to try to halt this process. By definition, however, the loss of currently rare species is unlikely to produce unstable ecosystem: it is more probable that large uniform strands of very common species will produce that particular effect.

Man's manipulation of ecosystems

We turn now to providing an introduction to the major part of this book, the consideration of man's uses of the world's environments, considered in ecological perspective. Each of the themes briefly reviewed here is taken up at greater length in a section of its own.

At some stage in the last million or two years, all the ecosystems were unaffected by man: they were pristine or virgin. A few such places remain, but much of the rest of the terrestrial surface has been altered in some way. The nature, intensity and antiquity of his transformations enable us to compile a simple ranking of the land-use systems according to their degree of manipulation away from the pristine condition. At one end are ecosystems which are either left untouched or which it is desired to keep in as 'natural' a condition as possible; at the other man's structures obliterate entirely those of nature. Thus the lowest manipulation is typified by the open oceans or the high peaks of the Andes, the most alteration by Manhattan or Europoort.

At present, our knowledge does not enable us to compile all the data about these systems that we would like, such as their primary and secondary productivity, their response to increased demands for their components, and the history and type of man's alterations. Many of these features are known about at a quantitative Level and the work pursued under the aegis of the International Biological Programme (IBP) has yielded a great deal of knowledge about the productivity of ecosystems. For the present we are mostly confined to more general accounts.

Apart from completely unused land, areas used for *outdoor recreation and nature conservation* represent the condition of least alteration. Where there is a single-use management aim for these, the object is generally to disturb the *status quo* as little as possible, and to maintain the ecosystems as they are at the time of designation or even to restore a former condition. This concern may even lead to the confusion of trying to stop natural processes such as fire or succession in order to preserve particular biota or landscapes even though under natural conditions they would be transient. Limiting factors in this system usually relate either to the numbers of a particularly favoured organism or to the populations of the observers, which if large enough diminish the quality of the crop. In this case it is mainly aesthetic satisfaction: the difficulties over the management of the National Parks of the USA illustrate this point. Most manipulation in these systems involves the control of animal populations which exceed the carrying capacity of the Reserve or Park because its legal limits are unrelated to the true boundaries of the ecosystem of which they are a component.

Requiring little more manipulation is *water catchment*, and indeed in many wildland areas it is successfully combined with the preceding land use. Apart from the major landscape changes induced by dams and the subsequent replacement of terrestrial by aquatic systems, the watershed area is often little altered, especially since there is not complete agreement about whether trees are beneficial to water yield (in terms of steady release) or harmful (by transpiring a goodly proportion of the system's water back into the atmosphere). Once caught, the water may be managed to effect far-reaching alterations such as irrigation or industrial growth, but the areally larger watersheds are not often subject to large-scale manipulations.

The longstanding but gradual system modifications represented by a pastoral economy are paralleled by one type of *forestry* which 'grazes' through virgin forest, allowing natural regeneration in its wake. On the other hand, modern forest plantation in Europe for instance is more like contemporary farming, exhibiting controlled inputs of nutrients, careful thinning out, lavish use of pesticides and orderly harvesting. So forestry is difficult to place on a gradient of manipulated systems since it can occupy most places from the barely altered to the totally artificial. A reasonable position seems to be to regard it as more manipulative than grazing and less so than agriculture. In biological terms forestry represents the removal of nutrients from the local ecosystem and the disappearance of the dominant of the community. Even if only temporary, the effects the therefore strong: careless forestry can lead to drastic soil erosion and consequent river silting, to quote two well-attested examples. As far as plant nutrients are concerned, the practice of removing and often burning leaf-bearing branches *in situ* means that many of the nutrients essential for tree growth find their way back into the local ecosystem and only the lignified wood is borne away, with its smaller complement of elements likely to be in short supply in the system, whereas soil erosion inevitably carries away mineral nutrients.

Towards the highly manipulated end of the scale must be placed *agriculture*, which is of three basic ecological types. The first of these is shifting agriculture where after the plot has been abandoned the system many revert to its original condition, there are suggestions that this is never totally achieved and that small differences in the secondary system always exist. As Geertz (1963) points out, the most interesting feature of shifting agriculture in tropical forests is that the crops planted in the clearing imitated the former community. The miniature forest of crops gives complete cover of the site, so that it is not exposed to the leaching effects of the heavy rainfall. The rapidly fluxing mineral cycling of the forest is not initiated, since the plot is cultivated until the nutrients are more or less exhausted and it is then abandoned to regain higher levels under the secondary forest. In the normal working of the system the critical points are probably the initial nutrient supply which, in tropical forests especially, is mostly in the vegetation and not in the soil (hence the practice of slash-and-burn) and the maintenance of soil structure. Under present conditions very little is usually done about either of these. The overall critical point comes when population levels make it necessary to re-utilize a plot before its nutrient levels and soil structure have been built up again, and diminishing yields coupled with soil erosion are the frequent consequence.

Pastoralism may also produce a high degree of manipulation. Traditional pastoralism moved over a very large area and so aimed at a minor degree of alteration of the ecosystem, but the growth of herds and the relative confinement of pastoralists (e.g. by political boundaries, the provision of constant water from tube wells, vetinerary centres, and the presence of government-provided markets) have brought about strongly changed ecological systems which in many dry places have been at the point of breakdown and erosion.

The most highly manipulated of all the non-urban systems is sedentary agriculture. Here, the food chains are subordinated to the monocultural production of a crop which either comes to man as a herbivore or is fed to herbivores for subsequent animal protein yield. Thus all competitors are removed and this necessitates either high inputs of energy, as in weeding and bird-scaring, or of matter, as in the form of chemical pesticides. The monocultures are usually prone to epidemics, so that the high yields of the intensive agriculture areas of the sub-tropics (where usually water is a limiting factor and irrigation is the means of lifting its threshold) and temperate zones are achieved only at the price of high inputs by man. This is especially true of mineral nutrients, for the use of the crop may be thousands of miles from the area in which it is grown, and few of the metabolic products find their way back to the fields: most are dumped in the sea. Although much of the energy gained in this system can be cropped by man, there is nevertheless a great reduction of fixed energy compared with he system which preceded it: compare for example the amount of photosynthetic tissue per unit area of a forest with that of a typical row crop, growing for perhaps five to six months of the year.

In the process of manipulating ecosystems for his own ends and in recovering and processing his resources, man may well affect the systems by the material which he returns to them. Very often there are unwanted and even toxic products and they are generally referred to as wastes and pollution. This topic along with other nagative features of resource use will also be dealt with in Part II as will the study of energy and minerals which provide the power and the tools with which to carry out the manipulations which are described.

Further reading*

ARMSTRONG, P. H. 1974: Some examples of the use of current ecosystem models as frameworks for land use studies.
BARBOUR, I. G. (ed.) 1973: *Western man and environment.*
BOULDING, K. E. 1966: Ecology and economics.
CLAYRE, A. (ed.) 1977: *Nature and industrialization.*
COOPER, J. P. (ed.) 1975: *Photosynthesis and productivity in different environments.*
COTTRELL, A. 1978: *Environmental economics.*
FIREY, W. J. 1960: *Man, mind and land.*
GLACKEN, C. J. 1967: *Traces on the Rhodian shore.*
HEWITT, K. and HARE, F. K. 1973: *Man and environment: conceptual frameworks.*
HILL, A. R. 1975: Ecosystem stability in relation to stress caused by human activities.
HUGHES, M. K. 1974: The urban ecosystem.
HUTCHINSON, G. E. 1978: *An introduction to population ecology.*
JANSSON, A.-M. and ZUCCHETTO, J. 1978: *Energy, economic and ecological relationships for Gotland, Sweden—a regional systems study.*
JONES, G. 1979: *Vegetation productivity.*
LEARMONTH, A. T. A. and SIMMONS, I. G. 1977: *Man-environment relations as complex ecosystems.*
LECOMBER, R. 1975: *Economic growth versus the environment.*
MITCHELL, B. 1979: *Geography and resource analysis.*
ODUM, E. P. 1975: *Ecology.*
ODUM, H. T. 1971: *Environment, power and society.*
—and ODUM, E. C. 1976: *Energy basis for man and nature.*
PARK, C. D. 1980: *Ecology and environmental management.*
PASSMORE, J. 1980: *Man's responsibility for nature*, 2nd edition.
PEARCE, D. W. and ROSE, J. (eds.) 1975: *The economics of natural resource depletion.*
SANDBACH, F. 1980: *Environment, ideology and policy.*
SKOLIMOWSKI, H. 1976: Ecological humanism.
WHITE, G. (ed.) 1974: *Natural hazards.*

* Full publication details are given in the bibliography.

Part II

Resource processes

2

Unused lands of the world

In this section we are concerned with those parts of the world where the terrestrial ecosystems appear to be virtually unaltered by man, either because he is present but does not exert any manipulative effect on the ecology, or because he is, in increasingly fewer places, absent. Men may be settled at a particular point and rely on outside sources for nearly all their materials and hence by their self-sufficiency not exert any resource-using pressures on the environment, as in the case of service camps like the US base at the South Pole which is supplied by air. In another category are the settlements which produce within their own area most of their needed resources and obtain the rest by trade along routes where self-sufficient transport is used. Desert oases with their accompanying trade (traditionally carried by the frugal camel) are examples of this category. At the point of settlement there is admittedly a more or less complete obliteration of the original ecology, but it is limited in area, and there is no tentacular reach into the environment for resources.

Man is also said to function only at an animal level in some ecosystems where he is subject to the same competitive and predatory pressures as other mammals. Only hunting, fishing and gathering cultures can be regarded as belonging to this group, and we shall look at examples of such people to try to assess the validity of the supposition that they do not in fact manipulate the ecosystems to which they belong. A schematic layout of these categories is given in Fig. 2.1.

Hunting, gathering and fishing

Hunting and gathering was once the dominant mode of existence of our species, but by AD 1500 only 15 per cent of the world's land surface was inhabited by such peoples, at a time when the ecumene was more or less fully occupied. Today only a few pockets of

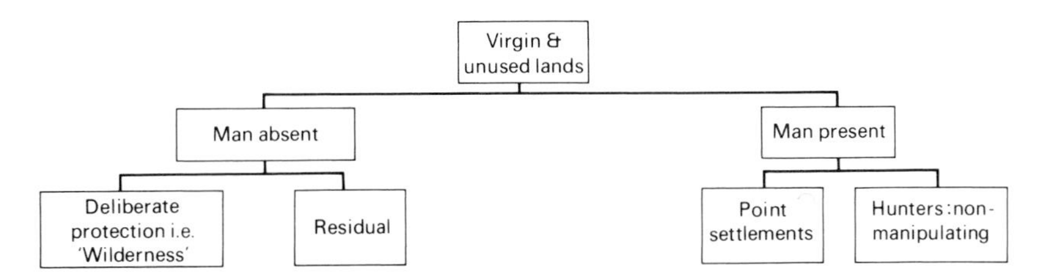

Fig. 2.1 A scheme of the various types of unused land and wilderness referred to in this chapter.

hunters and gatherers survive, and their culture has almost always been altered by Western contact. Some Indian and Eskimo groups in North America exist largely on hunted products, although contacts with whites are at a high level; less influenced perhaps are people such as the 5,000 Bushmen of southern and south-west Africa who pursue a very largely aboriginal form of life. In such a category also may be placed some of the non-agricultural groups of the 170,000 Pygmies of central Africa such as the Aka, Efe and Mbuti, the Aborigines of Australia, and some Indians of the interior of Brazil and Venezuela. Together with a few other groups, they form possibly about 0·001 per cent of the present population of the world (Murdock 1968).

Such a sample is not therefore the best material for estimating whether or not they exert long-term effects on their ecosystems. To add to the difficulty, most of the studies done on them have been on their cropping as it relates to features of their culture rather than long-term investigations of subsistence patterns and the ways these might have been affected by the hunting and gathering patterns of the people themselves. Most of what follows is therefore inferred from anthropological studies designed for another purpose.

At a very simple level of subsistence are the !Kung Bushmen described by Lee (1968) who live in the Dobe area of Botswana. Here, 466 of them lived in 14 independent camps, each associated with a water-hole. Around this focus, a 9·6 km hinterland (i.e. within convenient walking distance of the water hole) was exploited for subsistence materials. The !Kung practise very little food storage, for rarely more than 2–3 days' supply is kept at hand, and the main foods are vegetable in origin, comprising 60–80 per cent by weight of the diet. The main item is the Mongongo nut, the product of a drought-resistant tree, which contributes about 50 per cent of the vegetable intake and which yields 600 kcal per 100 g edible material and 27 g protein per 100 g. Although many kilograms of these nuts are eaten, many more are left to rot on the ground, so apparently the resource is not fully used up. Of the other sources of vegetable foods, 84 species of edible plants are recognized (including 29 species of fruits, berries and melons, and 30 species of roots and bulbs), but of these 75 per cent provide only 10 per cent of the food value and thus the use of them must be light.

There is a parallel situation with regard to meat-eating. The Bushmen name 223 species of which 43 are classed as edible, but only 17 are hunted regularly. It is not said whether this has always been the case or whether such a selectivity represents the end point of a long period of shifting numbers of animal populations. The adjustment of these Bushmen to the resources of their semi-desert environment is successful, for even though 40 per cent of the people do not contribute to the food supply, signs of malnutrition are absent and Lee (1969) estimates that they receive about 2,140 kcal and 93 g of protein per capita per day when an adequate diet would consist of 1,975 kcal and 60 g of protein. No starvation, even of the unproductive people, is necessary because of the basic dependence on reliable vegetable foods adapted to the local ecological conditions. It would seem therefore that no manipulation of the environment is necessary (nor indeed may it be possible) provided the population does not over-use the vegetable resources, and this appears not to happen, though Lee does not comment on any conscious attempts at population control.

A recurrent theme in the ecological relationships of recent hunting and gathering peoples is the amount of seasonal movement they undertake and their variation of the size of exploiting groups according to the availability of resources: G/wi bushmen, for example, split into smaller bands during the off season which practice presumably diminishes their impact on fauna and flora. Equally, such groups will eat a wide spectrum of organisms: the Mistassini Cree of central Labrador consume not only 'staples' like moose and caribou but also bear, beaver, hare, muskrat, porcupine, mink, squirrel, marten and otter (Bicchieri 1972). We read also that some hunters imposed upon themselves certain restrictions of a 'protective' type: Great Lakes and Vancouver Island Indians alike know to strip bark from only a portion of an individual tree; the Kaska allowed trapping of some species only at widely spaced intervals and the Iroquois spared female animals during the breeding season (Heizen 1955). But as Ray (1975) points out, some North American Indians were reluctant to accept any sort of conservation policies which were proposed by the Hudson's Bay Company for fur-bearing mammals. In the case of the Aleuts, it appears that by over-exploiting the sea otter, they nearly eliminated a key species in the littoral ecosystem. Without predation by these marine mammals, the near-shore herbivorous sea-urchins and molluscs eliminated the large algal kelps (Simenstad et al 1978). But despite the imprecision of the evidence, we may perhaps infer that the survival of such peoples until the advent of the Europeans is symptomatic of their success in managing wild populations even though there was some habitat and prey population management, albeit at a small scale because the population density of the hunter-gatherer cultures was so low.

A contrast between past and present is offered by the native inhabitants of the arctic regions of North America, of which the Netsilik Eskimos who inhabit the tundra regions north and west of Hudson Bay are a sample (Balikci 1968). Their base is essentially a marine-winter (seal) and inland-summer (caribou, salmon trout) movement with elements of substitution should one resource fail. For example, caribou would be hunted across the open tundra with bow and arrow should the normal method of killing them from kayaks at a river crossing fail. The critical period is the long winter, and caches of food for such a period are necessary. Over very large areas, the population density was quite low: in 1923, the Netsilik numbered about 260, in a region 480 × 160 km in size. In such a context, the harshness of the environment must have meant that the gaining of subsistence in the short term was a considerable struggle and the conscious manipulation of the ecosystem for the long term was scarcely likely. The advent of Western culture has altered the living conditions of most Eskimo, for they can now rely upon imported food from both trade and welfare programmes, and also have access to fossil fuel sources. In a group studied on Baffin Island, Kemp (1970) calculated that in one year the people gained 12,790,435 kcal of consumable food from wild sources. This was backed up by the outside world in the form of 7,549,216 kcal of store food, the equivalent of 8,261,600 kcal in ammunition, and 2,815 litres of gasoline for snowmobiles and outboard motors. In this area hunger is now virtually unknown, although it was in any case rarer than in the harsher terrain of the Netsilik.

In general, hunting and gathering seems to provide a very adequate way of life which can persist for centuries without radical alteration. This suggests that manipulation of ecosystems on a conscious level is rare; unconscious manipulation disturbs the system so little that fluctuations in yield or sudden adaptations to different food sources caused by catastrophes are virtually unknown. Thus obvious, and perhaps even borderline, malnutrition is rare. Even the minimal dietary requirements provided by hunting and gathering often allow such people to fare better than neighbouring groups who subsist on primitive agriculture or who are urbanized. Thus starvation is commonplace only in extremely rigorous climates where wild animals form practically the entire diet. Elsewhere, if there are any selective effects of shortages, they lie in the differential mortality of the old and infirm in the face of the water shortages which are more likely and more frequent than lack of food. Mortality from other causes is not particularly high among hunter-gatherers. Chronic diseases, traumatic deaths, accidents and predation vary in their incidence but are generally low (Dunn 1968). But in most such societies, social mortality has been part of the population-resources equation, with infanticide the most widespread practice. If, therefore, we have a tendency to regard hunter-gatherers as the members of the original affluent societies then it is because they did not populate their territories up to the limits of the carrying capacities of the lands (Birdsell 1968). Under such conditions, manipulation of environments, though not by any means impossible, becomes much less likely.

The evidence from modern hunters is so scanty and of such an indifferent kind for our purposes that it is scarcely possible to come to a firm conclusion about whether lands occupied by them are virgin lands, i.e. have unmanipulated ecosystems. In practical terms, since they are such small groups, it does not perhaps matter. Theoretically, however, there ought to be a point where the shift is made from man as an animal component of an ecosystem to man the changer of the ecosystem. The factors involved are very diverse, but two have major significance: the population numbers and the fragility of the ecosystem. The importance of the first of these is seen in the population control mechanisms exerted by many groups, although it has been argued that the adoption of agriculture was forced on hunters because of an inexorable use in their populations (Cohen 1976). Quite possibly experience at a past time had shown them that to grow too large in numbers was to exert an influence on the local ecology which was inimical to future food supply. The fragility of the ecosystem is another variable: semi-arid and arctic areas are particularly vulnerable, whereas temperate zones tend to have greater powers of recuperation after man-wrought alteration. The overall picture presented is of a mosaic in which the presence solely of hunters cannot automatically be taken to imply unaltered ecosystems; but most of the evidence which would enable more precise investigations to be made has now been overlain or wiped out by later changes.

Unused lands today

The 'unused' or 'virgin' lands of the world at present can only be found where human settlement is sparse, Not only is absence of man, or his presence at a very low density, necessary but also his use of either a simple level of technology or a 'point' settlement which is self-contained or relies on outside sources for its materials.

Let us differentiate between two types of 'virgin' lands: those which remain wild because they are residual lands of no value to contemporary resource processes, and those where there is a deliberate preservation of the wild for its own sake or for some other non-exploitative purpose. This second category is often described as wilderness: a term which although equally apt for the first set of lands is perhaps more useful where a particular concept of the role of wild terrain in relation to human purposes is envisaged. We shall deal with these two sets of virgin terrestrial ecosystems separately; the most pristine set of ecosystems of all, the sea, is treated individually later in the book.

Sparsely settled biomes

The largest more or less continuous single area of 'unused' land not deliberately preserved belongs to the arctic and subarctic zones of Eurasia and North America. These lands are mostly in the possession of the USSR, Canada and the USA (Alaska), with contributions from Finland, Sweden and Norway. The region has only relatively recently emerged from glaciation and is often characterized by immature soils and poor drainage. The characteristic vegetation type is the tundra, although many wetland and mountain communities also exist. Southwards, a belt of hardy deciduous forest like the birch scrub of Lapland may intervene between the tundra and the coniferous woodlands of the boreal forest biome. The animals are all adapted to survival over a long and very cold winter.

Like the high arctic and polar lands of perpetual snow and ice beyond them, these northern territories are sparsely settled. Modern incomers mostly live at 'point settlements' supplied by air or sea; the native peoples have always been few in number, although more wide-ranging at ground level both inland and across the ice-covered seas than immigrant whites. Both groups have had some effect on the ecology of the northern wilderness, though of different types. The native populations were largely dependent upon animals such as reindeer, caribou and seals for their subsistence. The number of people was small and the herds of animals immense so that probably no permanent change was effected in the natural ecosystems unless the biota were at a low level: the decline of the musk-ox in the Canadian tundra may have been due to native overhunting, although protection has now allowed it to regain higher population levels.

The advent of industrial man to the north has meant both direct and indirect change. Direct alteration of areas comes from mining, trapping and logging, although the last two need not wreak permanent change on ecosystems if carried out with due regard for sustained yield; inevitably such policies have often been neglected. Indirect changes come from incidental operations: fires from logging which run up onto lichen-clad slopes and destroy the winter feed of caribou, for example. The explosions from seismic testing for petroleum scare away the arctic fox, according to many Eskimo, and there is now the possibility of large-scale oil contamination from the finds on the arctic slope of Alaska and Canada's North-West Territories. In such a cold climate, the longevity of spilled oil is not known and might persist for decades. The same development has produced the Trans-Alaska Pipeline, and the effects on the tundra of spills of heated oil, and on the migration of the caribou, are not fully known.

At point settlements such as drilling rigs, considerable erosion of the tundra can be caused by vehicles and temporary dwellings if the permafrost is not protected by a thick pad of insulating material such as gravel. The wastes from such places are also apt to attract scavenging animals such as polar bears which then tend to end up as rugs. The northern lands, therefore, are still vast wildernesses of mostly natural systems, but exploitation (especially rapid in the case of Alaska) is eroding these wild areas, and agricultural expansion in their climatically more favourable parts is still an aim of some governments. The ecosystems are generally so simple and specialized that it takes very little to cause their breakdown: this can be seen to a minor extent in Iceland, where there has been no impact of heavy industry on the land to disturb the sparsely vegetated surface, but where sheep grazing has nevertheless caused shifts in vegetation towards more open communities and hence an increase in the surfaces susceptible to sub-aerial erosion processes. We must conclude that the fragility of these northern areas needs careful consideration in any proposals to develop them, whether it be for minerals, biotic resources or tourism.

Similar in many ways are the high mountain areas of the world, such as the Rockies, Andes and Himalayas. In general the human impact has been very much less than in arctic and subarctic regions because high relief is added to harsh climate, but apparently unsettled lands may have undergone grazing by domesticated animals. This may produce the symptoms of overgrazing because the ecosystems are fragile and easily changed: so in the nineteenth century John Muir was moved to write of sheep in the Sierra Nevade mountains of California as 'hooved locusts' and to point out the soil erosion that followed in their trail. Today, recreationists may outnumber sheep, but their effect can also be strong. Numerous boot-clad feet can wear away mountain vegetation and allow erosion from which recovery is likely to be slow. In winter, animals may be frightened away from their feeding grounds by motorized incursions on vehicles such as the snowmobile, which may also be used for hunting; by extending the range of hunters in winter it may help reduce the populations of some animals since they are likely to be in poor condition, although if predators like the wolf are killed then species like deer may expand beyond the carrying capacity of their habitat. In northern Ontario, some isolated lakes have been virtually fished out because of improved access during the winter by tracked vehicles.

Some forested areas of the world are also lacking in any density of settlement. The largest of such areas are the boreal forest zones of Canada and the USSR. The short summer inhibits the activity of the soil fauna and flora and so a considerable depth of humus is characteristic of many parts of this formation. Along with fallen branches and trees, this layer is potential fuel for wild fires, and when set by humans or lightning these can destroy large areas of forest. They tend to be even more damaging when a management policy of instant fire suppression has been carried out for some years, since this allows the forest floor material to build up. Some populations of animals, such as moose, may build up to extremely high levels when a man-induced predator control programme is effective, as has been that of the wolf in Canada. The herbivores may then seriously affect forest regeneration and growth because of the intensity with which they browse. Even minimal human occupance can therefore lead to quite far-reaching ecological change.

In the remaining stretches of tropical forest considered 'natural', such as those of Amazonia and the highlands of New Guinea, the most disturbing influence until recent times has been shifting cultivation (see pp. 165–6). After abandonment, the plots recolonize with secondary forest and if not eventually re-used for agriculture then they become climax-type forest again. It is not certain whether the secondary forest ever regains the exact composition of its virgin state, but if not then reserves may eventually contain the only truly natural tropical forest and all the rest reflect the influence of an unconscious management system, as do most temperate forests even if they are apparently 'natural'. These tropical forests are of course rapidly shrinking in area because of timber extraction, and conversion to pasture.

Possibly the most 'unused' land of all is in the desert areas of the world. Their biological productivity is low, and apart from some hunting and pastoralism human use tends to be confined to mineral exploitation and especially in North America, recreation. Extraction of minerals such as petroleum is a 'point' activity and pipelines do not alter the desert very much, although the arid climate means that any hardware left lying around will persist for a long time, witness the tank hulks in North Africa and the beer cans in Arizona. Recreational use of deserts in the USA, especially in California and Arizona, centres around activities involving four-wheel drive vehicles and, in sandy terrain, 'dune buggies'. The former do not change the ecology very much themselves but import people into another type of fragile ecosystem where, for example, the trampling of succulents or the gathering of dried cactus 'skeletons' for firewood disturb large numbers of niches for a very long time, since plant growth rates are so low. Dune buggies may easily change unfixed dune systems quite radically, but the wind soon rectifies the situation; on the other hand, partially vegetated dunes are soon converted into 'blow-outs' after a few vehicles have torn up the root systems of the grasses binding them.

Islands, particularly those isolated in large oceans, might be thought to be likely candidates for pristine status, but this is rarely so. As Elton (1958) pointed out, such islands generally have a depauperate biota, with the result that unstable ecosystems change very rapidly as a result of direct human activity or the man-induced introduction of alien plants and animals. The coming of the pig and goat to many tropical islands resulted in a revolutionary destruction of vegetation; likewise the rat caused a large faunal shift in small coral-based islands. The innate instability of the latter is underlined by the anxiety which resulted from the population explosion of the large Crown of Thorns starfish. Such islands as remain in an unaltered state are highly prized for scientific purposes: Aldabra in the Indian Ocean is one example, containing rare fauna as well as a low level of anthropogenic effects (Stoddart and Wright 1967). The plan to build an airstrip there caused a considerable outcry from the international scientific community; it was, however, financial rather than ecological instability which defeated it in the end (Stoddart 1968).

There remain for discussion a few special cases of ecological systems which can be regarded as natural because they undergo constant renewal. Unfixed sand dunes, in an early stage of succession, are one example. Coastal systems such as the Sands of Forvie (Aberdeenshire, Scotland) are examples of such a condition; inland, the White Sands of New Mexico (part of which are under protection as a National Monument) provide another. Any activity of man generally proves very ephemeral in shifting dunes, but once they become fixed by vegetation then they are much more vulnerable to human activity, especially the kinds of recreation which cause blow-outs. Tidal salt marshes which receive periodic inundation, and hence an input of silt and salt, might also be considered natural, although many of them are subject to grazing by domesticates; in the tropics the mangrove swamps occupy a similar niche. Large paludal areas are often close to their virgin condition but are increasingly becoming altered by outside influences such as the influx of pesticides or eutrophication agents from their highly manipulated watershed areas.

Wilderness

In this section, some examples of the conscious and deliberate protection of large wilderness areas are discussed. It is difficult to distinguish these from the national parks and nature reserves described in the next chapter, but the main criteria are the large areas (whereas parks and reserves can be of any size), the absence of deliberate management wherever possible, and the feeling of wholeness. The set of systems is preserved particularly because of its unity and not for any particular biota, landscape or recreational activity, as is so often the case with wildlife protection and landscape preservation areas (Plate 3, opposite).

Motivations

Ideally, wilderness areas should never have been subject to human activity which has resulted in manipulation, either deliberate or unconscious, of the ecology of the areas being considered. But in many cases our knowledge of land-use history is so fragmentary that we cannot honestly say other than 'it looks natural'. Research into Quaternary ecology and land-use history produces more and more evidence of the antiquity of man's imprint, and it becomes increasingly likely that some areas preserved for their 'natural wilderness' quality have in fact been altered by man at a pre-industrial or even pre-agricultural stage in the past.

The reasons given for the setting aside of wilderness areas vary from place to place and from culture to culture, but running through most of the legislation and regulations are a shared set of themes. There is often a declamation of the rightfulness of the independent existence of such areas: not all nature is to be perceived as a profitable or potentially useful resource and it should be allowed to persist on its own terms without any interference from man. In the last phrase is implicit the absence of any deliberate management of the areas designated and, like so many such hopes, this is often more pious than practical. This ethical strand of thought frequently stems from or is accompanied by religious motivation.

The most persuasive non-scientific argument in Western countries has been the advocacy of the spiritual aura of such lands. It is evident that their nature is changed by the incursion of large numbers of people, so their virtues reside either in their being experienced by small numbers of people as a special form of recreation which is basically spiritual refreshment, or in being symbolic. Then they do not need to be

Plate 3 A wild area under protective management: Mount Hood, Oregon, USA. Habitats such as this mountain meadow may experience recreational pressure severe enough to impoverish the biota and spoil the aesthetics; thus restriction of access becomes essential to prevent certain types of change. *(Grant Heilman, Lititz, Pa)*

visited, they are just there as 'the wilderness beyond'. At one extreme, wilderness advocates may insist that experience of great natural areas is a spiritual essential for every person. This is unlikely and, nowadays, impractical. A more balanced point of view regards wilderness as being just as much a state of mind as a condition of nature; the beholder's response becomes paramount and hence the type of person determines the necessity or otherwise of wilderness experience. Although a pristine condition is a requirement of consciously preserved wilderness areas, the desire of people to visit them soon leads to ecological changes, and this possibility brings about conflict with another major set of reasons for wilderness protection, which are scientific in origin. There are three main strands to these propositions. There is firstly the need to keep a gene pool of wild organisms, both plant and animal, to ensure present and future genetic variety. Circumstances under which new strains may be needed cannot be foreseen and by reducing diversity the possibilities for future choice are much restricted. Large wild areas ensure that a reservoir of potentially useful sources of breeding material for human use is preserved, a task which gains in importance as fewer and fewer genetic strains are used in major crops (Frankel and Hawkes 1975, Rendel 1975, Burley and Styles 1976, Siegler 1977, Bennett 1978, Frankel 1978) and as new discoveries of the properties of animals are made, such as the use of nine-banded armadilloes in developing vaccine against leprosy.

Secondly, there is an emphasis upon the maintenance of animal communities in their natural surroundings in order to facilitate research upon their behaviour. In changed habitats the animals act differently and the chance to study specialized patterns as the product of a particular ecological niche is lost. This argument is similar to that advanced for many national parks and nature reserves, but here the emphasis is on the large area of the wildernesses and the increased chance that the whole life-span and seasonal cycles of the animal are completed in undisturbed conditions. Thirdly, sharing the last set of motivations and constraints, is the absolute necessity of protecting undisturbed ecosystems of all kinds for research in ecology. Many of man's activities consist of tinkering with his environment, and the prime requirement of successful tinkering is keeping all the parts. Much useful research is done in smaller areas, but large wildernesses are essential not only for pure research but also for applied work. Efforts to increase the productivity of crops, for example, may gain much from an understanding of the functioning of natural ecosystems.

As some of the examples will show, different reasons have been responsible for the legal designation of wildernesses in different places. In some, such as Antarctica, scientific reasons have been dominant. In the case of the National Wilderness Preservation System of the USA, the recreational-aesthetic reasons won the day. Also, some wildernesses may be called by names such as National Parks; in differentiating them from the areas to be reviewed in Chapter 3, we shall try to follow the criteria of size, lack of settlement, and absence of management policies and practices.

The USA and other nations

'The Wilderness' has always played an important part in the symbolic life of North Americans (R. Nash 1967); the rapidity with which the continent was settled and exploited meant the disappearance of most virgin land except in marginal environments. Even here, in remote forests, unyielding deserts and high mountains, the determination to log, mine or graze was often successful even if only temporarily so. The erosion of the wild led to a rising groundswell of public opinion which resulted in the passage through Congress of the Wilderness Act of 1964 (Simmons 1966). This immediately established a National Wilderness Preservation System and placed in it 54 wildernesses, wild and canoe areas, totalling about 36,450 km^2. All additions to the System have to be approved individually by Congress: by 1970 the System totalled about 77,000 km^2. The areas thus set aside are to have no economic uses, except where mining has been previously allowed, in which case it has to be removed by 1983; the President may allow other economic uses but only in times of considerable emergency. The criteria for establishment include a normal minimal size of 205 km^2 and the Act's formal definition of ecological condition requires

> ... an area where the earth and its community are untrammelled by man, where man himself is a visitor who does not remain ... [and] without permanent improvements....

The wilderness areas are to contain no roads, nor use of motor vehicles and aircraft except in an emergency; wilderness users were divided in one survey about the desirability of telephones for emergency use, and none have yet been installed. The terrains designated are mostly in the west of the USA. This is where the wild country is, and where the Federal Government, especially the Forest Service of the Department of Agriculture, has its largest holdings. The wildernesses are mostly forested at their lower elevations but with an alpine terrain of high peaks, cirques and snowfields at the upper levels. Another kind of area is the Boundary Waters Canoe Area in the Quetico-Superior area of Minnesota-Ontario. Here a maze of natural waterways intersects coniferous forest, and canoe travel is the only method of transport permitted.

The Act makes it quite clear that the wilderness areas are recreative in purpose as well as scientific, for it speaks of 'outstanding opportunities for solitude or a primitive and unconfined type of recreation' as well as 'features of scientific, educational, scenic or historical value'. Thus, although wilderness and ecological purity are equated, there is a considerable amount of human incursion. The convention of wilderness recreation seems to be that the area is crowded if you see people other than those of your own party, and so the constraints on carrying capacity are narrow. Even with such small numbers of users, management of the wildernesses may present problems. For example, natural lightning fires are probably a regular feature of the ecology of western coniferous forests, but these can of course engulf people. Should such forest fires (as distinct from man-set ones, if the distinction can easily be made by a dragged-out-of-bed Ranger who has to decide whether to call in the fire-fighting services) be suppressed or let alone even if there is risk to human life? These and other difficulties, such as refuse and sewage disposal, pose problems for those whose task it is to keep the wildernesses wild, for even low levels of visitation. Only where the use is scientific in purpose, and for example the helicopter used to ferry people together with their gear and food, can the USA wildernesses be likely in the long run to remain truly inviolate in an ecological sense. The inevitable consequence of these management problems has been the introduction of rationing and pre-booking of visits to the more popular parts of the wilderness system.

Some other examples of the conflict between designated wilderness areas and the need to manage visitors are provided by some of the mountain National Parks of Sweden and Czechoslovakia. In Sweden, the northern wild Parks, administered by the Forest Service, are probably not totally natural since their lower areas are Lapp territory. Nevertheless an information brochure states,

> they must not be visited in such a way that . . . the land loses its unique character, and that any of what we can perhaps best express by the word 'mood' is lost.

Inaccessibility is a feature of Parks such as Sarek (1905 km^2) where provisions for a week have to be carried, while access is forbidden to parts of the forest land and mire complex of Muddus National Park (492 km^2), especially to protect bird life. In the Tatra National Park, jointly managed by Poland and Czechoslovakia, the high innermost zone can only be entered on certain paths and these must be kept to. Again, such areas were set aside with proper wilderness intentions, but their very designation has probably attracted some of the people whose numbers make management necessary.

Antarctica

The world's southern extremity is perhaps the best example of a deliberately protected wilderness (Plate 4, opposite). For many years following the early explorations it was an unconsciously preserved wilderness owing to its hostile nature, although the establishment of an increasing number of scientific bases by many nations meant that some destruction of biota ensued. The great impetus given to international scientific co-operation by the International Geophysical Year (1957–8), when research and observations in Antarctica played an important part, led the nations with territorial interests in Antarctica to conclude the Antarctic Treaty in 1959 (Heatherton 1965, Roberts 1965, Holdgate 1970). Even before the treaty the basic conditions for wilderness were in fact met: all settlement is 'point' settlement, supplied from outside, and much transport is by air; none of the local resources is used for food or building purposes except in an emergency.

The Treaty itself affirms the principle of international co-operation in occupance of the continent for scientific purposes. There is to be no military use or training, although military equipment may be used to aid research; no nuclear explosions or dumping of nuclear waste are to take place. No specific provision, however, is made for any terrestrial commercial resources should they be found: coal is the most likely, although this is highly baked and of marginal quality. The protection of the ecosystems comes in the Agreed Measures which are annexed to the Treaty and which aim at minimizing any disturbance to the plants, mammals and birds of the Treaty Area. No mammal or bird may be killed, wounded or maimed without a permit, for example; dogs must be inoculated against, *inter alia*, rabies and leptospirosis, and are not allowed to run free. Helicopters are not to approach large concentrations of animals, and all the signatories agree to try to alleviate pollution near the coasts and ice-shelves which are

Plate 4 Antarctica: the last great wilderness. Even this remote region is not immune from residual pesticides and its research-orientated function is being complicated by the advent of tourism. (*Aerofilms Ltd, London*)

the sites of most of the bases. Particularly noteworthy is the prohibition of the importation of non-indigenous species, except for laboratory use. There are also Specially Protected Areas where no vehicles are allowed and where even plants may not be collected 'except for some compelling scientific purpose', and Specially Protected Species, such as fur seals and the Ross seal, which are thought to be especially deserving of preservation. Overall, therefore, the Measures aim at the perpetuation of the variety of species and the maintenance of the balance of the natural ecological systems. The extent of their success is difficult to estimate but is probably high, since no large-scale ecological changes in the Antarctic appear to have taken place.

This position may not continue since the isolation of Antarctica from productive resource processes appears to be ending. Tourism has become a feature of the summer months, with cruise ships landing visitors to the point where Signy Island has now been agreed to be left alone by such parties, and the pelagic resources of Antarctica, formerly confined to whales, are now coming under increasing scrutiny, especially krill (see p. 212). Most likely of all to cause future disharmony between the Treaty nations and conflict with measures of environmental protection is the discovery of undetermined but considerable reserves of coal, oil and natural gas under the Antarctic continent and continental shelf. Even now, environmental contamination has found its way to these far southern latitudes in the form of DDT residues in the body fat of marine animals.

A wider context

It is unequivocally clear that the amount of 'unused' land in a 'natural' state is declining, and the rate of its disappearance is quite fast although precise data are lacking. The preservation of virgin land must therefore be a deliberate act, thus bringing it into the cultural-perceptual phase of resource allocation. The 'wilderness' idea is obviously culturally relative: many cultures see no virtue in the preservation of a large area of wild terrain just for its own sake. In many cases it is clearly a 'fringe benefit' for richer countries which poorer ones cannot afford if there are any resources at all in the wild areas. Are there, then, any strong reasons for the protection of wilderness areas other than those which are extensions of the arguments put forward for the designation of land as parks and nature reserves? Apart from the genetic pool idea, a suggestion which seems to have a world-wide validity is the role of wild areas, especially those with a high biological productivity, as 'protective' ecosystems which counterbalance the less stable 'productive' systems, a set of ideas (Odum 1969, Odum and Odum 1972) further discussed near the end of the book (pp. 366–8). A similar role has been postulated for the oceans. The exact role of these biomes is not known, but if there is any possibility that these wild ecosystems might be crucial then no deleterious action should be taken until more knowledge is available. The firm pressures of the numbers of people and their demands for materials militate against such action, for few nations or indeed individuals subscribe to Henry David Thoreau's motto for the preservers of wilderness, that,

> . . . a man is rich in proportion to the number of things he can afford to let alone.

Further reading

BENNETT, C. F. 1975: *Man and earth's ecosystems.*

BICCHIERI, M. G. (ed.) 1972: *Hunters and gatherers today.*

BRYAN, R. 1973: *Much is taken, much remains.*

FRANKEL, O. H. and HAWKES, J. G. (eds.) 1975: *Crop genetic resources for today and tomorrow.*

GOODLAND, R. J. A. and IRWIN, H. S. 1975: *Amazon jungle: green hell to red desert?*

HENDEE, J. C. and STANKEY, G. H. 1973: Biocentricity in wilderness management.

IVES, J. D. and BARRY, R. G. 1974: *Arctic and alpine environments.*

LEE, R. B. and DE VORE, I. (eds.) 1968: *Man the hunter.*

MEGAN, J. V. S. (ed.) 1977: *Hunters, gatherers and first farmers beyond Europe.*

MEGGERS, B. J. 1971: *Amazonia: man and nature in a counterfeit paradise.*

NASH, R. 1967: *Wilderness and the American mind.*

ODUM, E. P. 1969: The strategy of ecosystem development.

— and ODUM, H. T. 1972: Natural areas as necessary components of man's total environment.

SATER, J. E., RONHOVDE, A. G. and VAN ALLEN, L. C. 1972: *Arctic environments and resources.*

SIOLI, H. 1973: Recent human activities in the Brazilian Amazon region and their ecological effects.

STERNBERG, H. O'R. 1975: The Amazon river of Brazil.

STONE, E. C. 1965: Preserving vegetation in parks and wilderness.

UNESCO, Programme on man and the biosphere (MAB); final reports of expert panels:

1972: no. 3: *Ecological effects of increasing human activities in tropical and subtropical forest systems.*

1972: no. 5: *Ecological effects of human activities on the value and resources of lakes, rivers, marshes, delta, estuaries and coastal zones.*

1973: no. 11: *Ecology and rational use of island ecosystems.*

1973: no. 12: *Conservation of natural areas and of the genetic material they contain.*

1974: no. 14: *Impact of human activities on mountain and tundra ecosystems.*

3

Protected ecosystems and landscapes

In the previous section we considered that change was an accidental, if widespread, feature of wilderness areas and unused land. Since management was reduced to a minimum the ecosystems were insulated from anthropogenic changes. We now discuss certain classes of ecosystems where change is acceptable and management a necessary feature but where there is protection from certain kinds of alteration. Such action generally results from a desire to keep an ecosystem or set of systems in a valued state, and so is very much dependent upon cultural factors. The aim may be to preserve a species, assemblage or habitat which is rare, or typical, or symbolic, or perhaps to keep unchanged a traditional landscape because it occupies a high place in the values of those who see it or are responsible for it. Such ecosystems are usually wild rather than obviously man-made. Plants, animals, water and soil are, or are thought to be, relatively undisturbed and the terms frequently used of such areas are 'natural' and 'semi-natural'. In fact, a significant number of protected landscapes are cultural landscapes which exhibit distinctly the work of man, but this need not prevent them from being valued highly in the same manner that old buildings and other reminders of our cultural history are preserved (Newcomb 1979).

The desire to protect wild species and favoured landscapes usually resolves itself into two practical elements. The first group is nature protection (for which wildlife protection/preservation/conservation, nature conservation/preservation are synonyms) in which it is desired to perpetuate either a taxon or a group of taxa, or a particular habit. Species protection has a tendency to concentrate on the rare and unusual and habitat conservation upon the typical, but many exceptions can be found. The desire for protection is especially strong if the species is endangered (IUCN 1966–, Fisher *et al.* 1969), an attitude which has been growing with the realization that the extinction rate of animal species is now about one per year, compared with one every 10 years from 1600 to 1950. Perhaps as many as 20,000 plant species (mostly in the tropics) are at risk, a figure which may rise to 50,000 by 2000 A.D. if current trends in land use continue (Myers 1976, Raven 1976).

The second element may be termed landscape protection where a particular view or piece of country is protected from undesired despoliation or simply where change of any sort is restrained. These areas are sometimes used also for outdoor recreation, but that set of activities often requires much more manipulation and is considered separately in Chapter 4. As with wilderness, scientific and educational reasons are often advanced for the protection of biota and landscapes, partly for their study value as relatively undisturbed systems from which we may learn the better to manage intensively used systems, and partly for their role in maintaining the stability of the biosphere. Spiritual and aesthetic, even ethical, considerations also lie behind many protectionist movements: there is no denying the pleasure which many people gain from seeing wild rural landscapes, from Kinder Scout through Yosemite Valley to Fujiyama, and the popularity of bird-watching is such that in some crowded parts of Europe the great migration stopovers sometimes attract as many people as birds. Neither can economics be omitted: the public will line up to see rare fauna, as demonstrated by the nest of the Loch Garten ospreys in Scotland and by the African game parks, and in scenically beautiful areas they spread a lot of cash around the periphery in payment for souvenirs, film, food, lodging and public toilets. In a few instances, such as the ČSSR, the right of the people to enjoy unspoiled nature and to see beautiful scenery is enshrined in the constitution of the Republic. In the capitalist world the statement of national purpose is rarely as explicit, but is implied in many legislative acts to create nature reserves and protected landscape areas.

Nature protection

This protective process stems from a desire to perpetuate certain biota indefinitely because of recreational or aesthetic values, because of undiscovered economic potential, because of the possible key role of a species in keeping an ecosystem stable, or because they represent a form of biological organization which has demonstrated survival capabilities and therefore has lessons for mankind. Wild organisms may act as environmental datum lines and monitors of change and may thus be useful also as raw material for the restoration of damaged ecosystems. Lastly, there is the recognition that the extinction of a species is irreversible (Dorst 1970, Ehrenfeld 1976).

The object of concern may range from a single species living in a very restricted areas to a set of complex ecosystems covering a large tract like a mountain range. Such a continuum can only be divided arbitrarily, but for this discussion a differentiation will be made between species protection, where the resource manager aims to keep a viable population of a particular taxon, and habitat protection, where the whole is more important and fluctuations in the populations of component species or in features of the inanimate environment are not considered detrimental. A capricious feature of wildlife conservation is nomenclature: nature reserves, game reserves, national parks, nature parks and several other terms are used by the private and governmental organizations which manage land and water resources with the purpose of perpetuating wild nature. For example the term National Park covers systems managed for many diverse purposes, all the way from wilderness to intensive outdoor recreation. In this account the purpose of management will be stressed rather than the details of designation (IUCN 1971).

Species preservation

Nature reserves are usually designated in order to give protection to a species of plant or animal which is rare (on a variety of scales from the regional to the global) and which it is thought can be preserved in this way. Animals, especially birds and large mammals, are most often thought of in this connection, but plants may qualify for the same treatment, especially in densely settled lands where many of the wild mammals have disappeared. Typical early instances of such reserves are the nesting sites of rare birds, although general legislation preventing the killing of particular species is commonly passed before the setting up of reserves. In Britain, for example, a law protecting seabirds was passed by Parliament in 1869 and wild birds generally in 1880, but there were no National Nature Reserves until after the founding of an appropriate government agency, now the Nature Conservancy Council, in 1949. Privately owned preserves such as those of the Royal Society for the Protection of Birds were earlier in the field (Fitter 1963). Single-species conservation in Britain is dominated by plants and there are reserves owned or leased by the Nature Conservancy or bodies such as the County Conservation Trusts or the National Trust which are dedicated to protecting, for example, one of the last 20 good fritillary meadows or a particularly rare buttercup (Stamp 1969). In the field of animals, the Farne Islands off the coast of Northumberland are a reserve for the grey seal: these islands have a long history of management for the benefit of wild creatures, starting with St Cuthbert in the seventh century AD. National legislation (e.g. the Endangered Species Act of 1973 in the USA) is now beginning to protect wildlife both inside and outside reserves in the developed world, although the EEC cannot agree on the conservation of migratory birds because of the intransigence of France and Italy over the right to shoot practically any species: Italian hunters regularly kill 150–200 million birds/yr of all types, along with a few of their own species (3 in 1975, e.g.) on the opening day of the season.

Elsewhere in the world, animals tend to dominate single-species preservation mainly because they dominate worldwide concern, largely of a sentimental nature. In northern Alberta and the North-West Territories of Canada, for example, the Wood Buffalo National Park preserves in a wild and almost roadless terrain one of the few remaining herds of North American bison. The herd is culled regularly in order to keep it within the carrying capacity of the National Park area and also to try and keep down the diseases to which the herd is subject. Elsewhere, many other sanctuaries exist and it is often the first step in protection measures taken by undeveloped countries: two small islands in Sabah, Malaysia, protect frigate birds, for example; in north-east India the Kaziranga Sanctuary preserves the Great Indian rhinoceros; and the last 25–40 Javan rhinos are in a reserve at Udjung Kulon-Panailan in western Java, established in 1921 (Talbot and Talbot 1968).

Management of such reserves is usually limited, because of institutional and financial constraints. Where management is practised then it is usual to manipulate either the habitat or the population of the protected species. At Havergate Island in Suffolk, England, the RSPB have created pools in stretches of shingle in order to extend the conditions favourable for the breeding of the avocet, a bird rare in Britain. Further north, the National Trust periodically culls the grey seals of the Farne Islands in order that mortality said to be due to overcrowding should be reduced, and to keep down damage allegedly done to local salmon fisheries. The government of the UK has also undertaken to cull the grey seals off northern Scotland but in 1978 abandoned the event under public pressure, there being controversy over whether the population was in fact increasing rapidly and whether it had the alleged effect on the already declining fish stocks of the North Sea (Coulson 1972, Cherfas 1978, Summers 1978). An example of far-reaching management to protect a species is the Indian 'Project Tiger' which has increased the number of tigers from 1,827 in 1972 to 2,484 in 1977. Eleven tiger reserves were established and more are planned, and the government has removed and relocated villages from the core areas of the reserves: 4,000 people from the Kanha reserve were rehoused.

The protection of assemblages

Rather more common are reserves to protect an assemblage of species. These merge imperceptibly with those in which a whole habitat or set of habitats is preserved, but a distinction can perhaps be made on the arbitrary grounds of size: habitat reserves are large relative to the assemblage ones. The assemblages may have some linking affinity: wildfowl refuges, for example, which cater for the nesting or migration of many species of ducks, geese and waders (Plate 5, page 74). Islands may have endermics which are very vulnerable to extinction, as on the Galapagos or Aldabra. High mountain reserves often protect a very diverse suite of alpine plants, sometimes with their attendant fauna and where forested reduce the magnitude of floods in the river basins below. Although wild terrain is often sought for such reserves, the behaviour of wild animals often makes it necessary to set aside man-made habitats and quite frequently suitable places are of commercial value. The case of wetlands which are sought for reclamation as industrial land, airports or garbage dumps can be seen in many parts of the world: tidal estuaries are also among the most productive ecosystems in the world (as in Table 1.2) and clearly deserve special attention on this account, as do many coastal wetlands such as salt marshes, mudflats and mangroves, and especially coral reefs. All of these protect coasts from storms as well as being nurseries for fisheries, in addition to their wildlife value which is especially high for migratory birds. The USA has a system of National Wildfowl Refuges which are placed on or near the major flyways which run north–south across that continent (Fig. 3.1). They are designated in concert with Canada, on whose northern lands (along with those of Alaska) many of the birds nest. Their management is designed to provide maximum cover and food supply for the migrating birds and to provide refuge from hunting so that while sufficient numbers reach the waiting guns to satisfy the hunters (who pay for 'Duck Stamps' and a tax on all equipment in order to finance research and land-acquisition programmes), there are sufficient escapees to return to breed in the following year. Some limited hunting may be allowed on the Refuges along with a restricted variety of other recreational activites. Experience has shown that most populations of ducks and geese can replace hunting losses quite satisfactorily provided the cropping levels are planned in accordance with the breeding success. This necessitates a fairly sophisticated hunter who is able to distinguish between one species and another and between males and females, which surveys by management authorities have often shown to be not the case.

Plate 5 Wildfowl. Unpriceable but invaluable? *(Grant Heilman, Lititz, Pa)*

Plant assemblages are often protected in reserves, especially where some of them are rare on a regional or national scale. In the Upper Teesdale area of northern Britain, an assemblage of arctic-alpine plants, most of which are found relatively frequently in the Scottish Highlands and commonly in the Alps and Scandinavia, but not elsewhere in England (Piggot 1956), was designated first an SSSI (Site of Special Scientific Interest) under the authority of the National Parks and Access to the Countryside Act of 1949, and then a National Nature Reserve. Because of other interests in the land (especially grouse shooting), management will consist mostly of the traditional practices which include regular firing of heather (*Calluna vulgaris*) which is the food of the grouse. Experimental enclosures which relieve grazing pressure by sheep and rabbits first give a tremendous surge of the rare flora, but this then tends to be suppressed by the vigorous growth of the accompanying plants, often very common grasses. Land-use practices

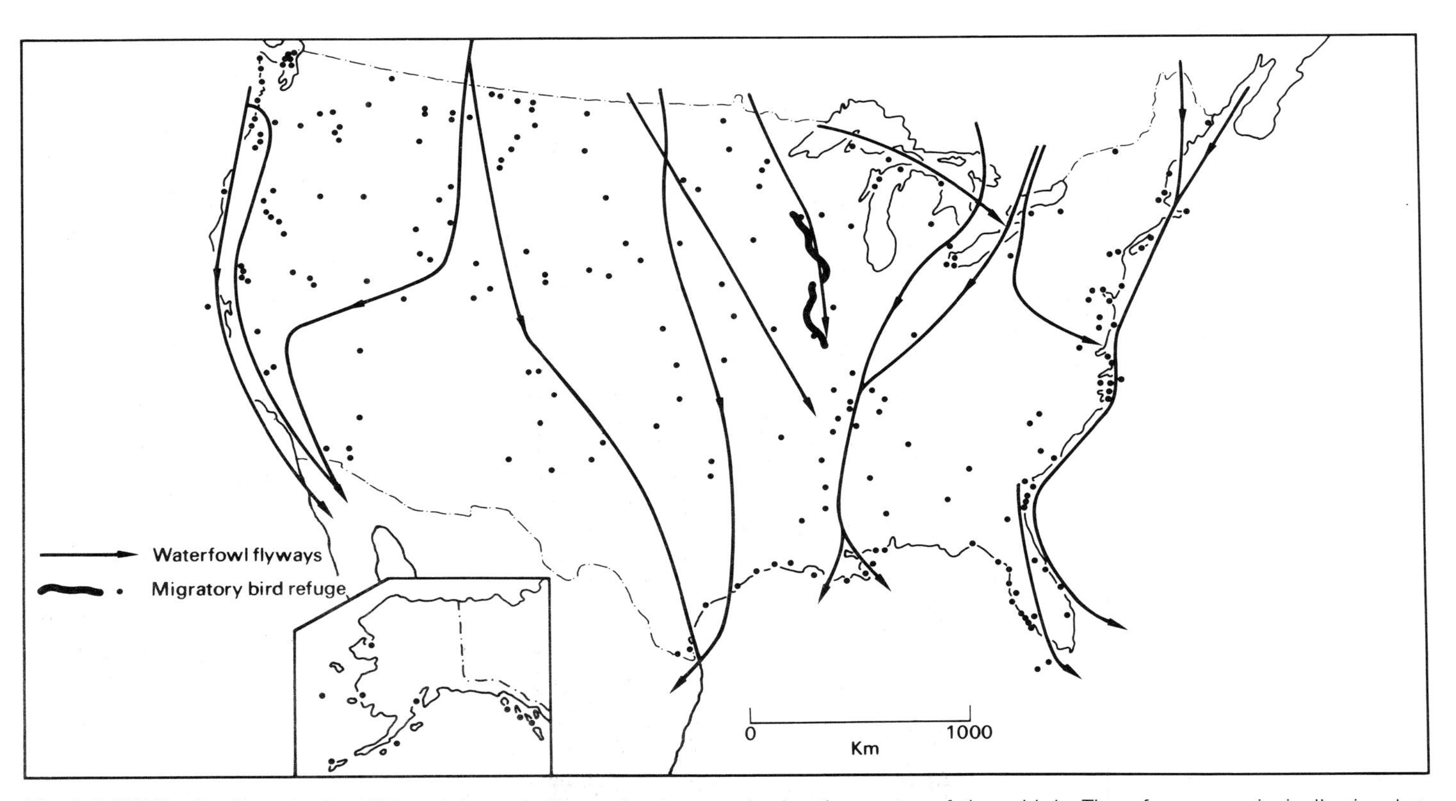

Fig. 3.1 Wildfowl refuges in the USA and the main 'flyways'—the annual migration routes of these birds. The refuges are principally aimed at perpetuation of a particular assemblage of avian species.
Source: Dasmann 1972

may therefore have been strongly influential in the continued existence of some elements of this flora, and complete protection by withdrawal of other land uses would be likely to entail the loss of some plants by competition (Squires 1978). A piece of woodland thought to be typical of a region is also a candidate for preservation, especially if it is thought to be substantially unmanipulated or even 'primary woodland', i.e. the ecosystem is quite strongly manipulated but there is evidence that there has been woodland on the site at least since the Medieval period. This is particularly so in Europe, where most woodlands have been either removed to make way for agriculture or intensively managed for centuries. In southern Bohemia, for example, the ČSSR has designated as a reserve the last fragment of beech-spruce *urwald* at Boubinsky Prales. A fence has been erected to keep out deer; the result has been rapid regeneration of the beech but not the spruce, and so other management devices are clearly necessary. The difficulties of ensuring regeneration in small woodland fragments are also illustrated by some of the relict pinewoods of Scotland in the care of the Nature Conservancy Council. Regeneration of the dominant *Pinus sylvestris* is confined to woods on bouldery slopes; elsewhere the raw acid humus appears to inhibit the seedlings. Here again is a case of *laissez-faire* being unsuccessful in perpetuating a particular plant.

Habitat preservation

Reserves which are large and diverse enough to protect whole sets of ecosystems which are either rare on a national or world basis, or thought to be especially typical of the country where they occur, may be designated as National Parks, although nature protection is often not the sole purpose of such places.

Few reserves in Britain are able to qualify for this category. The possible exceptions are some National Nature Reserves in Scotland such as the Cairngorms, Beinn Eighe and the Isle of Rhum. The first two are mainly mountainous terrain, with grasslands, mountain heath, and relict pinewood on the lower slopes. The red deer is the largest mammal and is in general too high in numbers for the amount of winter feed. But the Nature Conservancy Council is unable to carry out effective management policies since its leases do not confer complete control over animal management and because animals transgress the boundaries of the Reserves. This is not true of the Isle of Rhum, where it controls the grazing: the sheep which were overgrazing the open moorlands have been removed, and planned management of a herd of red deer substituted.

Some of the National Parks of the USA are devoted mainly to conserving a series of wildlife habitats. This is particularly true of the Everglades National Park in Florida, which comprises a complex of wetland areas such as coastal mangroves, tropical sawgrass marshes and 'hammock' forests on the slightly drier areas a few centimetres above flood level. The 4,856 km^2 park supports a wide variety of animals also, such as the rare Everglades kite, the roseate spoonbill, alligators, manatees and the many fish that are the basis of a sizeable industry. Until recently management was confined to the channelling of the ever-increasing numbers of visitors, and to measures like the protection of alligators from would-be makers of handbags. But the large size of the Everglades and its proximity to the popular holiday area of Florida mean that pressures to alter the ecology are intense. The main threat has been the loss of its freshwater input and thus the possibility of the wetlands drying out. The main water source is Lake Okeechobee and the water in this lake has been cut off from the Park in the name of flood control further north; surplus water was released into the ocean instead of into the Everglades.

On a still larger scale are some of the National Parks and Game Reserves of eastern, central and southern Africa, in such countries as the Republic of South Africa, Tanzania, Zambia, Uganda and Kenya (Fig. 3.2). Here the desire to ensure the survival of the fauna, especially the large mammals, is reinforced by two strong economic considerations. Firstly, most of these countries (especially the newly independent republics) make a great deal of foreign currency from tourism; for some it is their largest earner, and the visitors nearly all come to see the animals. Secondly, it has been shown that controlled harvesting of wild animals not only helps to ensure their preservation, since their ranges are being restricted by various forms of development, but also that the wild game can yield more protein per unit area than domesticated animals and with less damage from overgrazing. This is a fairly revolutionary concept and is not widely acceptable at grassroots level. Groups such as the Masai of Kenya, for example, are reluctant to give up their tradition of cattle-herding since the animals are not only meat and milk but also money to them (Myers 1972a, 1972b). The potential of controlled cropping is discussed in Chapter 7.

Asia too has many parks devoted to protecting mountain and forest areas, along with the associated fauna, in a natural state. Even countries like South Viet-Nam had two National Parks. The largest one, of 78,000 ha at Bach-ma Hai-Van, near Hue, was of virgin monsoon forest (Nguyen-van-Hiep 1968). In India, about 0·6 per cent of the country has been established as parks and sanctuaries but population pressure causes many management problems, one of which is the linkage of conservation with a growing tourist industry (Singh 1978). Newly emerging and undeveloped nations such as Papua New Guinea have enormous potential, as well as some difficulties, in conserving their flora, fauna and natural environments (Schodde 1973, Winslow 1977). The course of shifting cultivation and logging have injured many park areas in Asia where insufficient park personnel and wild terrain mean that control of these inimical land uses is scarcely possible. A further reason for preservation of upland forests is found in their protection of watersheds: if deforested they add to the flood hazard of urban and agricultural lands downstream and increase the silt burden as well as has happened in the Ganges Valley with deforestation of the mountains in the southern Himalaya.

The preservation of habitats is also the main function of the natural areas of the USSR which are withdrawn from enconomic utilization for scientific research and for cultural-education purposes. In 1966 there were 68 of these *zapovedniki*, totalling 4,300,000 ha, and the predominance of scientific research as their main purpose was well established, in such fields as the breeding and propagation of animals that are rare

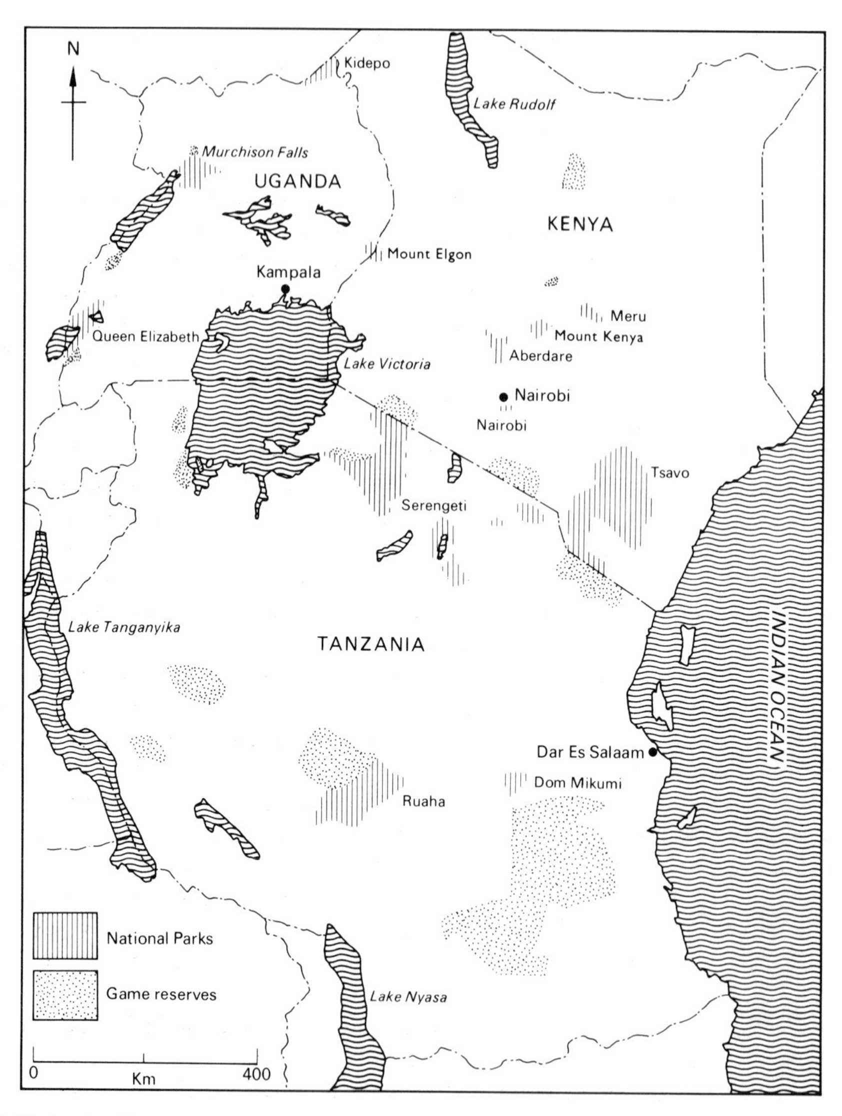

Fig. 3.2 The major National Parks and Game Reserves of Kenya, Uganda and Tanzania in 1970. The large areas which they occupy are symbolic of their role in the economy of the nations concerned as well as environmental considerations.
Source: Morgan 1972

or threatened with extinction, the study of both unusual and typical vegetation types, the investigation of the total ecology of particular regions, and the preservation of unique geological or archaeological features (Isakov 1978). Certain 'open' *zapovedniki* accommodate tourism, but they are limited in number and recreation appears to be a secondary consideration. A long-range plan for new reserves was presented in 1957, but Pryde (1972) reported that few of its recommendations had been implemented.

Management

The preceding paragraphs have given particular instances of difficulties associated with the management of ecosystems for nature protection, and some of these have a more general applicability. As with wilderness areas, deliberate designation is necessary but mere legislative enactment is insufficient. Apart from the obvious protective measures such as the prevention of poaching or picking, there are more subtle ecological interactions. For example, the removal of a particular type of pressure from a desired species may cause it to 'bloom' in an explosive fashion and crash thereafter. Similar difficulties have been experienced when the enclosure of rare plants has been followed by the discovery that they have survived because grazing pressure has kept down competing species. An example of this is given by some species of orchids in the chalk and other limestone grassland of southern England. In the absence of sheep this plant association soon becomes long grassland and then scrub, so that the orchids characteristic of the short-grass stage disappear very quickly once succession starts to take place. In this case, management measures for the preservation of orchids consist either of re-introducing sheep or of mowing the nature reserve.

A classic instance of the imperfections of designated reserves occurs when an animal spends part of its yearly cycle outside the reserve. A large mammal may spend its winters in a reserve where predators are protected but the rest of the year on a range outside the reserve which it shares with domestic livestock and where predator control, as of the wolf or the coyote, is maintained. The populations of the mammal are thus liable to build up to levels which cannot be supported by the winter feed, and the predators are incapable of trimming the population sufficiently. In such cases, management must be undertaken, either by culling a number of beasts or by feeding them artificially. The elk of Yellowstone National Park in the USA have had to be treated in this way: in this instance culling by the Park staff has been undertaken, much to the chagrin of local hunters.

Such problems basically arise from the fact that legislative boundaries rarely coincide with the natural spatial limits of the ecosystems which it is desired to protect; this may also be manifested in the operation of watershed influences which originate beyond the limits of the reserve but which penetrate into it. An obvious case would be the spraying with a persistent insecticide of agricultural land upstream from a protected lake. Similarly, the addition of large quantities of chemical fertilizer would probably result in the eutrophication of the lake. If the protected system is a forest, then commercial logging up to the boundaries of the reserve may possibly mean that the remnant area is too small to function as a natural unit—e.g. there may be insufficient territories for a viable population of a particular species. As a further example, in connection with the establishment of a Redwoods National Part in northern California in 1968, the Secretary of the Interior is authorized to enter into agreements and easements with the owners of the watersheds whose ecology affects that of the Park, in order to protect certain features of the Park itself (Simmons and Vale 1975).

Apparently destructive influences such as fire and flood, whose origin is frequently outside the reserve area, often pose problems for the managers of protected ecosystems. Fire may in some places be a part of the natural ecology, especially in coniferous forests where it seems to have an important role in mineralizing organic matter piled up on the forest floor, and in aiding tree regeneration since several species of conifers have cones that only open and release their seed after subjection to very high temperatures. The giant redwood (*Sequoiadendron giganteum*) of the western slopes of the Sierra Nevada in California is a case in point. Nearly all the remaining specimens of this tree are in State or National Parks, but regeneration has been sparse since the seedlings of the tree were shaded out by faster-growing competitors (Vale 1975). Following the realization that lightning-set fires had been a normal component of the local ecosystem, managers began programmes of controlled burning. Along with some restrictions on access to prevent trampling, these appear to have encouraged some regeneration of this rare and impressive tree.

A noteable development in reserve designation and management has been the elaboration of a body of theory about reserve size and shape with its origin in the study of immigration and extinction on islands (MacArthur and Wilson 1967). Biologists have noted that many nature reserves are 'islands' in a sea of more intensively manipulated terrain and so some of the principles of island biogeography apply, involving the mathematical relationships between taxonomic diversity, size of 'island' (i.e. reserve) and its shape (Terborgh 1975, Wilson and Willis 1975, Diamond 1976; Fig 3.3). Some

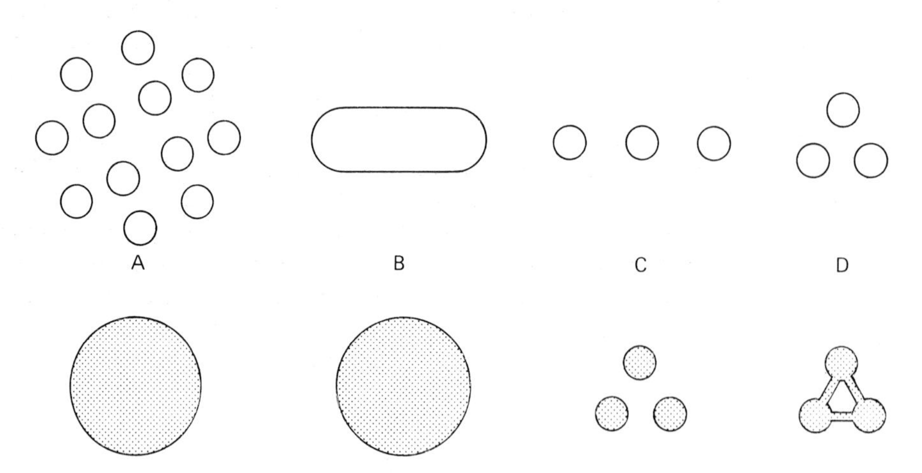

Fig. 3.3 Geometrical rules of design based on current theories derived from the study of island faunas. Both upper and lower hand figures have the same total area and represent reserves in an homogenous setting, but the lower configuration leads to a lower extinction rate than the upper. Rounded and continuous areas are preferable to fragmented and elongated areas, (A) and (B); distance effects are minimized by clumping rather than linear arrangements (C); and corridors between isolated areas are even better (D).
Source: Wilson and Willis 1975

of these ideas are being applied in the designation of reserve 'clusters' in the USA (Franklin 1977) but are only implicit in the reviews of all potential nature reserves carried out in small nations like the UK (Ratcliffe 1977).

Although the presence of wild nature is becoming more and more valued, it does not yet take precedence over more directly economic uses of land and water resources, and so conflicts over the conversion of use arise. The most common of these is the encroachment of industry on formerly wild lands which may have been under protection. Freshwater wetlands are especially vulnerable when surrounded by agricultural land, owing not only to substances dissolved in the drainage water, but also to the need to drain the agricultural land. If a 'perched' water-table is to be achieved in isolation from the surrounding drier lands, then considerable engineering works are a necessary prelude to management. This has been achieved by the Nature Conservancy at Woodwalton Fen in eastern England, where a remnant of original fen vegetation is maintained as an island in a sea of intensively farmed agricultural land (Duffey 1974). This theme is echoed in Japan where Numata (1974) records that most of the near-recent extinctions in the flora came from marshy habitats.

The protection of landscapes

The placing of whole landscapes under some form of legislative or regulatory protection comes from the desire to maintain or enhance a scene which is of great value, often for various types of recreation or for aesthetic purposes. Emotive and symbolic values are of considerable importance and so highly valued landscapes do not have to be 'natural'. Some of them are mostly virgin terrain, for example the National Parks in the western cordillera of North America like the Yellowstone and Rocky Mountain Parks in the USA and the Banff-Jasper Parks in Canada. Others however are landscapes which are thought to be natural but which closer investigation reveals to be man-made, even if they are extremely wild, like the moorlands of England and Wales. Lastly there is the frank preference for the man-made landscape which exhibits aesthetic qualities that makes it worthy of special attention. The combination of lake, wood, field and village in parts of Denmark, the small-scale neatness of rural England, the patterned richness of Lancaster County, Pennsylvania, the staircase of rice paddies flanked by cherry trees in rural Kyushu, all exemplify this type. The perception of the landscape, and the value then attributed to it, are obviously all very important in the decision to protect it. Management may (or may not) thereafter be carried out on scientific principles, but the initial motivation is inspired by a particular set of cultural values.

National Parks

One expression of the desire for protection is the designation of entire landscapes as 'scenic areas', 'areas of natural beauty' or 'national (state, regional county, country) parks' in such a way that the dominant purpose of management becomes the preservation of the qualities of the scenery. This 'total protection' is most often done with natural landscapes, since the lack of other uses produces fewer conflicts than in areas with more directly economic utilization. The legislation is therefore necessarily strong, since it must confer upon the manager the power to exclude all manipulations and changes which are deemed to be inimical to the perpetuation of the values of the landscape. In practice this usually means government ownership of the land and its recources, and the outstanding example of this category is the National Parks system of the USA (Fig. 3.4). Not only does it exemplify the way in which protection has taken place, but it also illustrates some of the conflicts which may arise within the framework of protection, even after the decision to remove the park resources from the ambit of commerce has been taken.

The first of the National Parks, Yellowstone, was designated in 1872. Others followed and the system grew to the point where an Act of the Federal Government was necessary to establish a National Parks Service to administer the system. The wording of the organic Act of 1916 enshrines the purpose of the Parks as being to:

> Conserve the scenery and the natural and historical objects and the wildlife therein and to provide for the enjoyment of the same in such manner and by such means as will leave them unimpaired for the enjoyment of future generations.

The system has grown to include not only the National Parks which are the chief concern of this section, but National Monuments, National Recreation Areas, Seashores and many other features. National Parks are, however, still being added (e.g. Utah Canyonlands in 1964, Redwoods and North Cascades in 1968), although it was intended to complete the system by 1972. In1978 there were 37 National Parks, with a total area of 63,000km^2 of which the Federal Government owned over 90 per cent, the difference being accounted for by privately held inholdings within the park boundaries. The types of terrain enclosed in the system are those generally acknowledged to be outstanding examples of the American scene: Yellowstone, Glacier, and Rocky Mountains National Parks in the Rockies, for example; Shenandoah and the Great Smoky Mountains in Appalachia; Grand Canyon, Bryce Canyon, Zion, and Canyonlands in the arid south-west; Yosemite, Kings Canyon-Sequoia, Lassen, Crater Lake and Mount Rainier in the far western ranges of the cordillera; and many more, including the Florida Everglades discussed earlier in this section. Hunting is not allowed, but the Parks are popular for many other recreational activities, of which the main one is the viewing of the scenery and its component natural elements which are the *raison d'être* of these Parks. Their popularity is such that visitor numbers rose by about 10 per cent each year during the 1960's.

In spite of these pressures, active manipulation of the Parks has been limited to relatively small areas. Forests and water-courses are little altered and there is neither grazing nor mining. In surroundings with such little change, wildlife is comparatively abundant, although sometimes the protection leads either to an overall superabundance, as with the Yellowstone elk, or to undesirable concentrations, as where a species turns to scavenging upon the leavings of the visitors and occasionally predating upon the visitors themselves, as has happened with grizzly bears (*Ursus arctos horribilis*). Because most of the visitors stay close to the roads, campgrounds and other 'developed' parts of the Parks, wilderness can also be a feature, and most of the Parks have large

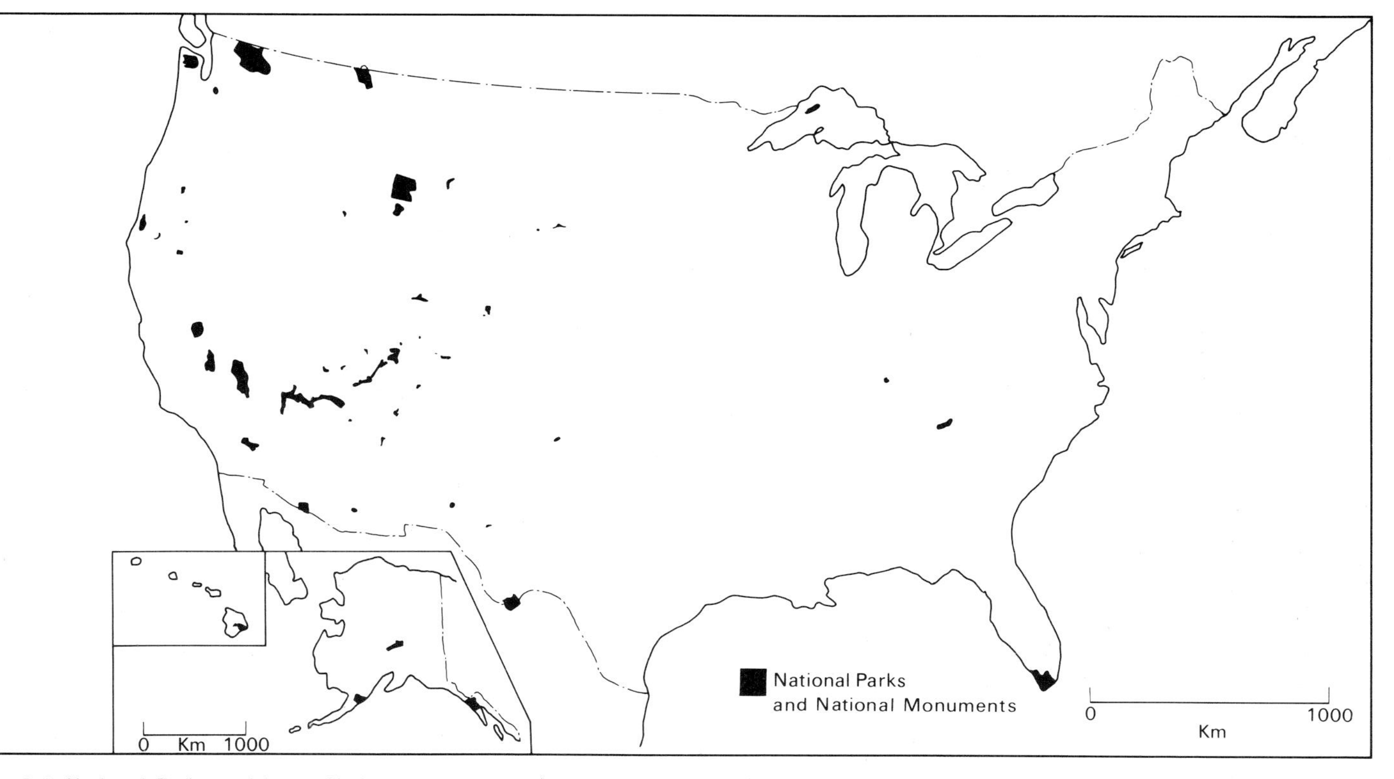

Fig. 3.4 National Parks and larger National Monuments (such as Death Valley) in the USA, 1970. The concentration in the west underlies the class of resource under this type of management, but the very small proportion of the nation so protected is evident.
Source: US National Parks Service 1970: *Areas administered by the National Park Service and related properties as 1 January 1970.* Washington, DC.

back-country areas which are only infrequently visited since they have to be traversed on foot or horseback. Inevitably, the more accessible areas are the most popular and here the demands of visitors have in large measure been met: there are car parks, campgrounds, stores, hot showers, cabins, nature trails and other interpretive facilities. Fishing is so popular that all the streams have had to be kept stocked. With so many functions for the park to perform it was inevitable that conflicts between them arose, especially under the relentless escalation of visitor numbers. Under the politically useful if ecologically contradictory slogan 'parks are for people', many developments were made which threatened wilderness values, such as roads into remote parts. More campgrounds intruded into wildlife foraging areas, sometimes those of easily angered animals such as bears, and the problems of sewage and garbage disposal from the compers and day-visitors in a few places resembled those of a small town. Such conflicts were the subject of a special advisory committee to the Secretary of the Interior. Their special concern was wildlife, but their report had far-reaching effects upon the long-term purposes of the Parks. They concluded that the management goal should be

> to preserve or where necessary to re-create, the ecologic scene as viewed by the first European visitors, ... in order to ... 'enhance the aesthetic, historical and scientific values of the parks to the American public, vis-à-vis the mass recreational values'. (Leopold 1963)

In line with these aims, a new set of policies appeared, designed to avoid land-use conflicts arising from confusions of purpose. Three major types of area within the National Parks system have been designated (natural, recreational and historic) and the administrative guidelines for the 'natural areas' emphasize the protection of the ecological systems. Some of the measures taken to achieve these ends included the formulation of control programmes for ungulate populations, the use of prescribed burning to pre-empt dangerous wildfire, and the gradual removal of service facilities outside the park boundaries. The motor car is to be restricted: in the summer of 1970 an area of Yosemite Valley was sealed off and a minibus service substituted and further plans envisage a 'de-development' at popular areas such as the Old Faithful geyser at Yellowstone National Park. It remains to be seen whether the public image of the National Parks as all-purpose scenic-recreation-resort areas is maintained or whether their avowed role as the jewels in the crown of protected lands with a special, even élitist, role can be substituted (Fitzsimmons 1976, Rowntree et al 1978). Their remarkable nature and distinctive purpose (it must be remembered that the USA gave the world the National Park idea) requires a farseeing intellectual effort on the part of managers and an understanding attitude from the visiting public. As Darling and Eichorn (1967) conclude,

> certain forms of decorous behaviour should be accepted and not questioned.... The national parks ... represent the glorious creations of nature and no expediency or misconception of their beauty must endanger the world heritage of which they are so shining a part.

It is perhaps not surprising that UNESCO has set up a World Heritage Committee, which has designated its first twelve sites which are held to be of world significance and hence in need of special protection. The list comprises cultural as well as natural sites so that for example Aachen Cathedral (BRD) is included as well as L'Anse aux Meadows National Historical Park (Newfoundland); Nahanni NP (Canada) and the Galapagos Islands (Ecuador), Cracow Historical Centre (Poland) and the City of Quito (Ecuador).

Protected landscape areas

The amount of land that can be placed under restrictions of the kind reviewed in the previous section is, by present cultural criteria, limited. Elsewhere, economic considerations may apparently dictate land and water use. Some of the landscapes created by economic usage are nevertheless highly valued, and the desire arises to prevent changes which threaten the areas which have evoked such reactions, for some landscapes exhibit in their aesthetic dimensions a quality which adds to their conventionally utilitarian worth. The type of protective measure which seeks to protect such landscape values while not interfering with the farming or forestry or other commercial use could perhaps be described as 'cosmetic', since it deals with the treatment of details of the surface, rather than with the underlying structure. A more conventional term is 'development control'. Development control seeks to minimize conflicts between the normal processes of economic use and the principally aesthetic appeal which the landscapes hold. Thus control conferred by legislation or by agreement seeks to preserve the familiar and well-beloved by preventing change which is deemed to be ugly, and by clearing up unwanted relics of the past. As examples, we may quote the banning of advertising billboards along roads in Denmark, the control exercised over certain types of agricultural buildings in parts of England and Wales, and the grants available in many countries for the camouflaging of auto junkyards in rural areas and on urban fringes. There are basically two ways in which such controls can be applied: by wholesale prohibitions and regulation over the whole of an administrative area; or by designating certain regions with a distinctive character and concentrating on the protection of that particular place. In the case of the latter, some distinguishing title, such as Protected Landscape Area, Area of Outstanding Natural Beauty, or even National Park, is given.

In the case of areas of a particular character which it is desired to protect *in toto* from certain landscape changes, the two main aids are development controls and the provision of governmental aid to clear up detrimental features (Plate 6, page 86). Thus one of the major aims of the National Parks and Access to the Countryside Act 1949 of England and Wales was to secure control over most types of development such as non-agricultural buildings, advertising, and caravan sites, in the areas which were to be designated (Fig. 3.5) as National Parks. Money for clearing up eyesores, mainly wartime remnants, was also available to planning authorities. Regions of slightly lower value, Areas of Outstanding Natural Beauty, have also been designated. As it has turned out, the major landscape changes in the National Parks to which many people

Plate 6 Derwentwater in the English Lake District. Apart from the lake this is largely a cultural landscape, but one so valued that it enjoys protection from certain kinds of change by virtue of its designation as a National Park. As can be seen from some of the mountains, afforestation is not subject to control. *(Aerofilms Ltd, London)*

have taken objection are exempt from the provisions of the Act. These were large-scale afforestation with coniferous trees, reclamation of open moorland for agriculture and the erection of large-scale farm buildings to take advantage of new developments in intensive farming. Because of the importance of retaining the economic life of the National Parks, the Countryside Act of 1968 did not provide for control of these changes, and the subsequent Sandford Report's (DOE 1974) recommendations that much more planning control be created over these uses was not accepted by the

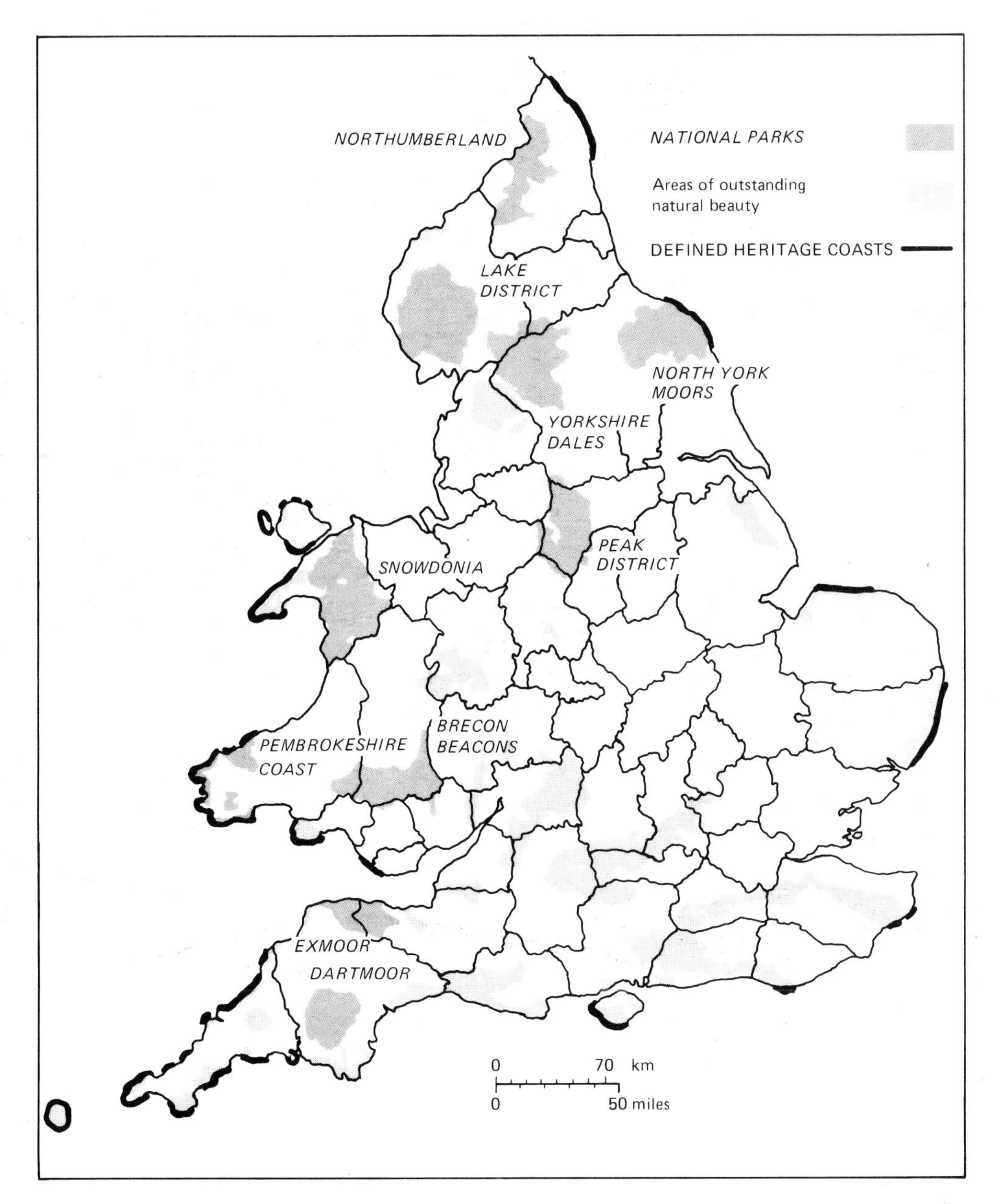

Fig. 3.5 Areas of England and Wales where strict development control is practised: National Parks, Areas of Outstanding Natural Beauty and Heritage Coasts. Economic activities continue within such areas. Source: Countryside Commission of England and Wales 1977: *Annual Report 1976–77*

government, one arm of which (the Ministry of Agriculture) was seeking to promote changes which another branch (the Countryside Commission) was trying to prevent.

In Denmark, the protected landscape may, if small in area, be part of the system of nature reserves. Larger units are called 'Naturparker', and the aim is to preserve a total environment or milieu with both cultural and natural features in which the Dane will feel at home scenically and in which there will be opportunities for outdoor recreation.

The formation of a 'Naturpark' takes place under the 1969 Preservation of Nature Act, amended in 1972 and 1975, which makes possible the designation of large areas (10–15,000 ha, for example) in all parts of the country. The chief aim will be the preservation of current characteristics, and to that end farming and forestry will continue, with new buildings and other developments strictly controlled. Management aims will allow the integration of recreation areas where these are compatible with the primary purpose. An early example of a Naturpark was the Tystrup-Bavelse area of Zealand, known as 'Sjaellands grønne hjerte'. It is an area of farmland, woods and villages comprising 12,000 ha around a lake, the Tystrup Sø. Until the 1969 Act it was not possible to pay compensation to landowners materially affected by the emparking, and the character of the area was maintained largely by voluntary agreements, but this legislation has made the position of both government and landowners clear.

Needless to say, there are many problems associated with the preservation of landscape values. Take for example the implication that a landscape should be kept as it is. Unless it is a totally natural landscape, then it has come to its present state in response to operation of past economic processes. If the needs now change, why should not the landscapes alter? What justification can be found for fossilizing a cultural landscape as it is now? The answer that present values demand it raises the question of whether such values are permanent, which they clearly are not. If we agree that uncontrolled alteration of landscapes in response to every economic whim of a *laissez-faire* economy is unacceptable, but that fossilization is also undersirable, then we have to find criteria for distinguishing allowable change from that which must be stopped (Simmons 1967).

Although development policies can be defined and adopted democratically (or autocratically), not every demand for change will fall neatly into an allowable or prohibited category, so that many *ad hoc* decisions have to be taken. In the face of pressures for a new industrial development such as a newly discovered mineral deposit, it is almost impossible to uphold a protective set of regulations, witness the controversy over potash in the North York Moors National Park and the concern over the copper and gold extraction explorations of RTZ in the Snowdonia National Park and the extension of limestone quarrying in the Peak District National Park in the 1970s. Sometimes the viewers of the landscapes create the greatest problems, especially if they are present in sufficiently large numbers (Duffey 1967). The unwitting damage which they then do, such as trampling and the scattering of litter, is made worse by the presence of numerous vehicles (Bayfield 1971, Streeter 1971). These not only damage the ecology unless properly managed: in open areas they represent a highly unaesthetic intrusion. In very fragile environments such as arctic and alpine areas, only a few people will cause considerable ecological change. Similarly small numbers of careless or malevolent visitors can create damage of a vandalistic kind. Examples of this are very familiar, and include direct economic loss to farmers as when stock are allowed to stray or are harassed by dogs; or when palatable but lethal rubbish such as some types of plastic are left behind for domestic animals like cattle to eat.

Ideally, the protected landscape is a non-consumptive resource: the crop is of a visual nature, and when this has been taken by the onlookers, the resource remains; the aim of management is to perpetuate this attribute.

Conflicts

The mention of landscape-protection areas which embrace forestry and farming provokes the discussion of the possibilities of multiple-use schemes which embrace a measure of wildlife or scenic preservation. Clearly, given goodwill and the possible loss of income from compromising with that aim, farming and forestry can be carried on along with nature parks, National Parks and the like. Any severe restrictions, such as the control of agricultural buildings or the enforced planting of hardwood screens around conifer plantations, need compensation payments if they are to be reasonably willingly adopted. Recreation and watershed management may also usually be integrated into scenery preservation if firm control can be exercised over the siting of developments. Reservoirs may of course come into the category of unacceptable alterations of the landscape: tastes vary from country to country. Nature protection often fits well, as when small reserves are embedded in a matrix of preserved landscape.

Some resource uses are decidedly incompatible with the protection of landscapes: military use almost invariably results in considerable damage, and industrial plants can rarely be given sufficient screening to render them unobtrusive. Large quarries are generally inimical developments, producing dust and noise as well as large holes. Processing plants are likewise highly intrusive and often sources of environmental contamination. So while landscape protection and nature conservation elements can and should be a part of every development scheme, rural or urban (as it is in the Netherlands where the Forest Service has a national responsibility for landscaping), the opposite case, where the valued scene and wildlife are the primary objectives of management, needs careful attention. The scope for technological solutions to the problems of management of protected ecosystems and landscapes is by definition limited. One possible avenue is that of the manipulation of habitats to accommodate rising numbers of people without damage, e.g., by deflecting them away from sensitive areas by the provision of alternative vistas or better roads. Another is the use of helicopters or boats to land people in an ecosystem while causing minimum change in it, especially if the visitors remain only a short time. Technological substitution is no doubt tempting at times: plastic trees and stuffed wildlife might not be recognized as such by every visitor to a national park.

The effects of increasing human populations are both direct and indirect. Directly, there are more people, usually with more vehicles either public or private, wanting to visit the protected places, and they are often concentrated at especially popular times like the summer weekends and holidays. This intensifies their impact upon the ecological systems which form the landscape, and in large enough numbers they become a feature, usually undesired, of the landscape itself. Indirectly, more people want more jobs, more housing, more water and the like. Even when regional or national planning is efficient, these pressures cannot be for long diverted from valued landscapes. The patterns of world trade and interdependence exert effects too, so that agricultural development in LDCs* stimulates demand for fertilizer production in DCs, whose industrial anabasis in turn requires more water and hence more reservoir construction in areas of high landscape value which inevitably will sometimes be in areas of considerable scientific interest, with an area of flora and fauna being threatened with damage or extinction (R. Gregory 1971). In this connection, work which shows that nature performs without cost services which otherwise would have to be paid for by extra resource use is of interest. For example, tertiary waste-water treatment and fisheries are both products of tidal wetlands: to duplicate these functions along the Gulf Coast of the USA would cost $205,000/ha; a 930 ha river-swamp forest ecosystem in Georgia yielded a minimum annual value of $1.8 million in ground-water storage, soil binding, water purification and streamside fertilization (Westman 1977). Despite considerable opportunities for creative planning and public education there remains among resource managers, in the West at least, no doubt that landscapes and nature can in the end be kept untrammelled only by restrictions on the number of people who come into their presence and by the slowing down or cessation of the competing demands for space created by higher populations with rising material expectations.

*Throughout this book the abbreviations LDC (less developed country) and DC (developed country) will be used.

Further reading

BURRELL, T. S. 1973: National Parks. The big three—conservation, recreation and education.
COSTIN, A. B. and GROVES, R. H. (eds.) 1973: *Nature conservation in the Pacific.*
DASMANN, R. F. 1964: *Wildlife biology.*
DOUGHTY, R. 1975: *Feather fashions and bird preservation.*
DUFFEY, E. 1974: *Nature reserves and wildlife.*
— and WATT, A. S. (eds.) 1971: *The scientific management of animal and plant communities for conservation.*
EHRENFELD, D. W. 1976: The conservation of non-resources.
FRASER DARLING, F. and EICHORN, N. 1967: *Man and nature in the National Parks: some reflections on policy.*
GREGORY, R. 1971: *The price of amenity.*
HELLIWELL, J. R. 1973: Priorities and values in nature conservation.
IUCN 1966—: *Red Data Book.*
IUCN 1971: *United Nations list of National Parks and equivalent reserves.*
JARRETT, H. (ed.) 1961: *Comparisons in resource management.*
MILLER, R. S. and BOTKIN, D. B. 1974: Endangered species: models and predictions.
MYERS, N. 1972: National parks in savannah Africa.
— 1976: An expanded approach to the problem of disappearing species.
— 1979: *The sinking ark.*
NELSON, J. G., NEEDHAM, R. D. and MANN, D. L. (eds.) 1978: *International experience with National Parks and related reserves.*
PRYDE, P. R. 1972: *Conservation in the Soviet Union.*
SCHOFIELD, E. A. (ed.) 1978: *Earthcare: global protection of natural areas.*
SHEAIL, J. 1976: *Nature in trust.*
SIMMONS, I. G. 1974: National parks in developed countries.
UNESCO 1974a: *Taskforce on criteria and guidelines for the choice and establishment of biosphere reserves. Final report.*
— 1974b: *Impact of human activities on mountain and tundra ecosystems.*
WALTER, H. 1978: Impact of human activity on wildlife.

4

Recreation and tourism

The use of resources for recreation is most apparent in rural areas, where land and water may obviously be devoted to that particular purpose; in urban areas the designation and preservation of open space is of the same nature. Indoor recreation, especially in built-up places, is scarcely distinguishable from other urban resource uses in terms of its utilization of energy and materials. The demand to use the countryside for recreation has, we may suppose, a very long history, probably as far back as the first cities but in pre-industrial times the inhabitants of even the largest cities could walk out into more open surroundings. With the urban explosion of the nineteenth century the working people became confined to the towns except for holidays but the steam railway soon began to provide a means of escape to countryside and coast. The working classes then began to acquire the taste for fresh air and rural landscapes which hitherto only the rich had been able to enjoy (Patmore 1975). The developments of the twentieth century have further entrenched the recreational desires of all socio-economic groups (especially in the West, and with rising expectations in the LDC's) to the point where Gabor (1963) listed leisure among the dangers to civilization along with war and over-population. The introduction of more automation with developments like the micro-processor, may in future create more structural unemployment to the level where traditional 'Protestant' valuations of work and leisure may be distinctly irrelevant.

Leisure, affluence and mobility in the twentieth century

At the turn of the century, the 60-hour working week was common in the west, but its reduction to an average of about 40 hours (In 1978 the UK Trades Union Congress was pushing for a standard 35 hour week; the number of statutory holidays in Britain is much lower than in continental Europe) has meant that individuals are now able to plan a very different time-budget from that common in earlier years. In 1900, according to Clawson (1963), 26·5 per cent of the time of US citizens could be classed as leisure (i.e. time after work, housekeeping and personal care); by 1950 leisure was at 34 per cent and likely to rise to at least 40 per cent by the year 2000 (Table 4.1). Shorter daily hours and flexitime have meant daily leisure periods as well as more long weekends and longer annual vacations. Only one counter-trend has been documented: the willingness of some industrial workers to trade their extra leisure for a higher income from 'overtime' or perhaps a second job. How far this will persist in the face of automation remains to be seen (Roberts 1978).

The relative and absolute monetary wealth of Western countries is not in doubt and

TABLE 4.1 Projected working hours and holidays

Year	*Weekly working hours*	*Weeks' holiday*	*Individual free days*
1976	35·4	2·8	8·5
2000	30·7	3·9	10·1

Source: ORRRC 1962b

it is mostly their populations who have the disposeable income to devote to recreation, especially those pursuits needing expensive equipment. Expenditure in western nations on recreation seems to run at 4–7 per cent of disposeable income and is surprisingly constant across income levels. The post 1974 minor recession and the increase in petrol/gasoline prices do not seem to have affected the demand for recreation to any considerable effect; this is the more remarkable since possession of a private motor car is a paramount factor in participation in most western nations although in East Europe and Japan, public transport is still the most important conveyor of recreationists even in rural areas. Even if the internal combustion engine becomes a museum exhibit within the span of projections now being made, possession of private transport seems to be very firmly tied into the pattern especially of outdoor recreation familiar in industrial economies.

Classification of activities and resources

As far as resource use is concerned, outdoor recreation will be given the most emphasis in this chapter, with tourism second; indoor recreation is treated little since its resource flows are not very distinctive. Outdoor recreation and tourism are difficult to disentangle, since the one may be the reason for the other and the other may incidentally participate in the first. Tourism is a leisure activity involving the short-term movement of people to destinations away from their usual residences and is often international in character. Outdoor recreation is leisure taken in relatively small groups usually in a rural setting, so while tourism may focus on cities, including their indoor attractions, it can also transform rural areas (e.g. for winter sport, or summer beach, complexes). In general there are two elements: the users and their accompanying demands for certain activities, and the resources in the use of which they will spend their time. Both are very high in variety and only a simplified account will be given here.

There has always been an element of this kind of 'non-productive' use in most places: in medieval England there were the Royal Forests, strictly preserved for the King, and areas of lower status like Chases and Warrens for the lesser aristocracy. Later, the great landscape gardens of the seventeenth and eighteenth centuries became backgrounds for the less lusty pursuits of the contemporary gentry. The ruling classes of most ages have had their resource-using pleasure gardens, whether rural or urban, as with the pleasure-dome of Xanadu, presumably a sort of early Havasupai City. By contrast with the requirements of the numerically insignificant rich, the demand to

TABLE 4.2 Major outdoor recreation activities (no ranking is implied by the order)

Activity	Notes
Driving for pleasure	
Walking for pleasure/Hiking	This is a useful distinction—the latter is the serious version.
Outdoor games	These should be of an informal nature, requiring little or no fixed equipment.
Swimming	Includes sub-aqua activities. It is implied that a 'natural-looking' water body is involved, not a 'pool'. Surfing also included.
Bicycling	
Fishing	
Nature study, archaeology	
Nature walks	Includes both guided walks and self-guiding nature trails.
Boating/Sailing	This category means motor-boats on inland waters and sea.
Canoeing	Inland water, and offshore.
Sea sailing	Sail only.
Sightseeing	Of cultural interests rather than appreciation of views, etc.
Caving	Restricted to limestone pothole country.
Hunting	This is an American term, and is used there mainly for deer-shooting. For use in this book it will be taken to be synonymous with shooting and to exclude fox-hunting.
Horseback riding	Including pony-trekking.
Camping	Including 'day-camping', not easily separable from picnicking; also caravaning: difficult to separate in North America.
Picknicking	
Ice skating	Outdoor only, on naturally frozen water.
Tobogganing	
Snow skiing	
Mountain climbing	Synonymous with rock climbing: fell-walking and scrambling under hiking.
Motor sports	Hill trials, motorcycle scrambles over informal courses.
Water skiing	
Wind-surfing	A surfboard with a sail.
Hang gliding	Powered hang-gliders introduced in late 1970's.
Flying	Both powered and gliding.

transfer resources into leisure use now emanates from a large proportion of the population in industrialized countries, some of whose activities are given in Table 4.2. Another way in which the dimensions of man-environment relations have altered with regard to leisure is the hardware involved. Once a pair of nailed boots and a woolly sweater (or solar topee in the tropics) sufficed. Now, trailbikes, dune buggies and snowmobiles extend mobility off the roads into most kinds of terrain; aqualungs allow penetration under water; and the humble tent is being ousted in favour of a caravan or an integral-chassis camper. In North America it is not uncommon to see a large integral camper with a boat on the roof, two trailbikes slung on the front, and a small Japanese motorcar being towed behind. The impact of all this technology upon some of the recreation resources can easily be imagined.

Plate 7 Perhaps most people's ideal of outdoor recreation? An isolated lake and a lone fisherman-camper, communing with nature. Cf Plate 8, page ▪▪. *(Grant Heilman, Lititz, Pa)*

The types of ecosystems which people like to be in for their outdoor recreation are very varied, according to both taste and availability, but certain preferences emerge. Overriding them all is the attraction of water (Plate 7, above). The pull of the seashore or the sandy edge of large lakes needs no stressing, but the presence of a small inland lake or river adds immeasurably to the value of a recreation area inland. There is, too, a preference for wild or seemingly wild vegetation such as woods, dunes and heaths, providing that the woods have ample openings or 'edge'. In climates with inhospitable tendencies some sort of shelter is demanded, so that the picnic tables at White Sands in New Mexico have reflective shades over them; in part of the Netherlands a very popular region is that where forest is found in the lee of sand-dunes so that people can move inland if the weather turns chill. It is not of course the actual nature of the ecosystem which attracts people so much as their perception of it, and varying degrees of alteration and change are acceptable for different recreation activities. While it is usually only possible to describe outdoor recreation use of resources in objective terms, therefore, it must not be forgotten that just as important are people's ideas about the sort of surroundings they choose, or would choose if alternatives were available. For example, Lucas (1964) showed that the two major groups of users (those with and those without motors on their craft) of a wilderness canoe area in Minnesota—Ontario have different perceptions of the amount of human interference with the forest and of the effect such alterations have upon their enjoyment.

One way of discussing both activities and recreation resources in the outdoors is the classification developed by Knetsch and Clawson (1967). It is basically a *user-resource* method in which the two major elements are the characteristics of the users in terms of the time whey they use the resource—whether it be afternoon, day or weekend use, or in their longer vacations—and the physical and ecological characteristics of the resource, especially its degree of wildness and of manipulation away from the natural state. The classification has three categories: user-orientated, dealing with the shorter periods of use; resource-based, treating the longer holiday periods; and intermediate, not unnaturally discussing the class in between—mainly the weekend and short vacation periods.

User-orientated recreation resources

The recreation areas deemed to be user-orientated are characterized by being as close to the homes of users as possible, and thus make use of whatever resources are available. The activities become the most important feature and a great deal of alteration is acceptable, and often necessary to protect the area against the damaging effects to heavy use. Activities such as golf, tennis, swimming and picnics, walking and horse riding, small zoos and model farms or railways and general informal play are the most usually found. The intensive use of what are often quite small areas, at most a few hundred hectares, with many smaller units, comes after school or after work and at weekends. Such areas are commonly owned by city or county governments or their various equivalents. The city park forms the inner-most element of the structure. Its nature is very variable and the present tendency is away from the starched formality of traditional city parks. The value of such areas of open space as Boston Common, Central Park in New York and the Royal Parks of London is probably best judged from the fact that they are still not built upon.

Examples of recreation areas near cities

Parks and recreation areas on city fringes are difficult to acquire because of the high cost of land which is also valuable for housing or perhaps industrial use; but some cities have happily acquired, been given or converted from dereliction suitable tracts of land and water. The Amsterdam Forest ('Het Amsterdamse Bos') on the south-east side of that city is a large (900 ha), totally artifical recreation area, converted from polderland. Since 1934 (Fig. 4.1) the major pattern is of woods with informal footpaths and cycleways interspersed with grass fields and water-bodies, some irregular and informal, others regular as in the case of a rowing course. The developed facilities include a sports stadium, riding school and openair theatre. There are roads and car parks, but access by public transport forms the commonest pattern of use.

The valley of the River Lee slices through the eastern suburbs of London and here a

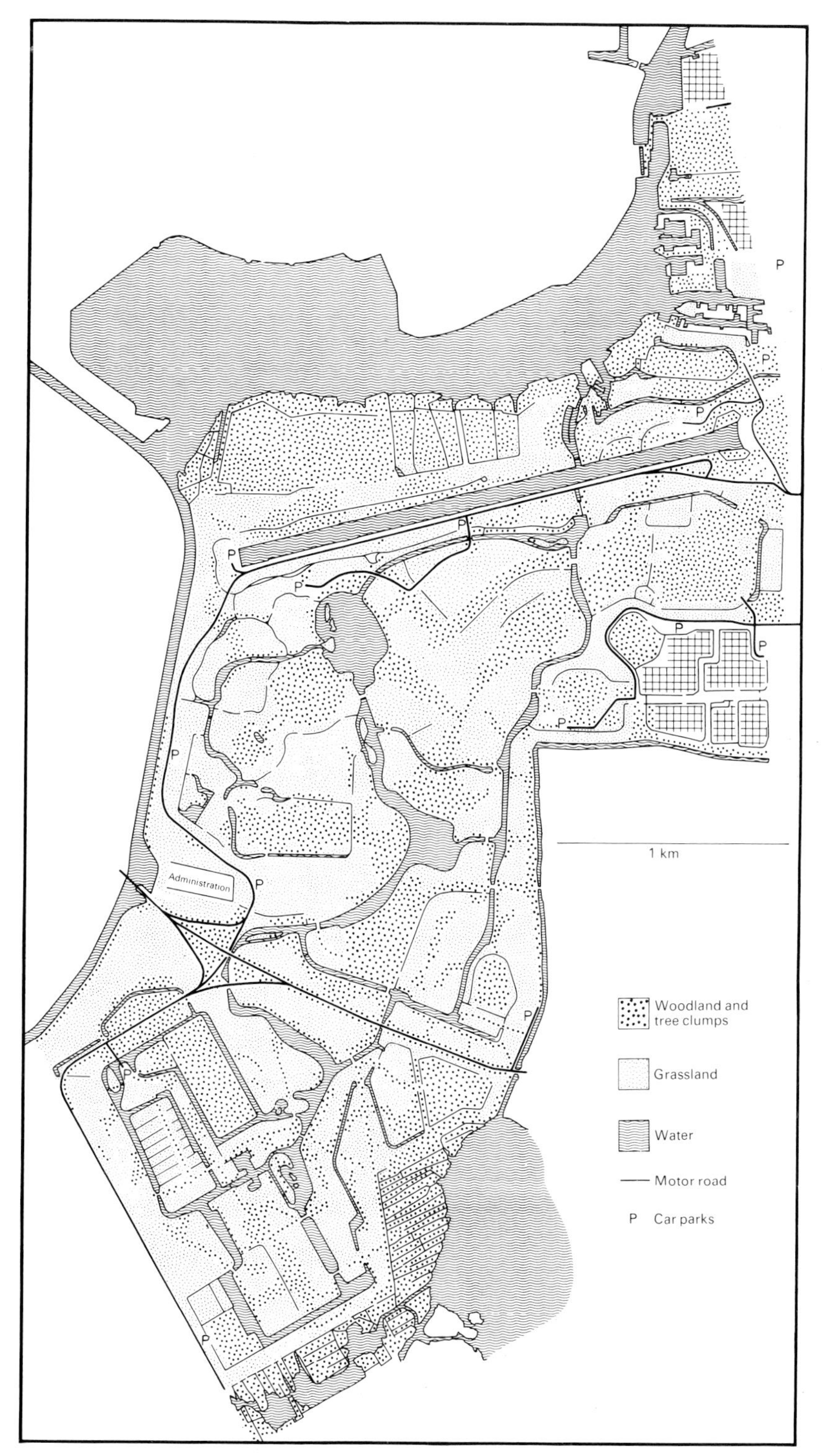

Fig. 4.1 The Amsterdam Forest: a completely man-made recreational facility which by use of trees and water creates a rural atmosphere in some parts. More urbanized and formal areas for games and a rowing course are also part of the area.
Source: Dienst der Publike Werken Amsterdam 1960: *Het Amsterdamse Bos*

scheme to develop a rather ill-assorted agglomeration of reservoirs, a canal, gravel pits, football pitches, refuse dumps and small industries like scrap-metal reclamation into a Lee Valley Regional Park has been put into action (Civic Trust 1964). The Park will have some nature reserves but will in general be highly developed, with playing fields, indoor sports halls, water-sport pools, restaurants, and possibly animal collections of special interest to urban children, like cows and horses.

These are more or less spatially isolated instances, and actual park systems on the edges of urbanized areas are rarer, though more and more public authorities are taking a strong interest in this zone. One example is the East Bay Regional Park District on the east side of San Francisco Bay in California. The hills behind cities such as Oakland, Berkeley and Albany are naturally grass- and chaparral-covered, with some redwoods, but since the nineteenth century eucalyptuses have become naturalized. The most user-orientated parks in the system are Tilden Park and Redwood Park. In the valleys nearest to the cities, development of features such as a swimming lake, steam-powered railways, golf course, a carousel, a miniature farm, and riding stables has taken place. Further in is a less developed zone with picnic sites, group camp sites and a botanic garden of a relatively informal kind. Beyond these the park is largely wild in appearance, with a few trails striking off into the chaparral and wooded areas. Here, deer herds exist probably along with their natural predator, the bobcat or mountain lion. The development of Tilden Park shows a distinct series of development zones aligned parallel to both the topography and the adjacent cities (Fig. 4.2).

Possibly the largest class of user-orientated recreation areas are beaches at the seaward fringes of seaside cities. These are very cheap resources since there is generally perpetual renewal of the essential elements by the sea. Where this action breaks down then sand has to be brought in and an artificial beach maintained, which is costly but usually essential for the viability of the resort. In Japan, for example, a first priority in open-space acquisition for public use is the provision of beaches which will be easily accessible to the people of the great conurbations.

Changes wrought by recreational use

The designation of a unit of land or water as a recreation area within this present category usually means that changes will occur in the local ecosystems. At a deliberate level, most changes have to be made in order that the area can withstand the intensive use which it receives. Thus circulation routes are inevitably hard-surfaced, and it may be necessary to plant trees and shrubs which can withstand compaction of the earth over their roots and the loss of a branch or two by vandalism. Picnic tables if installed have to be specially strong, and concrete frames, if not tops, are common. Biota rarely receive management in such places, although a pair of nesting swans can be inimical to other uses of water-bodies in spring, and in many countries the numbers of pigeons and starlings attracted to such areas may be undesirable. Beaches may of course suffer from polluted water, and enlightened municipalities endeavour to remove the source of pollution, which may be not only unaesthetic but a positive danger to health.

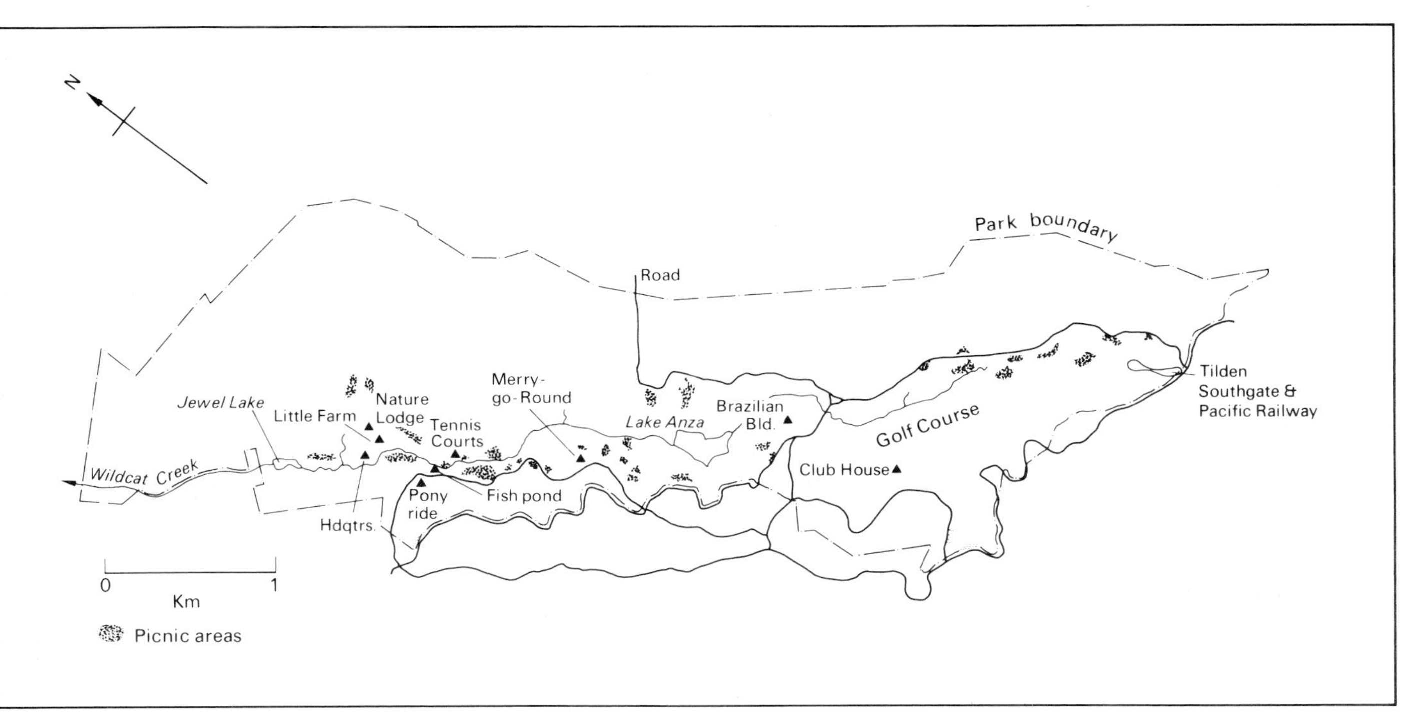

Fig. 4.2 Tilden Park, a component of the East Bay Regional Park District in California. To the east of the park is undeveloped watershed land, to the west the urbanized areas of the East Bay (e.g. Berkeley, Albany, Oakland). The concentration of developed facilities along the western edge is evident.
Source: East Bay Regional Parks District n.d.: *Charles Lee Tilden Park* (pamphlet)

Intermediate-type recreation resources

The zone of recreation resources available to the city-dweller for use on day outings and weekends is called by Knetsch and Clawson (1967) the Intermediate zone. The actual distance travelled by users will vary according to the road system, since transport by private auto is all-important for this type of use. Within the time—distance limitation, preference is shown for the best resources that are available, in terms for instance of scenery, water and forests. These are generally larger in size than the user-orientated areas, running from a few hundred to several thousand hectares. Emphasis is put on activities such as camping, picnicking, hiking, swimming, hunting and fishing, and there is also the pleasure of driving to get to the chosen location. Although public resources managed by governmental agencies are perhaps the most common feature of this category, the private sector is often involved. The degree of manipulation away from the natural state tends to be less than in user-orientated areas, and acceptance of quasi-urban features is lower. Insistance on a natural environment is of course impossible in continents like Europe and here high value is placed on areas which are obviously wild even if not natural, or on areas like heaths and moors which are popularly thought to be natural even though they are anthropogenic in origin. The pressures resulting from recreation may still be intense over a small area of the resource, but there is often a large area of back country into which the less gregarious can escape.

National Parks in England and Wales

Although dealt with once under the heading of protected landscape, the National Parks of England and Wales come under this heading too. People come to them not only to view the scenery, but for many active recreations as well. Driving for pleasure, walking and hiking, climbing, potholing, boating, natural history pursuits and many other activities are carried out in these 10 designated areas. Most of them are within the day-use zone of a major conurbation, yet several receive some usage of a resource-based category. Most of the valley land within the Parks is private and obviously so, which means that use is only by direct permission of the owner unless walking along a public right of way is contemplated. Even the owner may be under restrictions since, for example, the placing of caravans requires planning permission. Many of the farmers in the Parks do not welcome recreationists, however, because of the trouble caused by their ignorance of the working nature of rural areas. It is not surprising therefore that much recreation tends to be concentrated in the unenclosed zones which are altitudinally above the farmed lands. Such lands are often common land to which there is a *de facto* but not *de jure* right of public access (Fig. 4.3). It is here that most people like to set off on walks, to picnic or simply to sit in their cars. Since until the late 1960s there were few places where cars could pull off the usually narrow roads, such spots as there were frequently received very heavy pressure. The installation of car parks and toilets in many of the attractive villages that are found in the lower areas of the Parks has meant an alleviation of the congestion of narrow and picturesque streets by fleets of

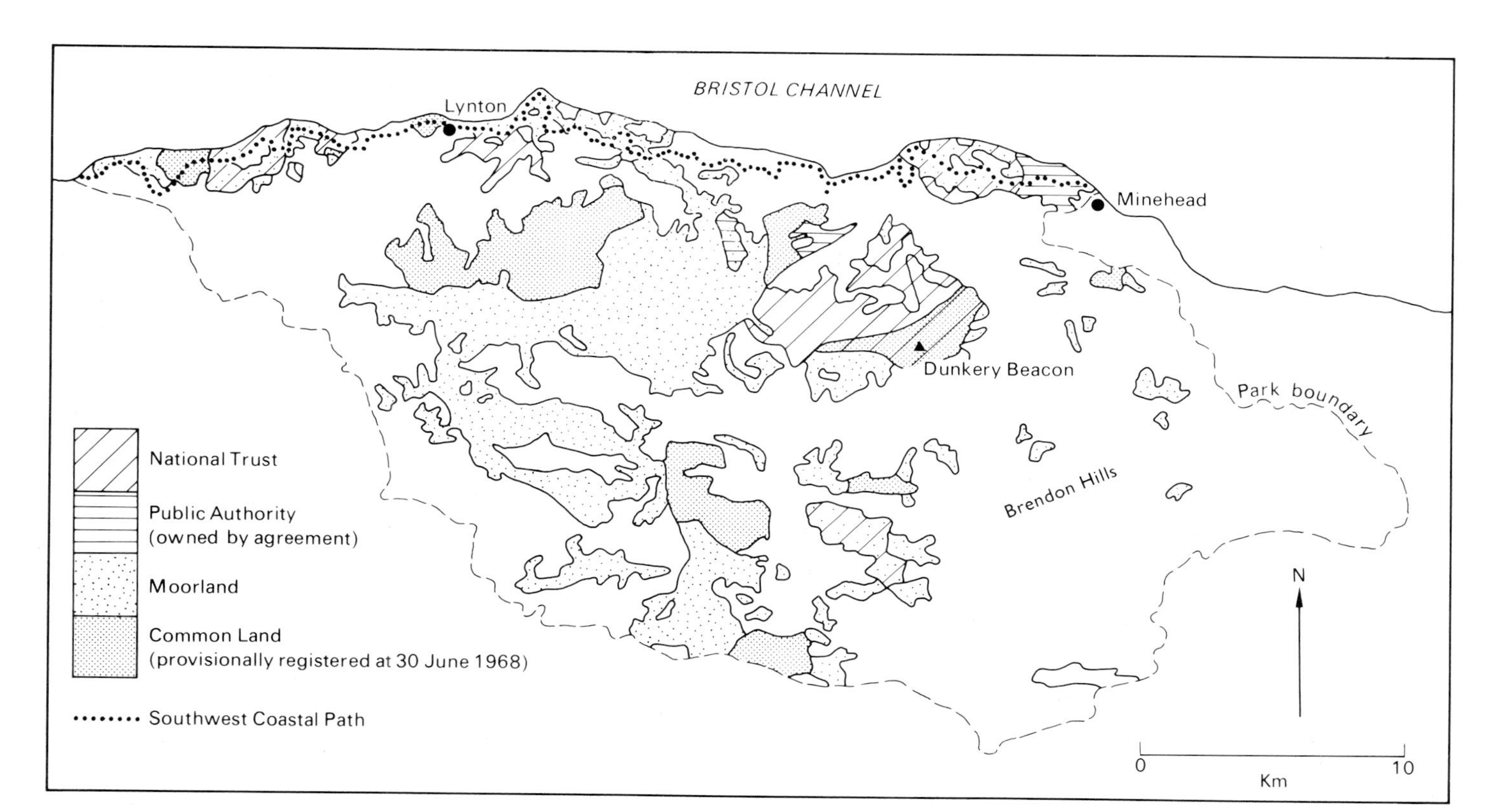

Fig. 4.3 Access land on Exmoor, one of the National Parks of England and Wales during the 1960's. *De jure* access is to National Trust and Public Authority land; *de facto* access exists to moorland and commonland but, strictly speaking, such access could be withdrawn. The Coastal Path is a public right of way for those on foot.
Source: Patmore 1971.

visitors' cars. In some of the Parks, the planning authorities have taken the lead in such actions as providing official but limited camp and caravan sites in order to prevent a rash over all the countryside. Experiments are being undertaken to control car access to some of the most popular but narrow roads in the interests of general amenity: the Goyt Valley in the Peak District National park is an example.

Even with such measures, it is difficult to avoid the conclusion that at peak times some of the National Parks (the Lake District and Peak District especially) are reaching their carrying capacity for outdoor recreation. This is especially so for people with cars, but queues to climb particular crags and to descend certain potholes suggest that the pinch is also being felt in other activities. Under the aegis of the Countryside Act of 1968, county councils are empowered to designate another kind of Intermediate-class recreation area, the Country Parks. The degree of development of these facilities can be varied according to the circumstances, and it is hoped that the National Parks will be relieved of some of their recreational pressures and be visited more for their unique attractions rather than for activities which can be pursued equally well in other areas. The Country Parks will amplify a role already played in some parts of the country by the plantations of the Forestry Commission. During the last 20 years there has been a revolution in the policy of this body regarding recreation, so that picnic sites, camping sites, trails and visitor centres are now often found, as at Grizedale in the Lake District, Allerston Forest in North Yorkshire, and in regions like the New Forest where they have been able to plan for the recreational use of a whole sub-region, to the point where in 1978, the Commission have called a halt to any more developments: 6 million visitors per year over 272 km^2 seems like a saturation level. As these areas have distinct boundaries, it is often possible to make a charge for facilities and this is usually done. In this context, recreation presents fewer problems as an element of multiple use than with farming: young forests are very susceptible to damage but at the same time are not very attractive for recreation.

The Dutch experience

The role of recreation in re-afforested parts of Europe is carried to a considerable length in the Netherlands, where 25 per cent of the area managed by the State Forest Service (Staatsbosbeheer, SBB) is managed primarily for timber production, and the remainder mostly for recreation and nature conservation, especially the former. Apart from the customary facilities, the SBB has created large informal swimming holes out of sandy lands, an example is to be found at Nunspeet. In at least one case the hole represents the further use of a borrow-pit resulting from motorway construction. Most of the State Forests of the Netherlands come into the Intermediate category although some are near enough to towns to be thought of as user-orientated and development is correspondingly intensive. In both situations, however, the Dutch love of food is exemplified in the provision of refreshment kiosks and restaurants in places where British and American experience and preference would find them a little intrusive.

State Parks in the USA

Most states have such a system of State Parks, but their effectiveness and range of utilized resources vary greatly. In general, the richer the state, the more complex the State Parks system will be: the states of New York, Michigan, Minnesota, Oregon and California are generally recognized to be the leaders. In all such systems these are areas of primarily historical or cultural interest (such as some of the Franciscan Missions in California), and in some the system embraces nature protection as well. For recreation, State Park systems usually enclose the best scenic and recreational resources which are not already within the Federal system and develop them to cope with heavy pressures, especially at summer weekends; winter sports may also be catered for at selected locations. Nearly all outdoor recreations are carried on in these Parks, with some emphasis on camping, fishing and hiking. Hunting is rarely permitted in them. Water is another attraction, whether on the few public beaches (those of southern California are very crowded in summer) or inland at reservoirs or natural lakes such as Clear Lake and Lake Tahoe, both of which have State Parks along part of their otherwise privately owned shores. The desert Parks are naturally popular in winter, as are skiing, tobogganing and other snow-based sports for which other parks are suited.

The second home

The possession of second homes in rural areas is a rapidly growing phenomenon in the West. In Europe it reaches the apogee of its development in Scandinavia, but it is burgeoning everywhere (Fig. 4.4). Weekend cottages, beach-houses, cabins, caravans and their kin are almost by definition part of the Intermediate zone; and because their owners wish to be in rural surroundings but also to have most home comforts such as electricity and mains drainage (or at least a cesspool or septic tank), they pose resource management problems. The aesthetic one of screening them to avoid the appearance of a rather tatty suburb is one difficulty, and where zoning laws do not exist it is often impossible to prevent a variegated scatter of bungaloid growth in favoured spots.

Ecologically their effects are most profound when a new area is opened up for cottages around a desirable place such as a lake. It is unlikely that regulations about waste disposal will be in effect and so raw sewage is discharged into the lake with consequent eutrophication and aesthetic damage. The actual placing of the cottages may have less effect since there is a desire to preserve trees and other biota.

Resource-based recreation

In contrast to the previous two categories, resource-based recreation takes place where the outstanding resources are to be found, independently of the distribution of population. This may well be some distance from the majority of users, although some people may live in the backyard of an outstanding resource which others travel thousands of miles to visit. It follows that the longer vacation periods are the commonest use periods and that activities are consequent upon the resource offered:

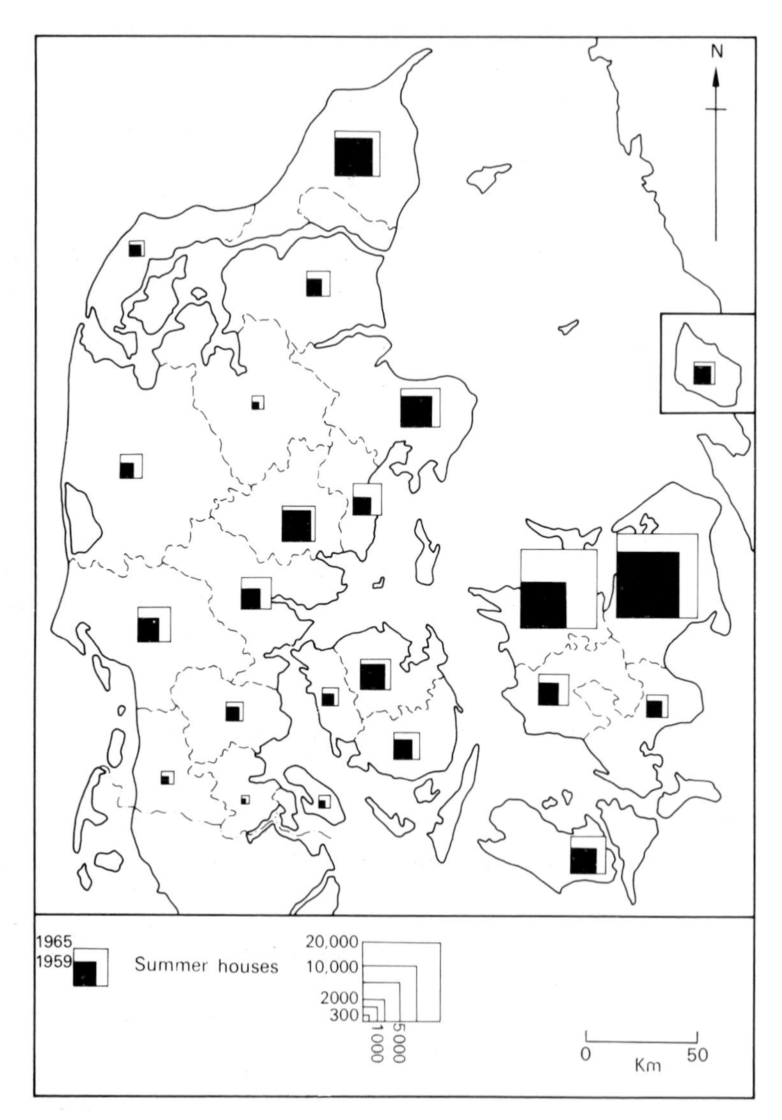

Fig. 4.4 Summerhouses in Denmark, by counties, 1959 and 1965. The concentration in Sjaelland (especially near Copenhagen) and in east Jylland is noticeable. Most of these summerhouse areas were not subject to planning control.
Source: Landsplanudvalgets Sekretariat 1965: *Strandkvalitet og Fritidsbebyggelse*

major sightseeing is perhaps paramount, along with scientific and historic interest, even if these latter two are very inchoately realized. Hiking and climbing, fishing and hunting may also be pursued because of the superlative qualities of the particular place; camping may be an end in itself or a cheap way to visit the resource area. The qualities of the attractive area are generally extensive: typically a resource-based recreation area may range from many thousands of hectares in size to ten times that magnitude. It is not surprising therefore that the resource is usually publicly owned and that the agencies of the national government are usually the managers. Exceptions to this are sometimes found in water frontages along large lakes and the sea.

As the environment is all-important, it should be a feature of the designated areas that human impacts resulting in alterations to the ecology should be minimized. This is generally so, except where visitor numbers are very high (or very concentrated as along the walking trails in the Himalaya) and require particular management responses. But highly developed zones are usually small in relation to the proportion of wild back-country.

National Parks in North America

The world's most famous examples of the designation of resource-based recreation areas are the National Parks of the USA and Canada, especially those of the western cordillera. The range of terrains is immense, but all the Parks share the property of being very wild and in most places natural or at any rate affected only by aboriginal economics in the form of hunting and burning. Economic activities on a large scale have never been permitted, although some early grazing may have left erosive scars, and there are numerous private inholdings which complicate management.

The major problems of the National Parks come from the great mass of visitors. In 1910, about 0·2 million people visited a US National Park system of about 9,308 km^2; in the 1970's about 140 million a year went to a system that was 63,000 km^2 in size. Such numbers initiate immense demands for roads (since most use private cars), for accommodation, food, gasoline, water and other services (Plate 8, page 106); in turn they produce sewage, garbage, litter, car exhausts and traffic jams (Cahn 1968). These effects, together with the presence in some National Parks of luxury hotels and golf courses, eventually led to the reappraisal of National Park policy in the USA described in Chapter 3, which has resulted in the designation of limited recreation areas within a park system whose function is primarily protective. Thus recreation is being de-emphasized, and eventually perhaps most developed facilities will be withdrawn beyond the park boundaries. Under the impetus of public opinion, the master plans for the National Parks of Canada include large wild areas and developed areas are being kept to a minimum (Fig. 4.5).

Even before such changes in policy, these National Parks were unequal to the recreation pressure they were being expected to carry. One result, analogous in function to the Country Parks in Britain, has been the development within the USA Federal system of National Recreation Areas. In 1971, there were 17 units comprising 14,164 km^2, representing outstanding opportunities for resource-based recreation

Plate 8 Perhaps most people's reality of outdoor recreation? Cf Plate 7. The area near Old Faithful geyser in Yellow-stone National Park. The Park Service is now 'de-developing' this part of the Park. *(Grant Heilman, Lititz, Pa)*

within the Federal system at areas not suitable for designation as National Parks. The selection criteria involve consideration of the size of an area, its ability to provide a high carrying capacity for recreation, and a location usually not more than 400 km from large urban centres. A natural environment is not required and many of the areas are primarily for water-based recreation along artificial impoundments such as Lake Mead, Lake Powell, and the Shasta Dam and Lake in California.

Multiple use with recreation as a primary purpose of management is represented in the USA Federal system by the National Forests. Most of their 752,742 km^2 are in the west and they often form the matrix within which the National Parks and National Recreation Areas are set. Since recreation is only one of the purposes of management (the others are wildlife, grazing, water and timber production), the multiple use tends to be a mosaic pattern, with developed recreational facilities and roads within a much larger area of undeveloped country. Even where campsites and picnic grounds are developed, they tend to be more simple than in National Parks and National Recreation

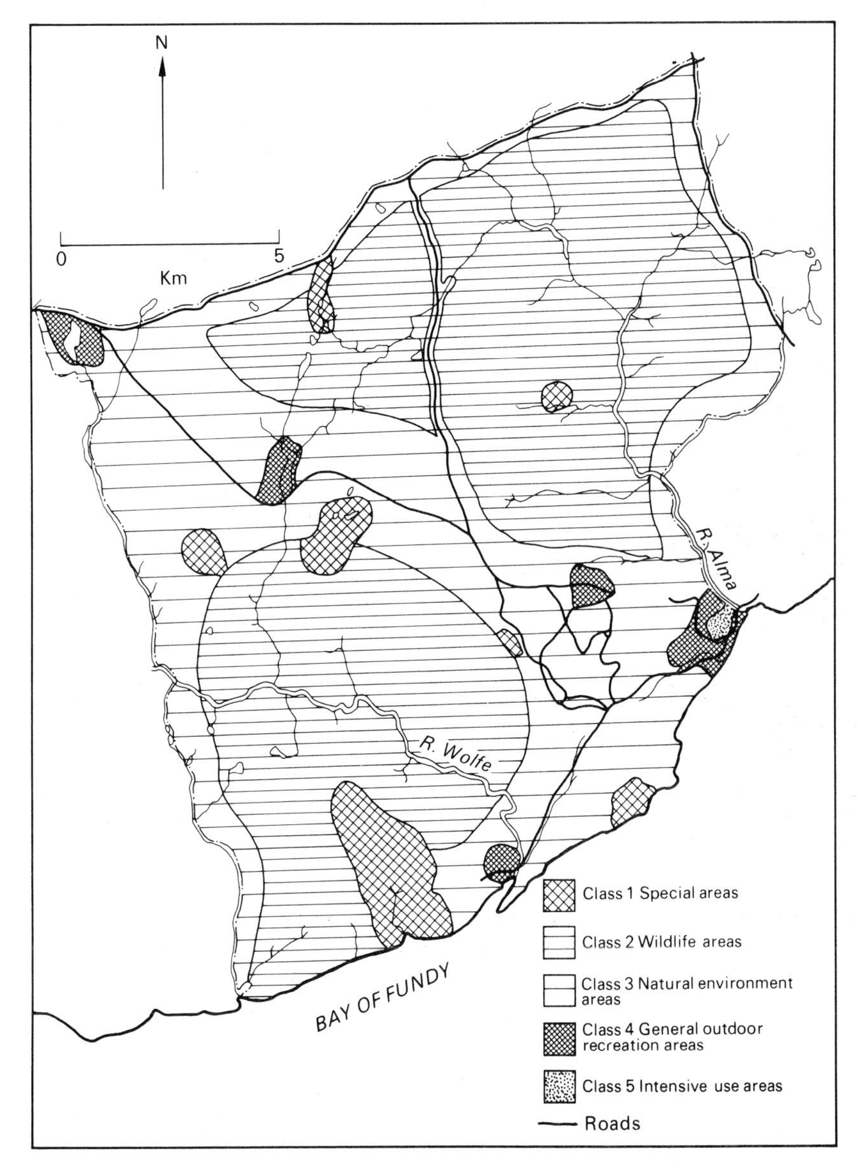

Fig. 4.5 A proposed master plan for the Bay of Fundy National Park, New Brunswick, showing the division of the park into zones. Developed areas are kept to a minimum consistent with the use of the park for recreation. Such a plan is a public document and the subject of public hearings, and will by now have been amended.
Source: National and Historic Parks branch kit for hearings on Bay of Fundy National Park Master Plan, Ottawa, 1968

Areas. Hot showers and launderettes are not found in National Forest campgrounds, and the deep-pit toilet replaces the flush type. Thus the local ecology is usually less modified than in the Parks and Recreation Areas, a feature which commends the National Forests to the more hardy campers. The National Forests in fact extend well beyond the forested zone in mountainous terrain and so enclose the alpine meadows and slopes and the high alpine country itself. Recreation activities are hence very varied in kind and range from the highly popular to the more eclectic such as river-rafting and wilderness back-packing.

Continental Europe

In more densely settled parts of the world like Europe, resource-based recreation areas are harder to find and are mostly confined to high mountain areas such as the Alps, Pyrenees, Carpathians and the Norwegian *fjells*. Each of these regions attracts many people both winter and summer for recreation. Management varies, but in general there are a few areas within each region which are particularly designated for the protection of wildlife (e.g. the Tatra National Park in ČSSR–Poland, and the Swiss National Park) and then development of recreation in the rest is subject to the ordinary planning laws of the country.

Within Europe, one area with potential for resource-based recreation is Lapland. At present relatively few people visit it except for such specialized activities as bird-watching and tourism based on the few roads. The same can be said of Iceland. Both these places have very fragile ecosystems with short growing seasons, and any development must recognize this if irresponsible damage is not to be done. On a European scale it would probably be best to make most of both areas into formally protected wildernesses.

An immense country like the USSR has enormous recreation resources; yet protected areas for recreation which have nationwide importance have been slow in coming. Pryde (1972) discusses the planned creation of two National Parks ('Russian Forest' south of Moscow, and Lake Baykal in eastern Siberia) which would be devoted primarily to vacation use by the Soviet people. Another 13 sites have been put forward as suitable areas, which would be at least 10,000 ha in size and would be divided into developed and undeveloped sections, with a central semi-wilderness having only trails and designated camping sites. The 'Russian Forest' Park, about 100 km from Moscow, will cater for more intensive use which will be combined with educative functions such as a demonstration forest.

Japan

In all essentials, Japan shares the West's tastes in outdoor recreation. Given a small country with a high density of population, particularly in the Tokyo-Osaka axis, there are large numbers of outdoor recreationists, whose quantity is increasing steadily with the advent of more leisure time and income. The only areas of the country which were specifically designated in 1972 as being available for outdoor recreation are the 26

National Parks and 46 Quasi-National Parks (QNPs), together with 286 prefectural parks; about 13 per cent of the surface of Japan is thus designated. The Parks function through all three user-resource recreation zones, but some especially are of the resource-based character. In these, the cultural attractions are generally less than in the intermediate zone, assuming not unreasonably that Fujiyama has a significance far beyond that of being a mountain. The resource-based National Parks are all distant from the major population axis of Japan but all the National Parks and Quasi-National Parks are probably accessible for weekend use if overnight trains or planes are used (and the Shinkansen trains will soon run to Hokkaido), and all are penetrated by public transport often to an intensity not now common in the West. In the case of Japan, the dilemma of development or preservation centres around two factors: the ownership of the land, and the application of zoning (Senge 1969). Where the land is in national ownership then it is frequently under the control of the Forest Agency, which places a high priority on economic timber production and tends to use clear felling as a method of harvesting; this conflicts at least temporarily with recreational activities as well as spoiling some of the visual attractions of the landscape. Thus the National Parks Agency is frequently at loggerheads with the Forest Agency over resource management in the National Parks. Where there are large areas of private land within the National Parks, the Parks Agency has little control over the development of hot springs, hotels and inns, restaurants, cafés and souvenir shops; these tend to be very frequent elements of the landscape at the nodal points of the National Parks, such as around Lake Shikotsu and Mount Showashinzan in the Shikotsu-Toya National Park in Hokkaido. Under the impact of the recreation-seeking element of the population of Japan, there seems at present to be an impetus towards development, and the appreciation of the virtues of the natural environment as a milieu for recreation which has characterized recent movements in the West is slow to penetrate: there is consequently considerable ecological change wrought by developments made in the name of the provision of recreation facilities, although moves to exclude cars from some scenic roads suggest that ideas of amenity are growing.

Changes caused by recreation

As with other types of recreation, the activities and development may affect the ecosystems within which the recreationists operate (Fig 4.6). At a conscious level, developments to cope with large numbers in popular areas are probably the most noticeable. In Yosemite Valley there has been a village catering to visitors; and in the Canadian Rockies earlier policies established two small towns (Jasper and Banff) which are currently undergoing expansionist pressures. In such Parks there is then the somewhat incongruous sight of sewage works and solid waste disposal areas, without which water quality would decline even to the point of containing dangerous pathogens. As pressure increases so more management has to be undertaken, necessitating more notices, more hardtop and generally less nature. Another deliberate policy stems from the desire of some managers to make as much country as possible accessible to as many people as practicable. The building of roads is therefore undertaken, with considerable

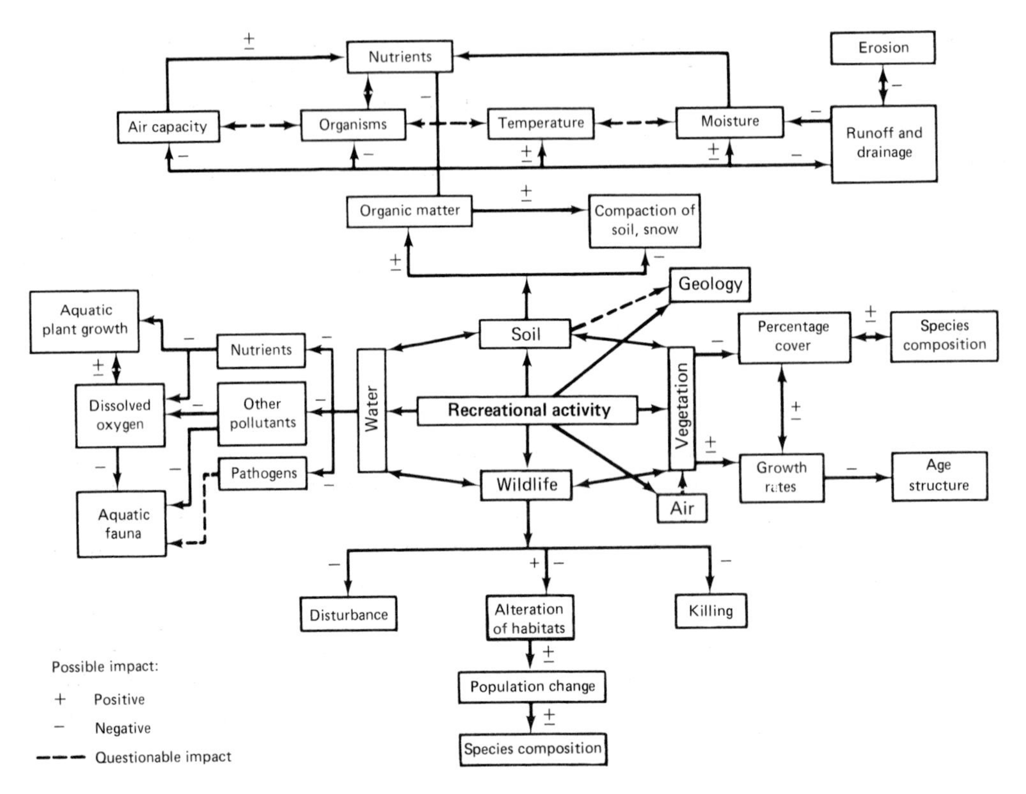

Fig. 4.6 A schematic diagram of the interrelationships between recreation and its ecological impact. Source: Wall and Wright 1977

effects on drainage lines, soil stability and many other features. A road through tundra may have unstable banks which are then seeded with a commercial mixture containing exotics and so the potential for very large biotic invasions may be created. Trees may often be felled to give a clearer view; conversely, insecticides may be used to preserve trees which have a special landscape significance or other associations. Hardware such as gondola lifts require the clearing of a swathe of forest beneath them like a fire-break; and fire roads themselves may be cut for the maximum travel efficiency of fire-fighting crews rather than with regard for the ecology.

Unconscious effects are numerous. The effect of human feet is perhaps the most noticeable, and numerous studies have been made of the effect of trampling upon plants and the soil along tracks and in campgrounds. In many popular resource-based areas, one animal develops the role of scavenger, to become dependent for its food on the leavings of the visitors: yellowhammers at Tarn Hows in the English Lakes, brown bears in western America. Bears can be dangerous, if surprised or with cubs; harmless-looking animals like ground squirrels and chipmunks may have rabies. At the other extreme are creatures which are driven out of their habitats either because of alterations brought about by visitors or simply by the presence of large numbers of our species.

Either way, the ecological patterns which are so valued in 'natural' resource-based recreation areas are distrupted. Even winter sports will affect vegetation because intensive use compacts the snow and reduces its insulative value. Off-road recreational vehicles can penetrate and damage all kinds of fragile ecosystems.

These few examples, and many more which could be quoted (Wall and Wright 1977), serve to highlight the basic dilemma of resource-based recreation: that too great a number of people destroy what they come to seek. The resolution of this in the short term lies in better management of the visitors and, where possible, in the transfer of additional resources from other uses.

Tourism

In his lively book *Tourism: Blessing or blight?* (1973), George Young suggests that the first recorded instance of travel abroad for curiosity was the visit of the Queen of Sheba to Solomon (I Kings 10 or 2 Chronicles 9). Since those days, travel abroad for pleasure, for conferences and on business has increased enormously: absolute figures are somewhat meaningless, but the 1970s saw 'overseas arrivals' reach the 200 million mark, having doubled in the previous decade. The mix of purposes (in one US survey, 42 per cent of 'visitor nights' were for pleasure, 35 per cent on business and 11 per cent for conferences, the remaining 12 per cent being inseparable combinations of two or more) suggests the same kind of social and economic background as in the other forms of recreation described in this chapter, with the added knowledge that travel on business and conferences is likely to produce more impact in cities than in the countryside.

The actual resource demands of tourists are naturally very variable: young people hitchhiking and the senior executive in the Concorde-caviar-Hilton syndrome are at different ends of a demand spectrum. The incidence of all kinds of tourism is uneven: Table 4.3 shows the world distribution of 'arrivals' and it is clear that Europe and North America dominate the scene, just as they provide most of the participants, though Japan is something of a rising star in the latter category at present. Perhaps even

TABLE 4.3 World arrivals 1970

Area	*Million people*	*Percentage share*
Europe	136·3	74·7
North America, Latin America & Caribbean	33·7	18·6
Middle East	3·4	1·8
Africa	2·6	1·4
Asia & Australasia	5·3	2·9

Source: Young 1973

more revealing are measures of 'tourist density' as measured by the number of visitors per thousand natives or per unit area of host country. A version of the latter measure (tourist nights/mi^2/yr) for the late 1960's, shows places like Monaco (800,000) and London (98,000) at one end of the scale, going through Bermuda (78,918), Singapore (53,876), Hong Kong (7,306), the UK (1026), and Spain (263) to places like Kenya (11) and Zaire (0·4) at the other.

The resources which attract tourists are very diverse: scenery, wildlife, history, traditions, friends and relations, architecture, theatres and museums, and 'nightlife' (covering, or in some cases uncovering, a multitude of sins) are all listed in official sources. Historic cities are a major magnet, and in rural areas, warm climates and the sea are another outstanding target. Thus in Europe the capital cities attract many people, as does the Mediterranean shoreline; in the Americas, Washington and San Francisco are parallelled by the Caribbean and Acapulco. The LDC's increasingly rely on sun and sand, as in the Seychelles, the West Indies and Africa (e.g. Kenya and Senegal) and a number of cities with an exotic (or even erotic) flavour like Bangkok. Minority tastes include wildlife safaris in Africa, offbeat places like the deserts of Jordan and Tunisia, cruises to Antarctica and winter weekends in Outer Mongolia.

In considering the impact of tourists on resources and environment, it is sometimes argued that the cities are already there, so the visitor functions like another inhabitant. To some extent this is true, but the tourist in large numbers is likely to be aggregated so that at any rate in some localities there is considerable extra noise, congestion, traffic, parking problems and difficulties in getting theatre tickets. At certain densities the permanent residents may feel like strangers in their own city like animals in the local zoo.

In rural areas the manipulation of ecosystems is more apparent. Agricultural land, for example, may be taken over for building tourist complexes, or farmers may divert their production to high-price commodities for the hotels, rather than local staples, a serious matter in small isolated areas such as the Seychelles. Buildings, roads, soil disturbance, water contamination, wildlife loss may all be found, and are especially serious in fragile areas such as high mountains and islands (UNESCO 1973, 1974b). Construction of hotels and associated developments along the shoreline will produce large quantities of silt offshore, disrupting fisheries and sometimes killing coral reefs as in Fiji (Baines 1977). At Sousse, in Tunisia, hotels have been built in dune slacks. The slacks are converted to gardens but the pumping of water has lowered the sub-regional water-table below the reach of small farmers. Residents walking across the dunes to the sea erode the vegetation and so the sand blows into the hotels. Wastes are another problem: it is commonest along coasts to pipe these out to sea untreated where winds and currents may bring them back again: parts of the Mediterranean have been dubbed the Costa del Mierde. With untreated sewage comes the threat of cholera as well as many other illnesses which may be foreign to the host locality.

All such findings suggest that the benefits of tourism to a receiving area need setting alongside a realistic estimation of costs, (especially social and environmental values), a calculation rarely undertaken even in advanced economies like that of Hawaii (Daws 1977). Beyond that, the development of techniques for the prediction of impacts is essential: the example of the Obergurgl model (Fig 4.7) in suggesting the limiting factors

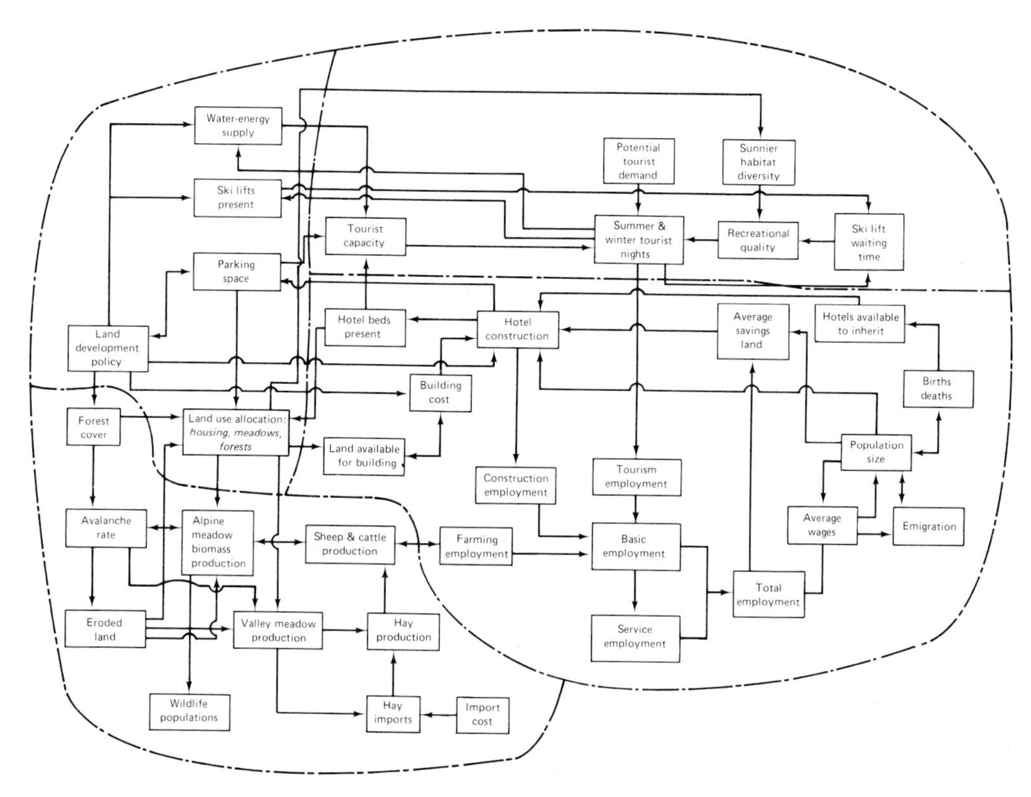

Fig. 4.7 Major components of the obergurgl model. This model aims to integrate both ecological components and social elements of the total system.
Source: Halling (ed) 1978

on the growth of tourism in an Alpine village is one which needs many followers (Holling 1978). At the risk of straying from relevance, however, we must stress the social impact of tourism, especially in poor countries (Smith 1978), and in small places as in the plan to build a tourist complex for 15,000 visitors on Lanai, population 2,000. No wonder that dissenters say, 'tourism is whorism'.

Effects of increasing populations

Although in the short term more recreation resources can be created by shifts of land and water from other uses, this process cannot be carried on indefinitely. Other more basic demands like housing, industry and agriculture cannot be deprived of their share. Rising populations exert an increasing pressure upon resources because demand for recreation in industrialized countries appears to rise at approximately five times the rate of population growth. An increase in the number of people hammering at the park gates is immediately caused to a great extent by increases in leisure, affluence and mobility, and in recreation areas they cause crowding with subsequent loss of satisfaction from the recreation experience; at popular times the peak loading causes tremendous management problems, and the intensity of use may lead to loss of biotic diversity. Environmental contamination may even result, since sewage and refuse disposal systems are unlikely to have been built to deal with such peak loads. People who come to enjoy wild places and rural environments seem inexorably destined to destroy them, even in vast nations like Canada and the USA (O'Riordan and Davis 1976).

In the medium term future the linkage of recreation and tourism to energy supplies is interesting. Recent increases in petroleum prices have had little effect, but if oil reserves diminish at the rates some people project (see pp. 246–7) then recreational travel may well be hard hit, and with it those economies that depend on leisure activities for their economic well-being. Interestingly this is one sector where nuclear power would not be much of a substitute since it comes in the form of electricity and so would only underpin train, tram and trolley travel and perhaps battery-powered autos; shifts in the pattern and magnitude of recreation and tourism would seem very likely under these circumstances.

In a more long-term perspective, recreation and the protection of wild ecosystems are essential components of what industrial countries call 'environmental quality'. Inasmuch as the demands for material resources are, per capita, the highest in the world, and that such desires seem to conflict with the maintenance of environmental quality, it is not surprising that population growth has been called the fundamental cause of the overloading of park and recreation systems. It is argued that just as LDCs suffer from certain material or 'quantity' deficits, because of their increases in population size, so even the relatively slow growth of population in the DCs is fundamental to the steady attrition of environmental quality. Such a train of thought is hard to prove quantitatively, since so much can be done to alleviate the problems by allocating more resources, by improved management, and by multiple use, though any matrix table will soon show incompatibilities and conflicts. But as populations and expectations continue to rise, it appears inevitable that competition for land and water resources, including those at present of value for recreation, will rise. The more basic demands are likely to prove victorious in any such conflict. Similarly, if the quality of recreation declines because of pressure, there are few substitutes that can be brought into this particular resource process. While it is scarcely possible therefore to argue that a cessation of population growth is essential for the persistence of outdoor recreation as a cultural activity, it certainly appears that slower growth with its consequent redistribution of age classes would make available a high quality of outdoor recreation experiences much further into the future than would be possible with a rapidly expanding population.

Further reading

ADAMS, A. B. (ed.) 1964: *First World Conference on National Parks.*

BARKHAM, J. P. 1973: Recreational carrying capacity: a problem of perception.

BRACEY, H. D. 1970: *People and the countryside.*

BURKART, A. J. and MEDLIK, S. 1974: *Tourism. Past, present and future.*

CLAWSON, M. and KNETSCH, J. L. 1966: *Economics of outdoor recreation.*

COPPOCK, J. T. and DUFFIELD, B. S. 1975: *Recreation in the countryside. A spatial analysis.*

DOUGLASS, R. W. 1975: *Forest recreation.* 2nd edn.

GOLDSMITH, F. B., MUNTON, R. J. C. and WARREN, A. 1970: The impact of recreation on the ecology and amenity of semi-natural areas: methods of investigation used in the Isles of Scilly.

GOLDSMITH, F. B. 1974: Ecological effects of visitors in the countryside.

LAURIE, I. C. (ed.) 1979: *Nature in cities.*

LUCAS, R. C. 1978: Impact of human presence on parks, wilderness and other recreational lands.

NELSON, J. G., NEEDHAM, R. D. and MANN, D. L. (eds) 1978: *International experience with National Parks and related reserves.*

VAN OSTEN, R. (ed.) 1972: *World national parks: progress and opportunities.*

OUTDOOR RECREATION RESOURCES REVIEW COMMISSION 1962: *Recreation for America.*

PATMORE, J. A. 1970: *Land and leisure in England and Wales.*

PATMORE, J. A. 1975: *People, places and pleasure.* Inaugural lecture, University of Hull.

POLLOCK, N. C. 1974: *Animals, environment and man in Africa.*

SENGE, T. 1969: The planning of national parks in Japan and other parts of Asia.

SIMMONS, I. G. 1975: *Rural recreation in the industrial world.*

SPEIGHT, M. C. D. 1973: *Outdoor recreation and its ecological effects.*

WAGAR, J. A. 1974: Recreational carrying capacity reconsidered.

WALL, G. and WRIGHT, C. 1977: *The environmental impact of outdoor recreation.*

YOUNG, G. 1973: *Tourism: blessing or blight?*

5

Water

Water occurs naturally in gaseous, solid and liquid phases; man's use of it is nearly all concerned with the last state and is also dominated by his demand for water relatively low in dissolved salts, i.e. fresh water. Of the various conditions in which free water exists, salt water in the oceans claims 97 per cent, in absolute terms $1{\cdot}357 \times 10^6$ km^3. The remaining 3 per cent is fresh water, but three-quarters of this is virtually immobilized as glaciers and ice-caps. Of the last quarter, most is ground water, so that at any instant in time surface fresh water (lakes and rivers) accounts for only 1·5 per cent of all fresh water, and the atmosphere 0·8 per cent (Table 5.1). Since our demands, like those of nearly all terrestrial living things, are for fresh water rather than salt, we deal with only a tiny fraction of the total water volume of the planet; yet the absolute amounts are large, and the energy relations of the various phases are such that human intervention is often no easy matter and water management can be very expensive.

The water that is not locked up as permanent ice is continually moving through various pathways in the atmosphere, biosphere and lithosphere, and this set of natural flows is called the hydrological cycle. A pictorial representation of its qualitative aspects is shown as Fig. 5.1. From the point of view of water resources, the quantity of water in

TABLE 5.1 Water distribution on earth

Total volume of water	1,357,506,000 km^3		
Of which we find	1,320,000,000 km^3	(97·2%)	in the OCEANS
	29,000,000 km^3	(2·15%)	FROZEN
	8,506,000 km^3	(0·65%)	FRESH {ATMOSPHERE, LAND}
The fresh water is find as	4,150,000 km^3	(48·77%)	GROUND WATER > 0·8 km deep
	4,150,000 km^3	(48·77%)	GROUND WATER < 0·8 km deep
	13,000 km^3	(0·16%)	VAPOUR IN ATMOSPHERE
	67,000 km^3	(0·8%)	SOIL MOISTURE & SEEPAGE
	126,250 km^3	(1·5%)	LAKES, RIVERS & STREAMS
	[i.e. runoff = 0·009% of total world water]		

Source: van der Leeden 1975

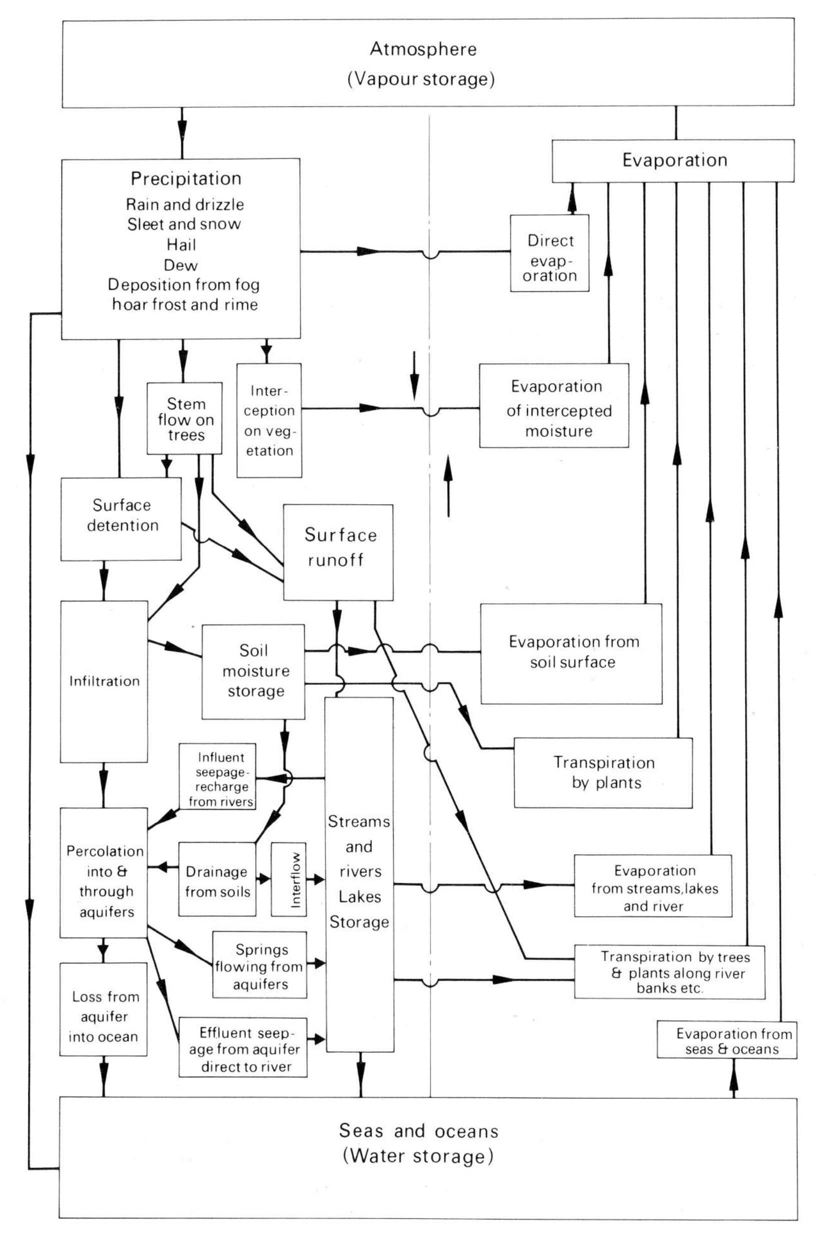

Fig. 5.1 A schematic diagram of the hydrological cycle. The boxes represent both the major storage zones and the transfers between the storages.
Source: Barry 1969

each major unit of the cycle and the rapidity of flux between each becomes the important consideration. The cycle can therefore be regarded as a series of storage tanks, interconnected by the transfer processes of evaporation, moisture transport, condensation, precipitation and runoff. The average water content of the atmosphere if rained out all at once would provide a global fall of 2·5 cm and *de facto* constitutes about 10 days' supply of rainfall (Barry 1969). Ten days is also the average residence time of a water molecule in the atmosphere. This points at once to a rapidly fluxing cycle of evaporation, runoff and precipitation: the annual precipitation over land surfaces alone is about 30 times the moisture content of the air over the land at any one time.

As far as most water resources are concerned, precipitation onto land surfaces is a critical component of the system since much fresh water falls onto the oceans where there is no possibility of garnering it. Over land the distribution of the various forms of precipitation is uneven and the natural flow of water once it has reached the ground is of importance from the resources point of view. Some is re-evaporated, some runs off by way of rivers to the oceans, and a third component enters the ground-water system (Fig. 5.1). If the precipitation falls as snow, then until melting occurs there is a lag between precipitative input and its re-distribution. Globally, water retention on the land surfaces is highest in March–April when there is extensive snow cover and freezing of lakes and rivers in the northern hemisphere. In October there is a rise in sea level of 1–2 cm, when an estimated extra $7{\cdot}5 \times 10^{18}$ cm^3 of water are present in the oceans. Water normally spends between 10–100 days on land, unless it enters ground-water circulation, in which case it stays much longer. In the great Artesian basin of Australia, some water in the aquifers has apparently been there for 20,000 years, but both younger and older sources have been found (Barry 1969).

Human use of water

These are two types of 'use' of water. In the first, water is used as a carrying medium in which materials or objects are carried either suspended in the water, or in solution, or by flotation. The last of these, theoretically, leaves the water in an unchanged condition, and from the former the water can be reclaimed by treatment. On the other hand, the use of water may result in its evaporating into the atmosphere and so passing at least temporarily out of a resource process, or being incorporated with some other material to make a new product. Then the use can be said to be consumptive, since the water is effectively removed from the resource process, even if not permanently. Water which is incorporated in a product is normally required to be very pure except where special contaminations are required as in the case of peat in whisky manufacture. Uses of both types are multifarious, since water is one of the most versatile as well as necessary of man's materials. The most basic need is our own metabolic requirement: men die of thirst long before they succumb to hunger. Since our bodies are 60 per cent water, those of politicians and professors included, a daily intake of about 2·25 litres is needed. To this must be added the residential demands of our species. In the LDCs this may not be very great (90 litres/day for the inhabitants of Karachi), but in industrialized countries the amount is much higher: figures of 263 litres/day in London

TABLE 5.2 Water requirements for selected industries

Industry	*Unit of product*	*Water required per unit of product (litres)*
Bread, USA	ton	2,100–4,200
Citrus fruits, raw, Israel	ton	4,000
Canned fruits, vegetables, juices, USA, av.	ton	24,000
Beer, UK	kilolitre	6,000–10,000
Beer, USA	kilolitre	15,200
Wood pulp, bleached, Finland	ton	450–500,000
Oil refinery, Sweden	ton of crude petroleum	10,000
Steel works, W. Germany	ton of output	8,000–12,000
Automobiles, USA	vehicle	38,000

Note: These figures are mostly for the 1960's and technological changes may bring about improvements.
Source: van der Leeden 1965

and 635 litres/day in the USA are commonly quoted. For health purposes, a supply of about 90 litres/day of high-quality water seems to be the minimum. Activities thus covered include washing, the preparation and cooking of food, and the disposal of household wastes and sewage. Such patterns comprise both consumptive and non-consumptive uses, with some emphasis on the latter. Industrial use of water reaches immense proportions (Table 5.2) and especially where high pressures are used, freedom from dissolved solids and silica is essential. Water is also much demanded for cooling: even though recirculation is practised, the generation of electricity in the UK causes a loss of 40 mgd to the atmosphere (Water Resources Board 1969).

Agriculture is also a very heavy water user. The transpiration from several cultivated plant ecosystems has been reported in the order of 500 mm/yr. The absolute amounts depend principally upon climate, and here the supply of water for irrigation may become important. In California, irrigation use may rise to 500–635 mm/yr, and even in Great Britain it has been estimated that 600,000 ha would benefit from supplemental irrigation (Prickett 1963). At the equivalent of 2·54 mm, this would require a water supply of 3,000 mgd which is similar to the demands of an equivalent area of a densely built-up city. Non-irrigated agricultural use of water is also very high: 0·45 kg of dry wheat needs 227 litres for its production; 1 lb (0·45 kg) rice between 200–250 gal (757–946 litres), and 1 quart (0·9 litres) of milk 1,000 gal (3,785 litres) of water. Neither can the water consumption of animals be dismissed as negligible. A pig of body weight 75–125 lb (34–57 kg) needs 16 lb (7–25 kg) of water per day; a pregnant sow 30–38 lb (14–17 kg) and a lactating sow 40–50 lb (18–23 kg). A lactating Jersey cow requires 60–102 lb (27–46 kg) per day in order to produce 5–30 lb (2·25–14 kg) of milk. By contrast sheep are very abstemious, for on a good pasture they need little if any free water, and on dry range only 5–13 lb (2·25–6 kg) of water per day; on salty range they can be kept happy with a daily input of 17 lb (8 kg) (US Department of Agriculture 1955).

Non-consumptive uses of water include the flotational use of fresh water, which is largest in timber-producing countries and on industrial waterways. Hence the rivers of Finland form 400,000 km of floatway for logs, conveying 3 million m^3 of timber annually; the St Lawrence Seaway is a major traffic artery in North America and in 1966 was used by 60·7 million tonnes of cargo, while the Rhine is another outstanding industrial waterway, converying 230·6 million tonnes in 1965. Such a use is theoretically non-consumptive, but the general contamination cause by shipping (sewage, oil, solid waste and garbage) generally renders the water unusable for other purposes without treatment. The recreational use of water is analogous to flotation in the sense that it is not consumptive at all except where pollutive contamination occurs. Again, many recreational activities involve floating (more or less) of people or boats. Only sport fishing involves a consumptive crop and even that is sometimes returned to the water. The use of water as a wildlife habitat is non-consumptive too when the wildlife is for scientific purposes or for observational recreation.

A genuinely non-consumptive use of water is the generation of hydro-electric power. This method of electricity generation is extensively used in some LDCs as well as in the West and the socialist countries (Beckinsale 1969b). In Peru, for example, 68 per cent of all electricity comes from this source, and in Colombia 63 per cent. Canada is very dependent with 81 per cent, but Norway heads the list at 99·8 per cent. Although the use is nonconsumptive, considerable man-directed intervention in the hydrological cycle at riverbasin scale is required (p. 254).

One theme runs through all these uses: the greater amount demanded in the DCs. It is in these areas that we expect to find the greatest manipulations of the hydrological cycle at whatever scales are manageable with the available technology. In the LDCs the demands are also high in particular sectors, but the development to match these requirements with the supply is often lacking so that large rapidly growing cities especially may suffer inadequacies of both quantity and quality. In rural areas, perhaps 1,200 million people are at risk from contaminated water supplies; the death-rate from water-borne diseases in the countryside of the LDC's being about 5 million/yr. The 1977 UN Water Conference considered an estimate that clean water for everybody by 1990 would cost US $60,000 million.

Ecosystem modification for water control

In order to divert water from the natural hydrological cycle to his own purposes, man must intervene at those places where technology makes it feasible and where the ratio of benefits to costs is deemed to be favourable. Most phases of the cycle are prone to intervention but naturally the runoff and storage phases of fresh water are the most usual. Others, however, need consideration.

Weather modification

The difficulties of manipulating the atmospheric phase of water are such that, compared with attempts to regulate water flow in other parts of the cycle, its practice is both very recent and small in scale. The methods employed have been dominated by cloud seeding. The statistical significance of results is hard to assess and there is a lack of undisputed evidence, although a series of careful experiments in Florida (Woodley et al 1977) suggested that cloud seeding was effective. But the most immediate difficulties are legal rather than ecological. If precipitation is induced at area A, then area B downwind is unlikely to receive the precipitation it might otherwise have had. At intranational levels this is a fruitful source of profit for lawyers (Maryland has made weather modification a crime, and Pennsylvania gives counties the option of making it so), but internationally it is conceivable that it could be used as a long-term weapon to subdue an enemy by desiccation (MacDonald 1968, Sargent 1969) or by inundation as was tried unsuccessfully by the USA during the war in Indochina (Westing 1976). Ecologically, the effects of one-shot 'weather modification' are unlikely to be serious, but persistent interventions come more into the category of 'climatic modification' and hence shifts in species are likely. Gross vegetation shifts are of course possible, but even more rapid would be those of rapidly disseminating species such as fungi and insects, some of which are bound to be 'pests'. So even meso-scale changes which are artificially induced are likely to produce serious and unpredictable biological consequences; should the technology for larger-scale changes become available, then the prospect is full of hazard.

Watershed management

Once precipitated, water in its liquid form is much more amenable to management. The portion not immediately re-evaporated goes largely through the pathways of the vegetation (transpiration), into water courses (runoff) and into ground-water storage at various levels (Fig. 5.2). In the first two categories, the most usual manipulations involve firstly the alteration of the vegetation of a catchment area or watershed, in order to produce more runoff, and secondly the increasing of the storage capacity in the runoff phase so that water may be held for use in dry seasons or to prevent floods.

The runoff characteristics of catchments are affected by many variables such as slope, soil type and depth, and vegetation cover, of which the last is the easiest to alter. In general, the higher the vegetative biomass the higher the transpiration and hence the greater the water 'loss'. Other factors such as the aerodynamic roughness of the vegetation may be important, but a forest normally transpires much more water than a grassland. On the other hand, a deforested watershed or a grassy catchment may yield greater quantities of unwanted silt than the forested zone, so that the optimal balance point between high water yield and high water quality must always be sought (Pereira 1973). Flood peaks are usually higher in unforested catchments unless the soil is unusually deep or unusually retentive as with blanket peat on British uplands. Forested mountains are often critical watershed areas because they may accumulate snow during the winter the melt from which forms a major source of water for lowland areas. Management aims at increasing the depth of snowpack but delaying the melt so as to produce a steady and prolonged yield. It has been found in the western USA that snowpack depth is lowest in the centre of dense coniferous forests and highest (by

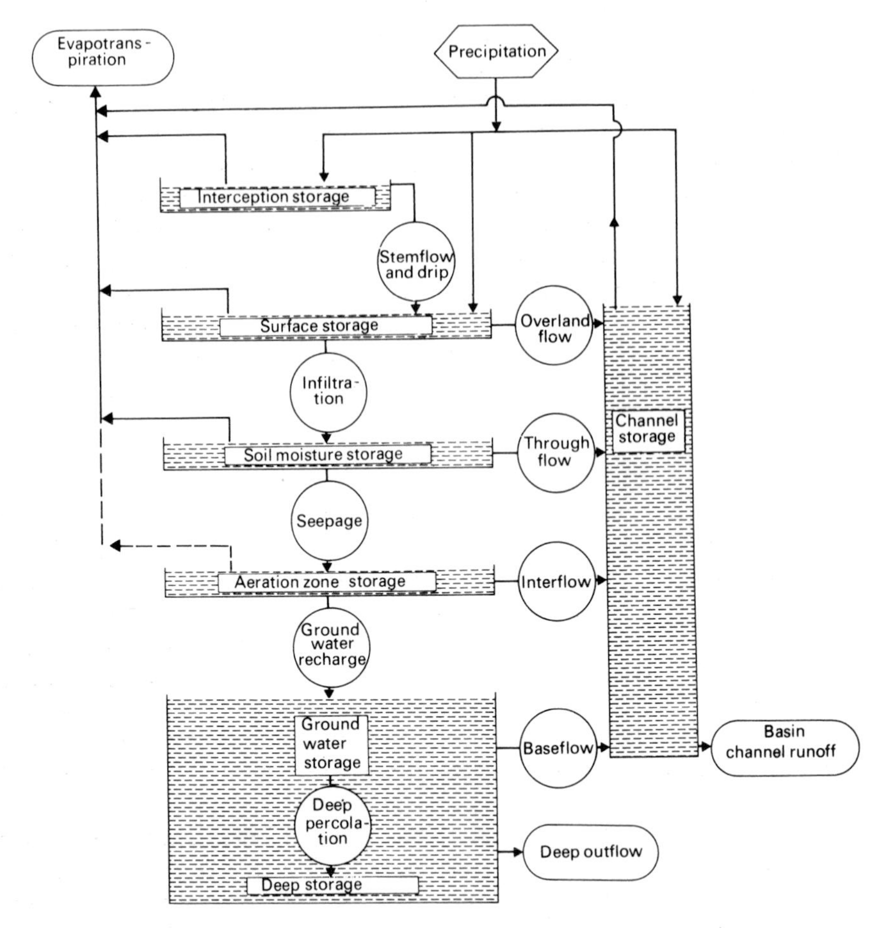

Fig. 5.2 Schematic relationships of the components of the hydrological cycle at the scale of a single basin. Interception storage refers principally to vegetation above the ground but might presumably also comprise buildings in an urban context.
Source: More 1969

increases of 15–30 per cent) in openings and lightly wooded areas such as aspen stands. Management techniques used include logging narrow strips on an east-west axis across the watersheds, which may produce 25 per cent more snow, and shading the trees over the narrow strips to delay melting; fencing on open watersheds may also increase the depth of snow. A major economic difficulty is that the benefits of such management may not accrue to the landowner but to the downstream user and so adequate costing is rather difficult; this makes the technique especially suitable for high terrain in public ownership (Martinelli 1964).

The removal of riparian vegetation can be a useful ploy which increases channel flow. On the San Dimas experimental watersheds of southern California, for example, a flow of $86{\cdot}3 \times 10^3$ m^3/day was produced during a wet season instead of $49{\cdot}3 \times 10^3$ m^3/day before the removal of riparian vegetation. The stream flow also continued all year round instead of drying up for the summer months. Such treatment is most effective in semi-arid and arid areas where phreatophytic vegetation (with species such as tamarisk) transpires large quantities of water (Hopkins and Sinclair 1960), but in some environments a riparian belt of trees is protected in order to trap silt from the watershed. Management of the water quality of runoff from agricultural watersheds is also feasible by manipulating the vegetation near the stream and channel morphology so as to affect sediment load, water temperature and nutrient dynamics (Karr and Schlosser 1978). Watershed management by vegetation control is not yet an exact science, but computer modelling and simulation are rapidly bringing about a situation where the optimum cover for a given mix of resource uses can be predicted and the major obstacles to successful implementation may then be institutional and social constraints.

Intervention in surface and immediate sub-surface flow

A primary aim of managing water flow over land surfaces is to get rid of any excess. Drainage is therefore important in questions of water supply, although it is not usually done with the object of providing a higher downstream flow. Nevertheless, different methods of drainage will affect many aspects of river regimes, as well as contributing to the quality of the water. Loss through evaporation must have been reduced by the advent of underdraining of fields, and large-scale drainage of wetlands such as the Fens of eastern England must have altered the regime of the rivers of the region. Urbanization has also produced its complex of effects, including the rapid runoff from impervious substances which is partly balanced by high potential evaporation from these surfaces.

Once the runoff from both rural and urban areas becomes committed to a flow channel or a lake, management and offtake become feasible. In effect, streams and lakes are moving storage cells of water, especially since many lakes are temporary widenings of rivers. On a world scale the amount of fresh water stored in lakes and rivers is dominated by the lakes of Africa, which amount to 30 per cent of the world's liquid fresh water at the surface; they are followed by the lakes of North America with 25 per cent, and Lake Baykal (USSR) with 18 per cent. Smaller lakes and the world's rivers comprise the remaining 27 per cent. River channels themselves are distinguished by a relatively rapid flow of water, but as storage cells they are held to be insignificant on a world scale. A world isochronal volume of water in river channels was estimated to be 1,200 km^3, compared with 125,000 km^3 of fresh water in the world's lakes and 124,000,000 km^3 on the world's land areas as a whole, a total which includes some saline water in inland seas. This emphasizes the minutes proportion of the planet's water with which man is concerned when he manipulates the river systems, although at regional and local scales the absolute amounts may be very important indeed. The storage capacity of rivers can be marginally enlarged by engineering works such as the deepening of channels and the regulation of flow: these are generally carried out as part of flood-control schemes but may have a secondary effect. Rivers are often the immediate source of water for the world's irrigated area of about 202,500 km^2, although

Plate 9 The traditional way of garnering water for industrial and urban purposes. In this Welsh example, the effect of the reservoir in drowning the sheltered valley lands of the farms can be seen, as can the apparent lack of recreational facilities which might have been a minor consolation. *(Aerofilms Ltd, London)*

pumped ground water is a significant component as well. Manipulation for irrigation is dominated by mainland China, followed by India, the USA, Pakistan and the USSR, and indeed some 80 per cent of all irrigated lands are in Asia (excluding the USSR). Thus few irrigated lands at present occur in very arid regions. Here, and to a lesser extent elsewhere, evaporation losses from the open channels and reservoirs used in most irrigation schemes can assume very high proportions (Beckinsale 1969b).

Water storage

There can be no doubt that the dam is the most popular of all devices for controlling water supplies on a large scale. The effect is to create a lake whose discharge can be controlled according to the demands placed upon the resource manager (Plate 9, opposite). The lake itself may create secondary benefits in the form of fisheries or recreation space to offset the value of the land which is drowned in its creation. The earliest impoundments, such as the tanks of ancient India and Ceylon, were built solely for water storage, so that water collected during a rainy season could be stored and used during dry periods for irrigation or human consumption. Such a function remains one of the primary purposes of the large-scale control schemes much in evidence today. Even in relatively wet climates, there are advantages in smoothing out river regimes, for example so as to avoid a summer low-flow period when the river may be unable to carry its downstream effluent loads without serious disturbance of its biological communities, and when its dissolved oxygen levels may also be very low. Where a river forms the major source for an industry then it is vital that a year-round supply is ensured so that the processes are not interrupted; the same applies to domestic supplies. The evening-out of the flow may also contribute to the solution of flood-control problems.

A steady release of water for irrigation is one of the greatest benefits conferred by storage. In seasonally dry climates it may be possible to utilize year-long sunshine for multiple cropping on land formerly rendered agriculturally unusable by aridity. The control of the Sacramento River in the Central Valley of California by the Shasta Dam and associated works of the Central Valley Project has facilitated treble and even quadruple cropping of some fruits and vegetables in the southern half of the Valley. Storage requirements per unit area of land increase in regions of climatic variability, so that per acre of irrigated land Australia stores 6·9 acft (1 acft = 1,234 m^3), the USA 31 acft, and India only 1·7 acft (A. B. Costin 1971). The generation of electrical power from falling water is much enhanced by the construction of large dams. In industrialized countries there is the frequent advantage that the power can be sold to industry in order to pay for the cost of construction and hence agriculture is made more efficient by its linkage with industry. It is one of the less happy features of some major dam projects in LDCs that no market exists for the power that could be generated.

The benefits to be gained from manipulation of the river flow usually overwhelm any suggestion that significant secondary costs might be incurred; indeed much water power is very cheap because the social and ecological costs are rarely reckoned up in the accounting of projects. The effects of large impoundments are wide-reaching and affect the hydrology, the terrestrial system and the aquatic systems of the basin in ways which have not always been beneficial (Lagler 1971, Ackermann et al 1973, White 1977). Basin evaporation rates rise, especially in arid and semi-arid countries, leading to the trial of surface films which will transmit oxygen and carbon dioxide, resist wind, be self-restoring after disturbance and be non-toxic. When the initial flooding takes place there is a biogeochemical enrichment of the water, usually resulting in an explosive growth of phytoplankton and other producer organisms. Some of these are floating water plants such as water hyacinth (*Eichhornia crassipes*), salvinia (*Salvinia auriculata*) and water lettuce (*Pistia stratiotes*). These weeds transpire water which would otherwise remain in the catchment: they and ditchbank plants together transpire $2{\cdot}3 \times 10^9$ m^3/yr, worth about $40 million, in the 17 western states of the USA. The cost of clearing them is likewise high: only the expenditure of $1.5 million/yr on herbicides has enabled the Sudanese to prevent the spread of water hyacinth from the White to the Blue Nile. In this region, the loss of water from *Eichhornia*-covered lakes is 3·2–3·7 times that from a free surface. The weeds also prevent algal photosynthesis and hence lead to serious depletion of fisheries. On Lake Volta, water lettuce serves as a habitat for the larvae of several mosquitoes, including the vectors of encephalomyelitis and filariasis (Holm *et al.* 1969). Submerged weeds may create difficulties, especially in irrigation systems, as with the introduction of *Myriophyllum spicatum* into North America from Europe. Control by herbicides is often used, but it is very expensive and a constant vigilance is necessary: the cessation of spraying of the Zaïre after the post-independence wars allowed water hyacinth to clog it again very quickly, and some herbicides are also toxic to fish. Biological control depends upon finding an animal with a voracious appetite for a particular plant: some snails are the subject of experiment, and the sea-cow or manatee is a possibility if enough of them can be found, since conversion into pelts is currently a profitable exercise. The white amuo fish and a wingless aquatic grasshopper (*Paulinia acriminata*) are also specialized feeders, the latter eating only *Salvinia*.

The hydrology of the basin is most altered downstream from a dam, especially if the dam is completely closed during the filling period. Thereafter, the volume of discharge and the current velocity downstream are most obviously altered, with a concomitant reduction in river turbidity. The water temperature may be affected severely if water is drawn off below a thermocline, and the quality of the water will also be changed if the impoundment is chemostratified and draw-off comes from only one depth in the lake. Silting reduces the life-span of a large impoundment and where possible a silt trap is built into the design; if the sediment piles against the dam, it may be periodically flushed out, with disastrous effects on the ecology of the river downstream. Silt removed by an impoundment may represent a loss of nutrient input to lands lower down and have to be replaced by chemical fertilizer; projected dams on the rivers of northern California would deprive some of the coast redwoods (*Sequoia sempervirens*) of their periodic injection of nutrients and probably prevent them reaching their enormous height (up to 112 m) which is their major aesthetic and commercial attraction. Control of river regimes may also lead to a net loss of soil moisture downstream, although it is sometimes predicted that lateral percolation from the impoundment may increase the groundwater supply to quite distant points. The stabilized stream flow favours the survival of sedentary and rooted organisms but disrupts species of fish which spawn in flood water or whose eggs or fry depend upon the occurrence of a nutrient-enriched zone of inundation. Adult fish may be catered for by fish passes round dams, but these have not been universally successful, and young fish going downstream sometimes find them difficult to traverse. Lastly, there is good evidence that large dams may trigger seismic movements, especially in highly faulted regions. An earthquake near Konya Dam in India cost 200 lives in 1967 and yet there was no history of tremors in that area (Rothé 1968).

A concatenation of the unforeseen effects of river impoundment has been observed on the Nile as a consequence of the construction of the Aswan High Dam. The removal of silt has taken away a natural source of nutrients which must be replaced by buying chemical fertilizers, and off the delta, the stoppage of nutrient input into the Mediterranean has caused a decline in fishery yield to Egypt since the sardines have disappeared. In 1962 the Egyptian fish catch was 30,600 tonnes, but this was reduced by 18,000 tonnes by 1965. The fishery of Lake Nasser was expected to reach 10,000 tonnes by the mid-1970s, but this figure probably represents an initial 'bloom' and will stabilize well below that figure. The productivity of the delta lakes has fallen owing to fish kills caused by biocide runoff, and by accelerated eutrophication (George 1972). The Nile delta itself is now in retreat due to lack of building material. The extension of perennial irrigation is instrumental in extending the range of blood-fluke diseases such as bilharziasis which now infects 100 per cent of the population of some areas; it is virtually impossible to cure, and control measures such as improving sanitary conditions, drug therapy or snail control have all been ineffective because of their expense or their incompatibility with cultural patterns (Van der Schalie 1972). These examples show that the damming of rivers may have many beneficial results, but that the consequences of the alteration of the many ecosystems to which the river acts as a common thread have rarely been explored and scarcely ever incorporated into the reckoning of costs and benefits.

Floods

One normal condition of the hydrological cycle has been perceived as abnormal by man and labelled as floods. Poor watershed practices have often exacerbated and in some places have been the cause of floods, but it remains true that most rivers have floodplains which get inundated from time to time with varying depths of water and at low levels of predictability. Human adjustment to flood hazard (White 1964, 1974, Burton et al 1978) may take the form of moving smartly away, lock, stock and barrel, or *inter alia* of pressuring the appropriate governmental agency to remove the threat by environmental intervention. Flood-control dams are one remedy as are channel widening, channel straightening and deepening, and the construction of by-pass channels. Apart from the three-dimensional aspect, the problem is rather like traffic engineering, but new freeways only create more traffic and concrete riprap does not reforest an urbanized watershed. In sum, flood-control measures rarely control beyond certain well-defined limits at certain places and do not look at the cause of the flood which are usually to be found in ecologically unsuitable land-use patterns as in the deforestation of the mountains of the southern Himalaya which exacerbates floods on the Ganges.

For all-round ease, manipulation at this surface-water phase of the hydrological cycle ranks first. But the essentially limited amount of water at this stage, together with the fact that most of it is not of the desired quality unless treated, means that nations with access to advanced technology are assessing their deep underground aquifers more keenly.

Intervention at the ground-water phase

Relative to other sources of fresh water, ground water is an important phase of the hydrological cycle. One estimate suggests that about 7×10^6 km^3 of such water is recoverable which is about half the total thus stored (Nace 1969). Though ground water is essentially a storage phase, it appears to have two components: cyclic ground water, which passes in and out naturally within the space of a year, and inherited ground water, which appears to have a much longer storage period.

One of the most important roles of ground water so far as man is concerned is its role as the major contributor to the base flow of rivers. Without this regular input, the reliability of rivers as resources would be much diminished. Of equal apparent importance is the pumping out of water for all purposes and this is done on many scales: dense networks of pumped wells serve towns and industrial areas in many parts of the world, as they do some major irrigated areas where river storage is not feasible. On the other hand, a single winddriven pump watering a livestock trough or an isolated farm is a common enough sight in semi-arid parts of the world. Pumped wells have their disadvantages because it is often difficult to estimate the quality and quantity of water available, and, like oil, recovery may be uneven. Near the coast, overpumping beyond a certain level may allow the influx of saline water into the aquifers, rendering them useless for a sustained yield of water. Artificial recharge is one response: water is spread into the aquifer by forcing it down through pumps, but achieving a reasonably even spread of the water through the rock can present problems. Nearer the ground surface, the Dutch, for instance, spread polluted waters from the Rhine system over the coastal sand-dunes, which act as a filter, and the water collected beneath is to some degree purified. Deeper recharge schemes often have the aim of allowing the period of percolation through rock to remove contaminants. Purification depends largely upon porosity, and if a high rate of flow is desired then a high-porosity rock needs to be chosen, which will mean a low efficiency of purification. The inverse relation between flow and purification appears to hold for most types of rock, so that ground-water recharge is not always a great success. Nevertheless, many overpumped aquifers such as the Chalk under London are now subject to recharge schemes.

Intervention at the saline phase

The limitation upon the human use of 98 per cent of the world's water is its content of mineral salts. Comprising all except 1 per cent of that proportion, sea water commonly has 35 g/litre of salts, and there are many brackish waters inland with contents above about 2 g/litre which make them useless for most purposes except flotation. In view of the increasing demands for water which is low in mineral salts as well as free from particulate matter, harmful organisms and toxic substances, it is not surprising that

Plate 10 A small desalinating plant in the Channel Isles, UK. Such plants are very useful on small islands low in ground-water resources, especially where there is a peaking of demand caused by a seasonal influx of tourists. *(Senett and Spears Ltd, Jersey)*

attention should have been turned to methods of demineralizing salt and brackish waters (Probstein 1973). In the 1970's there were about 60 plants in use of over 1 mgd capacity, of which the largest group were in the Middle East, followed by the Caribbean: the biggest plant was in Oman and had a capacity of 7·2 mgd. Most such plants use MSF (multiple storage flash distillation), as does a smaller plant in Jersey (Channel Islands) which has been used for 'topping up' during the vacation season (Plate 10).

Other methods tried include freezing, using butane as a liquid refrigerant; exchange resins, which will work only up to a salt content of 5 g/litre; and solar distillation, the drawback of which is the large area needed to collect the sun's energy; and plants using natural freezing have been designed in the USSR (Pryde 1972). Taken together, there was a world growth of desalination capacity from 59·8 mgd in 1961 to 247·2 mgd in 1968 with the expectation of 1,250 mgd in 1975 (van der Leeden 1975), representing about

700 plants. One of the difficulties with such water is that its cost (already high by comparison with conventional supplies) escalates the moment the water is raised, i.e. is used virtually anywhere away from a shoreline. Water costing 600/1,000 US gallons at the plant might cost 90c/1,000 gall delivered to the field (Clawson et al 1969). The cost of lifting water is generally about 1 cent/100 metres of lift/1,000 litres of water: van Hylckama (1975) quotes the example of Lubbock, Texas, of an elevation of 1,000 metres which would have to pay $5 million/yr for horizontal delivery of a supply of 1,000 l/cap/day and another $5 million/yr for raising the water from sea level.

The type of areas which would both benefit and be able to afford desalination are therefore restricted in number. Urban and industrial areas in high-technology countries may be one of these, but it may be noted that a plant in southern California to be built under a low interest rate (3·5 per cent) and a guaranteed market for 90 per cent of its water and power (150 mgd and 1·8 million kWe) was deemed uneconomic. If nuclear plants are not viable in such places, where might they be so? The use of desalted water as a catalyst for agricultural and industrial development in poor and arid areas is an exciting prospect, but as the cost of demineralized water is at least one whole order of magnitude higher than present supplies, it seems unlikely to be a cornucopia at any rate until virtually unlimited supplies of extremely cheap energy are available a position approached to some extant by heavy users of desalinated water like Kuwait and Saudi Arabia (Beaumont 1977a, b); even such a project would have to reckon with the ecological costs of the return to the sea of immense quantities (3,500 t of salt per 100 million litres of water) of hot concentrated brine.

In from the cold?

Van Hylckama (1975) suggests that since icebergs contain so much fresh water they should be considered as resources. The average Geenland iceberg has an initial volume of 15×10^6 m^3 or enough to supply a city of 60,000 people for one year. Their shape would make them difficult to tow, but the flat-topped antarctic bergs might be easier: an iceberg of dimensions 3,000 × 3,000 × 250 m might be towed to Australia in 1 month for a cost of $US1.5 million. Even if half of it melted there would still be 1×10^9 m^3 of water or enough to supply 4 million people for one year at an average cost of $1.50 per annum. Even better figures might be obtained for the dry west coast of South America, where ocean currents might be used to aid transport.

Scales of manipulation

The simplest and earliest forms of intervention in the hydrological cycle were at a series of single points. Wells, shadoofs, Archimedean screws and similar devices tapped the water at one place and it was then borne to the site of use by pipe, container or channel. It is a measure of their effectiveness and of the recency of much water-control technology that they are still found around the world, though often with a motor attached to reduce the labour. At a high intensity, such as closely spaced pumped wells, the effect on the water storage can be very great; point manipulation in rivers affects flow rather less.

Another scale of manipulation involves much more intervention, but the control is confined to one basin. Dams for irrigation, flood control and the various other purposes discussed above, irrigation schemes fed from wells or impoundments, and underground transfer lines such as the *qanats* of the Middle East (especially Iran) are examples of this scale of diversion (Beaumont 1968). As the effectiveness of technology increases and planning sophistication soars, so the proportion of water under control in the basin becomes greater so that multiple-use schemes for whole basins of varying sizes become feasible. The Tennessee Valley Authority scheme (quoted still so often as if it were the only example of a multi-purpose basin development) is one example, and those prepared for the Jordan basin and the lower Mekong others; the last two are problematical because of the political boundaries which transgress the natural water-control unit. The creation of large fresh-water lakes behind barrages at coastal estuaries is another major intervention that is especially popular in low-lying countries without steep-sided inland valleys to flood. The IJsselmeer and Delta schemes of the Netherlands and the desk studies for Morecambe Bay, the Wash and the Solway Firth in Britain are contemporary examples.

A still larger scale is the transfer of water between river basins, sometimes involving an ascent stage which necessitates pumping. The earliest examples of such interbasin transfers were usually to ensure a good head of water behind a particular dam which was being used for power production or irrigation, or both. The Conon Valley scheme in Scotland (Fig. 5.3) is a scheme designed to develop power from the flow of tributaries in the same basin by tunnelling water across watersheds, and also bringing it by tunnel and surface aqueduct from other basins: water from the Ewe (which drains to the west coast) is tapped for Loch Fannich which drains to the Moray Firth. Glascanoch water is piped into Loch Luichart, its parallel eastward-flowing system. Of the scheme's total catchment area of 119,140 ha, 5,957 ha belong entirely to other basins (Aitken 1963). The road to the isles nowadays is accompanied by the skirl of a different sort of pipe.

A larger-scale set of diversions can be seen in schemes planned for Alberta and north-east England. In both there is relatively ample water in the north and a thirsty south, with parallel west–east rivers. Plans exist therefore to transfer water from the northernmost stream via intermediate rivers to the middle reaches of the southernmost artery. In the case of north-east England the transfer is from the north Tyne via the Wear to the Tees with a new regulating reservoir on the North Type. Alberta was the centrepiece of PRIME (Prairie Rivers Improvement Management and Evaluation), in which the great untapped flow of the Peace River (Fig. 5.4) was to be the northernmost source of a set of interbasin transfers. Water from the Peace was to be fed via Lesser Slave Lake to the Athabasca and thence to the north Saskatchewan; there was also a transfer planned further west between the Athabasca and the north Saskatchewan, and water was to be fed from the north Saskatchewan to the Red Deer River as well. This scheme has been subsumed into a much larger project for the whole of the Saskatchewan-Nelson basin which is the fourth largest basin in Canada (Saskatchewan-Nelson Basin Board 1972).

Such schemes are puny compared with some of the ideas put forward for large continents, particularly where a small number of political units is present. The USSR is

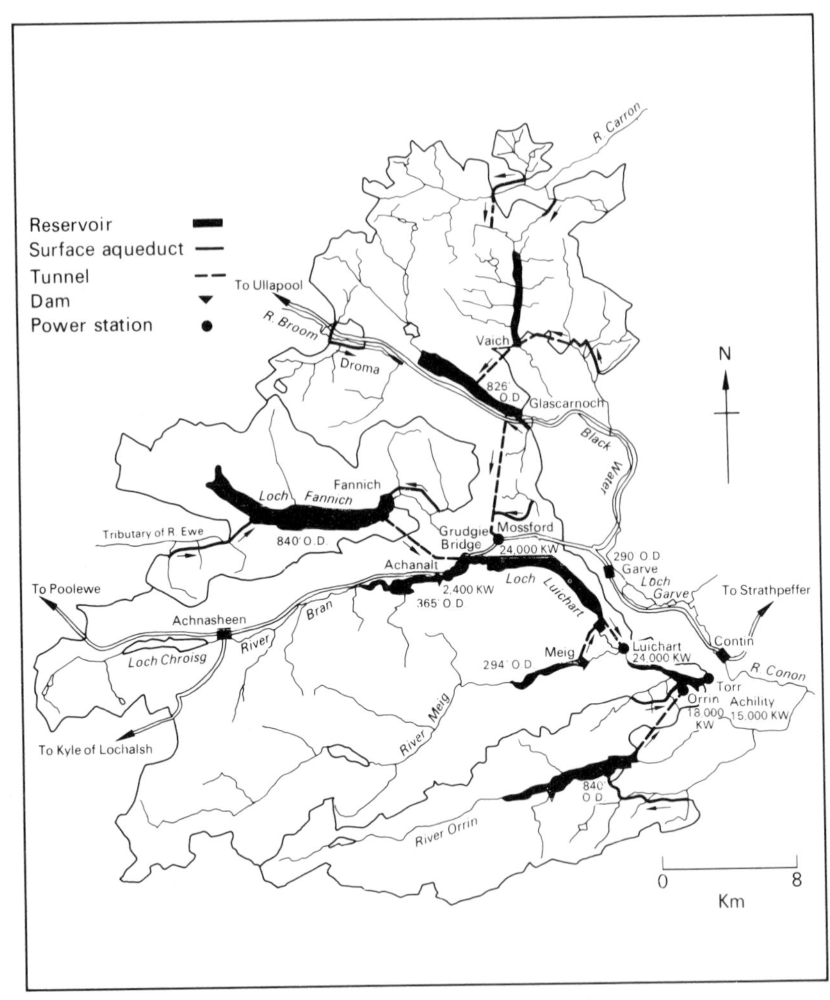

Fig. 5.3 The Conon Valley scheme in Scotland. Transfers within the basin (e.g. from Glascarnoch to Mossford) as well as from outside it (e.g. from tributaries of the River Ewe into Loch Fannich and from the headwaters of the River Carron to those of the Black Water) are used to provide the maximum head of water for hydropower generation.
Source: Aitken 1963

one instance and a number of plans have been put forward (Gerasimov and Gindin 1977, L'vovich 1978) which generally involve taking water from the northward flowing rivers and diverting it to the more arid but warmer regions between the Black and Aral Seas. In European Russia, the Onega, Sukhona, Pechora and Vychegda rivers would be used, and in Siberia the Ob, Irtysh and possibly the Yenesei (Fig. 5.5). Apart from the effects within the USSR, particularly on the moisture cycle in western Siberia, Kazakhstan and Soviet Central Asia, concern has been expressed for the wider

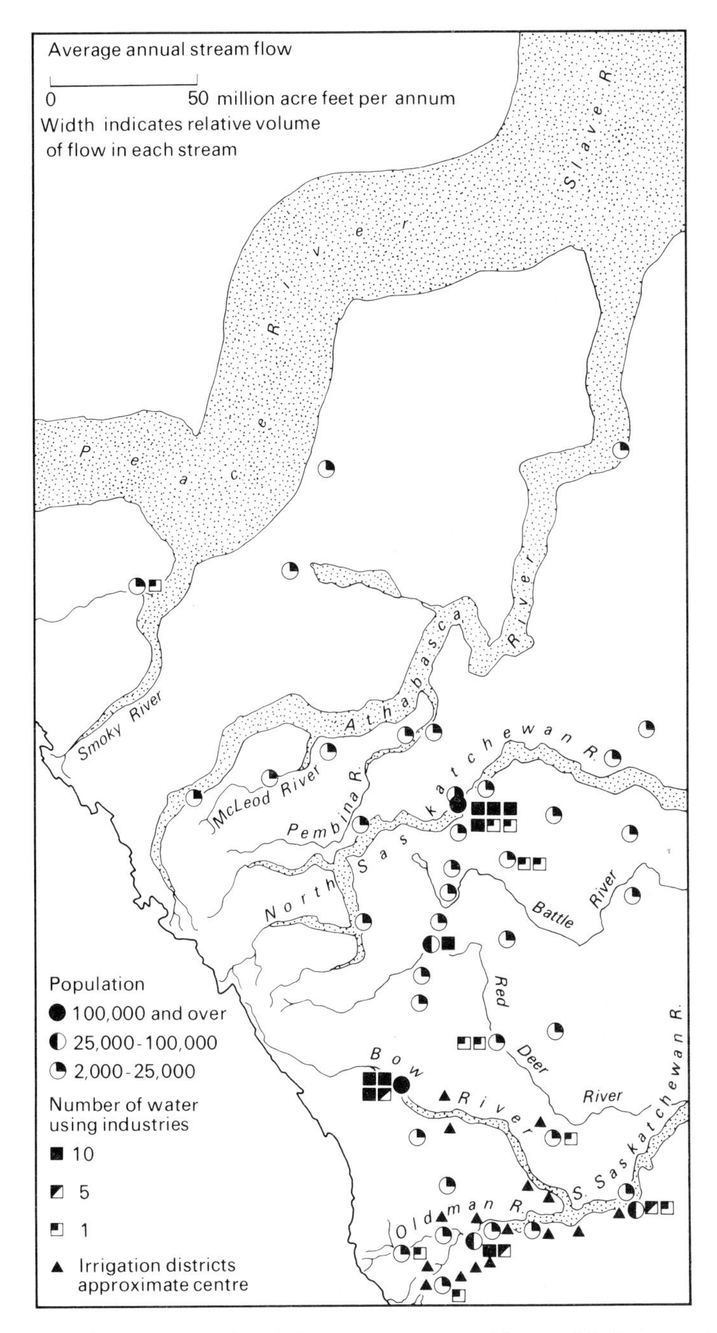

Fig. 5.4 The volume of prairie rivers in relation to user groups. The spatial discrepancies provided the impetus for a scheme of interbasin transfer, progressively transferring water southwards. This scheme (PRIME) has now been incorporated into a larger project. (Note: 1 acre ft = 1,234 m^3.)
Source: Province of Alberta 1969

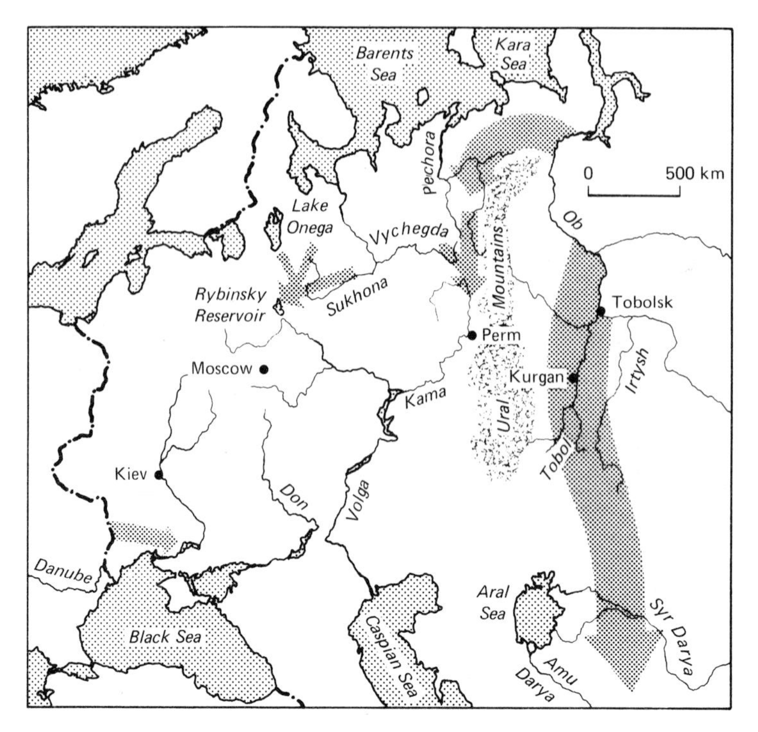

Fig. 5.5 Map showing the most favoured schemes for transferring water from the north flowing rivers to the south.
Source: L'vovich 1978

consequences, particularly on the Arctic Sea ice, if the volume of freshwater input diminishes significantly.

Another example of continental plumbing is the Parson's Company plan for taking 'unused' water from the northern rivers of Canada and feeding it as far as the Great Lakes, New York, Los Angeles (inevitably) and Chihuahua in Mexico (Fig. 5.6). Virtually all the western rivers of North America would be reservoirs or strictly controlled, and the centrepiece would be a flooded Rocky Mountain Trench in Montana and British Columbia, 914 m ASL. The objections range from the geological doubts about the ability of the Trench to withstand such a weight of water to the complaints on aesthetic grounds that the great playground of North America would not have a wild river left and that much wildlife habitat would be destroyed. At 1966 prices, the NAWAPA (North American Water and Power Alliance) scheme was estimated to cost $800 million; an even more potent objection, however, is the rising Canadian nationalism, which seeks to free Canada from economic dominance by the USA. Like many such schemes it presupposes that urban/industrial growth is an ever-expanding consumer which must be supplied, and that it is more necessary to take the water to the sites of use than *vice versa*. Apart from its engineering *folies de grandeur*, NAWAPA appears to be devoid of useful thought about water resources.

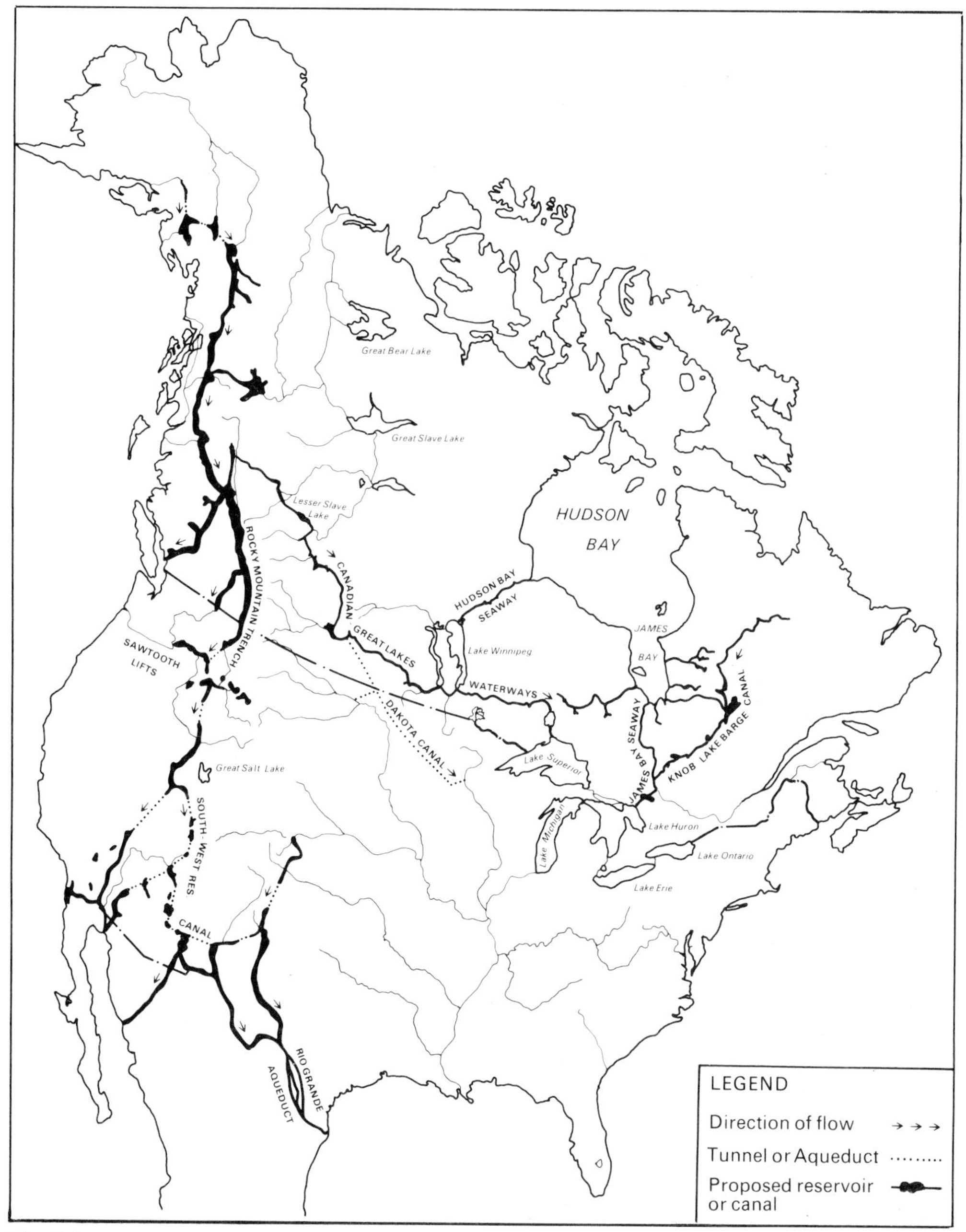

Fig. 5.6 The North American Water and Power Alliance (NAWAPA) scheme which would divert water from northern Canada and Alaska and use it to supply southern California, northern Mexico and the upper Great Lakes region. Navigation projects are also proposed. The effects of drowning the Rocky Mountain Trench upon wildlife, recreation, forestry, communications and earthquake frequency are hard to predict. Source: Province of Alberta 1968

Economic and social constraints

Economic pressures often derive from the status of water as a free good. It costs nothing itself, but manipulating it and treating it may be very expensive: a typical municipal scheme may incur costs of land acquisition and compensation at the site of a headwater reservoir, dam construction and maintenance, pipelines to the city, treatment plants, storage space near the city and distribution costs within the city itself. Inevitably, the greater the distance from source to user the greater the costs; not only is the initial supply system expensive, but its maintenance is likely to absorb large quantities of money. In the south-west of the USA water costs 5–15c per 1,000 gal (4,542 litres) per 100 miles (160 km) to transport, thus making some 'project water' sell at $1 per 1,000 gal, compared with the usual price of 50–70c for industrial water and 5–10c for agricultural supplies. Small wonder that many authorities try to generate power at their dams in order to sell the electricity to pay for the costs of the water-supply system. In the face of supply shortages and rising costs, the search for effective methods of cleaning effluents from water so that it may be re-used downstream is an important development, as is the technology for recycling water used at one site, for example as a cooling agent in an industrial plant or thermal power station. On the River Trent in England, water containing sewage effluent can be used by power stations for cooling, and it is successively re-used by power stations downstream at intervals of 16–32 km. The operations have to be regulated so that the river temperature does not exceed a prescribed limit of 30°C. Cooling towers themselves achieve re-use within a power station: the water is cooled partly by evaporation, which means a loss of just over 1 per cent of the amount of water being circulated through the tower. This amounts to 14 mgd for a 200 mW power station and is usually made up from a riverine source. The large bulk of water required for nuclear power stations has meant that coastal sites have been favoured, bringing them into conflict with recreational and wildlife resource uses.

Treatment of waste waters is more fully dealt with in the section on water pollution (pp. 294–7), but the economic constraints appear in the costs of the full treatment needed to return water from, say, domestic sewage to a high degree of purity. The necessity for the introduction of full treatment can be deduced from the fact that such processing is usually five times as cheap as the supply of new water. Here therefore is a major source of supply of potentially high-quality water: theoretically a town could recycle its own fluids on a closed-system basis. This argument can probably be applied to many polluted waters, particularly if the authorities responsible for sewage charge the true costs of treating industrial effluents.

The different values put on different uses of water or catchment areas often exert a social constraint on water manipulation. The most frequent instance of this is a controversy over the loss of land under a reservoir impoundment. Settlements may be lost, as in the Tsimlyansk reservoir in the USSR which necessitated the relocation of 159 towns, villages and hamlets (Pryde 1972); or the loss may be cultural as with the Abu Simbel temples which would have drowned beneath Lake Nasser in Egypt, but were saved by international action; it may be more narrowly economic as with the disruption of farm units when bottom land is drowned in upland Britain; it may be scientific as with the Cow Green reservoir in Upper Teesdale (northern England) where part of an assemblage of arctic-alpine relict plants was destroyed; or it may be scenic as with the drowning of Glen Canyon in Arizona, and the proposals to put dams in Bridge Canyon and Marble Canyon near the Grand Canyon of the Colorado River. In all these cases, and many others, there has been fierce opposition to the proposed water-management scheme, and the opposition to the building of dams especially has become vocal and well informed. In industrial countries, therefore, we are likely to see social pressures influencing large manipulation schemes to a greater extent than hitherto and hence accelerating the move towards closed-cycle re-use. If the benefits of wild country and natural ecosystems are considered at their true value, then the case against some impoundments may be stronger.

The effect of increasing human populations

The very vastness of the quantity of water on the planet gives it an aura of an illimitable resource. Regionally this is clearly not so, but is not the total amount so great that improved distribution would bring about plentiful supplies for every purpose? Since inter-continental transfers seem unlikely, and de-salting appears to be very expensive except for those with energy sources to spare, then problems have to be tackled on a regional basis. In industrial nations like the USA, vast quantities are needed to maintain the lifestyle. Wollman (1960) estimated that 68×10^3 litres/cap/day were used out of a total resource of 127×10^3 litres/cap/day, and projections for the future suggest a demand for very much more. Yet regional shortages are already apparent and are the moving force behind continental plumbing schemes like NAWAPA.

Estimates of future use such as those quoted in Table 5.3 show that needs are coming very near to the continental run off, especially if water is not re-used; proposed irrigation schemes are particularly heavy contributors to the consumption of water. Table 5.3 implies that in 20 years, Europe, Asia and Africa will demand more than 20 per cent of their total run off, at which level advanced water management measures will surely be needed, and indeed above 20 per cent water is likely to be a serious limiting factor in economic development. The problem looks especially serious in Asia since half the total run off means at least 65 per cent of the base flow of the rivers: effective storage of flood waters therefore seems to be an essential project. In Africa, development of the poorest regions may well need transfer from the wetter tropics and the use of deep ground water, which can turn out to be a non-renewable resource if recharge is not undertaken. In central Europe, pressures for large-scale transfer will develop and no doubt Scandinavia will be looked to as a possible supply area. Another critical area is Israel, which has a poor supply but a very high demand, especially for irrigation water (Wiener 1972). Sectorally, it looks as if irrigation requirements will be a very large identifiable need in order to keep with the increased need for food production (chapter 7); yet water will have to be provided in the right place at the right time and comprehensive management schemes for river basins look more than even necessary.

Such estimates of future use confirm what present data suggest, that human intervention in the hydrological cycle is now by no means inconsiderable. Such a view

TABLE 5.3 Estimated future water needs

	Needs in AD 2000				
	km³		*% of runoff*		*Est pop (millions)*
Region	*Alt A*	*Alt B*	*Alt A*	*Alt B*	AD *2015*
Europe	741	536	31	23	618
USSR	430	312	10	7	368
S and E Asia	4,286	3,465	51	36	4,684
Africa	1,044	742	25	18	1,154
N. Amer	437	317	7	5	380
S and C Amer	850	616	8	6	908
Oceania	46	33	2	2	43
WORLD	8,380	6,030	22	16	8,155

Alternative A: withdrawal needs without any re-use of industrial wastewater.
Alternative B: assumes 90% of the wastewater can be re-used.
Source: Falkenmark and Lindh, 1974.

is displayed in Fig 5.7, where the contribution of irrigation both to demand and to evaporation (with possible consequences for local climate and perhaps eventually for sea levels) is clearly indicated. So whatever the immense volume of the resource, it is within man's ability to change its flows quite markedly and thus to discover new bottlenecks in its supply. Within the envelope of these pathways, various developments may help to procure essential supplies. New technology will probably reduce demand in some industrial processes, as in the halving of water used per kWh generated in England between 1900 and 1965. Much pure water is now used for purposes which do not require such pristine conditions, as in carrying industrial effluents and domestic sewage for example; and the provision of separate supply channels for pure and not-so-pure water, as already happens for effluent in the Ruhr area of West Germany where one river is maintained clean as a recreational stream and urban supplier while a parallel water-course acts as the regional cloaca (Fair 1961), is a development which may have wider application. The purification of contaminated water would provide a major source of supply possibly equal to half that needed by industrial nations.

Such considerations need to be considered in the context of the energy linkages of water use. Granted that some water can be used to generate power (see pp. 254–5), its use is also a consumer of it: desalination apart, energy is needed to construct distribution and treatment systems and usually to run them as well, especially where the lifting of water takes place. Through the evaporative power of the sun and the pull of gravity, natural forces power much of the hydrological cycle but most human interventions require energy to divert the natural flows.

It follows that water may become a limiting factor in plans to increase the magnitude of resource processes. Rising expectations will create extra demand which management schemes will find hard to satisfy and there is no obvious technological fix: substitutes for water are hard to find and usually very expensive. Intervention at the largest scales brings with it other values also: what have we gained if we 'manage' so much water that every river is tamed into a series of lakes, if we lose the wild water of fast rivers and if

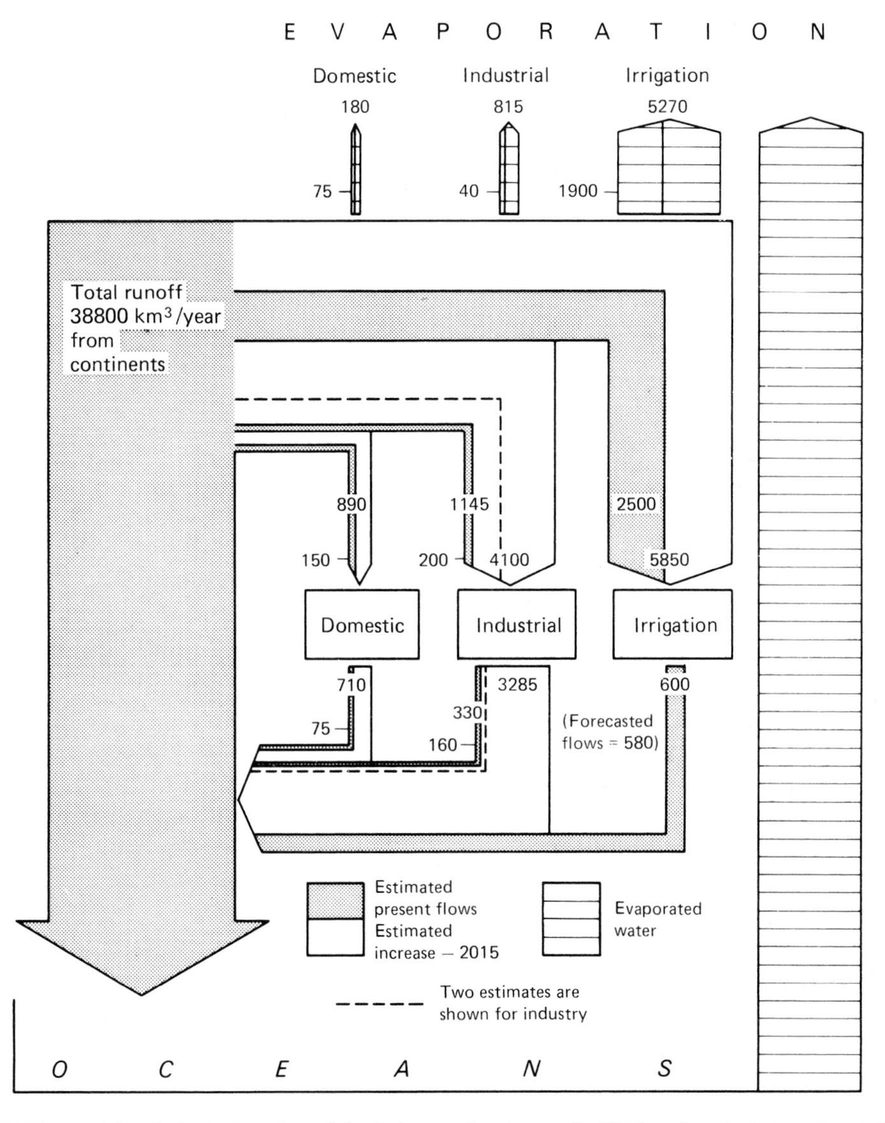

Fig. 5.7 The world hydrological cycle and the balance of water use in 1974 and projected use to AD 2015. Source: Falkenmark and Lindh 1974

every placed stretch disappears under another storage reservoir? Such attitudes involve a close look at how water is used and how it can be reused: how to keep the cycle 'tight', almost in the manner of nutrient cycles (pp. 13–14). We must at the very least disprove the validity of the first subject of the ancient Chinese proverb:

> Water and words . . .
> Easy to pour
> Impossible to recover

Further reading

BISWAS, A. K. 1974: Water.

BISWAS, A. K. (ed.) 1978: *United Nations Water Conference. Summary and Main Documents.*

CHORLEY , R. J. (ed.) 1969: *Water, earth and man.*

DUNNE, T. and LEOPOLD, L. B. 1979: *Water in environmental planning.*

FALKENMARK, M. and LINDH, G. 1974: How can we cope with the water resources situation by the year 2015?

FLÖHN, H. 1973: Der Wasserhauchalt der Erde.

GOLDMAN, C. R., MCEVOY, J. and RICHERSON, P. J. 1973: *Environmental quality and water development.*

GREER, C. 1978: *River management in modern China.*

VAN HYLCKAMA, T. E. A. 1975: Water resources.

HYNES, H. B. N. 1970: The ecology of flowing waters in relation to management.

JACKSON, I. J. 1977: *Climate, water and agriculture in the tropics.*

KORZUN, V. I. (ed.) 1978: *World water balance and water resources of the earth.*

VAN DER LEEDEN, F. (ed.) 1975: *Water resources of the world—selected statistics.*

MILLER, D. H. 1977: *Water at the surface of the earth. An introduction to ecosystem hydrodynamics.*

PROBSTEIN, R. F. 1973: Desalination.

SMITH, K. 1972: *Water in Britain.*

SPIEGLER, K. S. and BERGMAN, J. I. 1974: *Optimal expansion of a water resources system.*

UTTON, A. E. and TECLAFF, L. (eds.) 1978: *Water in a developing world. The management of a critical resource.*

UNITED NATIONS 1977: *Water supply and management.*

Special issue of *Ambio* 6 (1), 1977.

Special issue of *Ekistics* 43 (254), 1977.

6

Forestry

About one-third (4,028 million ha) of the world's land surface is covered with forests (whose continental distribution is outlined in Table 6.1), which we may intuitively define as ecosystems dominated by trees, reflecting the status of some of those organisms as being among the biggest living things in the world: both species of redwood are examples, and the bristlecone pines of California are certainly the oldest living organisms in the world at about 4,600 years. The size of trees subjects all the other elements of the forest system to a hegemony which is reflected in the organization of the layering of the forest plants: the canopy of the dominant trees may be penetrated by only a limited amount of light to be absorbed by shrubs; patches of light on the forest floor allow the growth of shade-intolerant elements of the ground flora. In deciduous forests a temporal element may be noticed in the herbs that flower, shed seed, and die down before the trees come into leaf.

The leaves of the trees themselves are the fundamental element in the system since they are the site of photosynthesis. Growing in Switzerland, spruce trees (*Picea abies*) need 2,300 kg (fresh weight) of leaves to produce 1 m^3 of fresh wood, and the average dry weight of leaves in measured forests is 7,000 kg/ha (Ovington 1962). These organs are also important in the food webs of woodlands, since the stem of the tree contains few nutrients and is physically intractable to many animals as food. The nutrient-rich leaves and twigs hence form the first level of most of the food chains of the forest, both predatory and saprophytic: defoliating insects thus have an important role to play in natural forests.

It follows that the biota of the forest floor are very important factors in the dynamics

TABLE 6.1 Forest lands (000 ha)

Europe	140,000
USSR	910,009
North and Central America	815,000
South America	908,000
Asia	458,000
China	76,600
Africa	639,000
Oceania	81,000
World	4,028,000

Source: FAO *Production Yearbook 25*, 1971. See also Fig. 6.1.

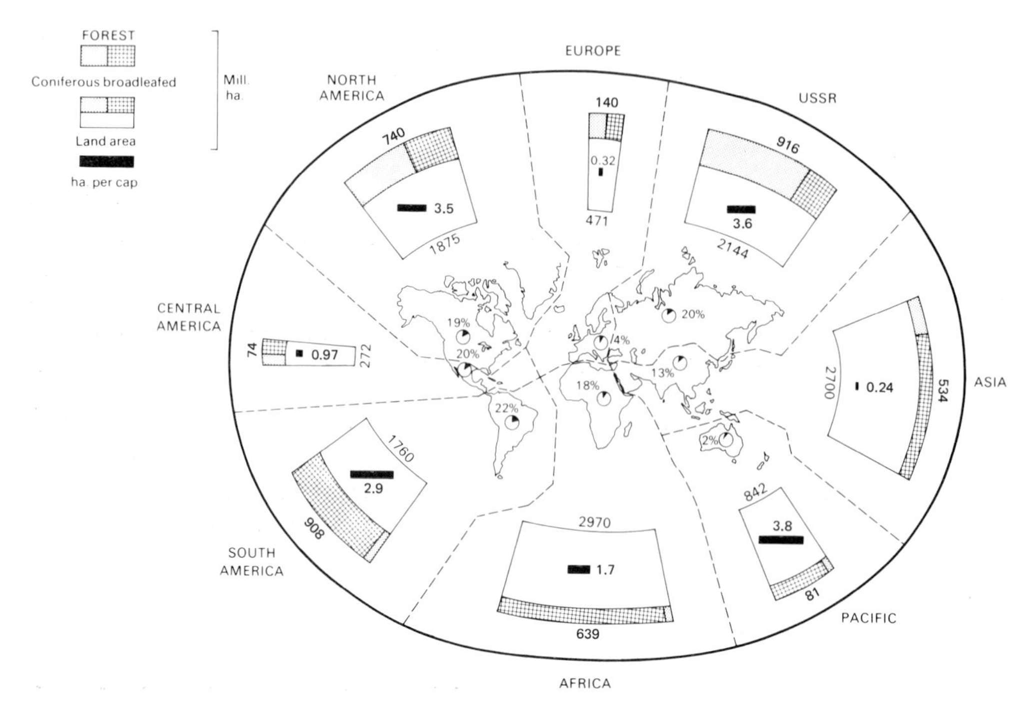

Fig. 6.1 A diagram of forest area and distribution in the world. The divided circle located in each major region indicates the proportion of the land area devoted to forest. The outer segments show (reading outwards) the total land area (million ha), the number of hectares of forest per person, the division of forest between conifers and broadleaved types, and the total forest area. The area of the segment itself is proportional to total land area, subdivided into forest and non-forest sectors. The predominance of the USSR and North America in terms of the coniferous trees so much in demand for paper and paper products will be noted. The dependence of Europe on outside resources in this field (as in so many others) can easily be inferred.
Source: FAO 1963, updated to 1976

of forests, for they exist mainly among the leaf and twig litter and are responsible for its mineralization and re-use by the tree and for providing some of the energy pathways through the ecosystem. In temperate woodlands, for example, they decompose 3,000–4,000 kg/ha of autumnal leaf fall by the next spring (Ovington 1965), and in evergreen tropical forests their activity keeps the ground surface virtually free of dead organic matter. Larger animals, too, may have vital roles in the ecology of the forest: the populations of small mammals, which feed on for example acorns and beech mast, will affect recruitment to the next cohort of young trees and so the populations of the predators of these animals such as owls can also be critical. An overpopulation of deer can damage and disfigure a whole generation of young trees by browsing off the leading shoots, while a population explosion of insects can strip a forest of its leaves and alter the ecology of a woodland for many years.

A forest is one of the most complex of natural ecosystems, and in man's use of natural forests and his attempts to imitate or improve on them these intricacies must be borne in mind, since manipulation of the ecosystem for resource purposes can easily bring about deleterious effects which were not foreseen.

Energy flow and nutrient circulation in forests

Because forests are so interesting and because the accumulation of organic matter in the tree stems forms the most important forest resource, more studies on woodland energetics have been done than on many other ecosystems. Forests are reputed to be very efficient users of incident solar energy because of the stratification of photosynthetic surfaces and the high density of chlorophyll-containing tissue. The maximum thermal efficiency of photosynthesis here, as elsewhere, is about 14 per cent, however, and of the net radiation received, a tree expends 60 per cent in transpiration, and the actual processes of photosynthesis account for only 1·5 per cent (Ovington 1962). The balance between photosynthesis and respiration determines the accretion of organic material in the tree, and this depends not only on the season but on the age of the tree. The rate of accumulation of organic matter is greatest at the dense pole stage of young forest, becoming slower in mature woodlands.

TABLE 6.2 Partition of primary production in two forests t/ha/yr.

	Gross production	*Tree respiration*	*NPP*	*Litter*	*Increase in biomass*
Mature tropical rain forest (Ivory Coast)	52·5	39·1	13·3	4·4	9·0
Immature beech woodland (Denmark)	23·5	10·0	13	3·9	9·6

Source: Varley 1974

Table 6.2 contrasts the energy partition of mature tropical forest and a young temperate beechwood: in spite of the higher biomass of the tropical forest, it is accumulating less matter than the young woodland. The range of biomass and NPP achieved by forests is shown in Table 6.3 which also brings together some of the structural and taxonomic characteristics of the world's major forest types. It can be seen that the tropical forests are among the most productive systems of the world, and even the least productive forest type is more productive than many other ecosystems such as the upwelling zones of the oceans, and average cultivated land (Whittaker and Likens 1975).

Because much of the energy of forests is taken up by the metabolism of the trees and the remainder either locked up as trunk material or shed to the litter layer, the animal production is much less conspicuous. Indeed, it might be said that much of the animal life depends largely upon the leaves of the trees, either *in situ* or as litter, though exceptions to this are not difficult to find, e.g. seed-eaters. An example of the energy

TABLE 6.3 Some characteristics of the world's main forest types

Forest type	*Regime*	*Structure*[1]	*Typical dominants*	*Spp/ha*	*Tree biomass (kg/m²) and [NPP] (g/m²/yr)*	*Typical fauna*	*Soil type*	*Uses*
Moist tropical forest (e.g. Congo; Amazon basins)	Evergreen broad-leaved	3TL 0SL 0GL climbers, epiphytes	Mahogany Ironwood *Morea*	200	45 [2,200]	Stratified: Canopy: flying squirrels, monkeys Trunk: martens, baboon Ground: tapir, ant-eater, deer	Variable. Nutrient-poor	Selected hardwoods, e.g. mahogany; shifting cultivation; clearance for agriculture; simplification, e.g. rubber
Subtropical broad-leaved forests	Deciduous broad-leaved	2TL 1SL 1GL	Teak Pyinkado	n.a.	35 [1,600]	Canopy: monkeys, civets Ground: tiger, elephant	Variable. Nutrient-poor	Dominants, e.g. teak, have commercial value; shifting cultivation; clearance for agriculture; watershed protection
Temperate zone coniferous forests	Coniferous needle-leaved	1TL 1SL 1GL	Douglas fir Redwood	1–5	35 [1,300]	Black bear	Leached brown-earth	Dominants have commercial value; clearance for agriculture; watershed protection; recreation; grazing
Temperate zone deciduous forests	Deciduous broad-leaved	1TL 1SL 1GL	Oak Beech Lime Chestnut	20–40	30 [1,200]	Deer; small rodents; owls; insect-eating birds	Brown-earth; leached brown-earth	Dominants have commercial value; clearance for agriculture; watershed protection; recreation; simplification for plantation, e.g. oak, beech; grazing of cattle, pigs
Boreal forest	Coniferous needle-leaved	1TL 1GL	Spruce Firs Pines	1–5	20 [800]	Moose; wolf; woodland caribou; reindeer	Podzol	Dominants have commercial value; recreation

[1] TL = tree layer; SL = shrub layer; GL = ground layer.
Productivity and biomass data from Whittaker and Likens 1975

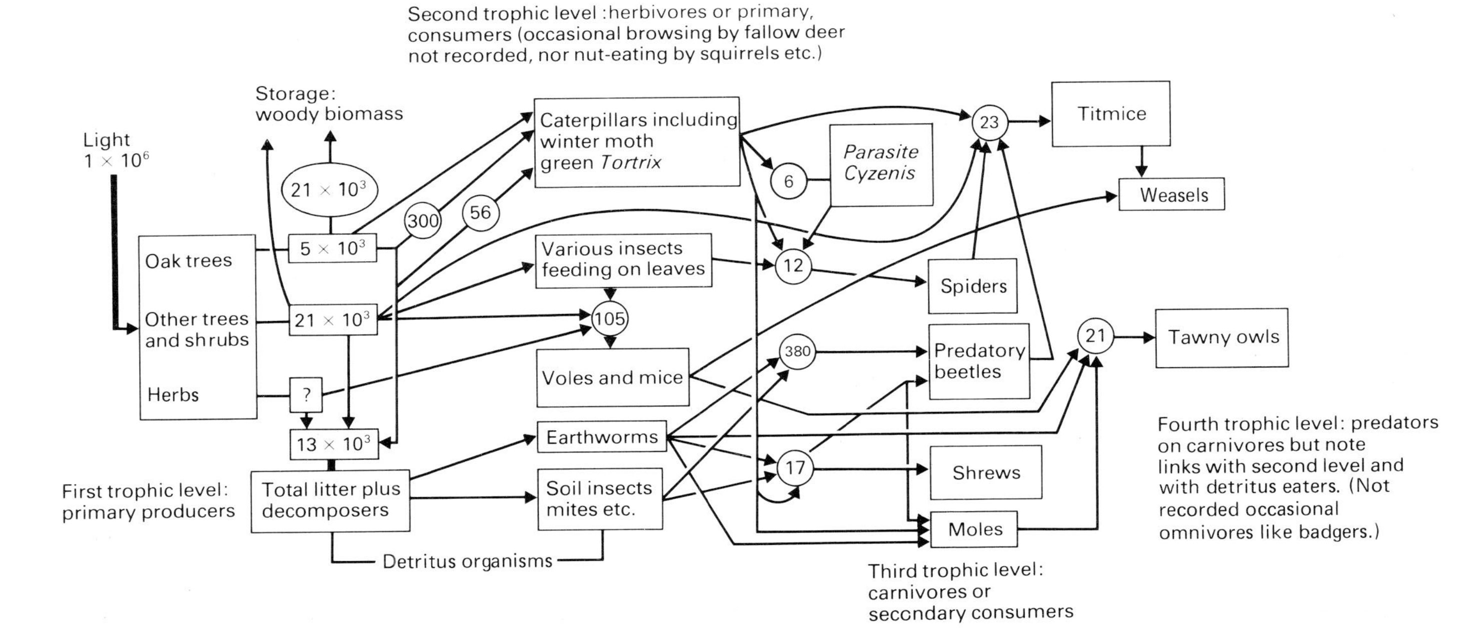

Fig. 6.2 Levels of biomass and energy flow in a woodland in lowland England. Figures in KJ/m^2 for biomass (rectangles) and $KJ/m^2/yr$ for transfer (circles).
Source: Learmonth and Simmons, 1977. After Varley, 1974

flow through a deciduous forest in England is given in Fig. 6.2. Note that here the equivalent of the NPP of the shrubs and trees other than the dominant oaks is going into storage as wood and that not very much less becomes litter. The quantity available for herbivores is not very large but note that some herbivores are omitted. The animal biomass is enhanced by the end-products of food chains leading from the litter layer, as with soil insects—beetles—titmice—weasels. But such a diagram goes some way towards showing why large mammals are not very densely found, even in natural forests.

The processes of the biota cannot be considered separately from those of the soil, since there is a reciprocal relationship. The importance of soil organisms is cor-

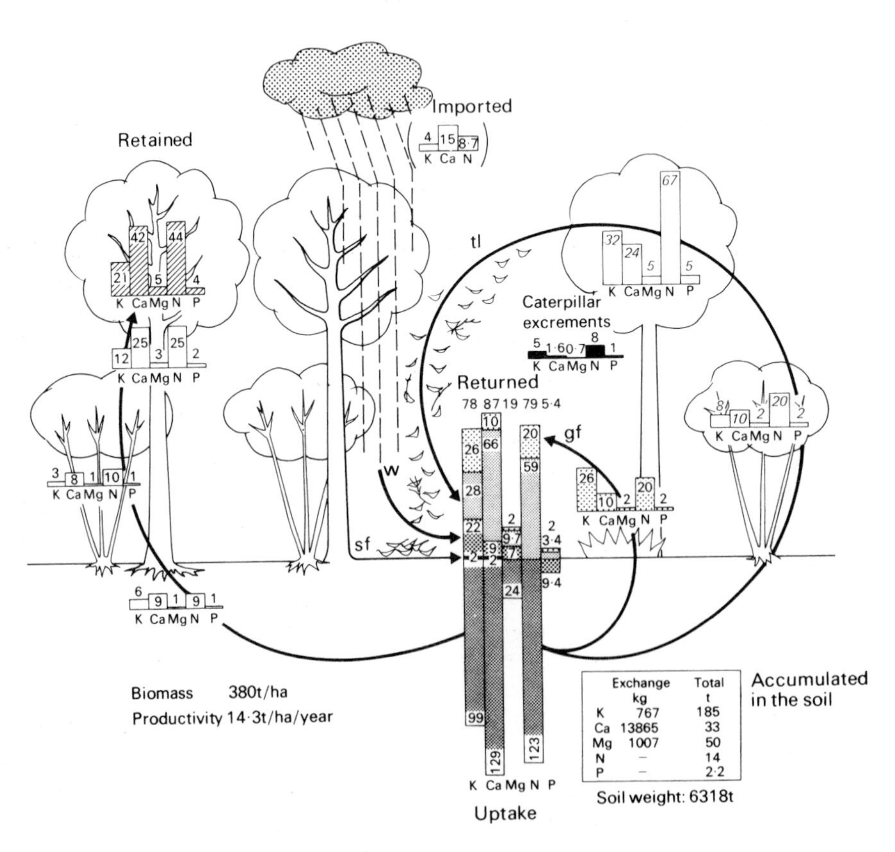

Fig. 6.3 The annual cycling of selected minerals (in kg/ha) in a Belgian forest of oak and ash with an understorey of hazel and hornbeam. The retention of the elements is in the annual wood and bark increment of both roots and above-ground portions of the trees and the 1-year-old twigs. The pathways of return are tree litter (tl on the diagram), ground flora (gf), washing and leaching of canopy (w) and stem flow (sf). Imported flow is from rainfall. The italic figures on the right of the diagram are for crown leaves at full growth in July, at which time defoliating caterpillars are an important part of the return flow. Uptake is equal to the sum of the retained and returned portions of the circulation.
Source: Duvignéaud and Denaeyer-De Smet 1970

roborated by the fact that the overwhelming energy use in the soil is in the breakdown of organic matter. Nevertheless, the soil humus is an important store of energy: in Devon, England, the annual leaf fall was calculated at 4,000 kg/ha (Ovington 1962), and in very wet climates the depth of organic matter in needleleaf forests (such as those of the west coast of North America) may be so great as to form a peat. Weathering processes in soil rank second in their consumption of energy, followed by the hydrothermic cycle and the transport of substances in the soil profile.

As with energy studies, quantitative investigations of the circulation paths of essential elements, such as calcium, potassium and nitrogen, have been undertaken in many different types of forest. The basic flow consists of inputs from rain and subsoil weathering, circulation within the forest-soil-animals system, losses by way of crops of plants or animals, and runoff in both dissolved and particulate forms. The exact quantities in each pathway vary greatly according to the species involved and environmental factors like precipitation. In wet climates, for example, there is often a high input of certain cations from rainfall, but rapid runoff removes much of this quite quickly.

By contrast, the cycle within the plant system is usually large in relation to both input and output; some idea of the nature and magnitude of flows in a deciduous forest in Europe is given in Fig. 6.3. Different rates of processing mean that essential elements need not be distributed evenly in both parts of the forest-soil system. In tropical forests, for example, the amount of calcium in plant shoots is 20 times greater than in the roots, and phosphorus 10 times greater. In temperate forests the order of difference is usually 2–3 times, and in a strand of Corsican pine in Scotland the nitrogen content of the trees was estimated to be only 0·4 per cent of the total nitrogen content of the ecosystem (Ford 1971). So the tropical forest is regarded as having evolved a circulation system which reduces nutrient loss from leaching and runoff by keeping most of the minerals locked up in the vegetation. The total nutrient budget is high in such places: tropical forests have been estimated to accumulate 2,000–53,000 kg/ha of minerals, whereas dry savannas reach only 1,000 kg/ha (Bakuzis 1969).

The long-term stability of the forest ecosystem depends upon a successful balance between output and input of nutrients. If there are high losses these need to be replaced, but it is more common to find that they are small in undisturbed forest. A small area of New Hampshire with an average slope of 26 per cent, which included some slopes of 70 per cent, lost on average only 14 at/km^2/yr of nutrients. Clear felling accelerated the nutrient drain by factors of $\times 3$ to $\times 20$ for various cations, and in the first year after felling an amount equivalent to a whole year's turnover of nitrogen was lost (Bormann *et al.* 1968). Within the forest itself the cycling of nutrients involves uptake by the plants, storage within the organisms, and return to the soil via dead organic matter. The quantities of nutrients at various stages are important in resource processes because they determine the quantities of minerals removed if the tree is harvested and taken elsewhere. Table 6.4 shows some comparative data for different European species, from which it can be seen that storage is the smallest of these stages and hence cropping of the stem, which in turn has the lowest proportion of nutrients, involves the least of these elements.

TABLE 6.4 Annual cycling of nutrients in forest stands (kg/ha)

Process	*Species*	*Nitrogen*	*Phosphorus*	*Potassium*	*Calcium*
Uptake	Pine	45	5	7	29
	Beech	50	13	15	96
	Oak	87	6·5	79	95·1
Return	Pine	35	4	5	19
	Beech	40	10	10	82
	Oak	55·6	3·1	58·6	82·8
Storage	Pine	10	1	2	10
	Beech	10	3	5	14
	Oak	27	3	13·9	0·8
Removed by thinning	Oak	5·1	0·4	6·5	10·5

Source: Bakuzis 1969

The nutrient flow of the forest may be summarized by emphasizing that under natural conditions it is cyclic and loss is nearly always in balance with gain. If forests are to be used as crops, then an ecologically sound resource process must ensure that this cycle is perpetuated; artificial forests must either establish their own cycles or be aided to do so by forest management. In all cases, the soil fauna and flora are a little-noticed but vital part of the mechanism of nutrient flow in the ecosystem: this is exemplified by the findings of Stark (1972) on the role of soil fungi as nutrient sinks for different kinds of forest (see p. 13).

Uses of forests and their products

Man's demand upon forests can be divided into two categories, of which the first comprises the direct uses of the wood itself, gathered almost entirely from the stems of the trees. The second category consists of indirect uses such as water yield and animal products which have come from the forested area and in whose ecosystems the trees have played an important part. The cropped forests of the world have an annual growth potential of 4,500 million t, of which 1,626 million t are estimated to be harvested (Weck and Wiebecke 1961). In Europe the deciduous trees appear to have the most efficient foliage, for 880 kg of fresh foliage will produce 1 m^3 of solid merchantable wood; even-aged spruce stands require 3,000 kg for the same production and Scots pine 1,200 kg (Bakuzis 1969). Table 6.5 shows the area and growth potential for the various forest types of the world. The dominance of the lowland equatorial forests is apparent, as is the subordinate status of the rest: possible exceptions are the summergreen (i.e. temperate deciduous) forests and mountain conifers, but an order of $\times 2$ separates them from the tropical rain forests. This growth potential is not equivalent to harvesting potential since species diversity and accessibility are very important too. The latter is self-explanatory; with regard to the former it must be remembered that the number of

TABLE 6.5 Estimated forest area and estimated growth potential[1] for different formation classes of the forests of the world

Formation class	*Estimated area* *Million ha*	*%*	*Estimated growth* *t/ha yr*	*Total million t/yr*	*%*
Equatorial rain forest, lower range	440	18	3·5	1,540	35
Equatorial rain forest, mountain range	48	2	3·0	144	3
Monsoon forests and humid savanna	263	11	1·8	474	11
Dry savanna and dry mountain forests in tropics	530	21	1·0	530	12
Temperate rain forests and laurel; precipitation below 1,000 mm	20	1	7·2	143	3
Sclerophyllous forests	177·5	7	1·0	178	4
Summergreen forests and mountain conifers	393	16	2·2	865	19·5
Boreal conifers	605·5	24	0·9	556	12·5
Total	2,477			4,430	100

[1] In tons of dry matter production per hectare and per year.
Source: Bakuzis 1969

species per unit area in tropical forests is so much higher than in the boreal conifer forests that selective harvesting is costly. By contrast, the immense stands of one or two species of pine and spruce in subarctic Eurasia and North America provide a uniform product much more evenly distributed.

Among the high number of direct uses, the use of wood for fuel has probably been the most significant until relatively recently and as Table 6.6 shows its proportion of the total wood harvest is only just under half. It remains an enormous consumer of forest and one in which there is little selectivity, for almost the whole tree can be used. It can be removed from its growth site, so any return of mineral nutrients to the soil is precluded, and although some trees are better fuel wood than others, in places of scarcity any tree or bush will be utilized (Openshaw 1974, Eckholm 1975). In many countries the wood is converted to charcoal before being sold, and even in a DC like Japan a large proportion of domestic heating in rural areas still comes from the charcoal-burning *hibachi*. Table 6.6 also states the magnitude of recent increase in total wood production and Table 6.7 the regional components of this production.

Of the uses of industrial wood, lumber is the most important, especially in the construction of housing, and the furniture industry is also a major market for both soft and hard woods. Other industrial uses include cooperage (a declining use due in part to

TABLE 6.6 Total wood and fuelwood production ($m^3 \times 10^3$) for the world

	1961–65 av.	*1976*
World roundwood (i.e. all wood) production	2,121,119	2,524,219
Fuelwood and charcoal	1,051,750	1,184,090
(and as % of roundwood)	(49·5%)	(46·9%)

Source: FAO *Yearbook of Forest Products 1976*. (Rome, 1978)

TABLE 6.7 Total wood production 1976 ($m^3 \times 10^3$) by regions

World	2,524,219
Africa	329,456
N & C America	517,750
S America	237,294
Asia	719,275
Europe	305,818
Oceania	30,092
USSR	384,534
Developed countries	772,884
Developing countries	1,050,625
Centrally planned	690,710

Source: FAO *Yearbook of Forest Products 1976*. (Rome, 1978)

the regrettable rise of beer sold in glass or metal kegs), veneer and plywood which consume about 10 times less wood than lumber but are a very large industry on a world scale, mine timber, railway sleepers or ties, posts, poles, fencing and various minor products such as cork, waxes, nuts, resins and bark. The uses of converted wood are dominated by the practice of converting wood to pulp in order to make paper. In 1976 a total of 112,151,000 t air-dry weight of pulp was produced. By far the largest proportion (94,897,000 t) came from the market economies of the West and Japan. The rank order of the major producers of pulp in 1976 was USA, Canada, Japan, Sweden, USSR and Finland. Most of it is consumed in the Western-type economies and converted eventually to ash and CO_2 and recycling has been little practised outside a few environment-conscious cities in Europe and North America. The profligacy of use is illustrated by the story of the little old lady in Idaho who came across a recently logged forest. As she walked among the debris of the former woodland, seeing the great stumps of decapitated trees, she wept. She cried so hard that she used up a whole box of Kleenex tissues.

Among the other uses of converted wood, the various kinds of particle board and building board have a very high growth rate. Not only is waste from lumber used up in this way but trees which have a relatively low market value as lumber can be used too, providing industry for places that might otherwise be economically remote. The whole question of waste utilization is important: sawing lumber produces large quantities of sawdust, and up to 35 per cent of total harvested timber may in fact be waste which at present is often used as a fuel at sawmills but which may be converted into a saleable product in future. The leaves of the trees are traditionally not used. They contain much protein, however, and experiments have been conducted into the feasibility of harvesting them for food, discussed in Chapter 7. The roots are never used and may either be left *in situ* or gathered up. The trunk of the tree is thus the source of nearly all the forest products which are harvested directly.

As the major consumer of wood, the West's tastes and technologies obviously affect the ecology of the forests which they use. Some changes in demand may eventually be reflected in the types of forest which resource managers try to produce. Modern housing, especially apartments, uses less sawn lumber than individual houses, so that the trend of the 1960s, when half the new housing in Europe (except the UK) was apartments, is towards a lower demand for sawn wood. Similarly, particle boards are absorbing much of the market for sawn wood in the housing and furniture fields. Mining timber output is falling, but the same kinds of wood are being absorbed into the enormous increase in pulp production. These trends have their repercussions in the move to the planting or manipulation of forests to bring about pure even-aged stands of fast-growing trees such as poplars, certain conifers, teak and eucalypts. Such forests are of course monocultures and, while possessing the economic cheapness associated with uniformity, are also subject to the ecological instabilities and aesthetic inacceptabilities which accompany a lack of diversity.

Crop ecology

The use of the forest for wood inevitably removes energy-rich material from the site on which it has grown. However, this loss to the local ecosystem is probably not of great consequence provided regeneration eventually takes place. In nature the removed tree trunks would eventually provide energy for decomposer organisms whose populations are probably linked to the available organic detritus and so fluctuate within wide limits. Thus the food webs based on the flora and fauna of the litter horizons must be deprived of energy by the practice of forestry. It has already been noted that forest trees utilize only 1–3 per cent of total incident energy for the production of matter; at harvesting a lot of this is left on site as litter, seeds and roots. The proportion of incident energy which is cropped is thus very small, and so forests are certainly not efficient ways of gathering winter fuel. Industrial users of wood are not of course interested in its energy content; but of the total assimilated matter of the forest, only about 32 per cent appears as usable wood even under very good conditions, since respiration accounts for 45 per cent, litter 16 per cent, roots 3 per cent, seeds 1 per cent, and 3 per cent is lost in logging and transport (Polster 1961). This last figure is much higher in places with a tradition of inexpert logging and in the Soviet Union has been placed at 33 per cent, including 50 million m^3/yr wasted at the cutting areas (Pryde 1972). So the final crop of wood from a mature forest (a fast-growing one is accumulating energy as organic matter more rapidly) is of the order of 0·1 per cent of the incident solar radiation (Plate 11).

Mineral nutrients in the natural forest are cycled, so cropping in this case may remove elements in short supply from the plant–soil system. Losses through harvest are highly variable, according to Duvignéaud and Denaeyer-De Smet (1970), and vary with the type of tree: 100-year-old stands of some European species had different amounts of certain nutrients in their trunks (Table 6.8). These data show broadly that a good proportion of the nutrients in the forest are removed by stem harvest but equally

Plate 11 Modern forestry has strong ties to industrial energy sources as seen in the heavy equipment portrayed in this photograph from the USA. Although most of the nutrient-rich parts of the tree are being left on site, the bark is being removed. The traditional suitability of winter for forest operations is confirmed here, as is the commercial suitability of relatively uniform stands of coniferous trees. *(Grant Heilman, Lititz, Pa)*

TABLE 6.8 Nutrients in 100-year-old forest stands

Kg/ha in trunks and roots, *and total for stand*						
	Calcium		*Potassium*		*Phosphorus*	
Deciduous hardwoods	257	*1,283*	121	*320*	20	*70*
Conifers other than pines	129	*676*	102	*375*	10	*70*
Pines	84	*283*	45	*138*	8	*30*

Source: Rennie 1955

that the leaving on-site of much of the crown provides a source of organic matter to be recycled. What is not shown by figures is whether any particularly scarce element in the ecosystem is removed which cannot quickly be replaced by subsoil weathering or precipitative input. There is at present no evidence to suggest that this is so, but the base status of the local rainfall and soils must inevitably be critical.

Forestry operations may involve the elimination of designated trees (selective logging) or the removal of whole stands (clear-cutting), although the so-called Montana definition of the former is, 'You select a forest and then you log it.' Clear-felling has been unequivocally shown to accelerate silt and nutrient loss into the streams of the forest. In a New Hampshire deciduous forest nutrient loss was accelerated by several methods (Bormann et al 1969). Transpiration was reduced so the amount of water passing directly through the ecosystem was increased, root surfaces able to remove nutrients from the leaching waters were reduced, and in some cases more rapid mineralization of added organic matter increased the loss. (Microbiological processes in the soil may also lead to an increase of dissolved nitrate in waters and eventually exceed levels of 10 ppm in the runoff, thus causing algal blooms, i.e. bringing about eutrophication.)

In many forests, however, the immediate losses of precipitative input in rainwater and the possibilities of luxury consumption by trees suggest that harvesting may not in the long term be deleterious to the ecosystem providing that the input in precipitation can be intercepted and stored in the soil, and that nutrient-rich but unwanted parts of the tree are left to mineralize on site; Likens and Bormann (1971) show that, for example, 36 per cent of the total calcium in the deciduous forest they investigated is incorporated in the stem bark of the trees. Forest practices which are inimical to such processes need to be eliminated, and in any case monitoring of the nutrient flows of all exploited forest is essential if sustained yield is to be maintained. This is especially so where the practice of the total utilization of tree-crops grown on short rotations (whole-tree harvesting or fibre farming) is being introduced, though exact predictions of its effects are difficult to make (Kimmins 1977). Existing forests may be very strongly manipulated in order to increase their yield of desired benefits, especially in the case of timber production where man-directed inputs of energy and matter may be of an almost agricultural intensity. Chief among these is the application of chemical technology against insect parasites and damaging fungi. Beginning with tar washes and Bordeaux mixture earlier this century, the advent of the aircraft and the organic pesticide has escalated warfare against creatures such as the spruce budworm to a very high degree. Resistance by the insects, however, has made the problem a chronic one usually met by more spraying with consequent effects upon ecosystems (and possibly humans as well) from the pesticide residues.

Fire control has also been ruthlessly practised on all forms of forest despite the fact that in many coniferous forests fires appear to be a normal part of the ecology. The cessation of fire due to a policy of suppression may produce two major effects in fire-adapted ecosystems. The first is to bring about a shift in species composition. This may favour trees less attractive from a commercial point of view, or, to take an example from the field of nature protection, the remaining trees of the fire-tolerant *Sequoia gigantea* more or less ceased to regenerate because their saplings were shaded out by competitors which were not fire-resistant and which would have been destroyed if fires had taken their normal course. A second consequence may be the piling up of humus and litter to great depths on the forest floor. When such a thickness is ignited the temperatures are very high, and 'jumping' to produce a crown fire which is the most destructive form of forest fires is likely. More regular burning of a thin layer of humus and litter does not lead to such damage. One result of these discoveries has been the careful introduction of prescribed burning into some sophisticated forest management schemes; such a practice may well be much cheaper than fire suppression. Even spatially minor forest exploitation works may have effects on the forests. Roads for logging are one example: if badly sited they act as channels for water and eventually as initiators of gulley systems in steep terrain. Together with careless logging and fire, enough silt may be shed into streams to render them aesthetically unattractive, kill the fish populations (especially salmonids), and contribute markedly to the silting-up of impoundments.

The net effect of manipulation of natural and semi-natural forests for timber yield is generally to simplify them by reducing undergrowth, eliminating 'weed' species and controlling herbivorous animals who feed directly upon the trees. The ecological analogy of an intensively managed forest is an agricultural crop like grass, and indeed cropping such a forest is very much like grazing with domesticated animals.

On the other hand, the new forests which are the result of afforestation of previously unforested land are analogous to crop agriculture. The ground is carefully prepared by ploughing and perhaps by applying chemical fertilizer; the trees are planted as seedlings from nurseries, weeded and carefully protected from grazing and browsing animals, sprayed with insecticides and generally pampered. The reasons for establishment of such forests are generally either production of cheap softwood or soil conservation, or both. Countries like Denmark and Britain which down the centuries had become denuded of forests undertook extensive planting of new forests, generally of native or imported (usually North American) conifers. In Britain the intention was principally to build up a strategic reserve of timber after the 1914–18 war, a reason only officially abandoned in 1972. In New Zealand about half the indigenous forests of the islands were removed in the century 1850–1950 (12·1 million ha to 5·7 million ha) and the establishment of large forests of exotic species, mainly pines, was undertaken after 1896 with the result that, by 1950, 360,450 ha of exotic forest had been planted, half by the State (New Zealand Forest Service 1970).

New forests are usually highly productive because the species have been chosen for their suitability for the prevailing climate, their ability to grow fast and their conformity to market demands. Some doubts about their long-term efficacy as wood producers still remain. One of these is the worry that the establishment of monocultures (necessary for the reduction of short-term costs) may make the forests prone to particular types of ecological instability as well as reducing genetic diversity (Kemp and Burley 1978). For example, small environmental shifts may allow new pests to spread rapidly and become rife. Their uniformity also means a constant fuel supply for an established fire. It is also feared that the short cropping cycle of 30–40 years may take away too many nutrients for production to be sustained, especially in cool temperate latitudes where podzolic soils are part of the coniferous forest ecosystem. In such cases chemical fertilization will become a standard procedure, showing these forests to be indeed a form of intensive agriculture.

The rapid growth of new forests, coupled with the high biological productivity of some natural and semi-natural woodlands, and with the knowledge of the high calorific value of wood (oven dry wood is 4·7; charcoal 7·1, cf coal 6·9, fuel oil 9·8), has led to the suggestions that an important future use of forests is as energy plantations. Earl (1975) calculates that each year a potential increment of forest of magnitude $7{,}700 \times 10^6$ tce is added photosynthetically, which is close to the world's total energy usage in 1970 of $7{,}435 \times 10^6$ tce. Thus there is here a renewable source of energy which would be especially valuable in LDCs as fuel (both as wood and especially as charcoal), to a large extent releasing them from dependence upon imported fossil fuels. Though particularly relevant to LDCs, work in the USA has suggested that 'energy plantations' there could easily substitute for imported oil and perhaps also for atomic power stations: 810 km^2 of forest in the Western States might supply 27 per cent of US energy consumption in 1974 (Livingstone and McNeill 1975, Burwell 1978).

Other uses of forests

The structure of the forest, and the relatively little damage that can be done to it (compared with field crops for example), have encouraged resource managers to think of multiple use as a normal aim of woodland management. Some of the possible combinations are set out in Table 6.9, which also attempts to assess the compatibility of the various components. Water is probably the most important accessory use, since undisturbed forest catchments have low silt and mineral yields and hence give water which requires little treatment. These considerations have to be balanced against the enhanced losses from transpiration, and the balance of benefits and costs probably depends upon the other used of the forests. Domesticated animals may be grazed in forests provided that the numbers are kept to a level which does not inhibit regeneration and that areas planted with young trees are fenced.

The use of the forest for recreation is one of the fastest-growing demands made upon the resource. A notable example of this is the Netherlands, where the State Forest

TABLE 6.9 Interrelationships among forest uses and attributes

Land use/ Environmental characteristic	*Scenic landscape*	*Recreation opportunity*	*Wilderness*	*Wildlife*	*Natural watershed*	*Wood production and harvest*
Scenic landscape		Moderately compatible; may limit intensity of use	Not inimical to wilderness but does not insure	Compatible to most wildlife	Fully compatible	Limited compatibility; often affects amount and methods of harvest
Recreation opportunity	Moderately compatible unless use intensity excessive		Incompatible; would destroy wilderness character	Incompatible for some kinds; others can tolerate	Moderately compatible; depends on intensity of recreation use	Limited compatibility depends on harvest timing and intensity
Wilderness	Fully compatible	Completely incompatible, can't tolerate heavy use		Highly compatible to much wildlife, less so to others	Fully compatible	Completely incompatible, precludes all harvest
Wildlife	Generally compatible	Limited compatibility; use intensity must be limited	Mostly compatible though some wildlife require vegetative manipulation		Generally fully compatible	Generally limits volume or conditions of harvest
Natural watershed	Fully compatible	Moderate compatibility; may require limitation on intensity	Not inimical to wilderness but does not insure	Generally compatible		Moderate compatibility; restricts harvest methods but does not prevent timber harvest
Wood production and harvest	Compatible if harvest methods strictly controlled	Moderately compatible	Completely incompatible; would destroy wilderness	Compatible if harvest methods fully controlled	Compatible if harvest methods fully controlled	

Service now manages about 75 per cent of the area of its larger forests primarily for recreation, nature protection and landscape enhancement (Staatsbosbeheer 1966). Small blocks of forest have also been planted in the Netherlands purely for landscape and recreation purposes in such places as the shores of newly embanked land in the Delta project and on the edges of agricultural land in the new polders of IJsselmeer. More commonly recreation is one element of multipe-use schemes. If it is to be compatible with the other functions, it seems essential that mass recreation should be managed as part of a mosaic of different uses of the forest rather than as one of several uses of the same stand of trees. Recreation affects forest ecology in many ways, the most noticeable of which are the increased number of fires and decreased incidence of regeneration. In spite of such difficulties one study indicated that a forest area in California could increase its recreation capacity 10 times and suffer a loss of timber production of only 13 per cent (Amidon and Gould 1962). If such a conclusion were to be true of most forests, the outlook for compatibility of these two resource processes is very good.

Established forests act as a reservoir of wildlife, both plant and animal. Although the animal biomass is not very high in relation to that of the plants, some of it is very visible in the shape of birds and of mammals such as deer. Natural and semi-natural forests with plenty of glades and 'edge' habitats are particularly valuable from this point of view. Forestry operations are rarely deleterious in the long term to such animal populations, unless very large areas are clear-felled. The exceptions are the rare cases where an animal cannot tolerate humans anywhere in its vicinity: the California condor now confined to the Los Padres National Forest in California is an instance, and special status has been given to the forest land around its remaining eyries. Reciprocally, wildlife populations such as deer must usually be managed if they are not to endanger tree growth. Browse-lines are a common sight in forests where there is insufficient predation, and in such cases controlled culling is essential.

Forest policies

The importance of forests and their products is such that many countries have adopted national forest policies, and even in the LDCs one of the better legacies of the colonialist era has often been a technically competent forest service. However, it is the DCs which have on the whole evolved the more detailed national schemes.

In the USA, for example, the Forest Service of the Department of Agriculture operates principally under the Multiple Use–Sustained Yield Act of 1960. A mosaic of uses is maintained, with accessibility by motor car the determinant of facility development. The Forest Service operates only on land belonging to the Federal Government: 75·7 million ha, out of a total of 186·3 million ha, is commercial forest area. The national resource process is thus dominated by private landowners, whether in the shape of giant timber corporations or woodlots attached to farms. Both state and Federal governments influence private owners however, through taxation structures and forest practice legislation. In such a large country, with complex patterns of ownership and fragmented levels of legislation, an overall picture is impossible except to say that in general the large forest landowners are very conscious of the importance of their management programmes, and that in this they are influenced by an articulate and concerned public which expects the Forest Service to set the standards for the nation (Clawson 1975, Stankey 1976).

In New Zealand, the national forest policy concentrates on the long-term management of both indigenous and exotic forests for timber production, recreation and water yield (New Zealand Forest Service 1970). In addition, however, the Forest Service is responsible for control programmes exercised on 'noxious animals'. The principal offenders are deer, whose large populations eradicate mountain vegetation and induce serious soil erosion. The Forest Service is also concerned with the reclamation of sandy areas, and co-operates with the National Parks Authority to preserve examples of natural forest ecosystems and scenic reserves. One sanctuary of 5,112·5 ha contains the rare kauri, and kiwi and blue-wattled crows are similarly protected by the Service. Countries with low population densities and large areas of forest need devote little attention to development for recreation. An example is Finland, where policy is to develop timber production and at present allow the recreation to look after itself. In the subarctic part of Finland, for example, there are 55 million m^3 of timber with an annual growth of 870,000 m^3. This growth is quite slow and regeneration uncertain in the northern part of Finland, so that investigations of climatic fluctuations, the encouragement of regeneration and the selection of suitable strains all become critical, especially as timber is the major export of the nation. Seeding using selected strains may become important even in wild terrain like Lapland (Mikola 1970). A different example is the Sudan where, apart from in the south, forests do not grow easily, where the overwhelming demand upon trees is for fuel, and where government policies must be directed at securing an orderly flow of trees for this use at a reasonable price. In 1976 the consumption of firewood and charcoal was 20925×10^3 m^3 out of a total production of 22371×10^3 m^3. Continued pressure on the woodlands and savannas of the Sudan is causing soil and wind erosion, sandstorms and the encroachment of the desert (Faris 1966). A major product is gum arabic, of which the Sudan is an important exporter. Fortunately, gum trees (*Acacia senegal*) are regarded as 'garden' trees and are mostly grown in rotation with crops, so they share the protection afforded to agricultural crops.

As a last example we may discuss the USSR, an industrial nation with strong and centralized control of resource processes. The forest lands belong collectively to a State Forest Reserve of 1,238 million ha, of which 738·2 million ha are actually tree-clad and 171·8 million ha are burned and unreforested logged areas. There are three categories of forest: group I (5·6 per cent of the total) which enjoys maximum protection and preservation and is often largely for amenity and shelter purposes; group II (7 per cent) which consists of forests covering important watersheds and woodlands in lightly forested areas of European Russia; group III (87·4 per cent) are the main productive forests subject to intensive timber harvest, and are mostly in northern European Russia, the Urals, Siberia and the Far East. The forests are controlled by the State Forestry Committee, and out of a resource of 76×10^9 m^3 an annual cut of 350–400 $\times 10^6$ m^3 is taken. According to Pryde (1972) there is a lack of effective sustained-yield harvesting practices, and overcutting is noticeable: a cut of 150–200 per cent of the annual growth is not uncommon in group III forests. In 1960 it was said that 10 million ha which has been cut in the previous decade had no regrowth. Forest land is also lost to fire (1 million ha/yr), insects, and reservoirs (a total of 18 million ha), as in many other countries. In spite of a centralized bureaucracy, the control exerted upon the field managers appears to be insufficient to inculcate modern forest management methods, especially in the face of what regionally must be perceived to be an inexhaustible resource.

A new task on the horizon for forest policy makers is the consideration of the energy linkages of modern forestry. The adoption of heavy machinery of various kinds coupled with rises in fuel prices has given impetus to studies of the energy balance of modern forestry. In 1972 industrial energy costs were 10 per cent of forest sales income in Sweden, where a solar input of 178×10^6 million kWh into the forest area was subsidized with 4,486 million kWh of industrial energy. Clearly, this is a very small proportion but energy accounting is likely to assume more importance in nations with a heavy fuel import bill (Nillson 1976).

Effects of increasing human populations

Population pressure on the world's forests is perhaps best seen from three viewpoints: those of the LDCs (especially in the humid tropics), the DCs, and the globe as a whole. First of these is the LDCs, where extensive disforestation results from the clearance of forest for fuel and for agriculture, and where it is subject to attrition from grazing animals. A principal result is soil loss and hence silt-laden waterways. Not only do reservoirs fill up rapidly, but irrigation and navigation systems are choked; some floods in Asia are 0·5 m high in rice paddies and overwhelm the short-stemmed high yield varieties, thus negating some of the effects of the Green Revolution (p. 187). The loss of fuel wood often means the use of dung for fuel, and the use of one tonne of dung for fuel rather than fertilizer could mean the loss of 50 kg of food grain which on a world scale could amount to 20 million t/yr. These losses are exacerbated by LDCs which offer their forest reserves to DC companies for commercial forestry: Indonesia for example has least nearly all of its moist tropical forest to such companies; Brazil's forests are being converted to grassland for cattle ranching, so that the tropical rain forests (the most productive terrestrial biome) are in danger of virtual disappearance and a great deal of concern is being expressed (Routley and Routley 1977, Eden 1978, Myers 1978, Stott 1978, Grainger 1980).

The relationships of the DCs to their forests at the national scale are rather different. Fuel needs are at present catered for mostly by fossil hydrocarbon and hydro-electric power, so that the forest becomes a supplier of timber, pulp and a desirable environment. In many places secondary forest becomes a desirable habitat for suburban houses: New England is a prime case. Virgin forest is even more desirable for second houses in the mountains or by lakes. Thus not only the trees but the whole forest ecosystem enters the economic-social-political realm and conflicts over priorities of use and manipulation assume large dimensions, especially since publicly owned forest lands become subject to pressures from all the citizens who consider they have a right to say how they should be used; perhaps a form of community-based forestry may help with this socio-political problem. Approximately 70 per cent of the world's forests are currently in public ownership (FAO 1963), although the extraction of their resources may frequently be in the hands, legal or otherwise, of individuals and companies. The pressures on the forests of DCs, therefore, will come mainly from the conflicts between recreational use of forests and the short-cycle uniform softwoods needed for pulping. If spatial zoning can be achieved, so much the better; but the size of management units is crucial, particularly where some of the nations of Europe are concerned. There seems to be little that technology can do in the way of substitutes for wood: for construction and other timber uses, another material might be feasible, but for paper and for forest recreation there appear to be none likely, though of course it is quite practicable to recycle most products made from wood pulp.

On a global scale, forests are responsible for about half the photosynthesis of the world, which means a much higher proportion of the terrestrial photosynthesis. If, as seems very likely, the balance of gases in the atmosphere is a biological artefact then the role of the forests in absorbing CO_2 and producing O_2 is very important. Just how the increase of CO_2 in the atmosphere (see pp. 282–3) and the decrease of forest area affect this balance is not known, but it seems likely that forests have an important place in the carbon and oxygen cycle of the planet. No suggestion is made here that a breakdown is imminent, but careful research into the global importance of forests is an obvious necessity. There seems to be no shortage of practical reasons for at least maintaining or at best enhancing the forest area of the planet; both at national and at global scale, W. H. Auden's line is apposite:

A culture is no better than its woods.

Further reading

BAKUZIS, E. G. 1969: Forestry viewed in an ecosystem perspective.
BROWN, C. L. 1976: Forests as energy sources in the year 2000.
CLAWSON, M. 1975: *Forests for whom and for what?*
COUSENS, J. 1974 *An introduction to woodland ecology.*
DUVIGNEAUD, P. (ed.) 1971: *Productivity of forest ecosystems.*
EARL, D. E. 1975: *Forest energy and economic development.*
GRAINGER, A. 1980: The state of the world's tropical forests.
GRAYSON, A. J. (ed.) 1976: *Evaluation of the contribution of forestry to economic development.*
HOLMES, G. D. *et al.* (eds.) 1975: A discussion on forests and forestry in Britain.
JORGENSEN, J. R., WELLS, C. G. and METZ, L. J. 1975: The nutrient cycle in continuous forest production.
MEGGERS, B. J. *et al.* (eds.) 1973: *Tropical forest ecosystems in Africa and South America: a comparative review.*
VAN DER MEIDEN, H. A. 1974: Forests and raw material shortage.
PESSON, P. (ed.) 1974: *Ecologie forestière.*
PRYDE, P. R. 1972: *Conservation in the Soviet Union.*
REICHLE, D. (ed.) 1970: *Analysis of temperate forest ecosystems.*
RICHARDS, P. 1952: *The tropical rain forest.*
WESTOBY, J. C. 1963: The role of forest industries in the attack on economic underdevelopment.
WHITMORE, T. C. 1975: *Tropical rain forests of the Far East.*
WINDHORST, H.-W. 1978: *Geografie der Wald- und Forstwirtschaft.*

7

Food and agriculture

Humans need food as a source of energy and for tissue replacement, like any other animal. Unlike them, our intake can be divided into metabolic food, necessary for the maintenance of the organism, and cultural food where preferences, taboos and excesses are manifested. The nutritional requirements of *Homo sapiens* vary according to size, age, and the kinds of activities which each individual undertakes: Table 7.1 shows the range of energy requirements for various tasks in an industrial society. If the individual is to function successfully an adequate energy intake must be complemented with sufficient protein to ensure the continuous replacement of tissues and, at certain times, to enable growth to take place. Some vitamins and mineral salts also appear to be essential, and water too is an indispensable part of the human diet. Table 7.2 sets out some recommended daily allowances for various ages and conditions. Such figures must

TABLE 7.1 Energy requirements (kcal/hr)

Writing	20	Cycling	180–600
Dressing	33	Coal mining	320
Ironing	60	Sawing wood	420
Walking	130–240	Walking upstairs	1,000
Polishing	175	Running quickly	1,240

Source: Pyke 1970a

TABLE 7.2 Daily dietary allowances

	Energy-yielding foods (kcal)	*Protein (g)*	*Others*
Man, 25 yrs	2,900	65	
Woman, 25 yrs	2,300	55	Calcium, iron
Woman, pregnant	2,700	80	Vitamins A and D, Thiamine, riboflavin, niacin, ascorbic acid
Children, 4–9	kg × 110	kg × 3·5	
Boy, 16–20	3,800	100	
Girl, 16–20	2,400	75	

Source: Pyke 1970a

be treated with caution, since it appears that some people are quite healthy even when receiving much lower amounts, especially of energy-yielding foods: nevertheless the figures are indicative of the level of intake considered necessary by Western nutritionists.

Cultural food is the translation of our metabolic requirements into such foods as are available and, for areas where food is plentiful, into choice between different kinds of food even to the point of the onset of diseases of obesity rather than dietary deficiency. Our species also ingests various organic substances derived from natural or man-made ecosystems but which are not strictly food, such as medicinal and social drugs.

Sources of food

The source much of the food consumed by man is terrestrial agriculture. This represents the most manipulated of all the non-urban ecosystems, in which the energy and matter pathways are directed almost entirely to man and where he maintains a high

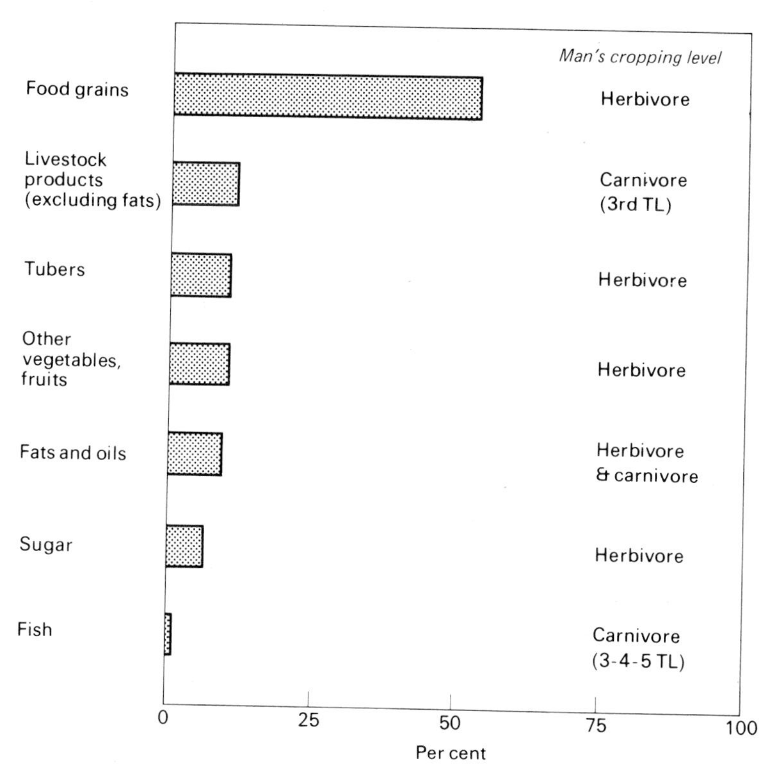

Fig. 7.1 Man's cropping level for food: his harvest is dominated by grains since wheat and rice each supply a fifth of his food energy, but offtake from other trophic levels is also common, including very high carnivore levels in the case of some fish. Yields are low compared with crops at the herbivore level. (TL = trophic level)
Source: L. R. Brown 1970

level of input of matter and energy to keep the system stable in order to yield his preferred crop. Not only is the ecosystem man-made, but the plant and animal components of it have usually been genetically altered by man in the course of their domestication. There are two main types of agriculture: crop agriculture, in which the plant production is harvested for use by man either directly or after processing; and animal agriculture, where a crop from a highly manipulated ecosystem is fed to domesticated animals. (Herein lies the difference from grazing systems, where the forage is more or less wild vegetation.) Ecologically, terrestrial agriculture presents man either as a herbivore or as a third trophic level carnivore, and Fig. 7.1 shows the energy harvest from different ecosystem levels.

Considering the importance of agriculture, it appears surprising at first sight how little of the surface of the continents is cultivated (Table 7.3). Closer inspection reveals the relatively small areas of land which can be subject to the manipulation required by agriculture and still remain as stable systems; furthermore the tolerances of domesticates are sometimes rather narrow, although increasingly accurate breeding methods may aim at widening them. The proportion of land under cultivation, which includes rotation grassland and fallow, is not an infallible guide to the production of agricultural crops since multiple cropping may be possible in tropical and subtropical latitudes.

Terrestrial agriculture is not our only source of food: grazing, the oceans (Chapter 8) and the so-called 'unconventional' foods (pp. 190–91) are all important, but the relative proportions vary enormously with locality. Together with terrestrial agriculture these sources also supply 'industrial crops', consisting of non-food material such as pyrethrum, sisal, ornamental flowers and pearls. This section is concerned entirely with food, and will concentrate largely upon terrestrial agriculture and 'unconventional foods'.

TABLE 7.3 Agricultural and grazing lands, 1975

	Thousand ha				
	Continental area incl inland water	*Arable land and permanent crops*	%	*Permanent pastures & meadows*	%
World	13,075,336	1,506,149	11·5	3,046,404	23·3
Africa	3,031,168	210,890	6·95	798,105	26·3
N & C America	2,246,465	292,148	13·0	320,725	14·2
S America	1,782,980	101,524	5·7	446,822	15·0
Asia	2,753,206	478,764	17·3	552,466	20·0
Europe	487,032	142,696	29·2	87,138	17·9
Oceania	850,934	47,920	5·6	469,148	55·1
USSR	2,240,220	232,207	10·3	372,000	16·6

Note: 'permanent crops' includes tree crops such as rubber, citrus and cocoa.
Source: FAO *Production Yearbook* **30**, 1976

The ecology of contemporary agriculture

Basic types

Agricultural systems today exhibit a major division into shifting and sedentary types. In terms of environmental impact (Harris 1978) total manipulation of the natural system is practised over a limited area but for only a short (1–5 years) period of time. Thus the agricultural path is spatially and temporally enclosed by wild vegetation. Sedentary agriculture, on the other hand, aims at a permanent replacement of the natural systems by the man-made ones. Partial reversions may occur in the shape of fallow periods, but a pioneer stage of recolonization is usually all that is achieved especially if domesticated beasts are allowed to use the fallow land for grazing. Grazing systems represent a modification of wild land vegetation rather than its replacement.

Shifting cultivation

Shifting agriculture today is largely confined to tropical forests, savannas and grasslands, although its demise in temperate zones is not particularly ancient. Its ecology has been conceptualized by Geertz (1963), who visualizes it as a miniature imitation of the closed plant community which it temporarily replaces, and this is especially so of the forest clearings. The crops are planted in a mosaic of different heights and times of fruition so that the plant cover of the soil remains as complete as possible throughout the year in order to reduce the leaching effect of heavy rainfall. The importance of mineral nutrients is emphasized by the burning which follows the clearing of the natural vegetation, for this mineralizes organic matter and allows its uptake by the crops. In virgin tropical forests most of the mineral nutrients at any one time are in the vegetation, and so 'slash and burn' provides a method of translocating some of these elements to the soil. The natural diversity of the forests is imitated by the variety of crops which are grown by some shifting cultivators. Conklin (1954) describes the Hanunoo of the Philippines as recognizing 430 cultivates, of which 150 specific crop types were found in the first year of a slash-and-burn plot. This cornucopian productivity is not maintained everywhere that shifting cultivation is practised: in the highlands of New Guinea, for example, there is a considerable dependence upon the sweet potato (*Ipomea batatas*), to the point where it usually provides 77 per cent of the calorie intake and 41 per cent of the total protein. Where a mixture of crops is grown then the food supply is abundant and varied, as Conklin (1957) says,

> over the first two years, a new swidden produces a steady stream of harvestable food ... from a meter below to two meters above the ground level. And many other vegetable, spice and non-food crops are grown simultaneously.

The major disadvantage of such a system seems to be the inability to store surpluses, even when partly processed to paste and flour form. The grains are more stable but highly vulnerable to insects, bacteria and scavenging rodents. Buffering against a poor season is made very difficult by the lack of technology and even worse for groups who have come to depend largely upon one crop. In normal years, calorie intake is sufficient but there may be protein deficiency, especially in newly weaned children (2–3 years) and in adolescents: the groundnut was introduced in order to help combat this problem, especially since animal protein (pig) usually forms only 3 per cent of their protein ingestion (Conklin 1957). Plots are normally abandoned when nutrient drain reduces fertility or when competition from weeds reaches an unacceptable level (Cassidy and Pahalad 1953, Walters 1960). Breakdown of this agricultural system appears to occur when plots are re-cultivated too soon or when the system is extended into less humid areas where trees cannot re-establish themselves and a grassland establishes itself as the fallow vegetation. In both cases the mineral nutrient cycles never build up to their former levels and fertility is lower when the plot is cultivated again. The breakdown manifests itself in either or all of three ways: in malnutrition of the people, in emigration from the district, and in ecosystem disintegration particularly in the form of soil erosion. The general conclusion is irrestible: that shifting cultivation is an ecologically well-adapted system in forested lands where the trees regenerate easily when the plots are deserted, and where an equilibrium population has been established. Given a rapidly expanding population, the system cannot cope with the more intensive crop production needed and will either break down ecologically and socially or undergo transformation into a basically sedentary system.

Grazing

This type of husbandry produces meat, fat, milk, blood and various other delicacies, together with hides, wool, and other minor items. They are taken by culling domesticated and semi-domesticated animals which have foraged off vegetation where there has been no deliberate return of nutrients to the soil. The animals themselves contribute nutrients by way of excreta and the occasional carcass but most of the ecosystem's nutrients that reach the body of the animal are removed from the site of intake; only in the very richest of countries is there any attempt to use chemical fertilizers. The traditional form of this system has been pastoralism of the nomadic and transhumant kinds; these are still carried on but have often been modified by at least partial sedentarization. This pactice has often increased the environmental impact of grazing by concentrating animals for long periods whereas under traditional systems the need for pasture and water made constant movement necessary and only a light pressure was exerted on the vegetation. Even so, as Fraser Darling (1956) observed, there have been shifts in vegetation caused by years of pastoralism, to the point where man achieved ecological dominance through the medium of domesticated animals, often accompanied by fire. The total area of 'permanent pasture' is given in Table 7.3; Lewis (1969) estimates that 28 per cent of the world's land surface is covered with forest which is used for grazing for at least part of the year, and 47 per cent is suitable only for grazing by domestic stock or by wild animals, and has no potential for more intensive use. As would be expected from ecosystem theory, very little of the energy trapped by the plants shows up as animal tissue, especially in the case of domestic stock which are very

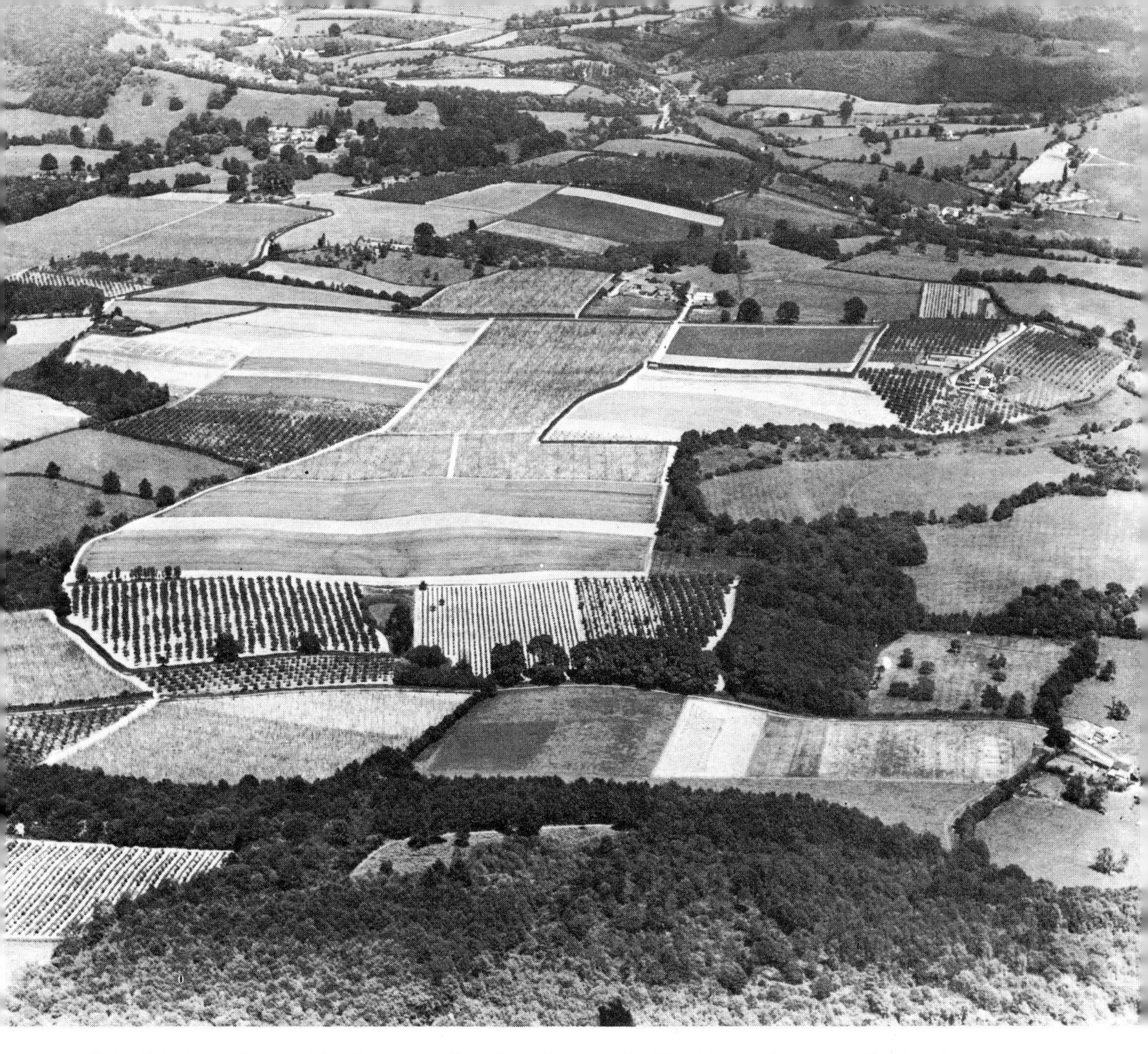

Plate 12 An agricultural landscape of high diversity near Cradley, Herefordshire, England. Fields, woods, orchards and hedgerows provide not only aesthetic pleasure but an ecological buffering system as reservoirs of predators upon the pests of man's crops. *(Aerofilms Ltd, London)*

selective in their consumption of range vegetation, (Macfadyen 1964). The efficiency of conversion of solar energy into domesticated animals may thus be of the order of 0·005 per cent on wild vegetation and 0·01 per cent on more intensively managed grasslands. However this carries the corollary that relatively small amounts of nutrients are lost from the ecosystem in the form of cropped animals, e.g. 0·08 $g/m^2/yr$ of phosphorus as sheep, compared with a soil pool of 302 g/m^2 on mountain grasslands in North Wales (Perkins 1978).

Taxonomically, the world's stock is dominated by cattle and sheep, with pigs, goats and buffaloes next in rank, followed by horses, asses, mules and camels. The small spectrum of taxa domesticated has led to examinations of the possibility of using other species and numerous experiments have been tried: red deer, musk ox, various

TABLE 7.4 A classification of farming systems

	Tree crops		*Tillage with or without livestock*	
	Temperate	*Tropical*	*Temperate*	*Tropical*
Very extensive Examples	Cork collection from Maquis in southern France **2**	Collection from wild trees, e.g. shea butter **1**	—	—
Extensive Examples	Self-sown or planted blue-berries in the north-east USA **2**	Self-sown oil palms in west Africa **2**	Cereal growing in Interior Plains of North America, pampas of South America, in un-irrigated areas, e.g. Syria **4**	Unirrigated cereals in central Sudan **4**
Semi-intensive Examples	Cider-apple orchards in UK; some vineyards in France **4**	Cocoa in west Africa; coffee in Brazil **4**	Dry cereal farming in Israel or Texas, USA **4**	Continuous cropping in con-gested areas of Africa; rice in south-east Asia **4**
Intensive Examples	Citrus in California or Israel **4**	Rubber in south-east Asia; tea in India and Ceylon **4**	Corn belt of USA; continu-ous barley-grow-ing in UK **4**	Rice and vege-table-growing in south China; sugarcane plan-tations through-out tropics **4**
Typical food chains (see p. 171)	A	A	A, B	A

Ecosystem type: **1** Wild, **2** Semi-natural, **3** Man-directed, temporary, **4** Man-directed, permanent.
Based on Duckham and Masefield 1970

antelopes, the warthog and African buffalo have all been considered as potential domesticates in various environments. Because of the ease with which overgrazing can occur, leading to ecosystem breakdown and erosion, it is often regarded as an unstable resource process but conversely if properly managed it can convert otherwise inedible cellulose to food and other useful materials, and the direct costs of harvesting the plants are low, even in difficult terrain (Spedding 1975).

Alternating tillage with grass, bush or forest		*Grassland or grazing of land consistently in 'indigenous' or man-made pasture*	
Temperate	*Tropical*	*Temperate*	*Tropical*
Shifting cultivation in Negev Desert, Israel **3**	Shifting cultivation in Zambia **3**	Reindeer herding in Lapland; nomadic pastoralism in Afghanistan **1**	Camel-herding in Arabia and Somalia **1**
	Shifting cultivation in the more arid parts of Africa **3**	Wool-growing in Australia; hill sheep in UK (sheep in Ireland); cattle ranching in USA **2**	Nomadic cattle-herding in east and west Africa; llamas in South America **1**
Cotton or tobacco with livestock in the south-east USA; wheat with leys and sheep in Australia **4**	Shifting cultivation in much of tropical Africa **3**	Upland sheep country in North Island, New Zealand **2**	Cattle and buffaloes in mixed farming in India and Africa **4**
Irrigated rice and grass beef farms in Australia; much of the eastern and southern UK, the Netherlands, northern France, Denmark, southern Sweden **4**	Experiment stations and scattered settlement schemes **4**	Parts of the Netherlands, New Zealand and England **4**	Dairying in Kenya and Zimbabwe highlands **4**
A, B, C, D	A (C)	C (D)	C

Sedentary cultivation

Sedentary agriculture represents the permanent manipulation of an ecosystem: the natural biota are removed and replaced with domesticated plants and animals (Plate 12, page 167). Competition by the remnants of the original biota or by man-introduced organisms may still remain, and considerable effort may be needed to keep these weeds and pests at an acceptable level. In dryland agriculture the soil assumes an importance which it did not have in shifting systems, for it now becomes the long-term reservoir of all nutrients and is constantly depleted as crops are harvested and removed. The nutrients must be replenished either by the addition of organic excreta or chemical fertilizers. The former have the additional advantage that they usually help maintain the crumb structure of the soil as well as adding the elements necessary for plant growth.

In contrast there is the important paddy-culture of rice, where the soil is very largely a mechanical rooting medium for the plants and the water supplies the essential mineral nutrients; it often contains blue-green algae which fix nitrogen, for example. Essentially this is an aquarium system with the boundaries made of earth instead of glass; it works particularly well when the catchment areas of the streams which feed the paddies drain from nutrient-rich rocks or soils. Thus paddy rice can be grown on a substratum which is very poor in essential elements and productivity can be maintained for long periods of time, since the cycle of cultivation practices ensures the replenishment of the mineral nutrients; the flooding provides a habitat for many kinds of fish, prawns and invertebrates which add to the local diet (Geertz 1963, Hecht 1979). The variety of crops grown under the various forms of sedentary cultivation is very high, and changing patterns of agriculture together with shifting trade flows and altered rates of consumption make a world kaleidoscope of infinite variety (Laut 1968). Table 7.4 reproduces one classification of farming systems, together with the degree of ecosystem manipulation they represent and the economic context (subsistence or commercial) in which they occur.

In more quantative terms in spite of an overall diversity of crops a very few of them dominate the agricultural production: the three major cereals comprise seven-ninths of the world grain crop and the three major oil-seeds likewise provide seven-ninths of the world output of this group. Secondly, meat is a very much scarcer product because it comes mostly from domestic herbivores which are inefficient users of energy, whether cropped from pastoralism or from more intensive agricultural systems although they have the advantage that the amino-acid balance is nutritionally better than many plants, and that animals are more adaptable to climate. Comparison with shifting agriculture of the numbers of people supported by sedentary agriculture is scarcely possible since so much trade is carried on. One of the most intensive of all, wet-rice paddy, appears however to be capable of supporting 2,000 persons/km^2 under subsistence conditions in favourable areas like Java. Odum and Odum (1972) calculate that the requirements for an American diet need 1·5 ac/cap, which would mean a density of 166/km^2.

Agriculture as food chains

General

A simple way of viewing agriculture ecologically is to use the model of a food chain with man as the end member. Four of these are delineated by Duckham and Masefield

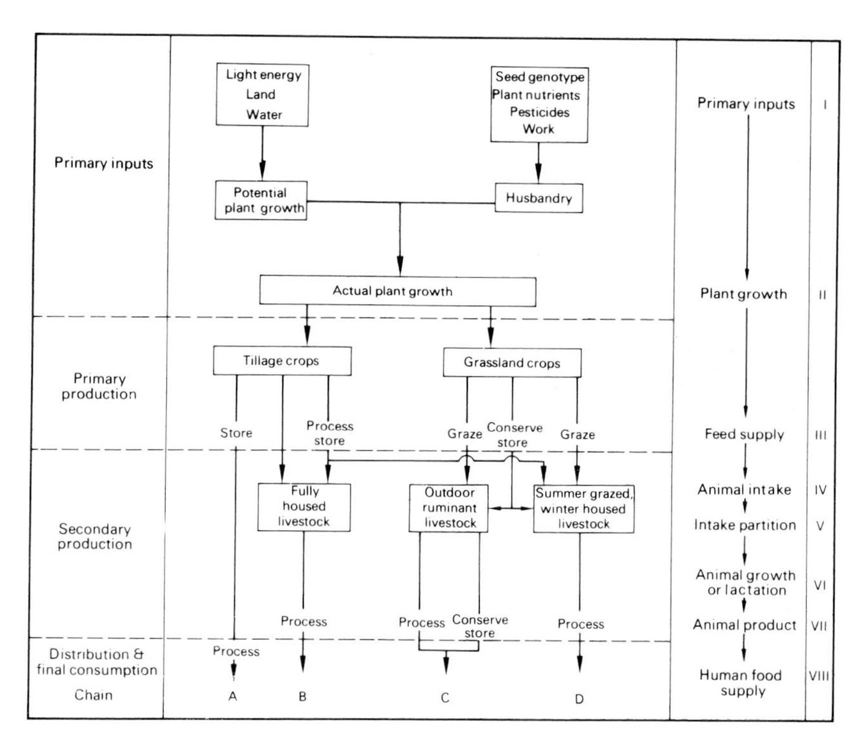

Fig. 7.2 The four food chains which characterize agriculture, with the basic stages set out on the right of the diagram. Where chain C derives its plant growth from grasslands not intensively managed then it falls into the class of the grazing system (Chapter 5). In the case of chain A, the absence of a box at the secondary production stage means that direct cropping by man at the herbivore level is practised.
Source: Duckham and Masefield 1970

(1970), which are chain A: tillage crops—man; chain B: tillage crops—livestock—man; chain C: the grazing system, which may be either intensive or extensive; and chain D: tillage crops and grassland—livestock—man. Fig. 7.2 shows the inputs of energy and matter for each of these chains and the alternative pathways through which the energy-rich organic matter comes as food to man. Examples of chain A are cereals, potatoes and sugarbeet; of chain B, bacon pigs and barley-fed beef; of chain C either a nomadic pastoralist group or an intensive beef herd on carefully managed pasture; and of chain D, a mixed farm with dairying as the main enterprise.

The economics and research input of agriculture have made it possible to compare the energetics and output of each of these chains, and some information is shown in Table 7.5. The general notions about trophic structure discussed in Part I are borne out here: man as herbivore in chain A has access to far higher quantities of energy and plant protein than when he acts as a carnivore in the other chains. In these, milk is the most 'efficient' product in terms of both energy and protein, and meat is obviously a great

TABLE 7.5 Relative outputs of food chains in UK

Chain	*mcal of food per 100,000 mcal solar radiation*	*mcal of food per ha*	*Edible protein kg/ha*
A (Tillage crop—man)	200–250	1,863–2,632	42
B (Tillage crop—livestock—man)	15–30	178–340	10–15
C (Intensive grassland—livestock—man)			
Meat	5–25	81	4
Milk	50–80	911	46
D (Grassland and crops—livestock—man)			
Milk	30–50	324	17

Source: Duckham and Masefield 1970 (1 mcal = 1,000 kcal)

waster of energy and a relatively poor source of gross protein, although the special nutritional qualities of animal protein make it particularly valuable. The table also hints at the basic problem of agriculture, which is that none of these chains is very efficient at energy conversion, and Table 7.6 shows some food outputs as a proportion of energy input from solar radiation; it is evident that even the herbivore chain is losing a potential energy yield, as would be expected, by respiration loss and by photosynthesizing for only part of the year. The harvest from indoor livestock and intensive milk production comes form the close attention paid to the feeding of pigs and their penning so as to avoid energy loss by movement, a feature also of intensive cattle-rearing, as is the avoidance of animal crop loss by disease.

Since man is acting as a herbivore, chain A should be the most efficient at producing

TABLE 7.6 Energy production as a percentage of annual solar radiation

Chain	*Production*	*Percentage of solar radiation*
A	Rice: Egypt	0·17
	Cereals: UK	0·16
	Potatoes and sugarbeet: UK	0·21
B	Pig meat	0·03
C	Fat lambs	0·01
	Summer milk on experimental farm	0·08–0·15
D	Summer and winter milk	0·05

Source: Duckham and Masefield 1970

food. The gross energy disposal of a potato crop in the UK, excluding disease and wastage, is as follows (Duckham and Masefield 1970):

Total organic dry matter formed per acre	24,750 mcal	100%
Respiration loss	9,000 mcal	37%
Unharvested vegetation	3,750 mcal	15%
Post-harvest loss	2,250 mcal	9%
Household waste	2,250 mcal	9%
Net human food	7,500 mcal	30%
(1 mcal = 1,000 kcal)		

The 30 per cent of the photosynthate which becomes available as human food represents about 0·22 per cent of the total solar energy received. With such losses it is scarcely surprising that agricultural development strategists for areas of nutritional stress concentrate if possible on multiple cropping as the sure way to increase energy uptake. The potato crop referred to above was used in the UK, and therefore a lot of cultural waste occurs which might be obviated in poorer countries: peeling the tubers is quite obviously wasteful.

Turning to chain C, a much lower efficiency is observed, as would be expected from the trophic structure of ecosystems. This example, also from Duckham and Masefield (1970), is an intensive grass crop grazed for beef production:

Total organic dry matter formed per acre	28,000 mcal	100%
Respiration loss	9,000 mcal	34%
Unharvested roots and stubble	3,000 mcal	11%
Uneaten grazing	4,000 mcal	14%
Faecal and urine loss	4,000 mcal	14%
Animal metabolism	5,000 mcal	17%
Tissue conversion loss	1,000 mcal	4%
Slaughter and household waste	500 mcal	2%
Net human food (1 mcal = 1,000 kcal)	1,500 mcal	4%

So this chain, cropped at carnivore level, yields 4 per cent of total photosynthate as food, which represents about 0·02 per cent of the solar energy received. Chains B and D show comparable efficiencies: pig meat from chain B may represent about 0·03 per cent of total solar radiation; milk from chain D, 0·05 per cent. All these are considerably more efficient than range ecosystems, where yields of 0·005 per cent of solar radiation as cattle are probably normal.

In attempts to increase the yields of food, inputs of energy from human and fossil sources and from abiotic substances such as fertilizers and pesticides are almost universal: Fig. 7.3 enables us to compare a simple farm with one in an industrialized country. The Ugandan farm has relatively few links outside its area, whereas the UK example is tied firmly to the rest of its economy by energy and matter flows from outside its boundaries. One result of the extra energy derived from fossil fuels ('energy subsidy' or 'support energy') is a greater intensification of production: whereas one family took 3·2 ha to feed itself in the Ugandan system, 8 workers on 178 ha of the

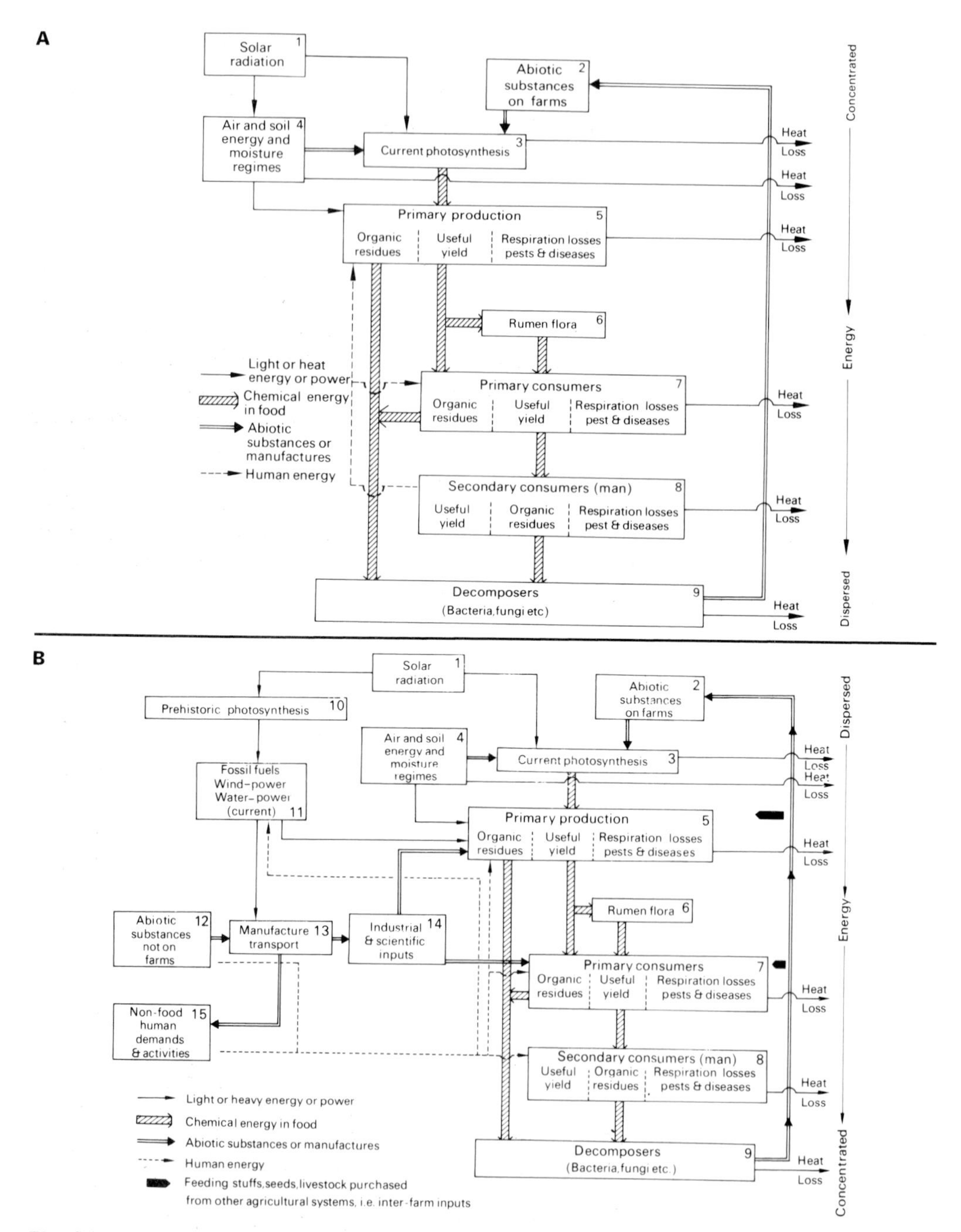

Fig. 7.3 Qualitative flows of energy and matter in a contrasting pair of farm systems. A is a simple livestock system in Uganda; B a livestock farm in the UK (equivalent boxes have the same numbers in them). Note the greater number of linkages in the industrial society example, and especially the connections to prehistoric photosynthesis, i.e. the use of fossil fuels. (In B inputs from other farms are represented by the two broad arrows on the right.)
Source: Duckham and Masefield 1970

industrialized farm produced enough food for 220 families, including a high proportion of animal products. Clearly the role of support energy is so substantial that it deserves further discussion.

Energy, agriculture and food consumption

The simplest relationship between energy and food is found on a subsistence farm where the energy input is limited to the solar source, which may be used immediately as photosynthate or in a stored form as wood. This latter is normally used as cooking fuel (often 1·0–1·5 t/cap/yr in LDCs) with an efficiency of conversion to heat of about 5 per cent (Leach 1975). In contrast, DC farms have access to support energy in a variety of forms. On the farm itself, the use of machinery is the outstanding example, including stationary motors for such purposes as grain drying or the ventilation of animal houses. A nation like Israel which has a lot of irrigated land uses about 35 per cent of its imported (pre-1967) boundaries) fossil fuel support energy in irrigation (Stanhill 1974). There are then 'upstream' uses such as the manufacture and transport of the machinery and of inputs like fertilizers, biocides and animal foodstuffs. 'Downstream' (i.e., beyond the farmgate), the output may be transported, processed, stored, distributed and cooked

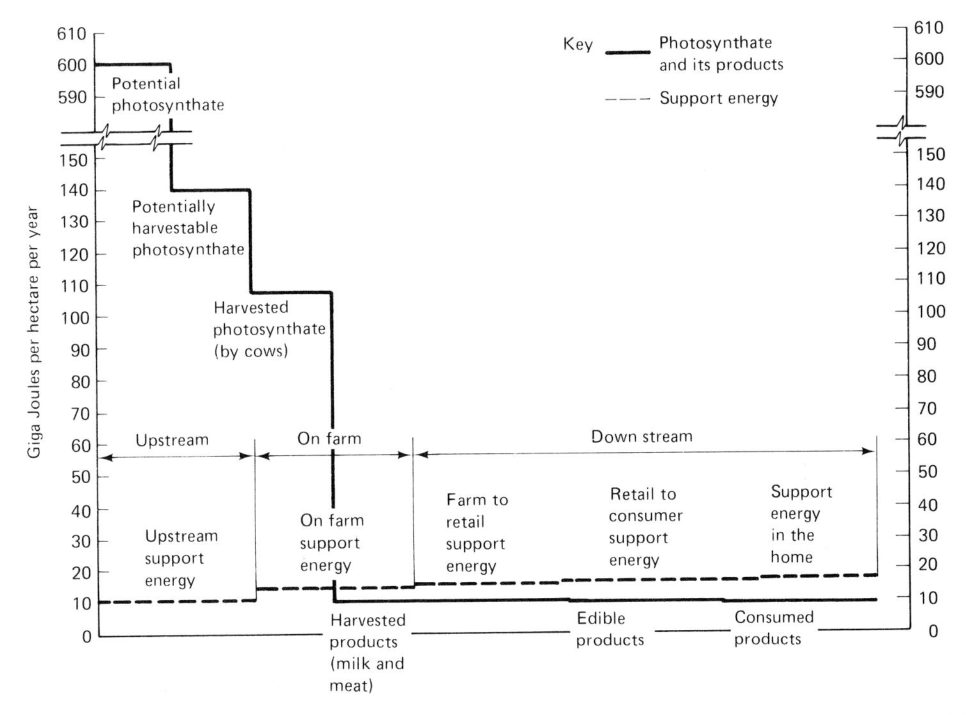

Fig. 7.4 The flow of photosynthetic energy and of support energy in the products of an English dairy farm.
Source: Duckham, Jones, and Roberts 1976

before finally being eaten by the consumer. The net food energy diminished at each of these stages (compare the potatoes and meat on p. 173) but the cumulative total of support energy increases, as can be seen for a dairy farm in Fig 7.4.

The outcome of the balance of the energy content of the foodstuff and the support energy may thus be positive at the farm gate in a market economy or at the table in a subsistence society, but it can easily become negative in an industrial context as the downstream energy uses are added to the reckoning. Table 7.7., for instance, shows the balance for white, sliced, wrapped bread on its journey from farm to shop. This negative ratio would fall even more if the bread were fetched by the consumer in a motor vehicle. Given this example, it is not surprising that many other DC foods exhibit negative ratios (Fig 7.5) compared with LDCs.

This price is paid for the variety in diet and for the high animal protein consumption in the industrialized nations and results in, for example, the UK 'eating' 30×10^6 t/yr of oil (0·53 t/cap/yr), and the inhabitants of the USA using more energy for food provision than the three times greater population of India use for all purposes (Slesser 1975 and Table 7.8).

Any debate centres on the question 'does this matter?'. It is pointed out that compared with other energy uses in the DCs, the share of agriculture is relatively small (currently about 13 per cent in the USA) and that it is worth paying for such a plentiful and varied cuisine and is in addition an amount that could easily be saved by conservation in other sectors. On the other hand, there is the danger that food production is tied very closely to fuel prices and availability, often fixed in a political manner. This is especially so for small, densely populated nations where fuel is substituted for land, especially in terms of protein output. So countries like the UK, Israel, the Netherlands and Belgium must secure their fuel supplies. There are, too, the side-effects of energy-intensive agriculture; the effects of heavy machinery on the soil and the runoff of nitrogen and phosphorus (see p. 181) are two examples. Much might be done to reduce energy use in the food system. A basic shift would be to less meat and deep-water fish eating in the DCs, and less processing and packaging of foods; nearly all of these are also changes which it is thought would benefit the health of many DC

TABLE 7.7 Energy budget for 1 kg white sliced wrapped loaf, UK 1970–71

		MJ
Inputs	To farm gate (fertilizers etc)	4·02
	Milling	2·68
	Baking	13·31
	Shops	0·71
	Total to point of sale (rounded)	20·70
Output	1 kg loaf	10·6
Ratio of energy out/energy in		0·525

Source: Leach 1976

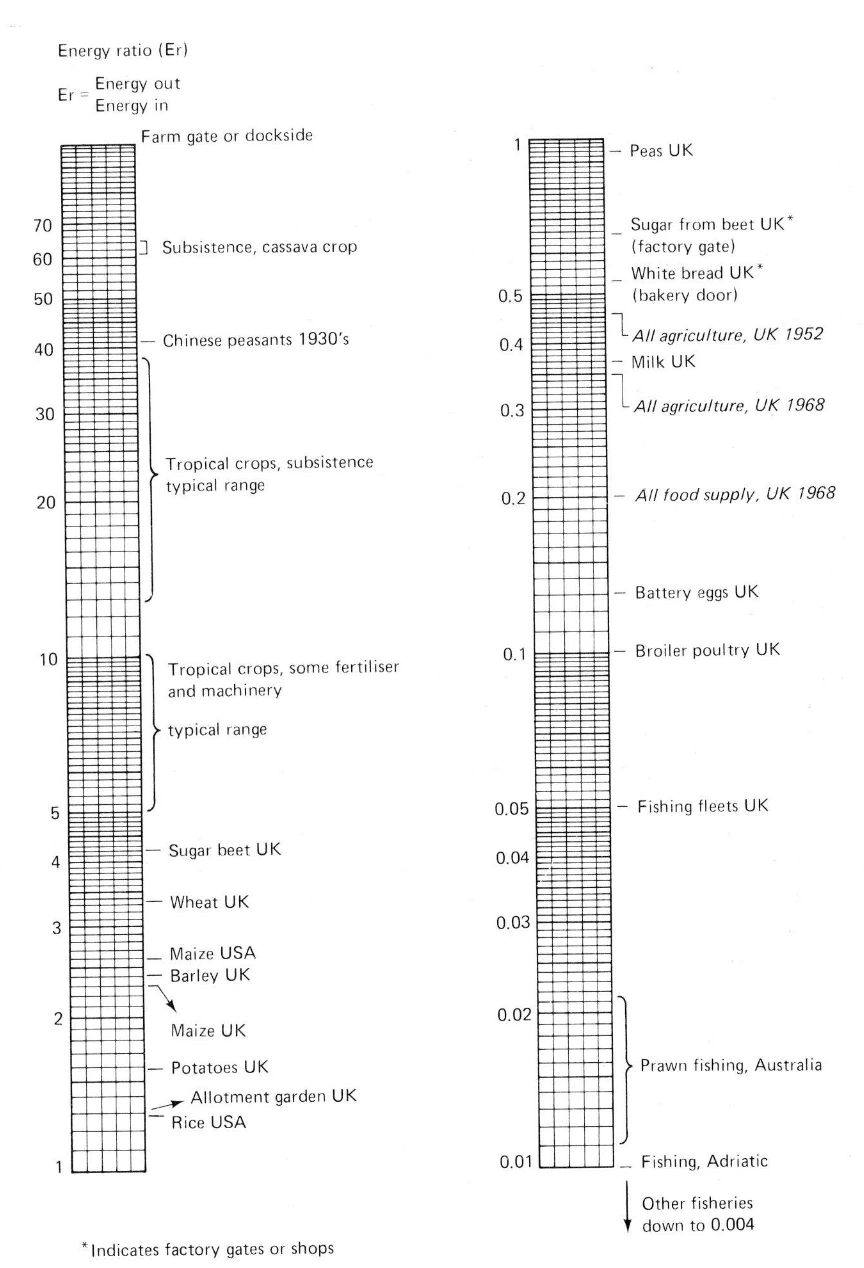

Fig. 7.5 Energy ratios for various foods.
Source: Leach 1976

TABLE 7.8 Support energy consumption in agriculture, ca 1972

	(All non-solar energy, J/yr)			
	World	*DC's*	*LDC's*	*CPE's**
Total	$204{\cdot}1 \times 10^{18}$ J/yr	$127{\cdot}1 \times 10^{18}$	$1{\cdot}6 \times 10^{18}$	$59{\cdot}6 \times 10^{18}$
(%)		(62·2)	(0·78)	(29·2)
Per cap	$54{\cdot}2 \times 10^{9}$ J/yr	$172{\cdot}1 \times 10^{9}$	$9{\cdot}6 \times 10^{9}$	$49{\cdot}9 \times 10^{9}$

*CPE = Centrally Planned Economy.
Source: Duckham, Jones, and Roberts 1976

inhabitants; on the farm many wastes (e.g. straw and faeces) might be processed to yield energy in order to make the farm more self-sustaining for energy. In some DC's, 'alternative' groups exist (like the Amish in North America) and they farm very efficiently without much access to energy subsidies. But this is largely because their whole life-style is geared to low consumption (see chapter 13), including major energy channels like the horse with an average life span of 20 years (Johnson et al 1977).

Nutrients in agriculture

Though fundamental to food production, energy is only part of the picture, for there can be no crop production without sustained supplies of plant nutrients normally held in the soil along with organic matter. Indeed, as Frissel (1978) demonstrates, it is possible to classify agriculture by its type of nutrient flow. The grazing system, for instance, relies on natural contributions to the nutrient pool in the soil. The animals are efficient gatherers of N_2 from the foliage but much of this is lost by the evaporation of ammonia from excreta. Conservation of phosphorus and potassium is good: they are lost from the ecosystem only when milk and meat are withdrawn and these quantities are low compared with the total pool, as can be seen from Table 7.9, 'E' examples. Improvement of pastures increases the output of animals and the proportion of nutrients taken off. An extensive arable system can also be characterized in which neither fertilizers nor manure are added to the crops which use the organic resources of the soil as in forestry. Fallow periods allow the accumulation of nitrogen, which otherwise has to be fixed by means of leguminous crops. A very important system is mixed farming where an area is devoted to N-fixing crops and another to non-fixing crops, and there are animal wastes to be added to the fields. The relatively low animal crop may thus compensate for itself by providing plant nutrients, of which a high proportion ('M' examples in Table 7.9) are removed by the plant agriculture. Indeed this proportion usually exceeds that taken off in intensive systems, where there is a continuous supply of chemical fertilizers. Offtake of nutrients in these systems ('I' in Table 7.9) is generally high (apparently 100 per cent of the plant pool in paddy rice) and the total ecosystem loses nutrients to the runoff. A last type of ecosystem, still to appear fully, will be based on large-scale recycling of nutrients from sewage and runoff, and possibly of water as well.

TABLE 7.9 Selected nutrient flows in agroecosystems

Type	Description	Crop	Nutrient flows: (1) Removed as crop (kg/ha/yr) and (as % of (2))			(2) Net uptake by plants from soil		
			N	P	K	N	P	K
E	Hill sheep, UK	(A)	1 (2)	0·2 (5)	trace (<1)	44	3·8	25
E	Hill sheep, UK (improved pasture)	(A)	3 (6)	0·4 (7)	1 (3)	52	5·4	38
E	Grazed bluegrass, fertilized, North Carolina	(A)	38 (25)	5 (25)	23 (15)	151	20	150
M	Mixed farm, Netherlands	(P)	49 (90)	9·5 (91)	34 (92)	54	10·4	37
	Mixed farm, Netherlands	(A)	1 (1)	0·2 (2)	0 (0)	71	8·6	63
M	Mixed farm, Brazil	(P)	113 (80)	n.a.	n.a.	141	n.a.	n.a.
	Mixed farm, Brazil	(A)	12·9 (24)	n.a.	n.a.	53·7	n.a.	n.a.
I	Intensive sheep on grass clover, UK	(A)	17 (9)	3·4 (17)	1 (1)	184	20·1	144
I	Winter wheat, UK	(P)	77 (81)	16·8 (79)	64 (44)	95	21·2	147
I	Corn for grain, Indiana	(P)	85 (67)	15 (68)	20 (18)	126	22	111
I	Soybeans for grain, Arkansas	(P)	90 (75)	10 (77)	22 (59)	120	13	37
I	Paddy rice, Japan	(P)	96 (100)	21·9 (100)	131 (100)	96	21·9	131
I	Tropical intensive farm, S. America	(P)	34·1 (38)	6·8 (37)	68 (38)	90·9	18·2	181·5
I	Intensive livestock, Netherlands	(A)	115 (25)	22 (38)	29 (8)	450	57	375

Notes: E = Extensive livestock; M = Mixed (self-sustaining) farm; I = Intensive agricultural system. (A) = Animal crop; (P) = Plant crop.
Data collected from Frissel (1978).

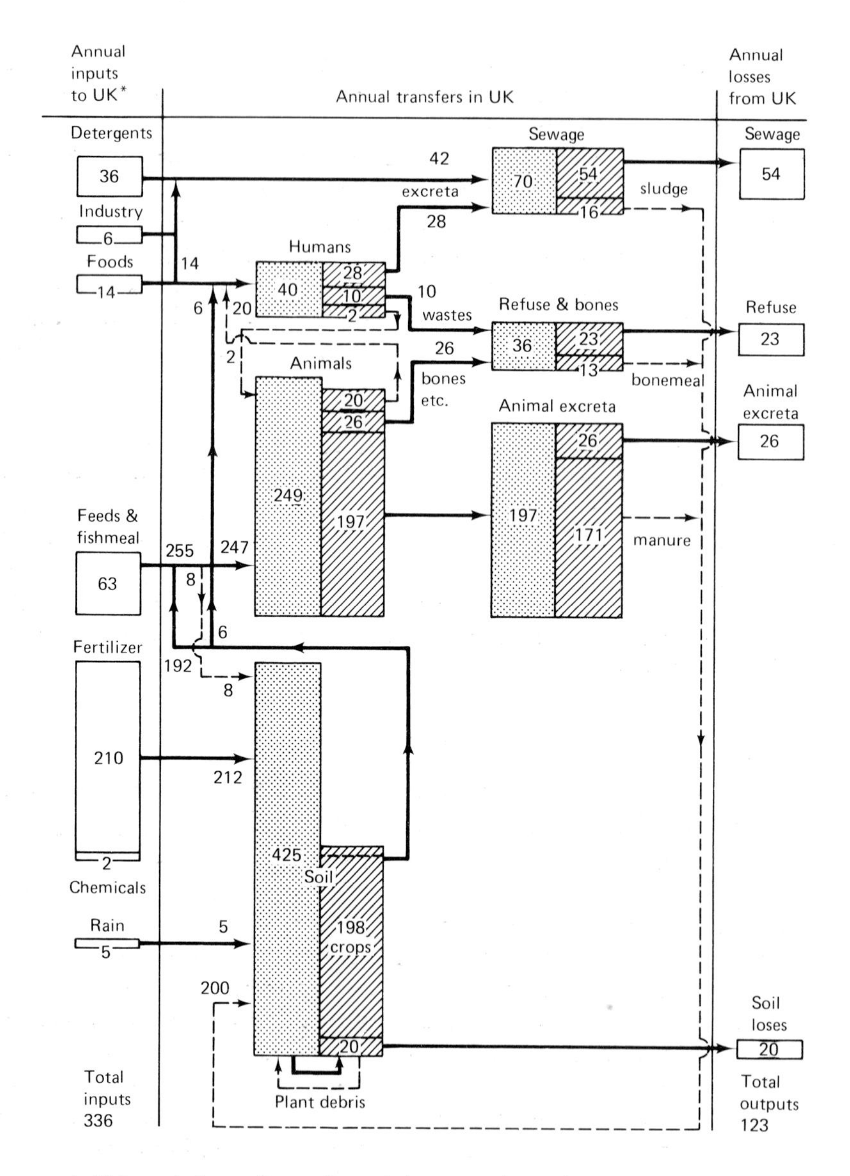

1 All figures in thousand tonnes P, rounded to nearest thousand.

2 Data based on 1973 except where otherwise stated in text.

3 * With the exception of rain, the annual inputs refer to imports of phosphorus.

4 [dotted box] represents inputs, [hatched box] represents outputs.

Fig. 7.6 The flow of phosphorus through the UK economy.
Source: Centre for Agricultural Strategy 1978

The quantities of nutrients in agroecosystems reflect to a great extent therefore the intensity of farming and at high intensities it is cheaper to buy chemical fertilizers than to handle bulky wastes with a relatively low nutrient content; industrially fixed nitrogen has now short-circuited the natural nitrogen cycle. There seems no danger of exhausting the supply of free N_2 but the energy cost of industrial N_2 fixation is so great that reliance on this source may have to be reduced. Phosphorus is in shorter supply since it is dependent upon an industrial supply from sources such as rock phosphate. Although immense quantities of phosphorus have been added to DC farms since the mid-19th century, much of this has either been removed or become unavailable to plants so that continuous additions have to be made. In countries like the UK, without any indigenous phosphatic rock, supplies are imported (Fig 7.6) and this raises concern about the best method for its conservation.

The Centre for Agricultural Strategy of the University of Reading in the UK (1978) points out that reclamation from sewage and runoff is very difficult and costly and that it would be better to prevent phosphorus use in another sector of the economy, i.e. in detergents. The environmental interactions of farms are today mostly controlled by short-term economic factors, but these glimpses of nutrient flows give us a clue that more attention may have to be paid to long-term ecological considerations including the conservation of non-renewable resources, which links farming neatly back to energy supplies once more (Snaydon and Elston 1976).

Constraints and buffers

Factors which hold down or diminish agricultural opportunity and production, and those which have the opposite effect, operate at all spatial scales, thus adding to the diversity of agricultural patterns in the world. Of all the constraints, probably the most important is climate, which forms an envelope for the tolerances of the plants and animals selected, and in some areas for the farm workers as well.

Climate is not necessarily static in its patterns, however, and some recent concern has focussed on variations in the values of climatic parameters which might be part of long-term climatic changes. Most forecasters suggest a cooling trend in global temperatures through the rest of this century, with a return to a climate rather like that of the 19th century. This might mean a shorter growing season in high-latitude growing areas like Canada and the USSR, more frequent failures of the monsoon of South and South East Asia, and the pinching of Chinese production by both cooling in the north and monsoon failure in the south; the growing season in Europe would be shortened. Significantly, the output of the USA would be little affected or even slightly enhanced, for as Thompson (1975) has shown, grain production there is rather higher in cooler years. A greater worry in the short term comes from the uncertainty of the weather at periods of climatic change: exceptional weather conditions (e.g., unseasonal frosts, warm spells, large storms, floods) are more numerous and are of course unpredictable in their effects on yields except in the general sense that they are normally adverse. The CIA (1974) suggested that such circumstances would lead to the need for large food imports into the USSR, China and South Asia, with considerable political impact since the USA might be the only area with a food surplus. One implication of this prognosis is, as stressed by Gribbin (1975) and by Schneider and Mesirow (1976), the need to set up a food bank to store surpluses generated in favourable years.

Daily weather is also of importance in determining the crop to be taken from an agricultural system, for a deviation from the expected at a critical time in the life-history of a cultivar may be instrumental in delaying growth or even causing death. The unusual occurrence of a spell of bad weather at the time of parturition in a free-ranging domestic animal is an instance. Except in technologically complex societies the means of forecasting such eventualities are lacking, and even in the industrialized countries some farmers prefer aching toes to inaccurate broadcasts.

Food losses significantly reduce the yield of many crops. World crop losses to pests are estimated to be about 35 per cent, with insects, pathogens and weeds dominating the causes. Overall post-harvest losses are of the order of 20 per cent, ranging from a typical 10 per cent in the DC's to levels of 40–50 per cent in the LDC's; the chief agents of loss are microorganisms, insects, and rodents which are held, for instance, to consume 20–30 per cent of India's stored grain crop (Pimentel 1976). One overall estimate suggested that if only half the post-harvest loss of food grains could be prevented, the additional food would be an adequate source of calories for 500 million people in the LDC's (US President's Science Advisory Committee 1967a).

Human constraints are of many types, most more appropriately discussed in economic or culturally oriented work, but infrastructural elements should perhaps be mentioned: if, for example, there are no roads to a farm, it is unlikely to produce cash crops or to make use of an industrial infrastructure which might make fertilizers and pesticides available. Simple economic problems may present themselves, as in the widespread case of the lack of purchasing power of the urban poor in underdeveloped countries. Thus not only do they remain malnourished, they do not stimulate increased production from traditional agricultural systems. Lastly there may be constraints resulting from the lack of knowledge of the farmer, his poor cognition of the limitations (or potentials) of farming in his locality, or simply prejudice against change.

To be put against these problem-inducing elements is a whole set of developments which buffer agriculture against such difficulties and stretch its capability. Energy inputs by man were the first of these and remain important, especially in some intensive systems such as paddy rice. Even here, however, the advent of the internal combustion engine has produced considerable change, and some of the crops of the 'Green Revolution' (p. 187) are heavily dependent, as is practically the whole of Western agriculture, upon fossil fuels. Storage is another buffer with an ancient history and one in which the technology is constantly being improved to prevent loss or deterioration of the crop. Not only man's food, but that of animals of lower stages in the food chains, can be stored against a rainy day or a very dry one. Dehydration, heat sterilization, radiation, fermentation, brining, smoking, freezing and chemically controlled ripening all spring to mind in this context. Continuous work on plant and animal breeding constitutes another attempt to improve the 'stretchability' of agriculture. Higher yielding strains, varieties higher in protein content, individuals more resistant to weather or pests, and even intergeneric hybrids like Triticale (wheat ×

rye), are all part of a post-Neolithic programme given added impetus by the discoveries of genetics.

The potential contribution of genetic engineering may be enormous in this field. For example, it may eventually be possible to breed into plants highly specific resistances to disease (thus eliminating the need for costly biocides with their unwanted side-effects) or even more usefully an ability even among non-leguminous plants to fix atmospheric nitrogen. Ways in which this latter state can be acheived in advance of genetic manipulation (e.g. by inoculation with appropriate bacteria) are probably the most important research area in agricultural biology today (Hardy and Havelka 1975).

The selection of a stable agricultural system for any one place must, as with all resource processes, satisfy three main conditions. Firstly, it must be ecologically feasible in the long term and must not lead to degradation of the ecosystem by means of such processes as soil erosion or structural breakdown. Secondly, it must be economically gainful to the operator: either he must make money at it, if it is a commercial enterprise, or he and his dependants must not die of starvation if it is a subsistence farm. Lastly, the system must be culturally acceptable both to the operator and to the society in whose context he farms. He may not adopt a new system because he fears the wrath of the gods, or the spirits of his ancestors, or the chilly disapproval of his neighbours. In these days of instant communication, neighbours may extend to non-farming people: witness for example the disapproval of 'factory-farming' systems which has sprung up in some Western countries; this may have persuaded a few farmers not to switch to such enterprises. He may also have to respond to technological imperatives such as the demand for machine-harvestable square tomatoes which became available in 1977. The sifting produced by each of these variables should in theory produce a stable, economic and culturally accepted agricultural system in each part of the world: perhaps the most remarkable fact is the number of places where it has done so.

Food shortages and agricultural responses

The performance of the world's agroecosystems appears to have been good. It has permitted not just the survival of the human population but its growth at overall rates of up to two per cent per annum. (See pp. 323–5) and food in abundance for a portion of it, although the absolute number of undernourished people increases from year to year. Increased use of science and technology has been evident in the rapid growth rate of output of major crops (Table 7.10). But such are the problems of distribution that there are areas of both over-nutrition and under-nutrition, for the overall increases in calorie and protein output are not spread evenly. Table 7.10 shows the much smaller rates of increase of calorie provision in Africa and Asia, compared with the world average, and the same for protein in Africa, Asia having made above-average gains in protein. In world terms, the DCs are consuming about 120 per cent of their metabolic energy requirements and LDCs only about 97 per cent. Protein consumption was 96·4 g/cap/day in the former and 56 g/cap/day in the latter during the early 1970s. There is thus in many LDCs a shortage of energy in the diet: in 1970, 24 out of 97 LDCs had energy consumption deficits exceeding 10 per cent and a group of 45 LDCs has been

TABLE 7.10 Output of major world crops

Crop	(rounded) 1969–71 average	1977	% increase
	$\times 10^6$ t		
Wheat	329·2	386·6	17·4
Rice	311·5	366·5	11·7
Maize	278·5	349·7	25·5
Total cereals	1,242·7	1,459·0	17·4
Potatoes	297·2	292·9	−1·5
Soybeans	46·7	77·5	65·9
Sugar cane	586·5	737·5	25·7
Coffee beans	4·2	4·3	2·4
Tea	1·3	1·7	30·8
Hen eggs	20,982	24,700	17·7
Beef & veal	38·8	46·2	19·0
		average =	21·0

Source: FAO *Production Yearbook 31*, 1977. (FAO 1978)

designated as the MSA nations (Most Seriously Affected). Of these, 26 are in Africa (e.g. Senegal, Niger, Uganda, Kenya), 9 are in the Far East 5 in the Near East, and 5 in Latin America. As the consumption of carbohydrates and fats is too low, protein is used by the body as an energy source and so deficiencies occur, especially among children less than 5 years old, together with pregnant and lactating women. In total, at least 400–455 million people (ca 25 per cent of the population of all LDCs) are undernourished with respect to calories or protein or both (Duckham et al 1967, FAO 1978) but it could be as high as 700 million. Certainly, in the MSAs, protein intake has been declining rather than increasing as elsewhere (Table 7.11).

In the face of these magnitudes of shortfall in the LDCs, the UN Food and Agriculture Organization (FAO) has indicated where extra production is and will be required, taking into account rises in population. The World Food Conference in 1974

TABLE 7.11 Nutritional intake

		Calories/cap/day	% gains (calories)	% gains (protein)	g protein/cap/day
World	1961–63	2,412			65·3
	1972–74	2,544	5·4	4·6	68·3
Africa	1961–63	2,075			55·4
	1972–74	2,111	1·7	1·2	56·1
Asia & Far East	1961–63	2,012			52·8
	1972–74	2,039	1·3	0·79	57·0

Source: FAO *Production Yearbook 31*, 1977 (FAO 1978)

heard that the demand in the LDCs would grow 3·7 per cent per annum compared with a world figure of 2·5 per cent per annum. This means a decadal increase of 31 per cent in wheat production, 30 per cent in rice, 48 per cent in vegetables, 56 per cent in meat. In absolute quantities, these figures mean an increase of 472×10^6 t/yr of cereals. Some ways in which the food-producing systems react to this are discussed below but it must be stressed that we are talking not only about a technical problem but a social and ecological one as well. There may well be a technical feasibility of acheiving growth rates for food of 2·5–3·5 per cent per annum but in the short term much will depend on how a multitude of societies behave under considerable pressure.

TABLE 7.12 Per capita total and (animal) daily protein intake

	grammes 1961–63 av.	1972–74
World	65 (22)	59 (24)
All DCs	91 (45)	98 (54)
All LDCs	53 (11)	57 (12)
MSA's	53 (7)	51 (7)

Source: FAO *The state of food and agriculture 1977* (Rome: FAO 1978)

The evidence points to malnutrition as a multifactorial problem whose causes have local variations. In some places the diet is well balanced but inadequate in quantity; in others there is a seasonal deficiency of protein; and in any of them endemic disease may prevent absorption of available nutrients. Elsewhere the social structure of a community may direct the available protein away from children and the poorer people. The complexities of malnutrition will be solved only by understanding the social and behavioural complexities of the consumers as well as the provision of more calories and protein (Payne and Wheeler 1971).

Extension of agriculture

The extension of the world's agricultural area seems an obvious way to resolve some of the problems. Improved machinery, irrigation, better roads, reclamation from the sea, transformation of other ecosystems, all ought to provide a greater agricultural area. Inspection of the potential reveals, however, a relatively small area of 'virgin' land suitable for modern agriculture. All the world has adequate sunlight and CO_2 for some form of photosynthesis, but when all other necessities like topography and soil are considered the proportion of the terrestrial surface suited to agriculture falls to magnitudes not far beyond the present total of 12·5 per cent. Regional estimates for the expansion of arable land in the LDCs are given in Table 7.13; in three regions the 1985

TABLE 7.13 Selected sub-regional totals for potential arable land

	Arable land 1962	*Proposed arable land 1985*	*Potential arable land*	
	Percentage of potential arable land	*Percentage of potential arable land*	*Million*	*% of total area*
Central America	64	76	46	19
South America	19	26	524	30
North-west Africa	100	100	295	6
South Asia (excl. Sri Lanka)	93	96	201	48
South-east Asia (incl. Sri Lanka)	44	57	47	47
Far East	81	97	4	28

Source: FAO *Indicative world plan for agriculture* 1970, **1**

total is at or near to the limit of potential and so further expansion will be nearly impossible. Again, the expansions suggested may need considerable cultural adjustments such as the resettlement of graziers and nomads as well as the relatively easier tasks of drainage or tsetse eradication. Irrigation will of necessity play a large part in agricultural expansion, and if schemes are not to end up choked with salt and silt then considerable technical skill in both planning and day-to-day management is required. Irrigation already accounts for 11 per cent of the world's cultivated land, and two-thirds of the world's population live in the diet-deficient countries which contain 75 per cent of the irrigated land, so that its importance to the malnourished is greater than world statistics imply. IWP envisaged a faster growth rate for new irrigation schemes than for non-irrigated harvested lands, together with equally important investments in modernizing existing schemes. All such expansions, in the words of the Indicative World Plan, are unlikely to be 'easy, rapid or cheap'.

Intensification

Most authorities agree that this process holds most promise of improving food yields from existing cultivated lands and that is more important than the extension of the agricultural area. It is a complex sequence of events which brings a low-yielding traditional agriculture into connection with the technology and economics of industry. Thus a great deal of capital and management skill are needed for success in applying developments such as irrigation, flood control, drainage, erosion control, mechanization, fertilizer and biocide use and the raising of improved varieties of plant and animal (Plate 13). The keys to intensification are energy availability, both on-site and in the places where tractors, pesticides, pumps and the milk are made; a steady effective demand for the products of the farming system; and good communications between the source of supply of the input, the farm, and the consumers. An estimate for doubling of agricultural production in the LDCs in the period 1966–85 required an increased application of plant nutrients from 6 million t to 67 million t, and from 120,000 t to 700,000 t of pesticides, representing capital outlays (in 1966 US dollars) of $\$17 \times 10^9$ and $\$1.87 \times 10^9$ respectively (US President's Science Advisory Committee 1967b). This assumed the availability of the appropriate materials: in discussing phosphates Eyre (1971) notes that 80 per cent of the world's phosphate fertilizer was used in western Europe. North America and the USSR in 1968–9, and that north German agriculture received twice as much as the combined systems of India, Pakistan and Indonesia. 'One must doubt,' he says, 'the feasibility of so expensive a commodity being made available in vast quantities to poor countries.'

The best-known intensification is the so-called 'Green Revolution', the development of new high-yielding and high-protein crops of basic cereals, especially the wheats bred by the International Corn and Wheat Improvement Center in Mexico and rice strains

Plate 13 An agricultural landscape of high intensity of use in the Noord-Ost polder of the IJsselmeer scheme of the Netherlands. This also shows one of the major sources of creation of new land for agriculture—from the sea. *(Aerofilms Ltd, London)*

evolved at the Institute of Rice Research in the Philippines (Harrar and Wortman 1969, Hutchinson 1978). Of these the development in 1966 of IR-8 ('miracle rice') was the most famous. The highly bred grain matures early after rapid growth and is insensitive to day length so that in the tropics and subtropics two to three crops per year become practicable; resistance to lodging is another important characteristic. IR-8 was developed from a cross between an Indonesian and a Taiwanese rice strain, and matures in 120–30 days. Its top yield average 1,067 kg/ha compared with the 368–405 kg/ha of its parents and 239–331 kg/ha of most local varieties in the Philippines. IR-5 was also an important new variety since it cooks dry and fluffy, whereas IR-8 tends to become soggy as it cools and thus is culturally less acceptable in some places (zu Lowenstein 1969). Such yields can only be obtained with careful cultivation. Fertilizers are the key element: IR-8 needs 13·0–16·5 kg/ha of nitrogenous fertilizer applied at particular times, together with a continuous water supply, and the use of biocides; the traditional criteria for harvesting time have also to be abandoned.

Continued breeding of high-yield varieties (HYVs) of cereals has continued at IRRI and CIMMYT, and more agencies have been created to carry on similar work applied to other regions and to develop other crops (e.g. pulses, cassava, potatoes, sorghum) and for livestock (Jennings 1976). The adoption of the HYVs has been very rapid, rising from plantings of 80 ha in 1965 to 32 million ha in 1973, and representing for example 50 per adoption HYV wheat in India and rice in the Phillipines within 6 years of the introduction of the new varieties (Wittwer 1975).

The impact of the new varieties upon production can be seen in Table 7.14 where they are given on a per capita basis for selected countries. All of them experienced very rapid rises during the 1960's, in some cases leading to self-sufficiency in basic cereals for the first time in decades. However, a combination of bad weather, renewed losses from pests, civil unrest and above all the inexorable and steady rise of population have, for two of the examples, brought no overall improvement; in Mexico and Pakistan, however, there has been some betterment although Pakistan is still classed as an MSA. The figures for the area planted to HYVs also conceals the fact that they have in some places taken the place of lower-yielding but protein-rich pulses; fears have been expressed that quantity of food has replaced quality.

Two biological dangers are inherent in the 'Green Revolution', both of which stem

TABLE 7.14 Impact of HYVs on cereals in LDCs

	Annual production of cereals in kg/cap			
	Phillipines rice	*India all cereals*	*Pakistan all cereals*	*Mexico all cereals*
1960	88	164	136	202
1965	82	136	142	265
1970	91	173	190	286
1973	88	165	180	272

Source: Brown and Eckholm 1975

from the lack of genetic diversity in new crops. Hitherto, individual farmers selected their variety according to their own idiosyncrasies and so a mosaic of different strains was produced. With the new types large contiguous areas are planted to one strain and so pathological susceptibility is multiplied. A small change in climate allowing the expansion of the range of an insect or a new strain of rust would cause a major disaster: the new wheats of India, Pakistan, Iran and Turkey might all fail at the same time, although more likely are less extensive failures such as the failure of 10 per cent of the hybrid corn in the Middle West in 1971 because of southern corn leaf blight (Dasmann 1972). Also, the breeding of the new strains means that the genetic diversity present in the old varieties is in danger of being lost,and a programme of conservation of the genetic characteristics of all types is essential (Frankel and Hawkes 1975). If such fears do not materialize then production will be very high indeed, and many modifications to current marketing practices, pricing structures and trade patterns must be made: what, for example, will become of the traditional rice-exporting nations of Asia? On the one hand the potentials unleashed are immense; on the other the price paid for the increased production is an enhanced risk of widespread catastrophe, particularly if the cornucopian aspects are used as an excuse to lessen the emphasis upon population control programmes in cereal-dependent nations.

The introduction of a technical revolution without understanding of its cultural context is always likely to be fraught with problems (Ladejinsky 1970). Socially, the introduction of the new varieties in a nation like India have swept away a great deal of conservatism on the part of farmers, but the selective impact of the agricultural changes has created unrest on the part of the many who want to be part of the new deal but cannot find the means to get started. Problems of land tenure have also been exacerbated, since rents have often risen as high as 70 per cent of the new crops and some owners would now like to get rid of tenants altogether. There needs to be effective demand from consumers: the penniless cannot buy all the IR-8 in the world. And beyond this there is the concomitant problem of how all the people displaced by even moderately efficient agriculture are to be employed: industrialization to give them all jobs would have to be on a totally unprecedented scale, although the development of a mechanized, industrially based agriculture could come to involve 30 per cent of the working population as it does in the USA (Paddock 1971). The general concensus about the Green Revolution seems to be that it has bought a little time in which to slow down population growth, a chance not always seized, and that it has, as Farmer (1977) puts it, had 'mixed social and economic effects, both creating and destroying livelihoods, both reinforcing and weakening exploitive relationships, both helping and depriving those whose lives are most precarious'.

New sources of food

Biological resources

Even if the rapid development of conventional agriculture is sustained, most authorities agree that nutritional deficiencies will persist. One strategy adopted is to work towards the improvement of the yields of basic crops on the one hand and the development of supplementary 'new' foods on the other. In the latter there is some emphasis on protein production because this has been perceived, not always correctly, as the main deficiency in areas of nutritional stress. But in any case animal flesh is usually very acceptable and its amino-acid profile is very similar to human requirements. Animal flesh has the further advantage that it is usually the more easily assimilable, since the plant proteins are locked away behind a cell wall of cellulose not easily broken down by the action of the human stomach. Animals have thus been a means of harvesting the plant protein in a digestible form, and so have considerable dietary advantages in spite of the energy losses due to their position in the trophic structure of an ecosystem: looked at economically, the livestock industries of DCs are gigantic welfare societies for domestic animals which return only 10 per cent of the energy invested in them by way of foodstuffs. If animals are to be avoided, a food source which will yield plant protein in a digestible form or from which the majority of cellulose has been removed is clearly attractive. Fungi appear to be easily assimilated and contain a good deal of protein by comparison with some other sources (Table 7.15). Other advantages are that they do not absorb much human or fossil energy in production, can be grown independently of environmental factors in places such as caves and abandoned railway tunnels, can readily be stored in dried form and require little sophisticated knowledge or technology (Pyke 1970b).

TABLE 7.15 Comparative yield potential of mushrooms

	dry protein · kg/ha/yr
Conventional methods of beef production	12·88
Fish farming	110·4
Mushroom growing in UK	11040–12880

Source: Pyke 1970b

Requiring rather more technology but readily available are leaves which are not normally cropped or which are fed upon by animals. Within their cells is a considerable harvest of protein if this can be separated from the fibrous material of the leaf and made palatable. Pirie (1969, 1970) notes that protein was extracted from leaves as early as 1773, but that the effort devoted to it by modern research is minimal. Leaves which are the by-products of another crop could be used and the fibre returned to the soil as a texture-maintaining essential, or otherwise unused grasses, shrubs or marginal aquatics might be harvested. Thus any leafy plant becomes a potential protein source, providing it is susceptible to propagation and harvesting. The potential of the tropics is especially high and the cropping of leaf protein would make attractive the retention of much of the forest cover, and help to prevent ecological degradation. Apart from harvesting, the major industrial input required is processing the extract to the point where it becomes palatable either by itself or as an additive to other foods. This need not be difficult but inevitably adds both to the cost and to the number of trained technicians who are needed. If such protein production could be allied to a labour-intensive technology then the social benefits would be high as well.

The potential is immense and probably greatly undervalued, since protein yields of 1,200 kg/ha/yr have been obtained with legumes in Britain, and 3,000 kg/ha/yr should be possible in the tropics; these harvests could be raised to 2,000 kg/ha/yr and 5,000 kg/ha/yr respectively if nitrogen fertilizer were added to the appropriate ecosystem (Arkcoll 1971). Algae are groups of plants which have received a good deal of attention as possible sources of food, especially the noncellular varieties which under optimal conditions have exceptionally high rates of primary productivity. 5 m^2 devoted to algae production could feed one man 10^6 kcal/yr, whereas it would take 1,200 m^2 of grain and 4,000 m^2 of pork to reach the same level. Since algae would be cultured in tanks, non-agricultural surfaces such as roof-tops might become food-producing. The potential yields have, however, been stressed at the expense of the drawbacks. Production of algae would be a very technical process, requiring stirring, bubbling of carbon dioxide, sophisticated machinery and skilled manpower (Pirie 1969). All things considered, the net energy input might be higher than the output and only if this were an acceptable price to pay for protein would the process become economically grainful; unquestionably it would depend upon the supply of cheap energy supplies. Algae are demonstrably not the panacea that has been claimed for them, especially in terms of dependence as in Nigel Calder's book, *The environment game* (1967).

Energy-wasters though they are, animals retain many desirable characteristics as cellulose-converters and as saliva-inducers. That so few species have become domesticated is often a source of wonder, and only recently has the potential of many wild animals been realized. If sustained-yield practices are adopted, together with minimal amounts of processing, many wild animals come within the ambit of possibility: most are eaten somewhere and it has been suggested that the best way to deal with locust plagues is to eat them. Birds such as young colonial seabirds, reptiles and amphibians are probably under-utilized, and a larger beast like the aquatic manatee which might feed on such nuisances as water hyacinths is also a feasibility. The large South American rodent *Capybara* (about 1·3 m long) which feeds on aquatic weeds is another candidate of Pirie's (1969). New domestications, among which the eland and the African buffalo rank as favourites (Jewell 1969), would be useful too, particularly if they ate plants which currently go unharvested.

Most domestic animals are very selective in their use of forage; by contrast, many wild animals of potential use for food, particularly ungulates, are much more thorough in their use of the available herbage; even if one species is very selective, it is likely to exist as part of a suite of animals whose combined cropping effectively utilizes the range. The faunas of the savanna and grassland biomes of east and central Africa are the outstanding example of this characteristic (Table 7.16) and a considerable amount of research has been done on their ecology (Fraser Darling 1960, Dasmann 1964a, Talbot 1966, Young 1975), although Parker and Graham (1971) point out that such cropping is more talked about than done, and suggest that the case for utilizing wild animals is not so strong as that for improving domesticates.

Cropping programmes require sophisticated knowledge of the ecology of the animals

TABLE 7.16 Comparison of biomass of wild and domesticated animals

Animals	*Range type and location*	*Year-long biomass kg/ha × 10⁵*
Wild ungulates	Savanna, east Africa	82·2–117·5
Domestic livestock	Tribal savanna, east Africa	13·2–18·8
Cattle	European-managed savanna, Africa	24·7–37·6
Domestic livestock	Virgin grassland, western USA	31·3
Bison and ungulates	Prairie, USA	16·4–23·5
Red deer	Deer forest, Scotland	0·14

Source: Talbot and Talbot 1963

involved and especially their social structure, so that discriminating culling can be practised: it is a different matter shooting young males of a pair-bonded species, for example, from culling spare males from territorially based harem groups kept together by one male. The role of hunter and predator has to be carefully evaluated, as has the contribution of species such as the elephant which, by virtue of its size and habits, strongly affects the development of habitats. On the one hand, they make forage from high trees available to other species by pulling them down, and they also dig water-holes: on the negative side, they will damage large areas of otherwise edible bush by trampling, will pull up trees which are the habitats of birds, for example, and at high densities wreak a general damage to the environment. Culling of elephants has often been a first step in good game management.

As well as the gains so far mentioned there is the income from tourism, which is the major source of foreign exchange for many east and central African countries. Here is a growing industry, and shooting animals with cameras does not conflict with the use of the game populations for meat. In spite of this, huge game-extermination programmes have been carried out in Africa in the hope of exterminating tsetse fly. Cattle are often then brought in and if managed inefficiently contribute to ecological degradation. Their concentration around water sources means puddling of soils, especially in the wet season, with subsequent soil erosion. This is exacerbated by the use of fire to encourage early growth at the start of wet seasons (Talbot 1972, West 1972). By contrast, native ungulates only degrade environments if their densities are too high, as has happened with hippo and elephants that have moved out of 'reclaimed' areas.

Fresh-water and brackish-water fish are other sources capable of development, especially in the tropics (Tables 7.17 and 7.18), where high yields are taken from fertilized ponds (milkfish in the Philippines are a good example). Israel and the USSR are also intensive raisers of fresh-water fish, whose protein content is very high (Hickling 1970).

In general, the scope for fish culture seems to be very wide since any unpolluted area of fresh, brackish or salt water is a potential space for rearing, and both waste water and waste organic matter can be used in a variety of trophic linkages with both terrestrial and aquatic ecosystems (In 1976 a trout and eel farm was started in a disused tram in

TABLE 7.17 Comparisons of milkfish (*Chanos chanos*) yields kg/ha

Country	*Milkfish total*	*Fish total*	*Total edible protein (dry weight) %*	
Java	33	52	15·1	Unfertilized
Taiwan	176	176	51·7	Fertilized
Philippines	55	55	16·2	Unfertilized
Agric. land: swine		83		
cattle		46		

Source: Walford 1958

TABLE 7.18 Estimated production of farmed fish for selected countries

	1975, tonnes
China	2,200,000
Taiwan	81,236
India	490,000
USSR	210,000
Japan	147,291
Indonesia	139,840
Thailand	80,000
Bangladesh	76,458
Nigeria	75,000

Source: Windsor and Cooper 1977

Yorkshire, England). Selective breeding is beginning to emerge, in for example, carp, trout and bivalve molluscs, and the criteria for successful species for aquaculture are now clear; these hinge around growth rate, conversion efficiency and hardiness. Given completely optimal conditions, with augmentation of water and food supply, 2 million kg/ha/yr of fish (wet weight) are said to be possible. One promising way ahead seems to be a polyculture which contains both herbivorous and carnivorous fish, bottom detritus feeders, and browsers, the whole being fed by wastes and perhaps calefacted water. With supplemental feeding, a polyculture of six species of carp in India gave yields of 8,200 kg/ha/yr wet weight of fish (Chaudhuri et al 1974). Aquaculture in the sea may initially use such areas as the lagoons of atolls and the structures such as offshore drilling rigs for organisms such as fish or turtles, and sedentary molluscs respectively. An output of 4 million t/yr of farmed fish in the 1980's would represent about 7 per cent of world fish supplies or about 10 per cent of fish for direct human consumption.

Industrial food

Where industrial technology is available, the choice of organisms for food can be widened, since close control of growth conditions and subsequent processing can be achieved. A first stage in industrialized food is the processing of otherwise unpalatable materials to yield either desirable food or a neutral substance which, if not exactly mouth-watering, is at least not repellent. For example, the soya-bean is high in protein and also highly adaptable, so that its content of plant protein can be disguised as (*inter alia*) turkey or pork sausages via flavouring, colouring and texturizing. It is also cheaper than another alternative, Fish Protein Concentrate. Many unmarketable fish can be defatted, deboned and dehydrated to a white tasteless powder that can be further processed or sprinkled on food as a powder additive. Its disadvantages include a complex and expensive processing procedure, but even so the marketed product is cheap if it can be got to places where it is needed without increasing its cost unduly.

Production of food by industrial processes which altogether sidestep contemporary photosynthesis offers new possibilities. Bacteria and fungi form the basis of the technique, with yeasts as the group upon which most work has been done. The substrates upon which yeasts can be induced to grow under industrial conditions include waste whey, some sugar waste, sulphite liquor and sewage. The cellulose in paper and sulphite liquor could be used to grow yeasts: one-third of the paper waste of the USA could supply one-third of its calories, but protein yields are less satisfactory (Mateles and Tannenbaum 1968). The most satisfactory substrates so far tested on an industrial scale have been the hydrocarbon by-products of petroleum. These are available in large quantities and not being agricultural products are relatively constant in supply and price. Yeasts grown upon them can be processed to a powder of dead cells containing 35–75 per cent of crude protein. Lysine content is adequate or even high, although methionine and tryptophan are relatively low.

The high price of oil since 1973–74 has however meant that most such processes have been in DC's and the product often used as animal feedstuff. Waste methanol from oil production can be used as a substrate for the bacterium *Methylophilus methylotrophus* which was due to achieve a commercial production of 50–70,000 t/yr in 1979 in England. Total production of yeasts, bacteria and similar sources of Single Celled Protein (SCP) is expected to be about 3 million t/yr by 1985, mostly in the industrialized countries (Taylor and Senior 1978). Industrially synthesized amino-acids may too become plentiful and cheap enough to feed to poultry and non-ruminant animals instead of conventional animal foodstuffs, soya-bean or FPC products (Pyke 1971).

Side-effects of agriculture

The extension and intensification of food production is not carried out without environmental side-effects. At a local scale there is the production of farm wastes (see p. 297), and the regional problems caused by the runoff of water containing high levels of nitrogen and phosphorus (pp. 297–9). More significant are the toxic effects of residual biocides (p. 285) and soil loss. This latter is a world-wide phenomenon but varies in its intensity. Accelerated erosion is commonest where cultivation takes place on steep slopes, where over-grazing leaves areas of bare soil, and where very intensive cultivation neglects to keep up the organic matter content of the soil. Thus soil erosion can be a problem even in highly managed farm systems of DCs as well as in peasant societies. In a few countries, the rate of loss is so great that it can be called a national problem of a high order; examples are El Salvador, Ethiopia and Peru. Soil is lost to agriculture as well through such processes as salinification of irrigated areas, soil compaction due to heavy grazing or the use of very heavy machines, and soil acidification after drainage (Kovda 1977). Drainage itself, both of wetlands and drier soils, produces many environmental changes, involving alterations in hydrology, channel form, sediment load, water temperature, chemistry, and aquatic biology (Hill 1976).

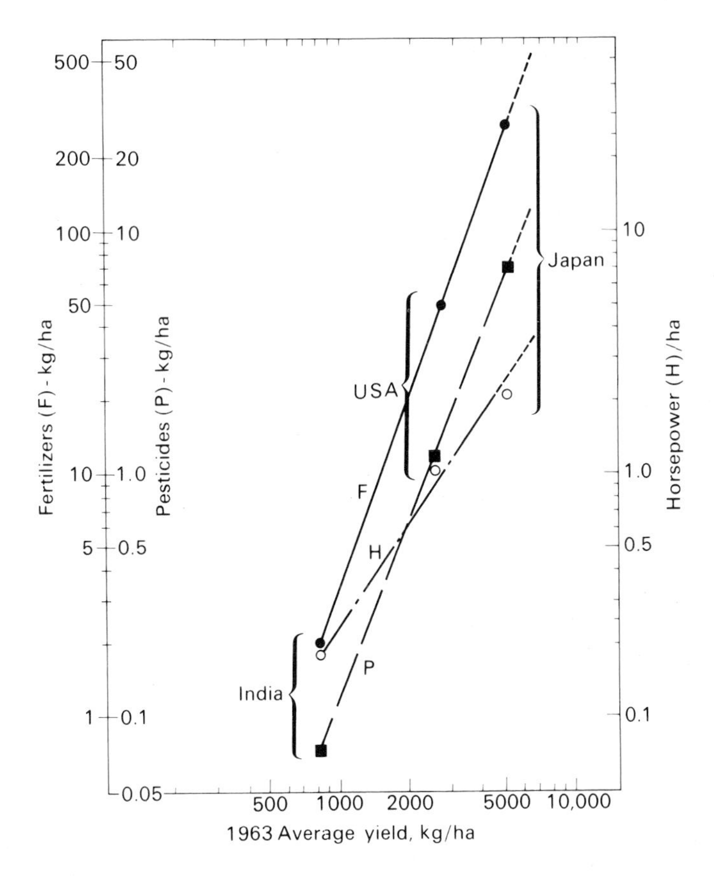

Fig. 7.7 Relationships between yield of food crops and requirements for fertilizer (F), pesticides (P) and horsepower (H) used in cultivation and harvest of crops. Doubling the yield of food requires a 10-fold increase in the use of fertilizers, pesticides and animal or machine power.
Source: E. P. Odum 1971

The intensification of agriculture inevitably reduces the quantity of wild life in rural areas: land reclamation, improvement of accessibility to machines, intensive grassland management, drainage, the contamination of freshwaters, hedgerow removal, modern harvesting methods and toxic residues have all made inroads on the populations of wild plants and animals (Davidson and Lloyd 1977).

The effect of increasing human populations

Population growth is more often considered in terms of the future availability of adequate nutrition than any other factor. The result is a complete lack of unanimity in forecasts, from famine in 1975 to enough and to spare. The effects of the Green Revolution and of industrial sources of food production cannot yet be assessed with any long-term meaning, and as C. Clarke (1967) demonstrates, if nutrition patterns are reduced from the US level to the Japanese standard then it ought to be possible to feed a lot more people. The surpluses of Europe and North America might now amount to 10 per cent of world food production and be used to make good some of the deficiencies in Asia, provided it was politically acceptable. Such a contribution, either as aid or trade, could only make an impact for a limited period given a rise in Asian populations of 2 per cent a year: as soon as 1980, four-fifths of the increase in world population will be in LDCs where the food situation is already bad.

To these additional people must be added an escalation of demand from developing societies whose better standard of living generates higher expectations, and of those whose aspirations are fuelled by worldwide electronic communications and better levels of literacy; Brown (1971) suggests these may be more potent than population growth. In the face of rapidly expanding food requirements, agriculture must be the main source of food in the foreseeable future, despite contributions from industrial and marine sources; but expansion of production is beset with difficulties both ecological and cultural, not the least of which has been the tendency to apply Gulliverian methodology to Lilliputian problems (Farmer 1969, Hendricks 1969).

The greatest reserves of potential agricultural land are in sub-Saharan Africa and the Amazon basin, if the successful management of tropical soils can be achieved and if these lands are not more valuable on a world basis as protective ecosystems. With intensification, the main process is the linkage of agriculture to the industrial world, and some results can be seen in Table 7.19. This comparison is to some extent invalidated by the more recent advent of the new varieties of cereals, but it stands as an example of the differences between an advanced agriculture and a simpler system; in more general terms, the statistics of Table 7.20 show that the production of a fossil-fuel-subsidized agriculture is able to support a much greater number of people than one in which there are no ties with the world of industry.

Such increases in output are not achieved without costs of various kinds; intensification is expensive for, as Fig. 7.7 shows, a doubling of agricultural output per unit area requires a tenfold increase in the inputs of fertilizers and pesticides. Apart from their monetary cost, such heavy use is likely to create dependence upon an advanced country for supplies which may bring political strings with them. Biologically, the new system

TABLE 7.19 Comparison of agriculture in Japan and India, 1960

	Yield (kcal/cap/day)	*Value of agric. output (US $)*	*Chemical fertilizers used (kg/ha)*	*% urban pop.*	*Tractors per 1,000 ha*	*Biocides applied tonnes*
Japan	2,360	961	303·7	64	1·55	150,000
India	2,060	91	2·3	18	0·21	10,000

Source: Dasmann 1972

TABLE 7.20 Outputs of agricultural systems

System	*Yield (kcal/m²/yr)*	*Persons/km² supported* On farm	*In cities*
Tribal agriculture	20	19	0
Unsubsidized agriculture	245	232	39
Fuel-subsidized agriculture	1,000	58	907

Source: E. P. Odum 1971

will be monocultural and hence prone to instability, and the runoff of surplus fertilizer and pesticides (not always used exactly according to the manufacturer's instructions) creates problems of eutrophication and toxification. In some LDCs the latter may reduce protein supplies, as has happened in Asia, where fish yields of 30–145 kg/ha from unfertilized rice paddies have been wiped out or made unpalatable by the use of γ-BHC to control the rice stem borer (Kok 1972).

Intensification also depends upon a series of inputs of the type described by L. H. Brown (1971) as 'non-recurring improvements'. For example, the ability of plants to respond to fertilizer has an upper limit, as does the capacity for faster growth conferred by hybridization; soya-bean cannot be hybridized and shows a limited response to nitrogenous fertilizer. An inevitable result of any intensification programme is, therefore, an S-shaped yield curve, sometimes for economic reasons like the cost of energy or input materials, sometimes the result of technical considerations like the genetics of a particular crop plant. As a context to the whole development, food prices become critical, especially in relation to the costs of energy, fertilizers, water and pesticides. The role of properly trained personnel at all levels is also an important part of intensification and can sometimes be a limiting factor, as can the gap between the promise of a crop in an experimental farm and its performance under the less controlled conditions up-country. Such considerations reinforce the view that in LDCs a labour intensive agriculture using 'alternative' technology (see p. 346) may be a better way of improving nutritional standards than hooking the people to a western-style agriculture.

In addition, Borgstrom (1965) has pointed out the possibility that water may be a limiting factor on agricultural production: it takes 132·5 litres to make a slice of bread,

TABLE 7.21 Water requirements in food production: temperate climates (kg H_2O per kg organic matter)

Millet	90–113
Wheat	136–227
Potatoes	272–363
Rice	680–907
Vegetables	1,360–2,268
Milk	4,536[1]
Meat	9,072–22,680[1]

[1] Includes water needed for production of foodstuffs.
Source: Borgstrom 1965

and as Table 7.21 shows, other food crops are high water users as well. To these amounts should be added the water needs of industrially based inputs such as fertilizers where 1 tonne requires 56×10^4 litres of water in its production, and food processing where 1 tonne of edible oil requires 35 tonnes of water (Paddock 1971). Where water is heavily polluted than agriculture may not be able to use it and national production of e.g. grain may be affected as in the case of the Vistula in Poland (Rich, 1979).

But even if we adopt every conceivable improvement in agriculture and the distribution of its products, there must still be an absolute limit imposed by the nature of the energy flux of the planet. A calculation of this sort has been provided by Buringh et al (1975) who, ignoring social, political and energetics factors and assuming that all areas can be made to yield at the levels of the best so far accomplished, suggest that 40 times

TABLE 7.22 Biological efficiency of food chains: world scale, 1972

	Parameter	*Energy/cap/day(MJ)*
(1)	Potential net photosynthesis on total cultivable land	2,126
(2)	Potential net photosynthesis on cultivated land cropped at least once in any year	682
(3)	Potential net photosynthesis in proportion of growing season used by such crops	477
(4)	Recoverable photosynthate actually formed in food production	59
(5)	Photosynthate recovered as potential food products	21
(6)	Food products which enter households	9·5
(7)	Food products actually eaten	8·6
Livestock excreta Fibrous waste from crops Human excreta		42

After Duckham, Jones and Roberts 1976, p 470.

the present cereal crop can be achieved, and at the very least some of the figures in rows 4–7 of Table 7.20 could be improved.

The perspective of ecology upon food production thus becomes less 'Can we feed the population we have and are likely to get, given also their rising expectations?' (to which the answer is 'Probably, yes'), but rather 'What are the ecological consequences of doing so?' (Holdgate 1978) Every move towards simplification of ecological systems produces higher chances of wider fluctuations and thus greater risks, many of which inevitably fall upon the LDCs, whose ability to cope with them is less buffered than that of the technologically advanced nations. In the DCs the intensive agriculture which is so successful has suffered from overcropping, and is a source of contaminants via animal waste, fertilizers and pesticide residues. And even assuming success in feeding immensely greater numbers of people (and writes Pawley in 1976, 'The ghost of Malthus, which Western economists had long thought to have chased away with ridicule, stalks the planet as never before'), there is a fundamental question of purpose: do we want the planet's management to be geared almost entirely to the production of food?

Further reading

ALEXANDER, M. 1974: Environmental consequences of rapidly rising food output.
BARNARD, C. (ed.) 1964: *Grasses and grasslands.*
BURINGH, P., VAN HEEMST, H. D. J. and STARING, G. J. 1975: *Computation of the absolute maximum food productivity of the world.*
DASMANN, R. F. *et al.* 1973: *Ecological principles for economic development.*
DUCKHAM, A. N. and MASEFIELD, G. B. 1970: *Farming systems of the world.*
DUCKHAM, A. N., JONES, J. G. W. and ROBERTS, E. H. (eds.) 1976: *Food production and consumption: the efficiency of human food chains and nutrient cycles.*
ECKHOLM, E. P. 1976: *Losing ground.*
FAO 1970: *Indicative world plan for agriculture.*
FRANKEL, O. H. and HAWKES, J. G. 1975: *Crop genetic resources for today and tomorrow.*
GEERTZ, C. 1963: *Agricultural involution.*
GREEN, M. G. 1978: *Eating oil. Energy use in food production.*
GRIBBIN, J. 1976: Climatic change and food production.
HAWKES, J. G. (ed.) 1978: *Conservation and agriculture.*
HENDRICKS, S. B. 1969: Food from the land.
LEACH, G. 1976: *Energy and food production.*
LEDOUX, L. (ed.) 1975: *Genetic manipulations with plant material.*
LENIHAN, J. and FLETCHER, W. W. (eds.) 1975: *Food, agriculture and the environment.*
MAKHIJANI, A. and POOLE, A. 1976: *Energy and agriculture in the third world.*
MELLANBY, J. 1975: *Can Britain feed itself?*
NATIONAL ACADEMY OF SCIENCES [of the USA] 1975: *Underexploited tropical plants with promising economic value.*
PIMENTEL, D. and PIMENTEL, M. 1979: *Food, energy and environment.*
PIRIE, N. W. 1976: *Food resources: conventional and novel.* 2nd edition.
Scientific American 1976: Food and agriculture.
SIMMONDS, N. W. (ed.) 1976: *Evolution of crop plants.*
SINHA, R. (ed.) 1978: *The world food problem: consensus and conflict.*
STEELE, F. and BOURNE, A. (eds.) 1975: *The man/food equation.*
SWANK, W. G. 1972: Wildlife management in Masailand, East Africa.
DE VOS, A. 1969: Ecological conditions affecting the production of wild herbivorous mammals on grasslands.
WITTWER, S. H. 1974: Maximum production capacity of food crops.
WORTMAN, S. 1980: World food and nutrition: the scientific and technological base.

8

The sea

The world's greater water bodies are perhaps less affected by man than any of the terrestrial ecosystems which have been treated so far. Byron could write:

> Roll on, thou deep and dark blue ocean—roll!
> Ten thousand fleets sweep over thee in vain;
> Man marks the earth with ruin—his control
> Stops with the shore.

and we can generally agree, with the proviso that if control stops with the shore, nowadays the ruin certainly does not, but it decreases quite quickly away from it. There are large areas of the oceans unfrequented by man because of their very size: approximately 71 per cent of the globe's surface is composed of the oceans together with the enclosed and fringing seas; volumetrically, this means about $1{\cdot}5 \times 10^{18}$ t (80 million km^3) of free water. The frozen water of the polar ice-caps forms some of the remaining land, although in this case the water is fresh and not salt.

A structurally important feature of the oceans is their depth. Whereas only 2 per cent of the land is over 3,000 m above the sea, 77 per cent of the ocean floor is more than that depth below sea level; the great trenches of the Philippines and the Marianas have a depth of 10,700 m and hence are deeper than the highest terrestrial mountain. Beyond the coastline there are three main zones: the continental shelves, descending gradually to about 200 m below sea level and the site of most human effects upon marine ecosystems; the continental slope, falling steeply from the edge of the shelf to about 2,500 m; and beyond that the deep ocean. Being most accessible to the land masses as well as the shallowest part of the ocean, the continental shelf is most often emphasized in studies of marine resources.

The salt nature of the water of the oceans appears to be derived from inwash off the land masses in which soluble minerals and particulate matter contribute to the salinity, which is thought to have been at a virtually stable level during the last 2,000 million years. The organisms of the sea must therefore play an important role in removing minerals from the liquid-soluble phase, otherwise a secular increase in concentration would be expected. The present-day average salinity at -300 m is 35 parts per thousand; nearer the surface there are regional effects such as the high evaporation rate and lack of freshwater inflow that produce salinities of 45‰ (parts per thousand) in the Red Sea, or the opposite situation which produces values as low as 10‰ in the Baltic. The chemical elements which produce this salinity are endlessly varied, since if an

TABLE 8.1 Selected chemical constituents of seawater

Element	*Concentration (μg/litre)*	*Residence time (yr)*
Hydrogen	$1{\cdot}1 \times 10^{8}$	—
Helium	7×10^{-3}	—
Lithium •	$1{\cdot}7 \times 10^{2}$	$2{\cdot}3 \times 10^{6}$
Beryllium	6×10^{-4}	—
Boron •	$4{\cdot}5 \times 10^{3}$	$1{\cdot}8 \times 10^{7}$
Carbon	$2{\cdot}8 \times 10^{4}$	—
Nitrogen •	$1{\cdot}5 \times 10^{4}$	—
Oxygen	$8{\cdot}8 \times 10^{8}$	—
Fluorine	$1{\cdot}3 \times 10^{3}$	$5{\cdot}2 \times 10^{5}$
Neon	0·12	—
Sodium	$1{\cdot}1 \times 10^{7}$	$6{\cdot}8 \times 10^{7}$
Magnesium •	$1{\cdot}3 \times 10^{6}$	$1{\cdot}2 \times 10^{7}$
Aluminium •	1	$1{\cdot}0 \times 10^{2}$
Silicon	3×10^{3}	$1{\cdot}8 \times 10^{4}$
Phosphorus	90	$1{\cdot}8 \times 10^{5}$
Sulphur •	9×10^{5}	—
Chlorine •	$1{\cdot}9 \times 10^{7}$	1×10^{8}
Potassium •	$3{\cdot}9 \times 10^{5}$	7×10^{6}
Calcium •	$4{\cdot}1 \times 10^{5}$	1×10^{6}
Titanium	1	$1{\cdot}3 \times 10^{4}$
Vanadium	2	$8{\cdot}0 \times 10^{4}$
Chromium	0·5	$2{\cdot}0 \times 10^{4}$
Manganese	2	$1{\cdot}0 \times 10^{4}$
Iron	3	$2{\cdot}0 \times 10^{2}$
Nickel	7	$9{\cdot}0 \times 10^{4}$
Cobalt	0·4	$1{\cdot}6 \times 10^{5}$
Copper	3	2×10^{4}
Zinc	10	2×10^{4}
Arsenic	2·6	5×10^{4}
Selenium	9×10^{-2}	2×10^{4}
Bromine •	$6{\cdot}7 \times 10^{4}$	1×10^{8}
Strontium •	8×10^{3}	4×10^{6}
Silver	0·3	4×10^{4}
Cadmium	0·1	—
Tin	0·8	—
Antimony	0·3	7,000
Iodine •	60	4×10^{5}
Barium	20	4×10^{4}
Gold	1×10^{-2}	2×10^{5}
Mercury	0·2	8×10^{4}
Lead	0·03	4×10^{2}
Uranium	3	3×10^{6}

Notes: 1 microgram/litre = 1 part per billion (10^{-9}) [1 ppb]; Elements marked • had concentrations valued at ≥$1.00 per 10^{6} gallons seawater ($3{\cdot}785 \times 10^{6}$ litres) in the late 1960's.
Sources: Cloud 1969, Goldberg and Bertine 1975

element is present on the land it will sooner or later find its way into the sea. There are, however, enormous differences in concentration, from chlorine as sodium chloride at $1{\cdot}9 \times 10^7$ μg/litre/down to gold at 0·2 μg/litre (Table 8.1). The commonest elements are of course the most important, since it is to their presence and concentrations that marine life has had to adapt, and it is they, together with offshore deposits of certain kinds, that constitute the inanimate resources of the oceans. Other resource processes for which the oceans are used include the harvesting of fish, shellfish and other marine life, including water fowl; recreational activities and the provision of aesthetic pleasure; navigation, the dilution and dispersal of wastes, and to a limited but increasing extent the extraction of a domestic and industrial water supply.

Mineral resources

The sea's mineral resources can be divided into three categories: those which are dissolved in the water itself; sediments present on the sea-bed at various depths; and those present at some depth below the sea-floor, beyond the sediments of relatively recent origin.

At present the utility of the dissolved elements is in direct proportion to their abundance and to the relative cost from terrestrial sources. Table 8.1 shows some of the commonest elements present together with an indication of those with the highest monetary values. Common salt immediately springs to mind as one of the resources that has been utilized since prehistoric times for its value in flavouring and meat preservation. At present only salt, magnesium and bromine are being extracted in commercial quantities and the sea does indeed seem to be inexhaustible for these elements: presumably replenishment is taking place at an equal if not higher rate. For most others, the chances for economically feasible recovery seem low. For example, 34×10^{12} litres of sea water, equal to the combined annual volume of the Hudson and Delaware Rivers, would yield 400 t of zinc. In 1968, 124,358 t of that metal were used in the USA alone (Cloud 1969). So at present only where concentrations exceed 1 part per million (= 1000 μ g/litre in Table 8.1) does commercial extraction look likely. In 1972, about 300 plants recovering sea-water materials (Table 8.2) were installed, producing salt, fresh water, magnesium metal and magnesium compounds, heavy water (deuterium oxide), bromine and small quantaties of calcium and potassium compounds. (Wang and McKelvey 1976). It may be possible to lower feasibility thresholds by investigating the capacity of marine organisms to concentrate desired elements (this is done *de facto* for nitrogen used in the form of fish-meal fertilizer and in sea-bird guano), and in the possible exploitation of zones along the sea-bed where fractures allow the escape of unusually high concentrations of mineral ions. As far as minerals are concerned the oceans are more like *consommé* than Clam Chowder, and the technology of handling the appropriate volumes of water is poorly developed, but presumably the economic perception of the minerals would be greatly changed by advent of cheap and ubiquitous energy should this become available.

Sediments and sedimentary rocks on the continental shelves are sources of certain materials. Placer deposits contain workable quantities of gold, tin and diamonds, and

TABLE 8.2 World annual production of minerals from oceans and beaches

Mineral	*1972 Production value (US$ × 10⁶)*	*Projected to 1980 (US$ × 10⁶)*
Subsurface soluble minerals and fluids		
Oil and natural gas	10,300	90,000
Sulphur	25	2,000 (incl. others not now extracted, e.g. phosphorite manganese)
Salt	0·1	
Freshwater springs	35	
Superficial deposits		
Sand and gravel	100	
Lime shells	35	
Tin	53	
Titanium sands, zircon, monazite	76	
Iron sands	10	
Precious coral	1	
Subsurface bedrocks		
Coal	335	
Iron ore		
Seawater		
Salt	173	200 (incl. others not now extracted, e.g. uranium)
Magnesium	75	
Magnesium compounds	41	
Fresh water	51	
Heavy water	27	
Others (potassium and calcium salts)	1	

Note: The 1984 projections assume a 30 percent increase in the value of minerals and an oil price of $10/bbl.
Source: Wang and McKelvey 1976

other sediments which are amenable to exploitation include sand, gravel and shells (e.g. in 1972, $9{\cdot}8 \times 10^6$ m^3 of sand and gravel were dredged up off the UK and 54×10^6 t of aggregates off Japan). The land-use problems created by their extraction from the land would largely be obviated by the use of the sea as a source, provided that the ecosystems of the oceans were not too greatly damaged by the recovery processes, which create great quantities of silt and also bring about imbalance in the sedimentary systems of the sea-floor; in the case of shell dredging in the Texas Gulf, the dredge plume was favoured by colonizing species because of the nutrient concentrations and shelter from predators in relatively turbid water. Phosphates are found as nodules and in crusts where the operation of natural concentration processes brings their recovery closer to economic feasibility, e.g. off Baja California, South Africa and New Zealand. A further resource of the continental shelves is fresh water: large quantities of artesian water may be found in certain aquifers, and although such supplies are currently costly compared with terrestrial sources, a demand may arise for their use in relatively humid lands, just as they are already tapped around some islands and being sought in the Mediterranean. Finally there are petroleum and natural gas, which are already exploited in many offshore waters up to depths of 2–2·5 km, which is the current limit of the techniques used.

The minerals of the deep ocean basins are difficult to appraise. The most extensive are the pelagic sediments, which are particles of biological, aeolian or chemical origin that have settled out on the ocean floor. They contain enormous quantities of certain metal elements, but only if they have been subjected to enrichment would they enter resource processes under current conditions. The most discussed of them is manganese (which in 25 years time might only be available from Gabon and South Africa), which forms in large nodules and crusts in which other metals like nickel, cobalt and copper are incorporated; they occur as a veneer with a mean depth of 4,000–5,000 m, coming up as shallow as 200–1,000 m off North Carolina, but their true extent is unknown, and large-scale methods of extracting the metals from the silica in which they are embodied are as yet undeveloped (Cloud 1969).

Their distribution is very variable but they are mostly found as pebbles with an average length of 5 cm, although a 750 kg boulder was lifted from 5,200 m near the Phillipines. Estimates for the Pacific vary from 9×10^3 to $1{\cdot}7 \times 10^{12}$ tonnes of manganese nodules and the area most favoured for commercial attempts at recovery is south-east of Hawaii where there is a 2–6 cm deep layer of nodules at 5,000–5,500 m depth. Phosphorite may also become recoverable in the future in areas far from other sources of phosphorus: it comes in nodules and slabs up to 1 m long, between 40°N and

TABLE 8.3 Environmental effects of marine mining

	Level of ecosystem affected							
Potential direct or indirect impact	Geomorphology	Water quality	Phytoplankton	Zooplankton	Benthos	Shellfish	Juvenile fish	Adult fish
Changes in bathymetry of sea floor	•							
Coastal erosion	•							
Removal of organisms					•	•		
Particle size change					•	•	•	•
O_2 depletion and free sulphides		•	•	•	•	•	•	•
High turbidity and reduced light in water column		•	•	•			•	
O_2 demand from sediment dispersion		•		•				
Release of nutrients		+	+	+	+	+	+	+
Release of metals		•	•	•	•	•	•	•
Release of pesticides		•	•	•	•	•	•	•
Rain of fine particles				•			•	

• = detrimental impact; + = beneficial effect.
Source: Baram et al 1978

40°S. In 1979 large deposits of sulphide ores of zinc, copper and iron at metal contents of 9–42 per cent iron, up to 6 per cent copper and up to 23 per cent zinc were reported from the East Pacific rise (Francheteau et al 1979) so that it looks as if the sea-floor spreading zones may have differentially high concentrations of minerals. However, deep-sea mining did not, in 1979, look as if it was likely to be at all important before 1990 because of the low prices of land-based nickel and copper (key elements in the economics of exploitation of managnese nodules), and the uncertainties over future laws of the sea relating to sea-bed ownership.

In 1974, about 5 per cent of the world's production by value of geological wealth came from the sources discussed above, mostly (*ca* 98 per cent) from oil and gas. That this proportion will increase is not in doubt, but the idea of an unending cornucopia is obviously false. The cheapest source, sea water, contains few elements demanded by modern industry in high concentrations, and access to the other materials is difficult. Energy costs, too are likely to be very high in any deep-water recovery programme, Lastly any discussion of the extraction of these resources must reckon with the external costs in terms of impact upon the ecology of the seas, for inevitably there will be a risk of destruction of biological resources of considerable value and perhaps greater indispensability, involved in the various processes of exploration, construction, extraction, processing and transport. The environmental impacts of ship mining the seafloor or if discharging tailings upon the various components of marine ecosystems is not quantitatively known but is bound to be mostly detrimental (Baram et al 1978).

Biological resources

It is commonplace to see calls for greater use of marine biological resources for food. Yet there are severe limiting factors on biological productivity in the sea. The euphotic zone in which photosynthesis can occur is only about 60 m deep, since from there to 520 m there is only blue light, the other wavelengths having been absorbed. There are very few terrestrial areas where the photosynthesizing zone is 60 m deep, but the primary producers of the sea are scattered very thinly through the water; if they were more concentrated then the euphotic zone would be shallower and so the level is self-limiting. Also limiting is carbon dioxide: as in fresh water, this tends to be scarce and the amount dissolved is dependent upon mixing at the interface between the water and the atmosphere.

In spite of such limitations, the sea supports a great diversity of living forms existing in complex interactive systems; there is in fact no abiotic zone. Fig 8.1 summarizes very generally, some of the main food-chain characteristics of which some need special emphasis. Firstly, the plankton are generally unable to control their regional movements and so their drifting nature leads to considerable patchiness in their presence and productivity. Secondly, food chains are generally simple and short-lived, according to Wyatt (1976). As animals grow, their diet changes so that food chains are continually forming and disappearing; the simple trophic level concept is therefore limited in its usefulness. Because the phytoplankton are so small, however, it is possible for food

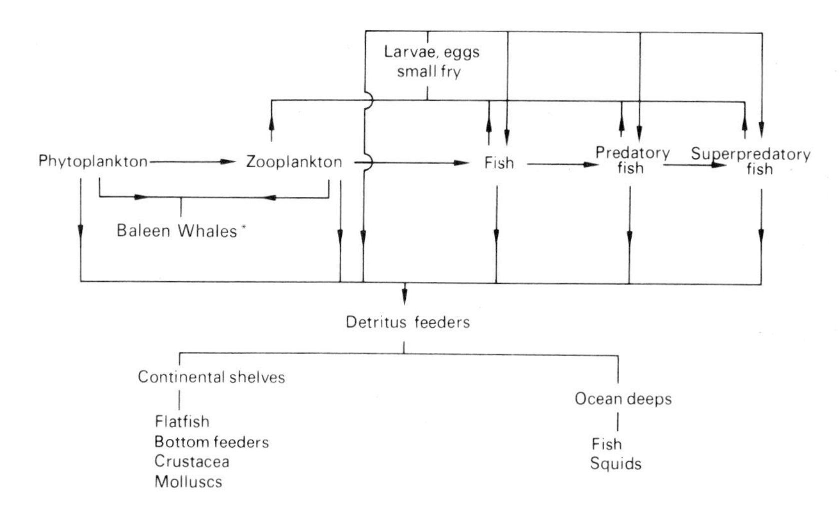

Fig. 8.1 Diagrammatic representation of the major pathways in oceanic food chains. The upper line attempts to show the possibility of a fish shifting its food source as it grows and also that the fry stage of superpredatory fish are vulnerable to consumption by animals at a lower trophic level, as if zebras ate lion cubs. The detritus chain is especially important on the continental shelves.
Source: Ricker 1969

chains with five or six links to exist in which the superpredators are a long way removed from the energy source. Their metabolic efficiency appears to be low, as is their fecundity and so their populations are limited by factors other than food supply. Detritus and bacteria are important food sources in shallow water and probably have a key role in deeper water since there are 10 g of dead and 100 g of dissolved carbon for every gramme of living carbon in the water column, and the dead organic matter can only be made available again via bacterial activity.

The primary productivity upon which the life depends is not uniform over the whole ocean area; the open oceans are the least productive and are in fact something of a biological desert, largely because of the small size of the autotrophic zone in relation to the heterotrophic zone in which the cycling of nutrients takes place. The distribution of productivity (Table 8.4) suggests that nutrients may be limiting factors in the ecosystem, which appears to have adapted itself by means of rapid mineral cycling with almost immediate uptake.

The most important producer organisms are probably the nanoplankton (2–25 μ in size) which have a short biomass turnover time and rapid nutrient cycling. Corals are also efficient at retaining, for example, phosphorus, presumably because it is recycled between the plant and animal components of the colony. The coastal zone has a higher

TABLE 8.4 Primary productivity in the oceans

	Area 10^6 km^2	NPP dry matter Mean $g/m^2/yr$	Total 10^9 t/yr	Biomass dry matter Mean kg/m^2	Total 10^6 t
Open ocean	332	125	41·5	0·0003	1·0
Upwelling zones	0·4	500	0·2	0·02	0·008
Continental shelf	26·6	360	9·6	0·001	0·27
Algae beds and reefs	0·6	2,500	1·6	2	1·2
Estuaries (excl. marsh)	1·4	1,500	2·1	1	1·4
Totals/avs	361	155	55·0	0·01	3·9

Source: Whittaker and Likens 1975

productivity because of its proximity to the sources of mineral nutrition, and upwelling zones share similar characteristics. Tidal estuaries and mudflats are among the most productive ecosystems in the world. The interpretation of estimates of yearly productivity must take into account the fact that the standing crop is very much lower: in the case of phytoplankton the biomass is probably about 1 per cent of the yearly turnover (Ryther 1969). This contrasts with the terrestrial ecosystems, where biomass of the standing crop may be roughly equal to yearly production as in grasslands and crops, or greater than that measure as with forests and desert shrubs. Even if all the phytoplankton were present in the top m of the sea, their average density would be 0·5 g/m^3 of water. Where the actual figure exceeds this greatly, as it does at certain places in particular seasons, direct harvesting by man is technically difficult and hence costs are high. The product is intractable not only on account of cultural factors such as taste and texture, which could be improved by industrial processing, but because of high salt and silica contents. The removal of organisms would also mean taking away nutrients and these would have to be replaced, just as if it were an agricultural system.

As in all ecosystems, productivity falls at higher trophic levels; zooplankton presents similar cropping problems to phytoplankton, and man harvests very little of it. Most of his crop comes at the level of secondary and tertiary consumers, a few species of fish and some molluscs coming from the first trophic level of consumer organisms. The third trophic level yields a great number of the desired species, such as flounders, haddock, small cod, herring, sardines and whalebone whales, while some strongly demanded species such as halibut, tuna, salmon, large cod, swordfish, seals and sperm whales are yet further along the food chain. At each stage there is competition for the production from taxa which are not important resource species for man: sharks, dogfish and seabirds, for example. A further harvest comes from detritus feeders which scavenge the sea-floor: many flatfish and crustacea belong to this group.

World biological production at the levels mostly used by man is estimated variously at about 200–325 million t fresh weight, mostly of fish. Estimates of the sustained annual yield of the seas vary from 55–2,000 million t of fish (Table 8.5); Ryther (1969) suggested that the length of food chains and the trophic levels at which man crops the

Plate 14 Although contemporary fisheries are dominated by the deep-sea fleets with modern equipment, the fish protein supplied to many nations comes from small inshore fishermen operating in a traditional fashion, unlinked to fossil fuel power, as in this part of the New Territories of Hong Kong. *(I. G. Simmons)*

TABLE 8.5 A calculation of fish productivity in the oceans

	Upwelling	*Coastal*	*Oceanic*
% of ocean area	0·1	9·9	90
NPP kcal/m²/yr	3,000	1,000	500
No. of trophic levels from plants to fish	1·5	3	5
Fish productivity 10^6 t/yr	98	122	1·6
% of fish production	44	54	0·7

Note: This table is calculated on the basis of various efficiencies of food transfer not given here.
Source: Crisp 1975

oceans will limit the yield to about 100 million t/yr, and this now commands a fairly general acceptance. The constraints imposed by economic factors probably mean that the upper range of the estimate will be difficult to achieve: of the actual capture 80–90 per cent is at depths of less than 200 m and it seems unlikely that commercial trawling could ever extend much beyond 1,000 m. Beyond this level the animals are so scarce in

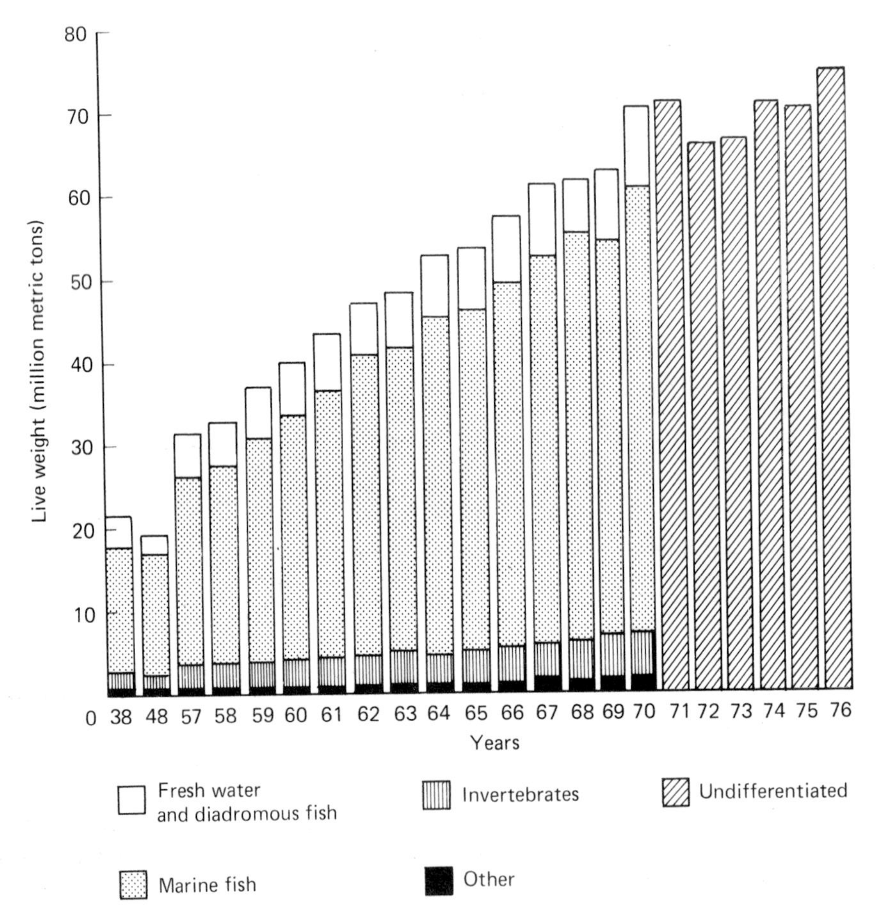

Fig. 8.2 The world fish catch 1938–76. In 1976, 69 per cent of the crop went for human consumption, the rest for livestock feed.
Source: Holt 1969, with later additions from FAO data, where the catch is not always subdivided by type and hence the columns from 1971 onwards are undifferentiated.

relation to the volume of water that it is probably better to harvest predators which go down to such depths to feed, as do sperm whales on large deep-water squids.

In 1976 world fish landings ran at 73·5 million tonnes (Fig 8.2), a slight rise on the fluctuations of the previous years which followed a steady increase of 8 per cent p.a. for the years 1948–66. As Table 8.6 shows, Japan led in absolute quantity, followed by the USSR, China, Peru and Norway, with most other nations some way behind, including the USA at 2995 $\times$ 10^6 t/yr and the UK at 1090 $\times$ 10^6 t/yr.

Consumption per person (Table 8.7) is also dominated by Japan, a fact as obvious to the viewer of the wax models outside restaurants in that land as to the avid *sashimi* enthusiast. The averaged figures also conceal the fact that several other Asian nations, such as Burma and Thailand, rely on fish for much of their animal protein (Plate 14,

TABLE 8.6 Catch size in 1976, rank order

	Million tonnes, live weight
Japan	10,619
USSR	10,133
China	6,880
Peru	4,343
Norway	3,435

Source: FAO *Yearbook of Fishery Statistics*, vol 42, 1977

TABLE 8.7 Regional consumption of fish, 1965 (kg/cap)

North America	14·5	USSR	13·1
EEC	12·1	Other central and eastern Europe	6·1
North-west Europe	20·5	China	6·9
Southern Europe	17·2	Latin America	6·0
Japan	53·7	Sub-Saharan Africa	8·2
Oceania	10·7	Near East and north-west Africa	3·2
South Africa	21·3	Asia	7·3
		World average	10·3

Source: FAO *Indicative world plan for agriculture*, 1970, **1**

page 209). The discrepancy between the landings of Peru and the low overall consumption in Latin America is in part caused by the export of most of the anchovy catch to Europe as animal feed. According to Dasmann (1972) the Peruvian catch could provide a minimal protein intake for 413 million people, although when the Humboldt Current is displaced by the warmer El Niño Current, as happened in for example 1957, 1965 and 1972, the fishery yield is relatively small (e.g. the 1971 anchoveta catch was 10·2 million tonnes, the 1973 catch 1·5 million tonnes; in 1974–6 it was steady at about 3·5 million tonnes) and so creates difficulties for Peru, since 40 per cent of her foreign earnings come from fish meal. Apart from particular regional and national situations, Table 8.7 shows that consumption per head of fish is highest in the developed countries, and since their livestock is also a major user of fishmeal, also that much of the world's catch is devoted to the industrial nations.

Extension of fisheries

The potential for extending fisheries comes from three sources: the utilization of untapped species, the cropping of hitherto unattractive areas, and the development of more novel methods of culture and harvesting. During recent years a number of new fisheries have started to flourish, such as Peruvian anchoveta, Alaska pollock, Bering Sea flatfishes and herring, and several more. The future for extensions of traditional fishing methods lies, for example, in the cool temperate parts of the southern hemisphere, which only produce about 10 per cent of the world's fish catch. There are difficulties, such as the small areas of continental shelf and the lack in some areas of suitable species, but it seems likely that considerable extension of fisheries could be wrought. Even in heavily fished northern areas there are abundant but little-used species: grenadiers in the north-west Atlantic, sandlance, anchovies and sauries in the Pacific, and small sharks like the dogfish, together with lantern fishes and other small oceanic fishes, especially in upwelling zones. Even where fish are found that are not very useful for direct human consumption, large-scale catches may mean that they are useful as 'industrial' fish. A protein-rich concentrate can be made that is currently used, for instance, in the broiler-chicken industry.

Other less direct ways of increasing the crop of marine resources could be more systematically investigated. When nutrients are limiting, fertilization by the addition of minerals to the sea may bring about higher productivities. Only this type of activity takes fishery management to a state much beyond the largely Mesolithic technology which, give or take a fossil-fuel-powered trawler or two, is still being used.

Krill

Antarctic whaling has reduced stocks to about one-tenth of their former size and so the presumably uneaten food of the whales is theoretically available to man. Approximately 80 per cent of the prey of blue and fin whales, and even more of humpbacks, is krill, the shrimp *Euphausia superba*, up to 60–70 mm long and 1 g in weight, with a net weight content of 7 per cent fat and 16 per cent protein (Moiseev 1970). Rather rough calculations suggest that their productivity is greater than 50×10^6 t/yr in Antarctic waters. How much is consumed by other predators is unknown, but no surges in the populations of seals, birds, fish and minke whales have been recorded (Mackintosh 1970); fish and squid may have been the chief benefactors. Since biomass and productivity are not yet known, the sustained yield cannot be calculated, but USSR and Japanese vessels are already catching and processing krill. The Antarctic powers are now in the process of trying to determine the sustainable yield.

Aquaculture

The first stage away from a hunting and gathering economy is that of herding, and this is being used in, for example, Hong Kong, the Philippines and Japan. Frameworks are lowered into shallow offshore waters and allowed to colonize with sedentary molluscs like oysters and mussels. With some species, the individuals grow on ropes that hang clear of the bottom so that they are out of the reach of predators such as starfish. Although productive (Table 8.8), such systems are very vulnerable to contamination, and since the organisms filter large quantities of water their ability to concentrate substances toxic either to themselves or to consumers is very high.

True aquaculture involves genetic manipulation of the chosen species by keeping them captive throughout their breeding cycle, a difficult though not impossible task.

TABLE 8.8 Aquacultural yields (fresh weight, without mollusc shells)

Location	*Species*	*kg/ha/yr*
USA	Oysters	
	(national average	9
	(best yields)	5,000
France	Flat oyster	
	(national average)	400
	Portuguese oyster	
	(national average)	935
Australia	Oysters	
	(national average)	150
	(best yields)	540
Malaya	Cockles	12,500
France	Mussels	2,500
Singapore	Shrimp	1,250
		500
Phillipines	Milk fish	5,000
India	Carp	8,200
USA	Sport fish, reservoirs	24

(Gathered from various publications)

The requirements are unpolluted sea water and a suitable coastal site with adjoining land. The most efficient plant would be large and would require buildings, stores, hatcheries and covered tank complexes on the land, together with enclosed tidal areas and tanks, and ponds or lagoons in deeper water. For preference, use of all the water areas would be possible by housing together algae browsers such as abalone, pelagic herbivores like the grey mullet, and bottom-dwelling carnivores (C. E. Nash 1970a, 1970b). The possibility of using waste heat to maintain constant water temperatures has been much discussed and tried, and in coastal temperate zones there exists the attractive possibility of raising tropical fish with a high productivity. Even the native species benefit from heated water, as experiments with plaice (*Pleuronectes platessa*) and sole (*Solea solea*) in Scotland have shown: most individuals attained a marketable size in 2 years, which is at least one year before the normal time for wild populations (C.E. Nash 1970b). More complex systems based on other waste products have been envisaged. For example, sewage and other eutrophicatory products might be used as the basis for algal production which forms the food of oysters which then filter the water as well. The oyster droppings are eaten by worms which are the prey of bottom-living fish, whose nitrogenous excretions nourish water weeds which oxygenate the water.

A general disadvantage of aquaculture seems to be the considerable skill needed for success, and it is therefore yet another competitor in the LDCs for scarce trained man-power. While we may assent to the principle of the Institute of Ecology's (1972) statement that money would be better invested in aquaculture than larger fishing fleets, the gloomier IWP (1970) statement, that even a five-fold increase in output from

TABLE 8.9 Utilization of fish potential by marine area

Marine area	Potential ($\times 10^6$ t)	Utilization (per cent)			
		1962	1965	1975 (proj.)	1985 (proj.)
Atlantic	53·7	26	32	46	63
Pacific	55·5	37	41	63	80
Indian	7·3	23	26	52	74
Mediterranean	1·7	50	54	61	76
World	118·2	31	36	54	72

Source: FAO *Indicative Plan for World Agriculture*, 1970, **1**

aquaculture by 1985 would be only marginal to the world situation (although perhaps being locally significant), seems closer to the reality of the near future. That total however might be, in the longer term, rather more important since farmed-fin fish alone totalled 5 × 10^6 t in 1975 (Windsor and Cooper 1977); a five-fold increase in these alone would equal one-quarter of the estimated sustained yield of wild fish. Any area of fresh, brackish or salt water is a potential space and both 'waste' water and 'waste' organic matter could be used in a variety of trophic linkages both within aquatic systems and with terrestrial resource processes.

Over-uses of biological resources

The harvesting of marine biological resources is subject to the same constraints as any other wild crop if sustained yield is desired: the population must not be overcropped to the point that its reproduction no longer provides sufficient individuals to constitute a resource. In view of the ecology of fish and sea mammals we might think that over-use is not likely, but the gregarious nature of many fish, the large size of whales and seals, and the product desirability of mammals such as sea otters have caused great inroads to be made upon their populations even in the traditional fishing systems of island LDCs (Swadling 1977, Johannes 1978). In seeking to expand commercial fisheries to their estimated potential (Table 8.9), rational management must be employed to ensure that depletion does not occur. In both past and present, particularly favoured fish species have exhibited considerable declines, as for example the east Asian sardine, Californian sardine, north-west Pacific salmon, Atlanto-Scandian herring, the Barents Sea cod; and a number of others, including the Newfoundland cod, North Sea herring, British Columbia herring and yellowfin tuna, are showing signs of strain (Holt 1971). International regulatory measures are sometimes applied to such species, regulating catch and net size, but these measures are difficult to enforce and once a species has been overfished it may not be possible for it to regain its place in the energy pathways of the ecosystem. The Pacific sardine (*Sardinops caerulea*) of the California current system was a major feeder on the zooplankton, and production of the fish aged 2 years and older was estimated at 4 × 10^6 t/yr. It was overfished in the 1930s and replaced by

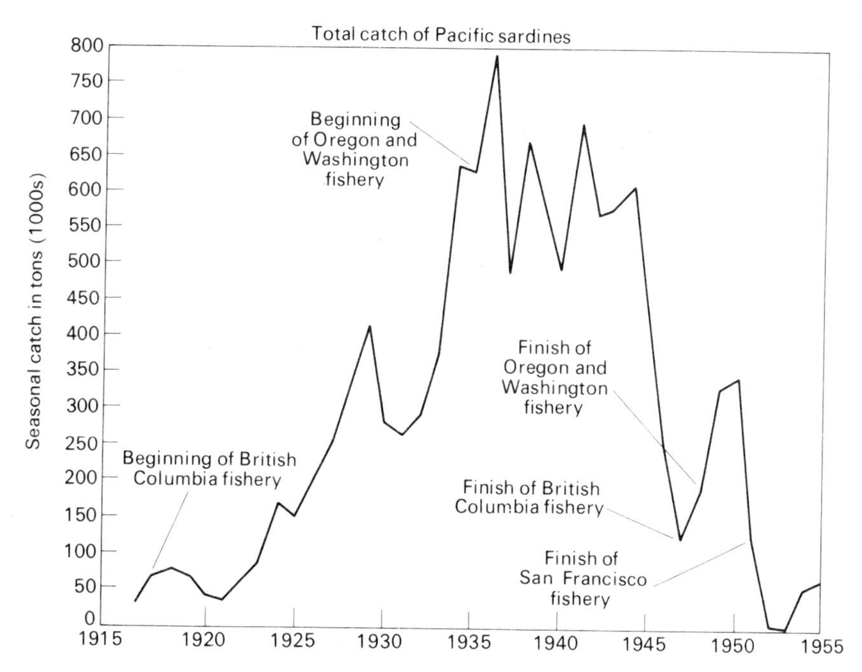

Fig. 8.3 At its height, the Pacific coast sardine fishery was the first-ranking fishery in North America in weight of fish landed, and third-ranking by value after tuna and salmon. The decline is attributed to overfishing and there has been no recovery in 1955–70. Presumably the niche formerly occupied by the sardine has been taken over by another organism or the breeding stocks are too low to enable the population to gain in size.
Source: Dasmann 1972

a competitor, the anchovy *Engraulis mordox* (Fig. 8.3). At the end of the 1950s the latter's biomass was similar to that of the sardine 30 years before and it has clearly ousted the former species, apparently irreversibly (Ehrlich and Ehrlich 1970).

The increasing number of incidents concerning fishing fleets in territorial waters are obviously indicative of competition for the sea's protein resources. A nationalistic attitude towards the use of the resource may promote ecologically sound harvesting of fish populations if the nation which is enforcing the fishery limit is a good manager, and if the extension of limits keeps out the overfishers, but there is no guarantee of such eventualities: the combination of heavy fishing and the intrusion of a warmer current (El Niño) caused a drastic collapse in the Peruvian anchoveta fishery in 1972–73.

Sea mammals other than whales have often been the subject of over-exploitative cropping. The porpoise family (Delphinidae) appears to be in no danger, although little is known about its worldwide status: only local populations have been studied. Japan, for example, is taking about 20,000 porpoises per year and the arctic porpoise (white whale or beluga) is extensively used in some northern regions. Fur seals are better documented: the Guadalupe fur seal was almost exterminated late in the nineteenth century, and the Northern or Pribilof fur seal was reduced from about one million

TABLE 8.10 Whale catch

	1966	*1968/69*	*1974/75*	*1975/76*	*1976/77*	*1977/78*
Number of animals	57,891	42,126	29,267	19,337	27,484	17,839 (quota set by IWC)
Oil production (000 t)	n.a.	206·8	90·5	n.a.	n.a.	

Source: UN *Statistical Yearbook 1977*; FoE 1978

individuals to 17,000 in 1910, since when careful management by the US Government has built the stock up to 1·5 million with a sustained yield of 80,000 animals per year. The true seals are also extensively killed, especially the harp seal of the north Atlantic. Canada, Norway and France share in this resource which, although the subject of controversy, is not in a great decline. The grey seal of the British Isles is also a controversial animal since it eats salmon, and some herds are culled in order to reduce its status as a competitor.

The most outstanding example of the over-use of marine populations is the history of whaling. The products of both baleen (plankton-consuming) and sperm (predatory, mainly on squids) whales have been highly valued in the past: oil, meat, blubber, skin and ambergris have all been used, although effective substitutes could now be found for most of them and whale products were forbidden in the USA in 1970. But a biologically depletive programme of whale cropping, mainly by Japan and the USSR, continues despite falling yields and obviously dwindling stocks. The decline of the whale resource is summarized in Table 8.10 and Fig. 8.4. the International Whaling Commission, which sets catch limits, is fully aware of the depletion of the whale stocks and has set out catch limits and preservation policies such as the complete protection of the Blue Whale (now numbering about 7,000) since 1965. Ehrlich and Ehrlich (1970) argue that exploitation to the point of extinction is occurring, whereas Gambell (1972) suggests that most stocks are stabilized at a sustained-yield level, with the exception of the overcropping of the Antarctic and north Pacific fur whales. Such a stabilization is presumably much below the level that could have been achieved if rational management policies had been followed earlier, for Gambell (1976) says that 'most stocks have been so reduced by the successive phases of the industry that there is only a small remnant of some of the major species left in the world's oceans ... yet this natural renewable resource could make an important contribution amounting to about 10% of the total yield of marine products on a sustainable basis'.

Effects of contamination

Large as the oceans are, they are not immune from the end-products of man's resource processes. The input of materials into seas is both deliberate and accidental, and only recently has there been much concern about the use of the sea as a garbage can, so that quantitative studies often lack long runs of time-series data. In general we can say that

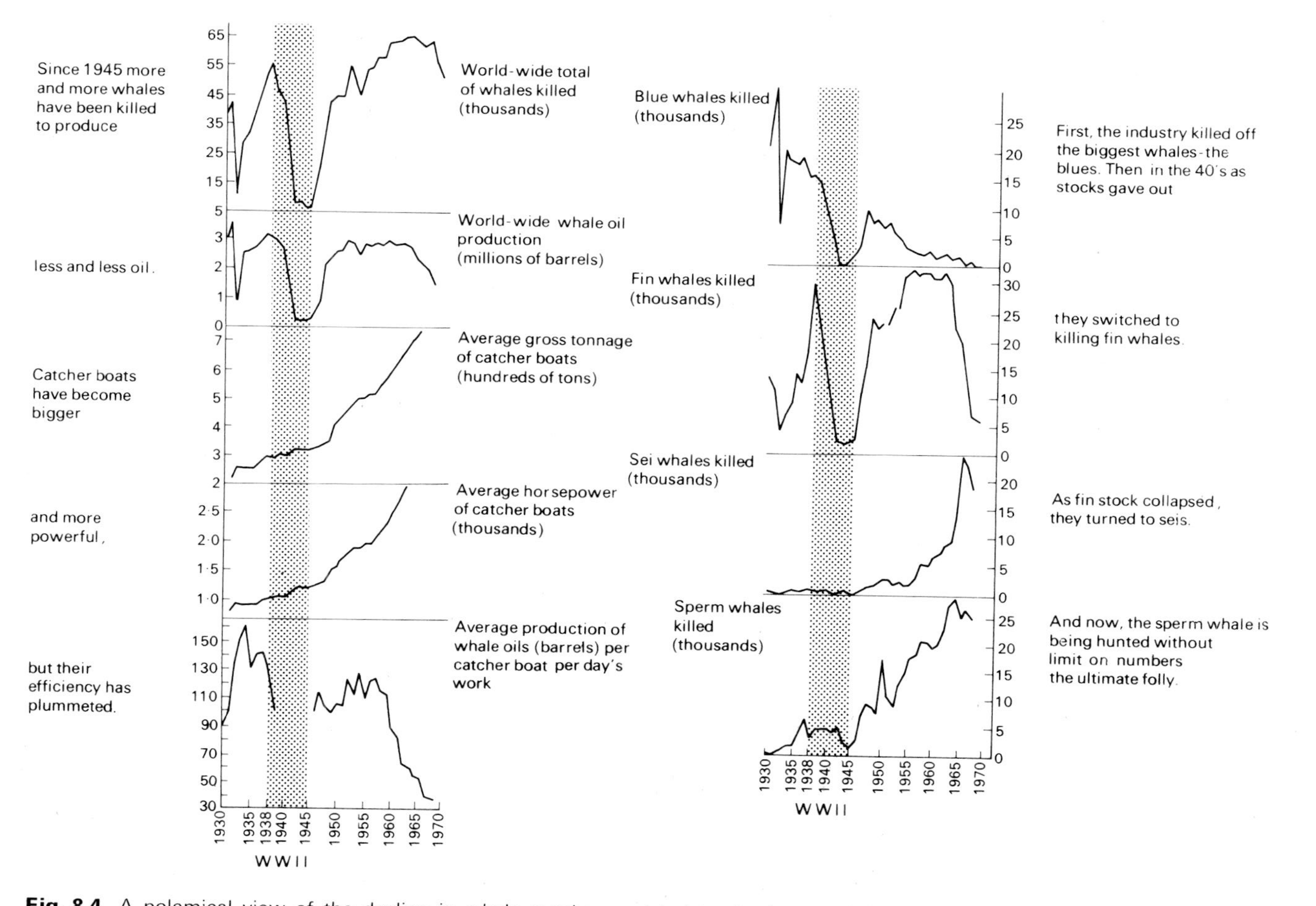

Fig. 8.4 A polemical view of the decline in whale numbers, related to the inputs of the industry. Some whale biologists would claim that the extinction of the blue whale has been averted and that most whales are now being cropped at a sustained-yield level. Source: Ehrlich and Ehrlich 1970

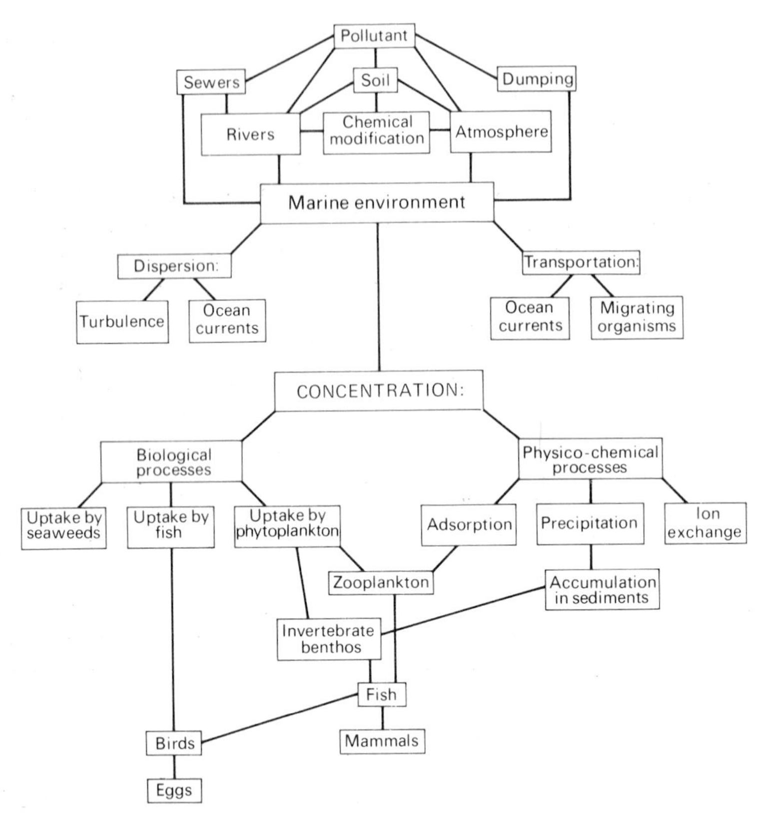

Fig. 8.5 A flow diagram of the pathways by which a contaminant can find its way into the oceans and the ways in which it can be concentrated, with various lethal and sublethal effects.
Source: MacIntyre and Holmes 1971

most substances present on the land will end up in the sea (Fig. 8.5) and that there is a danger that if will undergo concentration processes in the marine systems.

The nature and effects of individual contaminants of the biosphere are dealt with in Chapter 11 and here we will mention briefly only those which have been described as creating particular problems in the seas. They are: radio-isotopes, industrial effluents, oil, heavy metals, persistent pesticides, and eutrophication agents such as untreated sewage, fertilizer runoff and detergents. All of them affect coastal waters most markedly, but organochlorine pesticides such as DDT and its breakdown products DDD and DDE are particularly widespread because long-range atmospheric transport of these substances in aerosol form takes place. Where bans on DDT have been enforced, the residual quantities have fallen but in the interim the organochlorines have been implicated in the failure of birds such as the California pelican and the Bermuda petrel to reproduce successfully because of eggshell thinning. At one time it was feared that DDT was sufficiently concentrated in the surface layers of the oceans to cause poisoning of the phytoplankton and hence to reduce the primary productivity of the oceans. The laboratory experiments however which revealed this potential seem not to have been repeated in the oceans, although the effects of such contaminants is hard to measure because of the naturally wide fluctuations of the populations of small organisms (Longhurst et al 1972).

Even more widespread than DDT residues are those of the polychlorinated biphenyls (PCBs), a group of industrial chemicals introduced in the 1930's and escaping to the environment at a rate of about 28,000 t/yr in the 1970's, with much of this ending up in the seas. Unlike DDT there is no concentration along food chains but they too have been implicated in the poisoning of higher marine animals such as birds. Measures to reduce the loss of PCBs to the environment are now found in most DCs.

A form of eutrophication peculiar to the seas is the dinoflagellate bloom, common under natural conditions where there is an upwelling of nutrient-rich water or disturbance of bottom sediments by tides. The microscopic dinoflagellates may secrete toxins which directly poison the water, may use up all the oxygen in the water if in a relatively confined space, or if filtered through molluscs may build up to high levels. Eating mussels after a 'red tide' allows ingestion of a poison which affects the human central nervous system and to which there is no known antidote. Some recent blooms have appeared in the waters off-shore from known sources of untreated sewage, as off north-east England in 1968 and Nova Scotia–New England in 1972, and although proof is lacking, there is a strong suspicion that human activities can initiate the onset of such phenomena. At any rate red tides put up the price of shrimps by 15 per cent in North America during 1972.

Given these wastes, rational control of marine pollutants seems necessary. The identification of substances that might jeopardize marine resources (e.g. by reducing the diversity of primary producers) and the imposition of enforceable regulations have so far been confined to artificially produced radionuclides and a few other very hazardous substances such as mercury: one of the difficulties is that the effects may be long-term and low-level rather than obvious and crisis-provoking.

The future value of the oceans

Even under ideal institutional conditions the sea is neither an inexhaustible provider of food and mineral resources nor a bottomless sink for the end-products of resource processes. The ecology of the sea is inimical to the production of large quantities of organic matter per unit volume, as is the volume of water to harvesting of crop, and the sheer immensity of the quantity of water means again that mineral extraction is expensive. If minerals impose any constraints upon the primary productivity of much of the oceans, their removal for industrial purposes could possibly reduce marine harvests, a trend which is likely to be exacerbated by certain forms of pollution of the seas.

Man's 'ecological demand' upon the sea, as upon other biospheric systems, is increasing steadily. Over-use, particularly of fish, seems easy to attain: the adults are overfished and so reproduction is hindered and younger individuals are taken, reducing recruitment to the population. In turn, man's competitors for fish are perceived as pests and if possible their populations are reduced. The eventual effect is likely to be instability in the ocean systems, with large and unexpected outbreaks of 'pest' species and large fluctuations in fishing yields. These symptoms are likely to occur when yields of two to four times the present crop are achieved (Institute of Ecology 1972). Various estimates seem to agree, however, that the maximum sustainable world fish catch is about 100 million t/yr, out of a biological production of about 240 million t/yr. It is certainly unlikely to be reached if gross contamination of the seas by toxic substances takes place, since the coastal zones are especially vulnerable to pollution and from them come half the fish production and over half the money made on fishing as an occupation. Estuaries are especially fragile ecosystems and are much polluted and reclaimed, but are among the most productive ecosystems on earth, as well as often being important habitats for fish in their early stages of development. These seemingly barren places, haunts principally of wildfowlers and melancholic poets, therefore deserve special study and protection (e.g. H. T. Odum et al 1977).

In summary, the food resources of the sea can under the most optimal circumstances never be a panacea for all the nutritional problems of the world. Watt (1968) calculates that if we assume 100 g of marine food to contain 100 kcal of energy, and if the crop were to be multiplied 20 times, then about 9×10^8 people could be supported. With a population already at $3{\cdot}4 \times 10^9$ when he wrote, that meant only one-quarter of the present population could be thus fed. In some ways this is a misleading calculation, since the main use of marine food is for protein (9 per cent of the catch weight of fish may be edible protein), and the factor 20 for future cropping is obviously too high. Estimates of potential yield and role vary but are of the same order, and that of Ricker (1969) seems to reflect a general view. His opinion is that in the next 40 years the 1968 catch can be increased by about 2·5 times, giving an eventual total of 150–160 million t/yr, containing 20 per cent of usable protein. For a population of the order expected in AD 2000 this could supply about 30 per cent of the world's minimal protein requirements but only 3 per cent of its biological energy demands.

This all supposes rational exploitation of the marine resources, which regrettably does not happen (Larkin 1978). There are a relatively large number of international agreements about fishing rights and practices (such as the 1958 Geneva convention on Fishing and Conservation of Living Resources of the High Seas), but loopholes are not difficult to find, and the nature of fisheries has been changing more rapidly than the machinery to deal with them. The International Whaling Convention has sometimes disregarded the advice of its biologists and harvested well above maximum sustained-yield levels; pirate whaling outfits operate without regard to the IWC; Denmark refused to limit her oceanic catch of North Atlantic salmon in spite of the decline in its numbers. Numerous examples, some of them leading to international incidents like the Peruvian seizure of US fishing boats and the British-Icelandic 'cod wars' of the 1970s and the refusal of the EEC nations to accept reasonable conservation measures in the North Sea, can be found. At an FAO Conference in 1972, one fish-management scientist was quoted by the press as saying, 'If we wrote a book about our profession, there would be 20 pages of introduction, one page of results and 180 pages of excuses.' The sea, as Garrett Hardin (1968) has pointed out, is a 'commons' where every extra

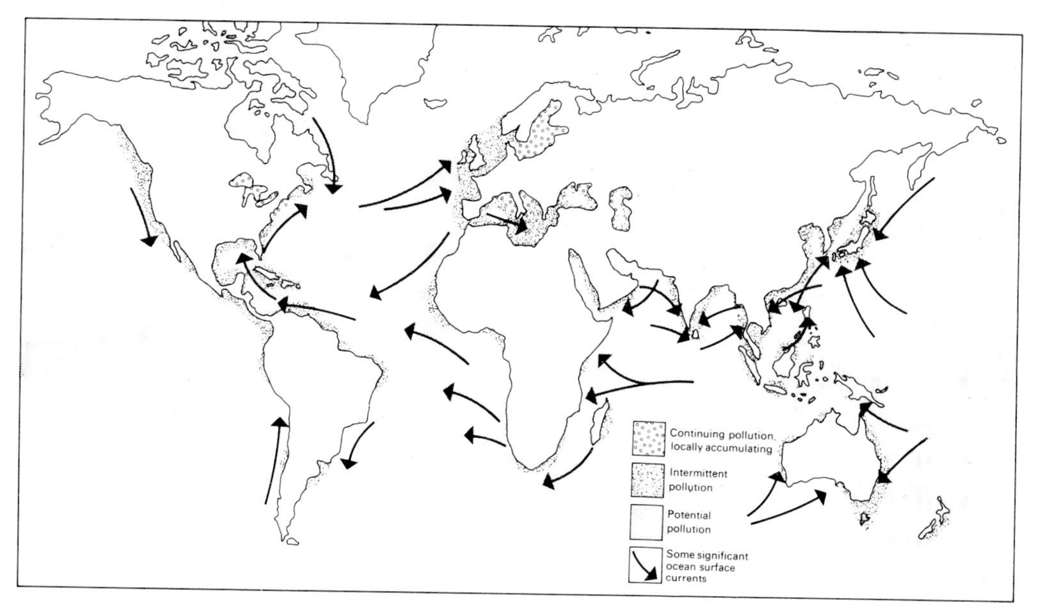

Fig. 8.6 Marine pollution around the world, both actual and potential. The latter refers especially to oil or noxious cargoes along the major shipping lines.
Source: M. Waldichuk and L. Andrèn, reprinted in *Ceres* **3** (3), 1970, 36–7

exploited unit beyond the ecologically permissible limit is of benefit to the individual cropper but a significant loss to everyone else. New concepts of the 'ownership' of marine resources are probably needed for rational management: the alternatives (Holt 1971) seem to be international ownership or the unprecedented extension of appropriations by nation states as has happened with the declaration of 200-mile exclusive economic zones. But even given substantial institutional agreement, no improvement in fisheries management is likely for at least 15 years. The first priority, however, is to reduce the already extensive pollution of the oceans (Fig. 8.6), towards which the 1972 agreements on dumping in the oceans was a first step. Thereafter long-term management of (a) the exploitation of fisheries and (b) inimical interactions between different uses for the oceans (Fig. 8.7, p. 222) become an absolute necessity; given the number of abortive meetings of the Law of the Sea Conference, perhaps the Papal Conclave technique should be applied: the delegates stay locked in until a result is achieved.

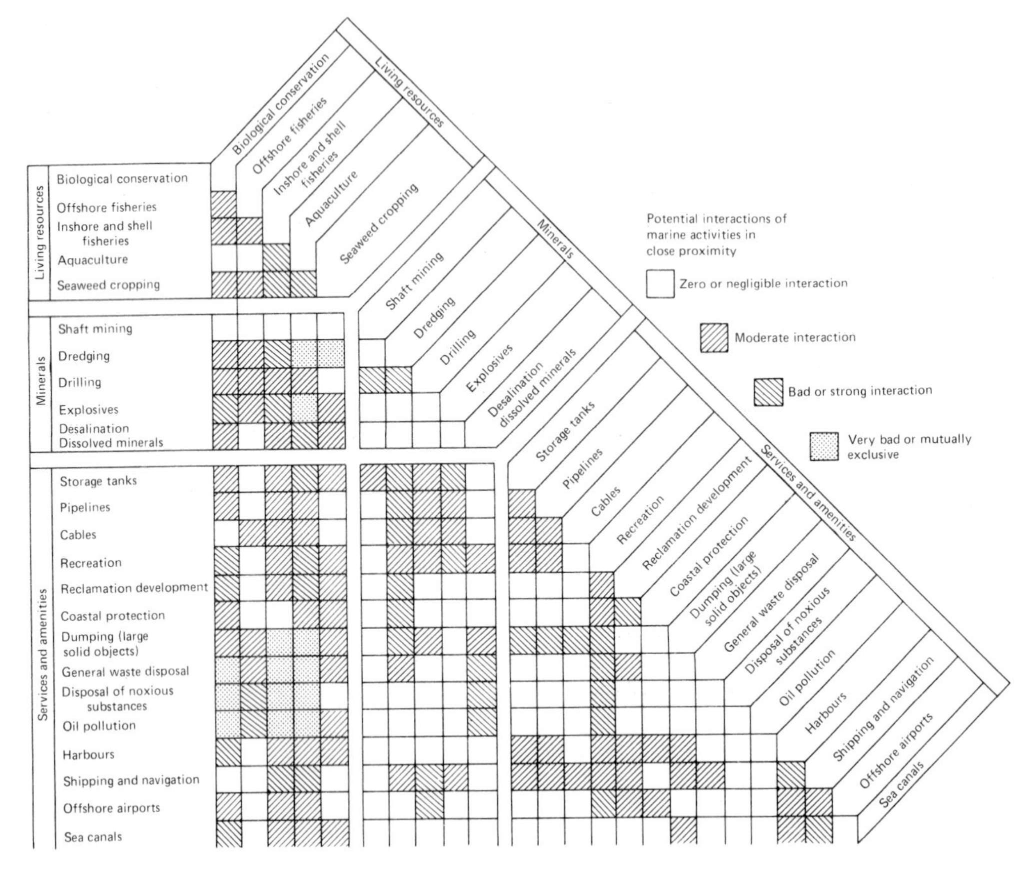

Fig. 8.7 Uses and interactions in the marine environment.
Source: Copper 1975

Further reading

BELL, F. W. 1978: *Food from the sea. The economics and politics of ocean fisheries.*
BOESCH, D. F., HERSHNER, C. H. and MILGRAM, J. H. 1974: *Oil spills and the marine environment.*
BURTON, J. A. (ed.) 1975: The future of small whales.
CALAPRICE, J. R. 1976: Mariculture: ecological and genetical aspects of production.
CHRISTY, F. T. and SCOTT, A. 1976: *The common wealth in ocean fisheries.*
CLOUD, P. 1969: Mineral resources from the sea.
CUSHING, D. H. and WALSH, J. J. (eds.) 1975: *The ecology of the seas.*
CUSHING. D. H. 1978: *Science and the fisheries.*
GAMBELL, R. 1976: *Population biology and the management of whales.*
GOLDBERG, E. D. and BERTINE, K. K. 1975: Marine pollution.
GULLAND, J. E. (ed.) 1972: *The fish resources of the ocean.*
GULLAND, J. E. 1975: The harvest of the sea: potential and performance.
HARDEN-JONES, F. R. (ed.) 1974: *Sea fisheries research.*
HOOD, D. W. (ed.) 1971: *Impingement of man on the oceans.*
INSTITUTE OF ECOLOGY 1972: *Man in his living environment.*
ISAACS, J. D. and SCHMIDT, W. R. 1980: Ocean energy: forms and prospects.
LOFTAS, T. 1972: *The last resource.*
MARX, W. 1976: *The frail ocean.*
MYERS, N. 1975: The whaling controversy.
POMEROY, L. R. 1974: The ocean's food web: a changing paradigm.
RAY, G. C. 1976: Critical marine habitats.
RICKER, W. 1969: Food from the sea.
RUIVO, M. (ed.) 1972: *Marine pollution and sea life.*
TAMURA, T. 1970: *Marine aquaculture.*

9

Minerals

All the minerals which we use to make 'things' come eventually from the earth's crust and their transformation requires some atmospheric elements as well, such as hydrogen and oxygen. The history of mineral use is also the history of man's material culture, dating from the first use of rock tools for hunting and coming through to the origins of present machine-oriented cultures which lie in the discovery of how to extract metals from their ores. Copper was first, its use dating from about 6000 BC but it was replaced by the much harder iron in the years after 1600 BC. It was the much later discovery (AD 1709) by Abraham Darby of Coalbrookedale in England of how to smelt iron using coke as a fuel, that set in place one of the structural girders of the industrial state; today's world can be said to be built upon a framework of steel.

Wealth from the earth

We demand a great variety of planetary inorganic materials. Chief among these are the ores which are used on a large scale to yield metals like iron, aluminium and copper. To them must be added elements which may not be needed in large quantities but which are indispensable in many modern industrial processes, as for example catalysts and hardeners: vanadium, tungsten and molybdenum are instances of this category. Finally there are non-metallic materials which are vital to industrialized nations such as sand and gravel, cement, fluxes, clay, salt, sulphur, diamonds, and the chemical by-products of petroleum refining. All these materials are characterized as 'non-renewable' resources: they can be extracted from the earth's crust only once as far as human time-scales are concerned, since their renewal takes place at geological time-scale. But the materials are not lost to the planet (if we except lunar module junkyards) and so ideally are available for re-use.

The distribution of minerals in the earth's crust is characterized by discontinuity; for a start, eight elements comprise nearly 99 per cent of the earth's crust (Table 9.1). There is as well spatial discontinuity in which deposits rarely coincide with the boundaries of nation states that wish to use them. North America is well supplied with the ore of molybdenum, for example, whereas Asia is not; by way of compensation Asia is rich with tin, tungsten and manganese. Between them, Cuba and New Caledonia have half the world's reserves of nickel, and industrial diamonds are dominated by Zaïre. Such discontinuities are emphasized by temporal patterns of use: the older industrial countries such as the UK are running out of their ore reserves and coming to depend upon imports, and heavy users like the USA face similar problems; in both cases iron ore stands as a good example.

TABLE 9.1 The most abundant elements in the earth's crust

Element	*Weight %*	*Volume %*
Oxygen	46·60	93·77
Silicon	27·72	0·86
Aluminium	8·13	0·47
Iron	5·00	0·43
Magnesium	2·09	0·29
Calcium	3·63	1·03
Sodium	2·83	1·32
Potassium	2·59	1·83

Source: Vokes 1976

Another type of discontinuity is exemplified by the richness of an ore. A few metal ores show a more or less continuous grading from the richest ores (which are usually worked first and are cheapest to extract) to the poorest: iron and porphyry copper are examples. On the other hand some ores are either very rich or very poor or both. A simple extension of extraction and refining techniques learned from the rich ores down to the poorest may not thus be possible: a whole new dimension of technology may well be essential (Lovering 1968). In spite of any difficulties of supply and processing, the industrial nations of the world have come to be very large users of minerals of all kinds. Iron and its products are perhaps the most important, and are a good index of industrialization. From 1957 to 1976 the world-wide rate of increase in steel production was 67 per cent, from 100 kg/cap to 167 kg/cap. In absolute amounts, the USA uses most steel (producing 116,121 $\times$ 10^3 t in 1976 out of a world total of 668,549 $\times$ 10^3 t) but per capita production seems to be highest in Japan: 535 kg/cap in 1976. All this is of course new steel to be added to that already in existence although there are losses from the system in the form of small particles produced for example by friction. Data for production and for losses suggest that during the 1960s each US citizen was supported by 9·4 t of steel mostly in the form of heavy structures, piling and sheet metal, but about 8 percent of which was in cars, trucks and buses (H. Brown et al 1963, L. R. Brown 1970).

Although figures for the USA are somewhat atypical, they nevertheless represent a level of aspiration in many places. Apart from the steel requirement discussed above, the US citizen is also responsible for the yearly use of 7·25 kg of lead, 3·55 t of stone, sand and gravel, 227 kg of cement, 91 kg of clay, and 91 kg of salt; in all representing about 20 t of raw material to be extracted from the earth to support each individual (L. R. Brown 1970). Demands upon the planet's resources and space caused by extension of such demands to many nations would be staggering, although under the present system of economics they are unlikely to occur.

The quantities of materials used and the anticipated demand (especially from LDCs desiring to 'modernize' their economies) has led to many attempts to quantify the amounts of minerals which are accessible to human use. The variables are manifold

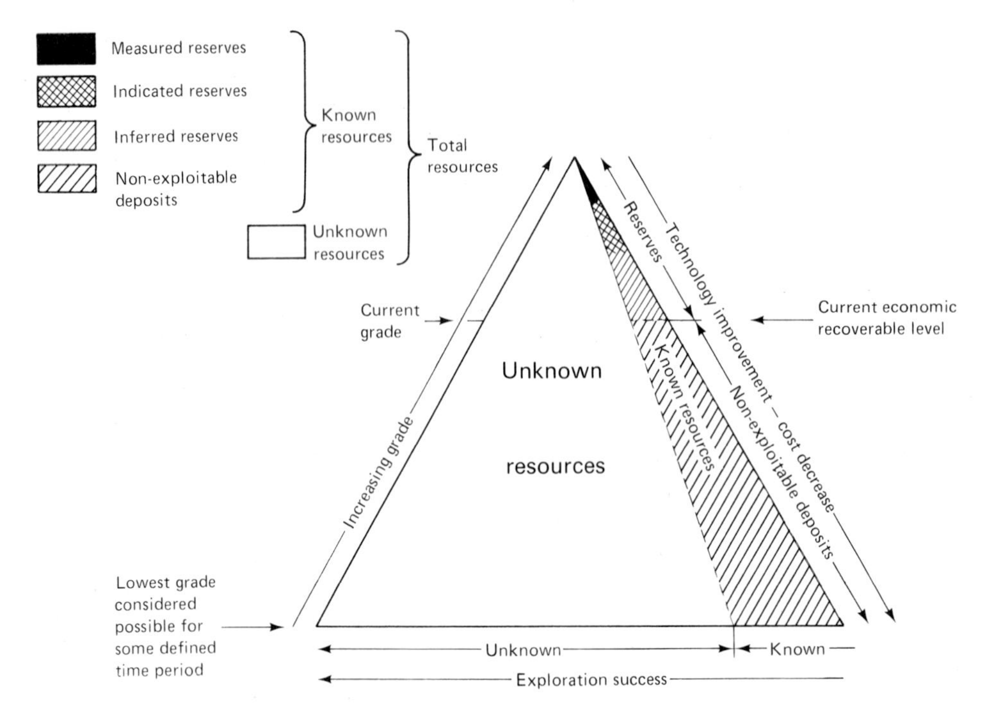

Fig. 9.1 Conceptual model showing relation between reserves and resources as a function of grade, technology, cost, and exploration.
Source: Govett and Govett 1976

since not only technological development but energy costs are difficult to predict as are possible substitutes; and political contexts are notoriously liable to change. A basic conceptual model which combines geology, technology and the state of knowledge about a mineral is given as Fig 9.1. All the values in such a model are liable to change, since the absolute amounts of an element in the earth's crust are rarely of any significance. Table 9.2 therefore represents an assessement at one point in time only of the relationships of the various categories of reserve and resource for some of the industrially most important minerals.

These apparently huge absolute amounts must be set against rates of use which, for instance, have brought about the consumption of more minerals since 1950 than in the previous recorded history of the world, and against projections of future demand. Given all the uncertainties of forecasts and projections (Table 9.3), the suggestion of a demand for minerals in AD 2000 of 2·5 times the 1971 level made by Just (1976) looks a reasonable estimate. It assumes that demand in the DCs will continue to rise but at a slower rate since some form of saturation will set in; but that the LDCs current proportion of 10 per cent of the consumption of minerals will rise markedly. Even this 'medium' projection of 2·5 × the 1971 level poses considerable challenges, especially in terms of the location of the minerals and the consequent political factors involved in their extraction and trade, and in terms of the energy costs of their extraction, refining,

TABLE 9.2 World mineral reserves and resources; $\times 10^6$ t unless otherwise specified

Mineral	*Reserves*	*Identified resources*	*Hypothetical and Speculative resources*	*Ratio of identified resources/reserves*
Bauxite	5,900	12–15,000	>9,600	2·0–2·5
Copper	273	312	653	1·1
Iron ore	251,000	779,000	enormous	3·1
Lead	141	1,500	n.e.	10·6
Manganese	6,500	7,700	10,000	1·2
Phosphate	n.e.	6×10^9	9×10^9	n.e.
Tin ($\times 10^3$ t)	10,100	10,400	17,030	1·0
Uranium	944	1,455	n.e.	1·5
Zinc	235	1,510	large	6·4

Source: M. H. Govett and G. J. S. Govett 1976. 'n.e.' indicates *not estimated.*

TABLE 9.3 Life expectancies of world reserves

Mineral Commodity	*1974 reserves (tonnes)*	*Life expectancy at growth rates of*			*Annual growth rate 1947–74*
		0%	*5%*	*10%*	*%*
Cadmium	$1{\cdot}3 \times 10^6$	27	18	14	4·7
Chromium (ore)	$1{\cdot}7 \times 10^9$	263	54	35	5·3
Copper	390×10^6	56	27	20	4·8
Iron	$87{\cdot}7 \times 10^9$	167	46	30	7·0
Lead	$145{\cdot}1 \times 10^6$	42	23	17	3·8
Manganese	$1{\cdot}9 \times 10^9$	190	48	31	6·5
Molybdenum	$5{\cdot}0 \times 10^6$	70	31	22	7·3
Nickel	$44{\cdot}4 \times 10^6$	67	30	22	6·9
Phosphate rock	13×10^9	128	41	28	7·3
Sulphur	2×10^9	44	24	18	6·7
Tin	$9{\cdot}9 \times 10^6$	42	23	17	2·7
Tungsten	$1{\cdot}6 \times 10^6$	42	23	17	3·8
Zinc	$118{\cdot}8 \times 10^6$	21	15	12	4·7

Source: Tilton 1977
N.B. These data are not necessarily the same as plotted in Fig 9.2 though the conclusions are not radically different.

transport and use in manufacturing. So the critical questions become, (i) can we sustain the production of minerals to meet the rising expectation of an increasing population? (ii) Can this be done without irreparable damage to the biosphere? (iii) For how long can such processes continue?

The attitudes to these questions can be divided into two groups. They are rarely as clearly separated as presented here but for clarity of exposition only they are set out as black and white without the intervening shades. The first group shares with Barnett and Morse (1963) the beliefs that 'the progress of growth generates antidores to a general increase in resource scarcity' and that 'technological progress is automatic and self-reproducing in modern economies'. Hence no major difficulties in mineral supplies are seen since atomic power will produce cheap and virtually inexhaustible supplies of energy, that the role of economics as a distributing mechanism will ensure substitution whenever necessary, and that as the grade of ore decreases arithmetically, its abundance increases geometrically so that it may eventually be possible to extract minerals from common rocks such as granite rather than from special deposits. Beyond the conventional deposits there are, too, the resources of the sea.

These premises of future abundance have been attacked by writers such as Lovering (1969) and Cloud (1975). They suggest that even atomic energy is unlikely to be cheap because of the capital involved and the high costs of safety and waste processing. They argue that the common rock solution ignores the contamination problems of extracting, treating and storing the high volumes of rock that would be required (e.g. the volume of rock increases by ca 40 per cent after it is mined), and that grade/abundance ratios may be continuous for metals such as iron and aluminium but for many (e.g. lead, zinc, copper, manganese) there are sharp discontinuities which might necessitate whole new technologies for recovery. As long as metals are mined for common purposes, limits to the useable grade will be imposed by energy requirements, which show hyperbolic increases as the grade of ore decreases (Page and Creasey 1975). Further, economics may only bring about the desired substitution if there is a long enough lead time for development, a condition made difficult by the rapid incremental rates of growth of consumption. As for the seas, as chapter 8 has shown, the various problems of extraction are likely to confine their usefulness to a few minerals and to limited quantities of them. The major question to be asked of technology is, 'can it raise the expected population of AD 2000 and beyond to DC levels of supply?' It might be possible for iron ore, for example, but not for the molybdenum needed to convert the iron into steel. In the case of lead, zinc and tin, such a demand would exceed all possible reserves of these metals.

Unless there can be created a technology based on universal and abundant elements only (iron, aluminium, magnesium, silicates, hydrogen, oxygen, nitrogen and sunlight) then the ideas of finite supplies and extraction cycles which will soon reach their peaks and will be back near to zero in the next century, to take on a more realistic air than the affirmation of cornucopian supplies (Fig 9.2). But the currently low prices of many minerals means that adjustment to such ideas will be slow in taking root.

Ecology of mineral use

As Flawn (1966) puts it,

> Man, like an earthworm, burrows into the earth and turns over its surface; like a bird, he brings material from elsewhere to build his nest; and like the pack rat he accumulates quantities of trash.

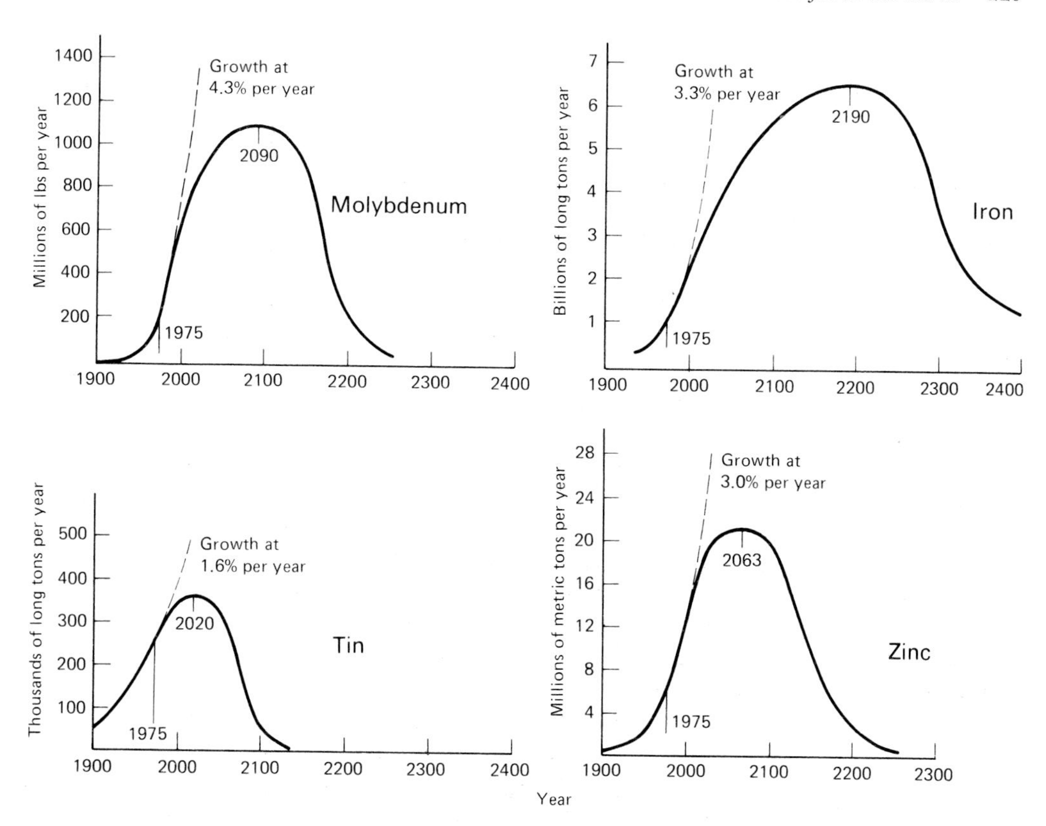

Fig. 9.2 Complete cycles of production of various minerals, given certain assumptions about abundance and rate of growth of consumption. Note the different units along the vertical axes.
Source: Read 1975

And most of these activities have side-effects of an ecological nature. Underground mining, for example, may include whole new towns among its surface installations: the San Manuel coppermine in Arizona necessitated 1,000 dwellings, 19 km of paved streets and three schools. Timber is cut in forested areas, often leading to soil erosion, and the tailings and mine-waste have to be discarded. Large-size solid wastes can be used as backfill or sold for aggregates, but tailings usually yield silt particles to wind and water and are often chemically unstable; the only suitable treatment appears to be to 'fix' them with vegetation. Mine waters are often heavily contaminated and have to be treated chemically and physically, or injected into 'safe' rock strata. Either method is expensive and the desire to forgo such outlays may overcome some mining operators, resulting in ecologically toxic effects. It is reported, for example, that the Bougainville mine in the Solomon Islands, extracting 0·48 per cent copper ore, was preceded by the removal of 40 million tons of overburden, including forest, and two-fifths of all material mined (over 400 million tons) will be dumped in a neighbouring valley. An oval basin

2·1 × 1·5 km will be an end result, as will the choking of the water courses and silting of the coastline. The lifetime of the project is estimated at 25–30 years and the profits to the developer at £20 million/yr (Counter Information Services n.d.).

Considered for the whole world, open-pit mining is more widespread than extraction by shaft; in the USA, for example, 80 per cent of metallic ores and 95 per cent of non-metallic minerals and rocks are extracted by this means (Plate 15, opposite). The technology has been facilitated by machines such as dumper trucks with a capacity of 137 t, coal haulers of 245 t capacity and draglines with 84 m booms. Thus the biggest operations can extract 100,000 t of ore per day. Waste disposal becomes a major problem to which backfill is the obvious solution; if the topsoil is saved then restoration to agricultural or recreational use is often possible. Where dredging is concerned, disposal of spoil is the main difficulty, together with changes in the equilibrium systems of water bodies. The extraction of sand and gravel from bays has caused the erosion of beaches, and in river beds has increased the downstream erosive capacity of the stream. The extra sediment suspensions created may kill flora and fauna, as with oyster reefs along the Texas coast. The processes of concentration, beneficiation and refining may all create biological change if various products are released into nearby ecosystems. Washing yields sediment and slime-charged liquid wastes, leaching produces spent acids, flotation is a source of tailings and contaminated liquids, and smelting gives rise to slags and gases high in elements like sulphur or even fluorine. Devastated areas like the lower Swansea Valley of south Wales or Copper Hill in Tennessee are familiar examples.

Nearly all these effluents can be treated or put to gainful use, usually at some extra cost, but they would be produced in unprecedented quantities if 'common rock' such as granite were to be used as a multi-mineral source, for the ratio of waste to usable rock would be 2,000 : 1. If underground leaching were adopted, large underground explosions would be needed to shatter the rock, followed by the drilling of input and recovery wells. Subsequent contamination of ground water with extractive chemicals would be difficult to avoid. One further aspect of mineral extraction and processing is the differential invasion of wild ecosystems where competition for other uses tends to be less and where processing is remote from concentrations of people who might be harmed. An example is the Canadian shield, a very heavily mineralized region, where the wild lands are highly valued for their unaltered state. Then controversy ensues, as with the constant pressure to open the Wilderness Areas of the USA to mining, or with the pressures to extend limestone quarries in the Peak District National Park in the English Midlands.

Each individual industrial plant, along with the energy use from which it is inseparable, produces local ecological change by placing a strain on the capacity of the systems of the biosphere to absorb the concentrations of elements which industrial processes create, and which economics dictate shall be discarded. Whether there is a

Plate 15 The skeleton of the industrial world is derived from iron ore. This pit in Minnesota, USA, exemplifies the attributes of mineral resource recovery in holes and heaps, high energy input, and a large land area for both the pit and its associated services and housing for workers. *(Grant Heilman, Lititz, Pa)*

worldwide capacity for wastes is difficult to assess, but monitoring of potentially toxic elements would seem to be a minimal step towards trying to avert any breakdown of the biogeochemical cycles of the planet on anything other than a restricted spatial scale.

Recycling

Increased recovery of mineral elements from scrap and waste is often advocated as a response to shortages and a way of minimizing the environmental impacts of the extraction of new materials. Success of the processes depends upon the quantities available in the recoverable unit (obsolete iron machinery has scrap value but not iron rods embedded in concrete foundations), and the resistance of the material to chemical and physical breakdown. Costs of collection will be high from dispersed sources: thus lead from exhausted vehicle batteries can be reclaimed but not that from lead bullets, except perhaps in Texas. The costs of recycling must be set against those of new materials and at present the latter are generally less. Many of those who hold neo-Malthusian or 'environmentalist' views advocate a minerals tax which would increase the price of new material *vis-à-vis* recycled substances, forcing a shift towards greater employment of secondary sources. Some notion of the sources and quantities of metal recycling in the USA is given in Table 9.4.

Minerals in the face of rising populations

In the near future, the patterns of mineral production appear to depend on whether the development of high-cost mines in the DCs will be economic, whether environmental constraints will allow the mining of low-grade deposits, and which countries will spend money on new exploration and development. It looks as if self-sufficiency in minerals for industrial nations will be even less common and thus increased levels of conflict are likely between producers and those who have to import to maintain an industrial economy.

In the longer term two wider perspectives may gain wider acceptance. The first is that many of the economic measures of mineral use (as of other resource processes) will be recognized as indicators of throughput and not of stock, so that they value decay and rapid obsolescence rather than longevity and thrift in materials. The replacement of such indicators is overdue. Parallel to this is the knowledge that one of the effects of mineral use is to accelerate by many times the natural flows of elements through the biophysical systems of the planet. One result is concentrations of mineral elements far beyond the adaptive capabilities of most organisms and so toxification results. Mostly now on a local or regional scale, any growth to a global extent must be strongly resisted, whatever the short-term economic gain.

Population growth underlies the extensions of demand described earlier in this chapter and its diminution would be welcome here as in other systems. But it is not the whole story for it is now very difficult to separate needs and demands in those industrial societies which have consumer-oriented economies. To take a single example, and readers should consider others for themselves, a private auto may seem unlike a

TABLE 9.4 Scrap metal in the United States

Metal	*Approximate annual recovery from scrap*[1] *(1 ton = 1·016 t; 1 oz = 28·349 g)*	*Remarks*
Iron	70–85 million tons[2]	In the iron cycle from mine to product to recovery, the loss of iron is 16–36 per cent. About half the feed for steel furnaces is scrap.
Copper	1 million tons	Secondary copper production from old and new scrap ranges from 900,000 to 1,000,000 tons per year, about half of which is old scrap. Old scrap reserve is estimated at 35 million tons in cartridge cases, pipe, wire, auto radiators, bearings, valves, screening, lithographers' plates.
Lead	0·5 million tons	Estimated reserve is 4 million tons of lead in batteries, cable coverings, railway car bearings, pipe, sheet lead, type metal.
Zinc	0·25–0·40 million tons	Zinc recovered from zinc, copper, aluminium, and magnesium-based alloys.
Tin	20,000–25,000 tons	Tin recovered from tin plate and tin-based alloys, 20 per cent; tin recovered from copper and lead-based alloys, 80 per cent.
Aluminium	0·3 million tons	Because aluminium is a comparatively new metal the old scrap pool is small, but it is growing rapidly.
Precious metals	gold = 1 million ounces silver = 30 million ounces	Precious metals including platinum are recovered from jewellery, watch cases, optical frames, photo labs, chemical plants. Because of the high value, recovery is high.
Mercury	10,520 flasks	Recovery is high. Nearly all mercury in mercury cells, boiler instruments and electrical apparatus is recovered when items are scrapped. Other sources are dental amalgams, battery scrap, oxide and acetate sludges.

[1] Includes old and new scrap. New scrap or 'home' scrap is produced in the metallurgical and manufacturing process; old scrap is 'in use'.

[2] Old and new scrap *consumed* rather than *recovered.*

Source: Flawn 1966

necessity in a densely-settled European city but in Los Angeles with scarcely any public transport, the case is altered. Whether future demands for minerals have to meet the demands or the needs of future populations may bring about considerable differences in most of the issues described in this chapter.

Further reading

BROWN, H. 1954: *The challenge of man's future.*
BROWN, H. 1970: Human materials production as a process in the biosphere.
FLAWN, P. 1966: *Mineral resources.*
GOVETT, G. J. S. and M. H. (eds.) 1976: *World mineral supplies. Assessment and perspective.*
REED, C. B. 1975: *Fuels, minerals and human survival.*
SCHURR, S. H. (ed.) 1972: *Energy, economic growth and the environment.*

10

Energy

Access to sources of energy mediates the whole of man's relationships to the planet and in particular the use of all kinds of resources, since their extraction, conversion and transport may depend upon the control of large quantities of energy (Foell 1978). Access to stored energy sources is the basis of industrialization with all its concomitants in terms of the manipulation of ecosystems by mechanical and chemical means. It also permits penetration to nearly all parts of the planet, making possible such activities as recovering minerals from under the sea-bed or living permanently at the South Pole. Together, energy and mineral use have also provided the means for man to escape from the surface of the planet and hence the chance to view it from outside, both personally as in the case of astronauts and vicariously as with remote sensing. This capability may on the one hand evoke our sense of wonder at its apparent singularity or on the other enable us to view it as an object capable of exploitation; time alone will tell if either of these views has become predominant.

All resource processes can be characterized by quantifying the flows of energy through them, and studies of energetics may be used to link their ecological and economic dimensions (H. T. Odum, 1971, H. T. and E. C. Odum, 1976). The role of energy in natural systems has been mentioned in chapter one, and solar flows dominate unused lands, wildernesses and protected areas. Where modern outdoor recreation is important then flows of energy from fossil fuels are superimposed on the solar fluxes, as in the construction of roads and other facilities, the use of powered vehicles, and the operation of installations such as ski-lifts. Energy dominates forestry in terms of the balance between solar capture as trees and fossil fuel expenditure in harvesting and processing; similar considerations are seen to pervade the food system, especially in industrial countries. Even water supply has its linkages to energy sources in, for example, the costs of demineralizing water and the price of pumping it up from the source in places where gravity feed cannot be used. So too with minerals, where energy costs are becoming ever more important in deciding which deposits can be exploited; a relationship which may apply *mutatis mutandis* to the processing and recycling of many kinds of wastes, where if energy were very cheap and ubiquitously distributed, there would be more economic incentive to collect, separate and re-use a whole spectrum of materials currently buried or burned.

The increase in organizational complexity of the industrial nations is only made possible by the understanding and application of energy flows, and it is not without significance that the term 'power' is used for energy obtained for urban and industrial purposes. The actual quantities of energy used are measured by the rate of flow of useful energy that can be made to do work.* Nearly every source uses different measures of energy and power, and conversions are bedevilled by different conversion factors based on different values for the thermal values of, for example, coal and oil. The older units sometimes found are calories and British thermal units; the SI units are the Joule and the Watt. Both of these are small and so the commonest multiples are the kilojoule (KJ), megajoule (MJ) and the kilowatt (as a rate: kWh). For very large quantities the gigajoule (GJ-1 $\times$ 10^6 J) and exajoule (EJ-10^{18} J) are seen, as is the gigawatt (GW-1000 kW) and the terawatt (TW-10^9 kW). American sources also use Q, for Quad, which equals 10^{15} Btu.

It is a one-way flow and degradation occurs, so that heat sinks are an inevitable consequence. Resource processes should therefore consider the ecology and economics of the whole process, from the 'capturing' of the energy source through to its dispersal into the ecosphere as heat. Table 10.1 shows the energy levels of various natural and man-made phenomena. The supply of energy to all resource processes which yield a tangible product is so important, and the escalation of use so rapid, that concern is being evinced over the relations between supply and demand both in the near future and in the longer term, even if not in the time-scale of William Blake's dictum that 'energy is eternal delight'. The various sources of power supply must therefore be examined, with the understanding that this, along with mineral science, is one of the areas where the development of technology may change the prognosis most rapidly, albeit on a time-span of decades rather than years. This must be set against a world consumption of energy that has recently been doubling approximately every 14 years.

TABLE 10.1 Energy levels of various processes

Process	*Energy (kilojoules)*
Green plants covering 1 m^2 of ground, per day	16,700
Food energy for 1 adult for one day	10,000
Combustion of 1 US gallon of petrol (gasoline)	135,000
Fuel consumption of Boeing 707, flight of 5,000 km	1,406 $\times$ 10^6
Average thunderstorm	160 $\times$ 10^9
H-bomb of 1 megaton	4,000 $\times$ 10^9
Industrial energy consumption of world, 1970	220 $\times$ 10^{15}
Solar insolation at top of atmosphere, per year	5,361,120 $\times$ 10^{15}

Compiled from various sources (1 KJ = 0·239 kcal)

* Some basic equivalents are:

1 MJ = 0·277 kWh
1 kWh = 3·6 MJ
1 kcal = 4·19 $\times$ 10^{-6}MJ = 1·16 $\times$ 10^{-6} kWh
1 Btu = 1·055 $\times$ 10^{-3} MJ = 2·93 $\times$ 10^{-4} kWh
1 mtce = 28·8 $\times$ 10 MJ = 8 $\times$ 10^9 kWh
(million tons coal equivalent)

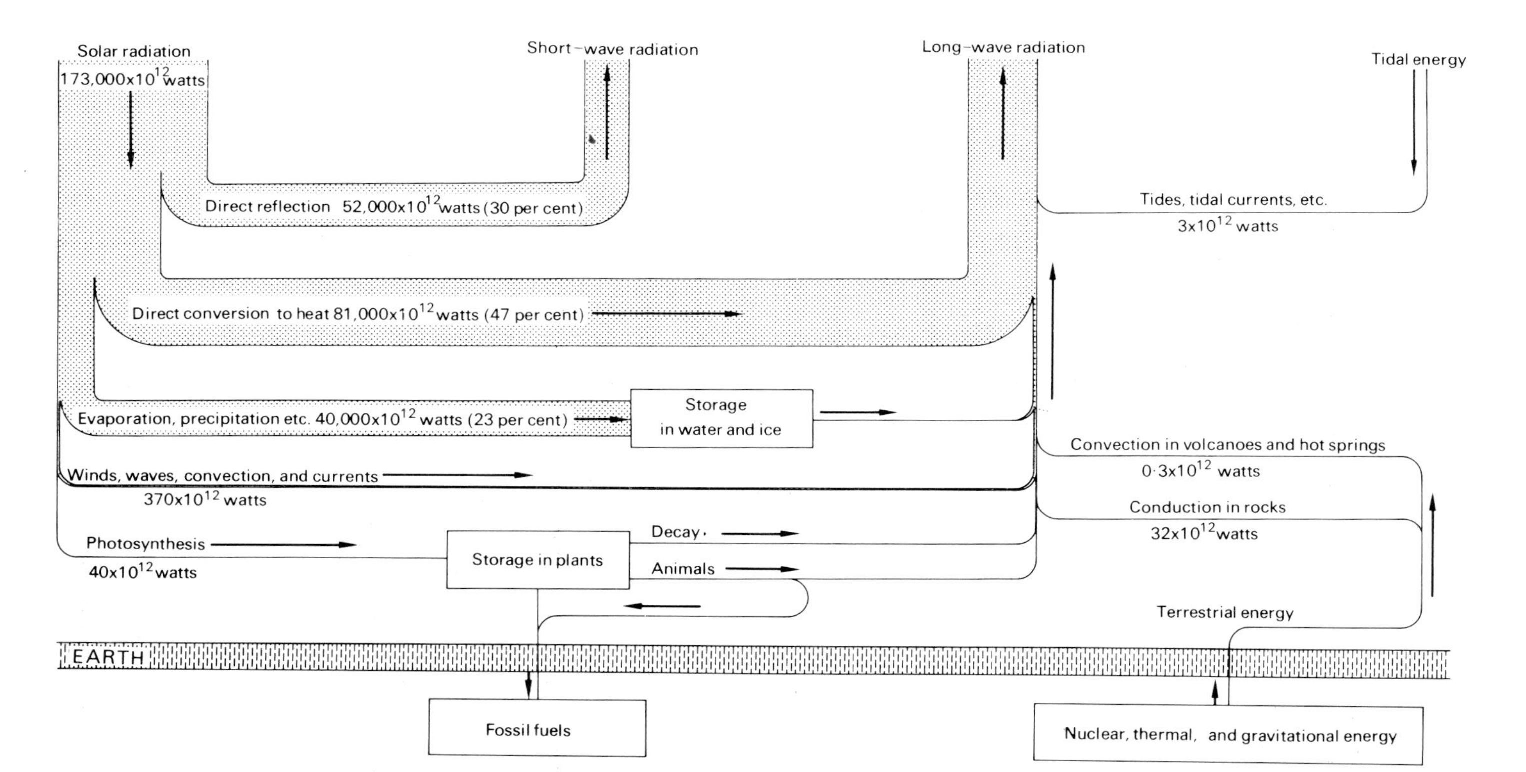

Fig. 10.1 The flow of energy to and from the earth. The overwhelming contribution of solar radiation can easily be seen, as can the small proportion of it which at present enters resource processes, especially via photosynthesis. The thin band leading to fossil fuels reflects their slow rate of accumulation.
Source: Hubbert 1971

Types of energy source

Human societies have access to two different kinds of energy source. The first is simply a diversion of the current flows of the earth's energy systems (Fig. 10.1) of which the solar flux is the most important and which we tap through the medium of photosynthesis and, less directly, through the action of falling water and wind. Gravitational energy can be harnessed via the tides, and the inherent heat of the earth's crust provides geothermal energy. These sources are termed equilibrium sources since the flows would be present even in the absence of man. The other type is the non-equilibrium set of energy sources which consists of the humanly-procured release of stored energy which would either not happen at all under natural conditions or would happen only at an infinitesimal rate compared with its discharge under the impetus of technology. In this category fall the fossil fuels (oil, coal, natural gas, oil shales, tar sands) which represent stored photosynthesis from the geological past and which yield energy upon combustion. Compared with the equilibrium sources they are very concentrated, a quality which applied most strongly to nuclear power, also a non-equilibrium source which liberates the energy contained in the nucleus of the atom. A measure of the significance of this concentration is the equivalence of the energy content of 1 g of uranium 235, 1826 kg of oil and 2700 kg of coal.

Energy uses and demands

Most societies can be characterized by their access to and uses of energy. A hunter-gatherer group traditionally uses only solar energy via food, at a level sufficient for

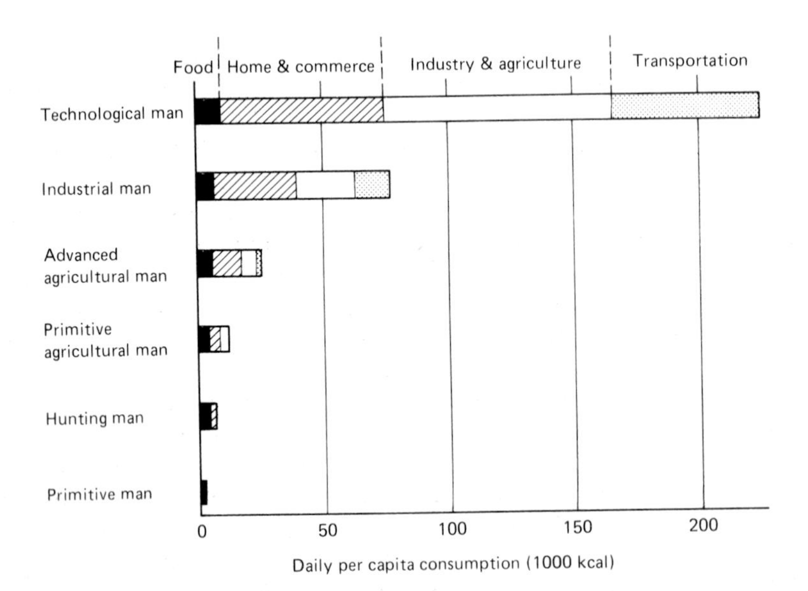

Fig. 10.2 Man's daily energy consumption at several levels of socio-economic development. (1,000 kcal = 4,200 KJ)
Source: Cook 1975a

TABLE 10.2 Industrial energy* consumption 1971

Country	*Consumption kJ/cap/day*
USA	929,632
Canada	771,628
Sweden	503,690
DC average	497,002
UK	455,620
Australia	450,604
Norway	431,376
USSR	375,364
Japan	270,028
WORLD AVERAGE	159,676
Spain	137,940
Iran	84,854
China	46,398
Brazil	42,636
LDC average	28,842
Egypt	23,408
India	15,466
Pakistan	6,688

*Does not include food, fuelwood, animal or human energy.
1 kcal = 4·18 KJ
Source: Cook 1975a

metabolic purposes, together with fuel-wood for fires. An agricultural society will tap the energy of the sun but add to it the labour of beasts and perhaps power from wind or falling water as well. Combined with the energy-intensive domesticated crops, a surplus may be produced which enables the society to support people who are themselves not farmers on a much larger scale than is possible in a hunting culture. Because of its access to stored energy, an individual group uses many times its own metabolic energy in order to produce food and manufactured goods, run transport systems and generally underpin many of the multifarious activities of a complex developed economy (Fig. 10.2). As we would expect, levels of energy use are not constant in the world: the United States citizen consumes 2·4 times as much as his counterpart in Europe and Japan, and 32 times the average for LDCs (Table 10.2).

In advanced industrial societies the pattern of use of energy is in most places similar to that of the USA, which is depicted in Fig. 10.3. Twenty-five per cent of it was used in the transport of people and goods, 20 per cent in houses (of which more than half was for heating), 12 per cent in the 'commercial' sector such as offices, hospitals and schools. The remaining 42 per cent was in the industrial sector, including agriculture with its very heavy use of energy (see p. 175). About half the ultimate energy use was controlled by private individuals, the rest in the hands of institutions and business firms. Of all the forms of energy available, a characteristic of the USA and similar societies is the high proportion of electricity: about 25 per cent of all primary energy

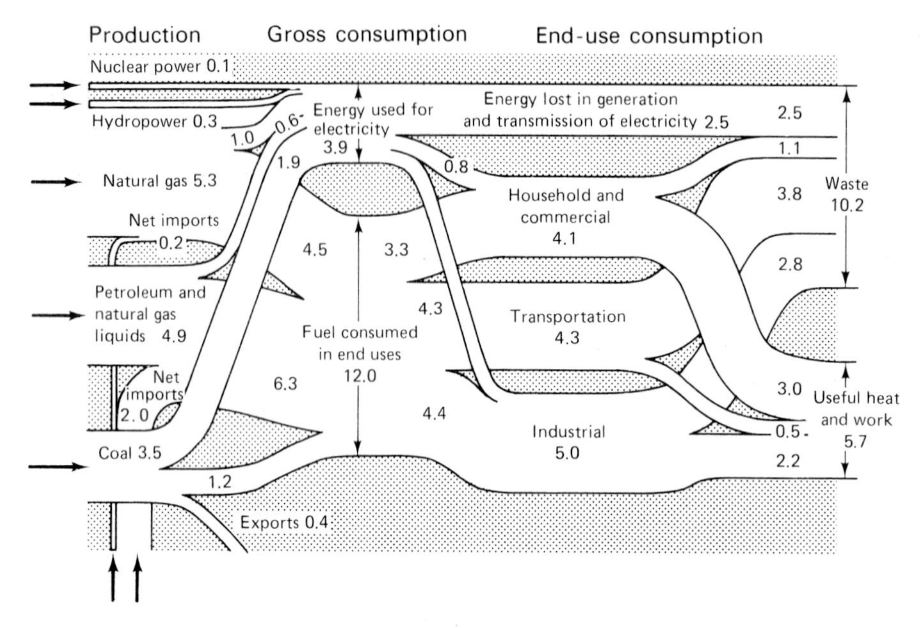

Fig. 10.3 Flow of energy through the US economy in 1971.
Units are 10^{15} kcal = ×4·2 EJ.
Source: Cook 1975a

sources (oil, natural gas, coal, hydropower, nuclear power) were converted to electrical power. A diagram for the UK (e.g. in Foley 1976) shows a similar picture except that the losses in conversion and distribution are 26 per cent of the end-use consumption rather than 50 per cent as in the USA (Holdren 1975).

This mention of losses implies that the various processes of conversion of energy from one form to another vary in their efficiency in terms of the ratios between the input of energy and the work done. The overall efficiency of the US system is about 50 per cent, that of electricity generation 33 per cent and of the average automobile 15 per cent. Some of these losses follow inexorably from the laws of thermodynamics but technological advance may decrease waste in some processes: for instance, the energy requirement for making pig iron has fallen from about 226 MJ/kg in the Europe of the 1800's to 30 MJ/kg today (Slesser 1978). Nevertheless, space heating is generally reckoned to be about 8 per cent efficient and air conditioning only 5 per cent.

The levels of use described above are the outcome of a period of rapid growth in energy consumption since the industrial revolution, with even faster growth since 1945. Table 10.3 shows the world growth of total industrial energy, and Table 10.4 the changing mix of the various sources. The growth has been spatially uneven and in the last decade the picture has been dominated by North America, with 33·9 per cent of world consumption, Western Europe with 20·1 per cent. The centrally planned economies of the USSR, Eastern Europe and China together consumed 29·6 per cent, Japan 2·8 per cent and the rest of the world 10·9 per cent (Smith and Toombs 1978).

TABLE 10.3 World energy consumption in recent decades

	1925	*1938*	*1950*	*1960*	*1968*	*1970*	*1974*
Total consumption millions of tonnes coal equivalent	1,484	1,790	2,610	4,196	6,306	6,821	8,104
Per capita consumption kg coal equivalent	785	826	1,042	1,403	1,810	1,893	2,060

Source: Darmstadter 1971; Foley 1976

TABLE 10.4 Percentage of world energy supplied by main fuels

Fuel	*1900*	*1940*	*1965*	*1974*
Coal & lignite	94·2	74·6	43·2	30·6
Oil	3·8	17·9	36·7	47·6
Natural gas	1·5	4·6	17·8	19·6
HEP and nuclear	0·5	2·9	2·2	2·4

Source: Foley 1976. Comparison with Table 10.2 will make clear that a relative decline may mean an absolute rise: e.g. coal production rose steadily from $1,367 \times 10^6$ t in 1929 to $2,652 \times 10^6$ t in 1974.

These levels of use represent world growth rates around 6 per cent per annum (a doubling time of *ca* 14 yrs) in recent decades, with a slower rate after 1971, exacerbated by a five-fold increase in oil prices in the period 1973–75.

Future demand

The amount of energy required by various nations in the future is unknown. Forecasts and projections abound but are all subject to errors and assumptions and indeed some may become self-fulfilling prophecies. A complex mixture of assumptions about economies, economics, the politics of supply, people's expectations of higher material standards and the environmental effects of power generation and use, is involved in considering future usages. The usual technique is to postulate alternative scenarios of life-style in a society and to calculate the energy needed to sustain that life-style. For example, Chapman (1975) puts forward three demand scenarios for the UK which result in annual demands in AD 2020 of $> 6000 \times 10^9$ kWht, 4250 kWht, and 2750 kWht respectively (cf 1970 base of 2300 kWht). The Ford Foundation (1974) also set out three scenarios using a 1973 base of 75×10^{15} Btu and arriving at AD 2000 demands for the USA of 187, 124 and 100×10^{15} Btu depending on the assumptions made. On a world scale, Häfele and Sassin (1978) present a world future scenario for AD 2030 with a population of 8000×10^6 and an overall economic growth rate of 3·7% per annum. With a 1971 base of 236×10^{12} MJ/yr (1·9 kw/cap/yr), they envisage a consumption of 4·4 kw/cap/yr resulting in a total demand of $108{\cdot}5 \times 10^{13}$ MJ/yr; in simpler figures, an expansion from 7·5 TW (terawatts) to 35 TW, where 1 TW $= 10^9$ kW.

Many demand projections, especially for DCs, have assumed that historical rates of growth must continue and the future needs are predicted on such growth. Increasingly, reaction to these assumptions has promoted 'low growth' scenarios, where efficiency of use and avoidance of waste are emphasized. The energy industries, however, are eager to sustain high growth rates and often suggest that unacceptable changes in life-style would result from them. We shall look closer at such possibilities near the end of the chapter, but we now turn to consider the energy supplies which will be needed in future to sustain various levels of demand.

Energy supply

As with minerals it is difficult to say exactly what constitutes a resource since the appraisal varies with factors such as price, technological change, and the political element introduced for example by uneven spatial distribution. Generally speaking, the terms used are *reserves* for materials whose location is known (*proved reserves*) or inferred from good geological evidence (*probable reserves*) and which can be extracted under present or near-future technological and economic conditions, and *resources* which are materials whose location, quantity and quality are not well known or which cannot be extracted under currently feasible technological and economic regimes. There is also the concept of the *ultimately recoverable resource* which is a crude estimate of how much of the material actually exists assuming that the conditions might one day be favourable for its extraction. It is thus important to remember that in the contect of minerals and energy, the word *resource* has a much more precise and specialized meaning than usual.

For non-renewable materials, we can envisage a production cycle (Fig. 10.4) with a

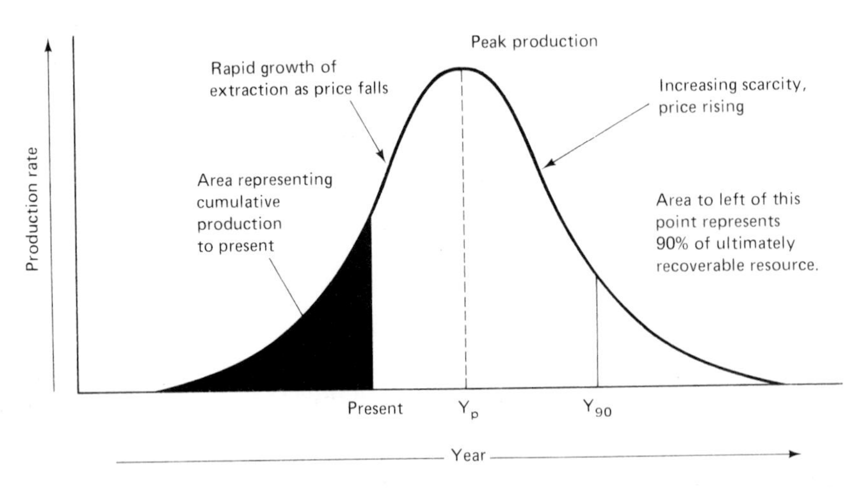

Fig. 10.4 The production cycle of a non-renewable resource, e.g. a mineral or a fuel such as oil or natural gas.
Source: Holdren 1975

rapid growth rate as demand rises and costs fall, then a levelling off of production as the reserves become scarcer and the price rises. This is followed by a decline as increasing scarcity keeps ahead of technological change. For each major energy source, therefore, we need an estimate of its production under different conditions and how far this will go towards meeting estimated demands. But no energy source can be treated in isolation since most countries (and especially the heaviest users) employ a 'mix' of sources. Because of doubts about the longevity of supplies, it is necessary to follow the currently most important fuels with an estimate of the present and potential development of 'unconventional' or 'alternative' energy sources, most of which are of the

TABLE 10.5 Conventional fossil fuels: reserves, resources and supply

	1 Proved reserves (EJ)	*2* Remaining URR (EJ)	*3* Recent annual consumption (EJ)	*4* Reserve lifetime (col 1 ÷ col 3)	*5* Resource lifetime at growth rate increase of 5% p.a. (Yr)
World coal	124,000	220,000	90	1,380	96
World petroleum	3,700	10,900	100 (136 in 1978)	37	37
World natural gas	2,100	10,600	40	53	53

Source: Holdren 1975 and later additions. Smith and Toombs (1975) give largely similar figures for proved reserves of petroleum and natural gas but differ on coal where they estimate only 15,250 EJ.

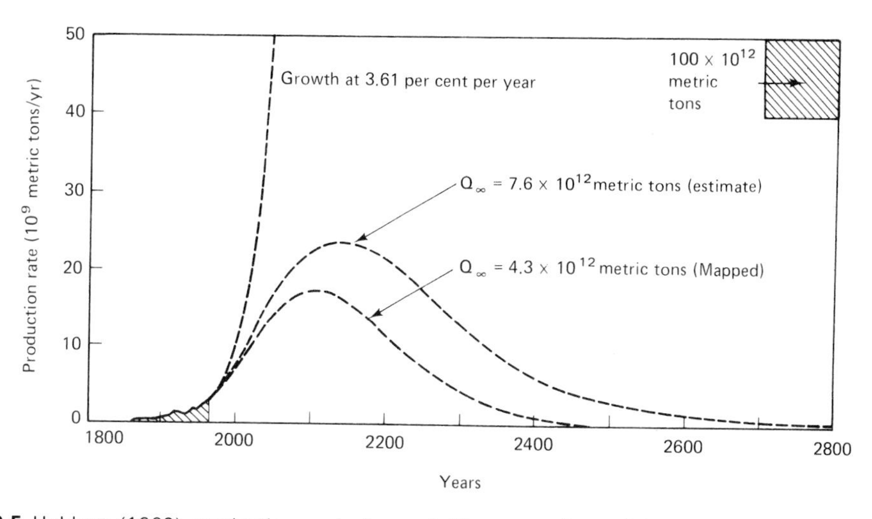

Fig. 10.5 Hubbert (1969) production cycle for coal. The two values of Q_∞ represent the ultimately recovered quantity of coal and the values are broadly similar to those of Table 10.4 since 76 × 10^{12} t ≏ 202,462 EJ and 4·3 × 10^{12} t ≏ 11,455 EJ. Note the longevity of the supply compared with oil in Fig. 10.6. Source: Hubbert 1969

renewable, equilibrium type (Pearce-Batten 1979). For both conventional and unconventional energy sources, large numbers of estimates have been made; those used here have been selected after comparing a number of different authorities.

Coal

Table 10.5 gives Holdren's (1975) estimates for coal reserves and supplies at various levels of use. In recent years, the rate of growth of coal use has been about 3·6 per cent per annum (Hubbard 1969) and so the global resource would last slightly longer at that rate than the 53 years at 5 per cent growth. The cycle of production for coal given by Hubbard (1969) is shown in Fig. 10.5, where it will be noted that his figures for reserves and resources are broadly similar to those of Holdren.

The reserves and resources of coal, therefore, if treated as a world patrimony might last well into the 21st century, provided that the rate of growth of use is diminished

TABLE 10.6 Minable coal and lignite

Region	*Estimated resources* $\times 10^9$ t	*Established by mapping*
USSR (including European part)	4,310	2,950
USA	1,486	710
Asia outside the USSR	681	225
North America outside the USA	601	70
Europe	377	280
Africa	109	35
Oceania	59	25
South and Central America	14	10
Total	$7{\cdot}6 \times 10^{12}$	$4{\cdot}3 \times 10^{12}$

Source: Hubbert 1969, Darmstadter 1971

TABLE 10.7 World coal production 1975

	$\times 10^6$ *tce*	%
USSR	614	23·6
USA	581	23·3
China	349	13·4
Poland	181	6·9
UK	129	4·9
W. Germany	126	4·8
Rest of world	617	23·7
World	2,597	

Source: Peters and Schilling 1978
2,597 million tce = 76 EJ.

considerably. But of course the distribution of coal is uneven, as shown by Table 10.6; its present production is uneven although naturally closely related to the distribution pattern (Table 10.7); although there is some long-distance export of coal (e.g. from Colombia to the USA, from Canada to Japan, and Australia to Japan), the quantities moved are much less than is the case with oil, although a continued rise in oil prices might, combined with very large bulk carrier vessels or slurry pipelines, increase the volume of trade substantially.

Forecasts of the future production of coal (e.g. Peters and Schilling 1978) suggest an increase in consumption of coal from 76 EJ/yr in 1975 to 245·5 EJ/yr in 2020, which represents a drop from 28 per cent of total industrial fuels (coal, oil, natural gas, HEP, nuclear) to 22 per cent in 1990. Most coal producing countries seem to be capable of meeting their own demands within such totals in the next ten years and thereafter, but the distribution of the resources make it clear that most of these are in the DC's and are dominated by the USA, USSR and China. Few of the LDC's have substantial coal resources which they might use for industrial and infrastructural development: India, Botswana, Colombia, Brazil, Indonesia, N. Korea, S. Korea, and Chile, are the only significant likely producers. Given that LDCs are unlikely to be major importers of coal, it seems unlikely to play a large part in fulfilling their energy needs.

The forecasts for coal production to AD 2020, given by the latest World Energy Conference (1977) are set out in Table 10.8 and hold few surprises. Of all the major fuels, coal's position seems the most secure and assured as far as the industrial nations are concerned, although the environmental side-effects are considerable (p. 263); nevertheless, the pressures to develop coal in order to substitute for increasingly scarce oil supplies will be strong.

TABLE 10.8 Estimated coal production for 1985 and 2020

		$\times 10^6$ *tce*	
	1975	*1985*	*2020*
USSR	614	851	1,800
USA	581	842	2,400
China	349	725	1,800
Poland	181	200	290
UK	129	137	173
W. Germany	126	129	155
Rest of world	617	914	2,071
World	2,597	3,798	8,689

Note: $2{,}597 \times 10^6$ tce = 76 EJ; 3,798 = 111 EJ; 8,689 = 254·5 EJ. At a URR of 220,000 EJ (Table 10.4), these would be 864 years' production.
Source: Peters and Schilling 1978

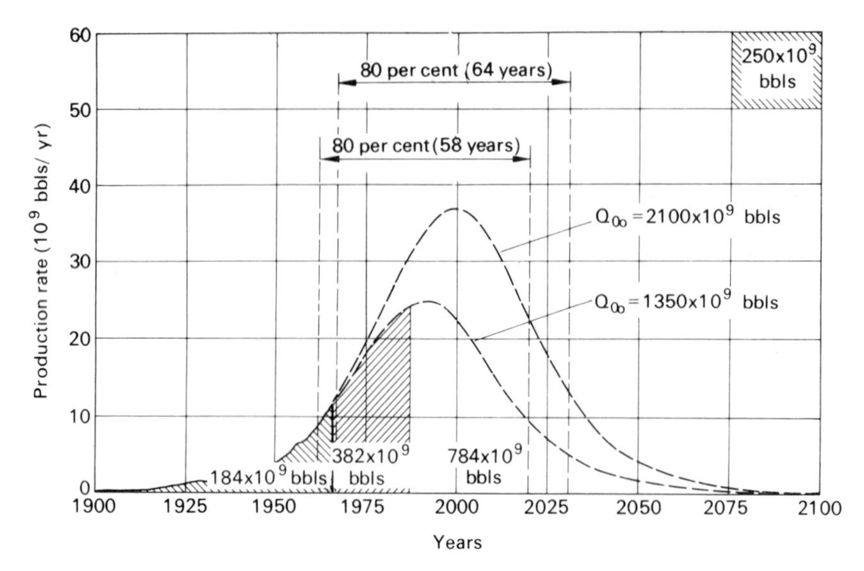

Fig. 10.6 A projection of the complete cycles of world crude oil production. Q_∞ represents the ultimate amount of the fuel recovered during the cycle and is given here for different estimates of the Q_∞ of crude oil. The difference between them makes little difference to the long-term situation. The quantities enumerated under the curve represent the amounts recovered during various time-segments of the recovery curve.
1 barrel of oil $\simeq$ 5·92 × 10^9 J. Thus 2,100 × 10^9 bbl $\simeq$ 12,432 EJ and 1,350 × 10^9 bbl $\simeq$ 7,992 EJ
Source: Hubbert 1969

Oil

Comparable figures for petroleum liquids are given in Fig. 10.4, the main feature of which is the estimate of a shorter life-time for oil than coal or natural gas. A production cycle for 12,432 EJ ultimate recovery is given as Fig. 10.6 (cf Holdren's estimate of 10,900) and compared with an ultimate recovery of 7,992 EJ. The actual quantity of the ultimate resource recovery is, however, more difficult to estimate than for coal since oil is present in such deposits as tar sands, bituminous rocks and oil shales (Maugh 1978), and it is difficult to forecast whether they will be economic to work even on a large scale; the same kind of consideration applies to off-shore oil fields in very deep water or difficult seas such as the Arctic Ocean.

The rate of use of oil has been, in aggregate, a remarkably smooth curve, rising at nearly 7 per cent per annum (i.e. doubling every 10 years) until the 1970's when the price rises of 1973–74 and the Iranian revolution of 1979 caused some diminution of the rate of increase (Fig. 10.7). In 1978, for example, demand in the USA rose by 2·5 per cent and in most other western countries and the USSR by 4 per cent. China's consumption, however, went up by 10 per cent.

The distribution of oil production from estimated reserves (Fig. 10.8) is unlikely to change greatly from the present pattern with the Middle East and North Africa as dominant producers but with a role for South America enhanced beyond the current position. But the Middle East and North Africa account for nearly half the world's

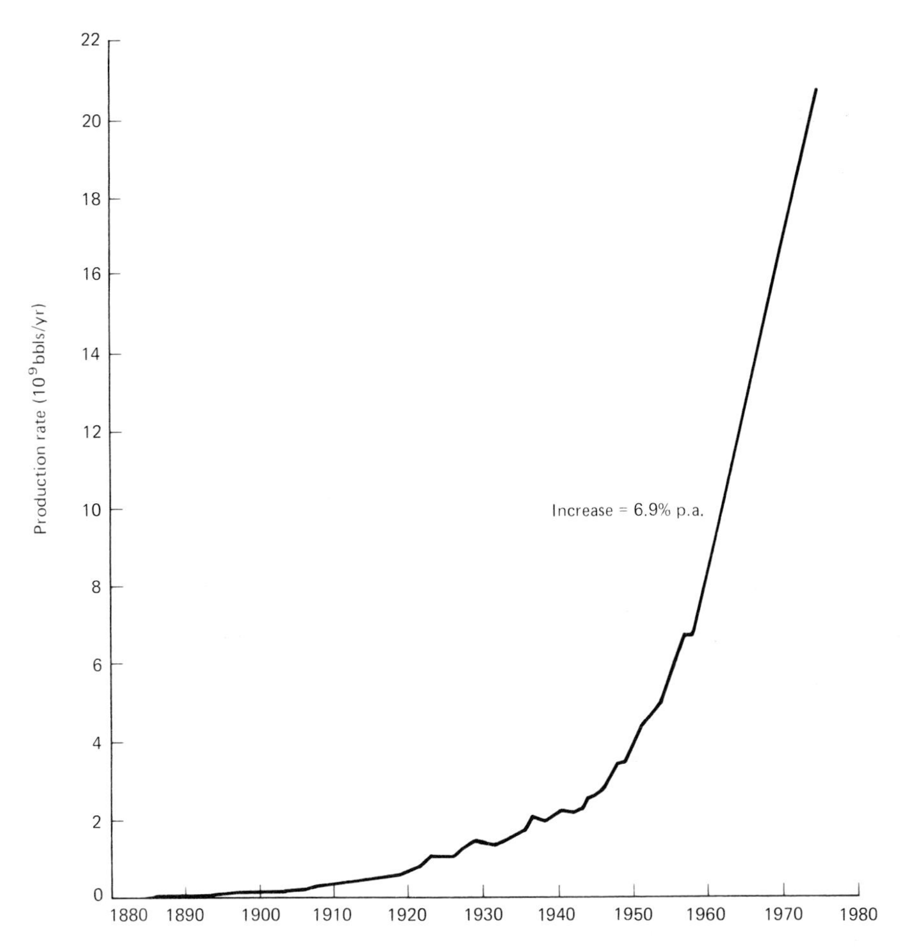

Fig. 10.7 World production of crude oil.
Source: Hubbert 1969 plus additional post 1970 data

apparent reserves, with the socialist countries constituting the second largest group (Desprairies 1978). Environmentally, some 45 per cent of the reserves are likely to come from off-shore sites. A degree of uncertainty about production rates is introduced by such factors as the costs of discovery and development, the percentage of oil recovered from the deposits, currently 25 per cent but expected to rise to 40 per cent by AD 2000, and the role of 'unconventional' petroleum (deep off-shore deposits, polar zone deposits, oil shales, tar sands, synthetic oils) whose price is likely to confine them to specific uses such as transport and chemicals rather than as general energy sources. Indeed, this latter points to a specific problem of oil: that it has many more uses than merely as an energy supplier and that once past the production peak it may become too precious to burn.

The role of coal, natural gas and nuclear fuels are the power sources most likely to be

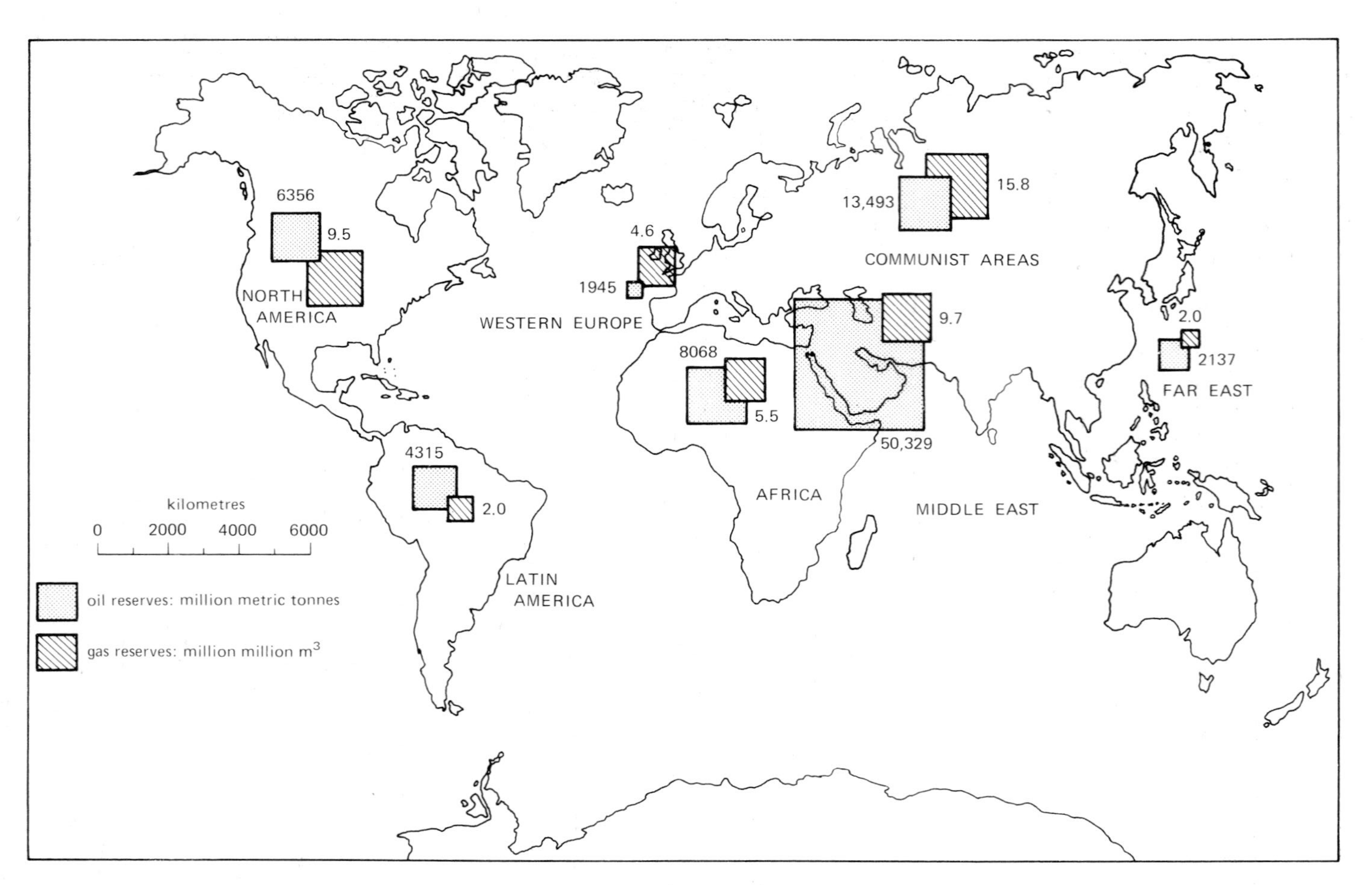

Fig. 10.8 At present, fossil fuel reserves are a key factor in the future of the industrialised nations. This is especially true of the hitherto cheap and easily-processed oil and natural gas. This map shows the estimated reserves of each in 1972. The key role of the Middle East (especially Saudi Arabia in fact) is emphasised, as in the question, 'If an LDC wishes to industrialise, where is it to get its energy from?'
Source: Desprairies 1978

enhanced should this come about. The production ceiling in 1990 was estimated by the World Energy Conference 1977, to be 176–220 EJ/yr, and 110 EJ in 2020, compared with 132 EJ in 1977. As far as trade is concerned, the dominance of the Middle East could cause its production rate to vary between 53 and 57 EJ/yr in the years 1985–90 when dependence on that region is likely to be at its height.

Whatever the marginal adjustments, the production of oil as a source of energy is likely to run down seriously by *ca* AD 2030, when several estimates suggest it can only be half of the 1970s rate. Flexible though oil is, it is not stretchable forever and the use of much of the remaining resource merely as a fuel may be a waste of very versatile material.

Natural gas

This requires little additional treatment since its distribution is closely linked to that of petroleum and its recoverable resource usually amounts to about 80 per cent of the oil discovered. Production of natural gas worldwide is now about 50 EJ/yr, proved reserves are estimated at 2,500 EJ and remaining reserves about 8,100 EJ. Cumulative production so far has used up about 40 per cent of proved reserves and 11 per cent of remaining resources. At a production of 100 EJ/yr, the resource base would last at least another 50 years, even if 'unconventional' gas supplies from coal seas, shales, and 'tight gas formations' are not considered, nor enhancement of gas recovery rates from present levels of 70–80 per cent. Regional estimates for future production are given in Table 10.9

General remarks on fossil fuel availability

It is clear that these fuels are not in danger of running out immediately, as some more alarmist views have implied. But equally, they are absolutely finite and economically

TABLE 10.9 Estimated natural gas production

	EJ	
	1976	*2020**
North America	23·0	7·5
W. Europe	6·4	1·6
Japan & Oceania	0·3	4·5
USSR & E. Europe	12·8	25·3
China	1·4	6·0
OPEC	3·9	61·0
All others	2·3	8·9
World	50·3	114·8

* This is a medium estimate which assumes that 75 per cent of the original recoverable resource base will be extracted.

Source: McCormick et al 1978

even more finite: given continued demand in the DCs and increased demand in the LDCs, it seems that the year 2030 will see the exhaustion of the readily available oil and gas reserves and a significant part of the coal reserves. The stretching of the resources can and doubtless will be undertaken, i.e. by intensifying the search for oil and natural gas on the one hand, and by reducing demand for them on the other. But to continue the industrial nature of the DCs and to improve material standards in the LDCs, access to other energy sources will be needed. Of the non-equilibrium sources, nuclear power appears to have the most potential, and then there is a spectrum of equilibrium sources, some with a highly developed technology like HEP, and others which are in a very early stage of development, like solar power. Some account of all of these must now be given.

Nuclear power

The physiochemical basis of nuclear power differs markedly from the combustion of hydrocarbons: in the processes currently used for generating heat, the nucleus of uranium 235 is bombarded with neutrons. It breaks into two fragments with the release of large amounts of energy, each fission producing $1{\cdot}6 \times 10^{-13}$J. This means that, if used for generating electricity at an efficiency of 30 per cent, 1 kg of Uranium 235 produces as much commercial energy as 2500 t of coal, and here lies much of its attractiveness. The source material is natural uranium, of which only 0·7 per cent is the fissile isotope ^{235}U, the rest being ^{238}U which is more stable. Natural uranium has to be enriched therefore before it can be used in an atomic reactor.

The sources of uranium are therefore of interest to nations engaged in nuclear power programmes: even though a typical reactor will fission only 10–20g of nuclear fuel for every kilogram of mined uranium, the absolute quantities required are not large, but continuity of supply is important. So is the price, since the economics of nuclear energy may be an important factor in determining the penetration of nuclear energy into a national mix. (They may, of course, be much less important if there seems no alternative to nuclear power, or if a commercial reactor is also to produce fissile material for military use). Thus estimates of uranium resources are quoted at a particular price and one such set of assessments is shown in Table 10.10.

To 1976, the world production of 470,000 t of uranium has mostly (68 per cent) come from North America, and the bulk of the rest from sub-Saharan Africa; present production is ca. 33,000 t/yr and present operations might yield 110,000 t/yr, above which level new sources of production must be identified. The sufficiency of these supplies depends entirely on the rate of introduction and growth of nuclear power installations, the type of fuel cycle used (and in particular the penetration of the 'breeder' reactor described below) and, naturally, the world demand for electricity which is the only type of useful energy that nuclear power can supply unless it is coupled to another process. Calculations akin to production cycles seem therefore to lack usefulness.

Another way in which nuclear energy differs from fossil fuels is in the nature of its waste products. Many substances harmful to human health and to ecosystems are

TABLE 10.10 Estimated world resources of Uranium at costs up to $130/kg U

	Tonnes	
	Reasonably resources assured	*Estimated additional resources*
North America	825,000	1,709,000
Western Europe	389,300	95,400
Australia, N. Z. & Japan	303,700	49,000
Latin America	64,800	66,200
M. E. and N. Africa	32,100	69,600
Sub-Saharan Africa	544,000	162,900
E. Asia	3,000	400
S. Asia	29,800	23,700

Source: Dufet et al 1977. Present (1977) world prices are in the order of $104/kg U. No estimates are available for the USSR and China.

TABLE 10.11 The growth of nuclear power

	1954–1974 actual, 1975–80 estimated	
	Number of reactors	*Capacity (mWe)*
1954	2	7
1960	24	1,304
1965	66	7,106
1970	98	20,015
1974	182	72,417
1975	219	100,058
1980	401	262,596

Source: IAEA quoted in Patterson 1976

produced by burning hydrocarbons (see p. 263) but the radioactive wastes of the nuclear industry are seen as comprising a different category altogether. They are only detectable at low levels by instruments, yet, even at such low levels, they may be deleterious to health and produce genetic effects: the aftermath of Hiroshima and Nagasaki is still with us. Further, many of the wastes emit radiation for a long period: strontium 90 has a half-life of 28 years, caesium 137 of 30 years, so that it takes, for example, about 300 years for their radioactivity to drop by a factor of 1,000; the half-life of plutonium 239 is 24,400 yrs. These are planned wastes: the releases from accidents are many orders of magnitude higher. (A more detailed account of nuclear wastes and their disposal is given in chapter 11).

Fission reactors of various kinds (usually identified by the coolant which surrounds the radioactive core and which conveys the heat to a turbine that generates electricity, e.g. Magnox-carbon dioxide; AGR (advanced gas cooled)—carbon dioxide; PWR (pressurised water reactor)—light water; FBR (fast breeder reactor)—molten sodium or sodium and potassium, are the basis of the world's installed nuclear capacity, the rapid growth rate of which can be seen in Table 10.11.

A prognosis for further growth is difficult to make beyond the immediate future for, apart from the social and political contexts, the development of other conventional fuels is relevant to the attractiveness of the nuclear option. The World Energy Conference suggested a rise in consumption from 1·5 EJ of nuclear power in 1972 to 13·3 EJ in 1980 and 67·1 EJ in 1990, where it would represent 15 per cent of world industrial energy consumption. Most of the present and predicted capacity is in the developed countries (Table 10.11); the LDCs with programmes include Argentina, India and Pakistan. At one point Brazil embarked on an ambitious programme of acquisition but in 1979 was cutting this back because of the cost of acquiring the reactors; Iran cancelled its programme in the same year after the Islamic Revolution. The LDCs are obviously interested in nuclear energy for development, since they are at present mostly dependent on oil and wood, but the barriers in terms of cost and capability are strong: the IAEA's forecast that the installed nuclear capacity of Third World countries will rise 200-fold over the next two decades seems optimistic and more development of indigenous hydrocarbons (especially oil) and 'alternative' energies seems likely (Hayes 1977).

TABLE 10.12 Nuclear-produced electricity by type of economy 1975

	$\times 10^9$ *kW (rounded)*	%
World	342	100
DCs	319·8	93·5
LDCs	5·7	1·6
Centrally Planned Economies	16·6	4·8

Source: UN, *World Energy Supplies 1971–1975*. Statistical Papers Series J no 20. New York: UN, 1977

Industrial nations dependent not only upon increasing their nuclear capacities but also on making more efficient use of raw materials such as uranium are investigating the breeder reactor. This produces ^{239}Pu from ^{238}U and ends up with more plutonium than there was enriched fuel originally: a non-breeder reactor will produce 0·86 TJ per kg U whereas a breeder reactor will yield 51·75 TJ/kg U ($1\ TJ = 1 \times 10^{12}\ J$). This attractiveness is offset by the high quantity of radioactive wastes produced, especially ^{239}Pu which is not only highly toxic to man but which is a fissile material capable of being made into a bomb, even with relatively unsophisticated equipment and knowledge. The economic benefits of the FBR are sometimes disputed as well, and international programmes may be necessary to develop the technology to an efficient and safe level (Chow 1977, Dombey 1979).

The future of fission-produced atomic energy is therefore the subject of intense debate in countries where such controversies are permitted by their governments. The proponents of nuclear power point to the favourable economics of nuclear processes (e.g. Rossin and Rieck 1978), especially with breeders, to the dangers of 'energy gaps' without their contribution to supplies, to their safety record compared with coal, and to the immensely complex precautions taken over power generation, waste disposal and the combating of terrorism. Their opponents suggest that 'demand' can be reduced by more efficient energy use and generation so that the 'gaps' are not necessarily real, and produce studies which suggest that for the whole of a nuclear cycle (the plant itself becomes 'waste' after its lifetime) the economics are not so competitive; they point to the potentially horrific consequences of accidents even if they were very rare, and argue that, if so much care is needed, it proves that the process is unlikely to be out of the reach of human error and unforseen technical problems, as was shown at Three Mile Island (USA) in 1979, an argument which applies very strongly to LDCs where technical skills are still scarce. They are unconvinced of the safety of the waste management schemes so far proposed and point to the immense contract which society would take on in keeping large quantities of waste plutonium isolated from the biosphere for perhaps 250,000 years. The anti-nuclear lobby also notes the 'loss' of radioactive materials, some of which are thought to have found their way to aspirant possessors of nuclear weapons. Nuclear energy from fission will be discussed again in this chapter; here it is necessary to stress that in comparing sources it is essential to consider the whole of a fuel 'cycle' from raw materials to waste disposal, including the energy 'investment' needed before there is any power yield.

Fusion

The drawbacks to fission energy have brought to the fore the possibility of generating electricity from nuclear fusion, in which two very light nuclei are fused to form a heavier one. The obvious sources are deuterium and tritium which are isotopes of hydrogen. The fusion of 1 g/hr of deuterium nuclei would generate electricity at the same rate as 27 t/hr of coal. A further advantage is that deuterium is present in sea water: the deuterium in 1·5 km^3 of sea water would yield as much energy as the world's recoverable oil, and there are $1{\cdot}5 \times 10^9$ km^3 of sea water: a more lasting source of energy (other than the sun) can scarcely be imagined.

A major drawback, however, is that theoretically the fusion of the nuclei can only take place at a temperature of 200 million degrees C, although in practice 20 million degrees C might be sufficient. At such temperatures all substances are gaseous and atoms are stripped of their electrons: the result is called a plasma. Such a plasma can only be held in an immaterial container made of magnetic forces, or possibly on a small scale by laser beams. It is possible that some constraints may be put on construction of reactors by the quantities of rare metals (e.g. vanadium, niobium and molybdenum) required, since the reactor core would have to be at least 100 times as large as a conventional power station. But again, it looks theoretically as if the wastes are much lower in quantity and longevity than with fission. But this conception of energy, although the subject of vigorous research, is not yet with us; indeed, nobody is yet certain that it will work at a commercial scale. The investment of money and energy needed in its development along with other engineering considerations, suggest that even if fusion power is available, as the optimists suggest, in 40 years' time, it will be expensive. (Steiner and Clarke 1978, Parkins 1978).

The problems associated with nuclear power, together with the likelihood of the high prices of fossil fuels in the later parts of their production cycles, have excited interest in other sources of energy, usually called 'alternative' supplies (with the exception of hydropower) and these must now be examined. They are, it will be remembered, 'equilibrium' sources of energy which add no extra heat burden to the atmosphere and some of them are said to produce much less environmental manipulation than the resources discussed so far, though they all produce environmental impacts (Hoare 1979).

Hydropower

The basis of harnessing the energy of falling water is simple and has been in use at least since 85 BC this almost infinitely renewable source creates power in relation to the volume multiplied by the drop and in modern plants the power is converted into electricity. Large hydroelectric plants date from the 1890s onward when the first was installed at Niagara Falls, and the method is generally associated with large scale installations (the largest, Krasnoyarsk Dam on the Yenesei in the USSR has a capacity of 6·1 GW) but in China there are 50,000 plants with an average capacity of 35 kW. Since the greatest falls of water are often distant from areas of demand for electricity, high voltage transmission may be needed and this now appears feasible for distances up to 1,600 km. In 1976 the world capacity was reported at 16 per cent of the estimated potential: 372 GW of installed plant was producing 5·7 EJ/yr. The major producers are

TABLE 10.13 Major producers of hydro-electric power 1975

Country	*Production* 10^9 *kWh*	*EJ*
USA	306	1·1
Canada	202·4	0·73
USSR	125·9	0·45
Japan	86·2	0·31
Norway	77·4	0·28
Brazil	71·9	0·26
France	59·8	0·21
Sweden	57·6	0·20
Italy	44·5	0·16
China	37	0·13
Spain	28·7	0·10

Source: *World Energy Supplies 1971–75*. UN Statistical Paper Series J no 20. New York: UN, 1977

shown in Table 10.13; several countries rely quite heavily on their hydropower: Brazil generates 80 per cent of its electricity this way, Spain 50 per cent, Italy 40 per cent, Japan 23 per cent and Canada 16 per cent, for example. The Yenesei-Angara scheme produces electricity equivalent to the total UK production of that form of power.

Estimates of the potential are typically for a AD 2020 capacity of 2–3 TW. The distribution of the extra installations is dominated by the LDCs, where perhaps half of the increase will be located: the World Energy Conference 1977 suggested that 28 per cent of the development would be in Asia, 20 per cent in South America and 16 per cent in Africa. Fortunately, regions poor in coal such as South West Asia, South America and Africa, seem to have the largest hydropower resources. Most of the DC's have already tapped their capacity, with the exception of the USSR. So hydro electricity may well help some developing nations to increase their electricity supplies and from time to time TVA is quoted as an example of what could be done on the Mekong or the Amazon, for instance.

Although falling water may have an infinite (though not invariable) supply, the rest of the process may be rather more mortal. Reservoirs behind dams are liable to silting and have a lifespan of perhaps 100–300 years and the creation of the impoundments increases evaporative loss so that the net volume downstream is less; this may be significant where irrigation is important. The creation of large dams has numerous side effects (p. 125) of geophysical, hydrological, ecological, epidemiological and social kinds and so the future in LDCs especially may lie with smaller installations. Overall, the contribution of HEP to world demand is not likely to be very high, although it may be regionally important; Häfele and Sassin (1978) talk of AD 2030 capacity of 2·9 TW out of a world total energy consumption of 35 TW (8·25%). Their total is optimistic but if hydropower is vigorously developed, then the proportional importance may be higher; but it is difficult to envisage it as a major element in the solution of energy problems.

Geothermal energy

The average thermal gradient of the earth's crust is 25°C/km depth but it is steeper in many places, especially where vulcanism and tectonic instability is present. In such localities, geothermal energy is in the form of hot water or steam; elsewhere it may be feasible to pump water down to dry 'hot rock' and use it back at the surface. In either case, the heat is either used to generate electricity (Table 10.14) or for other direct uses.

Underground water is used to heat a large block of offices and 13,000 dwellings in Paris (Garnish 1978), to heat greenhouses and hotels in Iceland, and for spas, tourist attractions ('hells') and crocodile farms in Japan; most of this category of use is residential and commercial rather than industrial. The first electricity plant using geothermal power was installed at Lardarello, Italy, in 1904, and it now has a capacity of 400 MW; the Geysers plant in California, opened in 1960, has a capacity of 12·5 MW. In 1976 the world installed capacity for electricity was 1325 MW and most countries with such power were developing their industry; a number at present without it are expected to have the capacity early next century if not before, notably the central American nations, Kenya, Indonesia, the Phillipines and Taiwan. There are a few

TABLE 10.14 Electricity generating capacity from geothermal sources

	1976 Installed capacity MW	*2000 Estimated capacity*
USA	522	20,000
Italy	421	?
New Zealand	202	1,400
Japan	70	50,000
Mexico	78·5	1,500–50,000
USSR	5·7	1,500–20,000
Iceland	2·5	500
Turkey	0·5	1,000
El Salvador	60	?

Source: Clarke 1978

environmental problems: noxious substances such as boron and sulphur may be brought to the surface in large amounts, and land subsidence may occur.

A considerable increase in the utilization of geothermal resources is to be expected but the contribution to world demand in, for example, AD 2030 is unlikely to be more than 1 per cent.

Wind power

The harnessing of the wind to provide power has been operated for at least 2,000 years in China and in Europe since the 12th century, so that the basis of the technology is well known. Modern interest in the wind focusses more on its potential to generate electricity and here a 70 per cent efficiency increases its attractiveness. But a structure will have to compromise between an ability to work at very low wind speeds and a strength to withstand the higher velocities, and it is likely that the blades of a mill would have to be feathered at very high windspeeds.

Since wind velocities are so inconstant, storage becomes an important stage in the process: batteries are one possibility, especially for small-scale decentralised installations. Alternatively, the electricity could be used to hydrolize water to provide hydrogen as a fuel, to pump up water for later hydropower generation, or to fuel storage heaters. The largest practical possibility seems to be of the order of 100–500 kW output per mill, with smaller units (10–50 kW) more likely. Another problem might thus be the land area needed: with units of 100 kW, something like 5,000 windmills would be needed to generate the same power as a 50 MW thermal power station; one calculation suggests that in the UK 1·0 per cent of the land area would be needed to generate 10 per cent of the current electricity consumption and, in the Netherlands, 1·1 per cent of the surface area. In general, the land area needed for a given electricity generation would be 4–5 times that needed for direct solar collection (p. 261). Thus there would be considerable intrusion into land use patterns and even if the design of the towers permitted multiple use of land, the diminution of visual amenity might be considerable.

Häfele and Sassin (1978) talk in terms of a 3–9 per cent contribution of wind to their AD 2030 / 35 TW scenario: the contribution of wind seems likeliest where it can be coupled to a grid where there is also hydropower, for complementary use, and in small-scale decentralized power installations. Thus LDCs might be expected to regard this as a favourable technology: maintenance problems have caused the abandonment of windmills in Uganda and Mali but successes have been reported with electric power generation in Ethiopia, Zambia and the Argentine. In overall perspective, Gustavson (1979) argues that wind energy offers a far larger potential than many other renewable energy sources.

Tidal power

The use of tidal mills for power has been known in Europe since the 13th century. The technology uses the energy of falling water and modern installations trap water at high tide and allow it to ebb through turbines, generating electricity. It appears that only very high tidal ranges will suffice and these are often found where the water is funnelled into an inlet. The technology is currently available: a few installations have been built and many feasibility surveys made (Table 10.15).

TABLE 10.15 Tidal power sites: potential and ACTUAL

Location	*Av tidal range m*	*Energy potential 10^9 kWh/year*
North America		
Bay of Fundy	5·5–10·7	255·0
Cook Inlet, Alaska	7·5	18·5
South America		
San Jose, Argentina	5·9	51·5
Europe		
Severn, Great Britain	9·8	14·7
France:		
Mont St Michel	8·4	85·1
Somme	6·5	4·1
Abquenion/Bancieux	8·4	3·9
LA RANCE	8·4	3·1
Others	6·4	1·65
USSR		
White Sea	5·7	126·0
Mezen Eshary	6·6	12·0
Lumbovskii Bay	4·2	2·4
KISLAYA INLET	2·4	0·02

Source: Clarke 1978. Hoare and Haggett (1979) list other potential sites in Ungara Bay, the Kutch Gulf and Cambay Gulf (India), Kimberley Coast (Australia) and the Inchon Gulf (S. Korea).

At an efficiency of 18–25%, the potential contribution of such plants to national electricity supplies might be from 5 per cent as high as 80 per cent, if transmission losses are not high in the case of remote stations. Tidal power plants might be combined with the regulation of navigation, provision of recreation, and the accumulation of fresh water; on the other hand, silting problems and the destruction of wildlife habitat are reckoned as costs rather than benefits. On a world scale, the potential in AD 2030 seems unlikely to reach more than 0·2 per cent of demand but regionally the schemes could be of considerable significance, especially as cornerstones of development for poorer areas of industrial nations.

Wave Power

At least 2 TW of energy are contained in the waves of the ocean at any one time but historically attempts to harness it have been few. Recent investigations revolve around the possibility of converting this energy to mechanical or electrical power and involve devices such as cylinders in which the waves compress the air sufficiently to drive a turbine, or the differential motion of floats to operate a mechanical engine, or cam-shaped 'ducks' which oscillate relative to a frame and drive hydraulic pumps. Most writers think that unless very large operations are undertaken with, for example, off-shore wave generation of electricity producing hydrogen which is used on site (i.e. also off-shore), then the potential is unlikely to be realized. As a contributor to conventional grids or as a decentralized source for regional or LDC development, this method is not generally regarded as a front-runner. Apart from the capital costs, the conflicts with other uses of the coastline (e.g. recreation, fishing and navigation), present immense obstructions to the widespread dissemination of wave-power generators.

Ocean thermal gradients

In the tropics, the temperature of the surface water is about 25°C but at 1,000 m depth, it is 5°C. This difference could be used in theory to drive an engine but it has not yet been tried on a large scale. It looks as if a large system would produce only a little power and, so, taken along with the economic and technical problems, it appears unpromising besides other methods of collecting solar energy directly.

Biomass

The idea of generating power from plants or from organic wastes stems fundamentally from the calculation that the energy content of the NPP of land plants is about 1,900 EJ/yr and that of the seas 1,100 EJ/yr, whereas the commercial energy consumption of the world in 1975 was estimated at 215 EJ (but if the United States' annual NPP of all types were used to generate energy, it would generate about 25 per cent of its requirements). The process can be no more efficient than photosynthesis (see p. 7) and optimistic estimates suggest that about 3 per cent of incident solar energy might be recovered as heat, using an average value of 20 MJ/kg plant material.

The raw materials for such fuels could be organic wastes such as straw, urban refuse, sewage, bagasse or pulp wastes, or plants grown especially for use as energy sources. These are known as energy plantations: guayule, pines, eucalyptus and sugar cane are all mentioned in this context. In the first category, the organic wastes are usually fermented to give off methane and are thus called biogas producers: there are already at least 25,000 such installations in India and others in, for example, Nepal and China. One advantage of them in LDCs is that they reduce the pressure to burn wood and dung. Wood, for example, is the most important fuel in the LDCs: it still constitutes about 15 per cent of the world's energy consumption and in countries like Thailand, Tanzania and Cambodia (Kampuchea) is 90 per cent of the population's fuel supply, with people using up to 1·5 t/cap/yr. At these rates, forests are a depleting capital stock rather than a renewable resource. Dung is frequently used as a substitute (since kerosene or paraffin is too expensive) and so its value as a fertilizer is lost: to the tune of 60–80 $\times$ 10^6 t/yr in India. Even the litter layer of soils may be used for fuel, robbing them of their organic matter and hastening soil erosion. Biogas plants may well therefore have a very important role in the LDCs, the more so since the technology involved is not particularly complicated. A communal latrine in Nepal produces gas for cooking, fertilizer, and improves health standards.

Energy plantations, on the other hand, envisage the conversion of plants to methanol which requires a high technology. Trees, grasses, water hyacinth and algae have all been considered: 34 t/ha/yr of dry matter of water hyacinth has been mentioned. The country with the largest programme at present is Brazil where sugar, pines and eucalyptus are grown for methanol: 1 per cent of the national area produces about 27 per cent of its energy supplies. To Brazil this is a small area but it is in fact about the size of Denmark. Similar proposals have been made elsewhere, including using coconut husks in the Phillipines. In the US it has been suggested (Burwell 1978) that unharvested wood trees and other biomass might be used for space heating on a local basis, perhaps especially for peak winter conditions. In scenarios for a solar powered Sweden (Johannson and Steen 1979), energy plantations based on fast growing willows (*Salix* spp) play an important role.

Municipal wastes are often high in energy content: in the USA the municipal wastes had an energy potential equal to 3 per cent of the national consumption and many plants burning up to 2,000 t/day were in operation. Much the same processes are happening in Europe, where the EEC nations produce 1·7 $\times$ 10^9 t/yr of wastes, growing at 5 per cent per year. Perhaps the butter mountain could be turned into fuel bricks and the wine lake fermented to car fuel? Countries with large quantities of process waste (e.g. from forestry and paper-making) might not only produce energy but possibly nitrogenous fertilizers as well.

Biomass energy, therefore, is of considerable potential as a decentralized low-cost power source for remote rural regions in DCs or for LDCs with wood supply difficulties. Energy plantations are more attractive to DCs (though competition for the land has to be considered) or to LDCs with a low population density and, preferably, a humid tropical climate. Overall the contribution of this potentially renewable energy source is one of the hardest to predict but underestimates of its usefulness seem more likely than the reverse (Lewis 1977).

Solar energy

In the face of the difficulties of supply or of harnessing many of the equilibrium sources discussed so far, the impulse to turn to direct use of the sun's radiation is very strong. Not only is this for all practical purposes constant and inexhaustible but its magnitude is immense. The world average radiation at ground level is 180 W/m^2, with 210–250 W/m^2 (equivalent to 18·22 × 10^6 MJ) in tropical deserts but diminishing to 80–130 W/m^2 (equivalent to 7–11 × 10^6 MJ) in northern Europe. The total incident on the US in a year is 650 times the total energy consumption, for example; the energy falling on 100 km^2 of the tropics would take care of present consumption and the Sahara Desert receives enough energy for the needs of more than any conceivable human population. But 100 km^2 is the area of the world's buildings and it sounds an unlikely task to build all the world's structures all over again in order to trap solar energy. Solar energy is, of course, very diffuse compared with fossil or nuclear sources and whether we are dealing with direct beam or diffuse radiation, we are considering a low energy density which has somehow to be focused or used as a converter to a more tractable store of energy such as electricity or hydrogen. A number of methods are in the early stages of research and development, and it is possible that the problem of storage of solar energy has been exaggerated (Metz 1978).

Solar-generated thermal energy concentrates the radiation by the use of mirrors, focusing on a single point to produce a solar furnace with temperatures up to 3,000°C. At Odeillo in France, such an installation is used for smelting of rare metals and at one time it was thought the simple versions of the solar furnace would be of use for cooking in LDCs, but not everybody wishes to eat at about noon.

Low temperature ($<$ 100°C) heat is the mainstay of the current market penetration of solar collectors. A flat plate collects the radiation and the heat is then transferred by liquid or gas to media for its end use in water or space heating. Water or solids may also be used for storing the heat. Rather higher temperatures, however, are needed for air conditioning. At present, this type of installation only makes overall sense if conservation measures such as insulation have already been carried out. Although locally useful, this type of collection seems unlikely to contribute more than 2–5 per cent of primary energy needs in the foreseeable future.

Solar-thermal electric conversion (STEC) uses an optical concentrator (a heliostat) which focuses on an absorber which then transfers the heat to an electricity generating plant. A 1 MW plant is installed at Odeillo but installations of 50–100 MW should be possible: these would use 12,500 heliostats on a tower tracking the sun. Each would be 40 m^2 and their output would drive a turbine for 6–8 hrs/day so that energy storage in some form would be necessary. A disadvantage of this development would be its size: to achieve the same output of electricity as a medium-size thermal power station, a STEC plant would need to be 5 km^2 in size, although some reduction in cost and complexity might be achieved by using parabolic trough reflectors.

Photovoltaic conversion uses the principle of the exposure meter and turns solar radiation directly into DC current via a silicon cell. Currently, the cost is 100–200 times that of conventional electricity but development in cell technology may make this more attractive on a commercial scale, especially since there are no moving parts.

Solar hydrogen production acknowledges that other methods have difficulty in storing the power output and proposes using solar-powered electrolysis of water to produce hydrogen, (Bolton 1978), which can be stored and transported, even using existing natural gas pipelines, for example. If coal is available then it can be used not only to generate more electricity but, along with the hydrogen, to make methanol which again can be transported using existing installations. But the technology for this sequence is as yet little developed.

In general, we should note that none of these systems is more than 20 per cent efficient, so that large areas are needed for solar collection, with the subsequent likelihood of intrusion into land use patterns. Uninhabited areas are usually a long way from demand centres and in any case may have high value simply because of the absence of technology. Energy accounting shows that most of the systems are so complex that their construction is highly energy-consumptive and that they must have a long life if they are to be yield-positive, although economic feasibility studies in US cities have shown that some systems are already competitive with existing fuels (Bezdek et al 1979). But we must note that the 'waste' products are confined to some concentrated heat and land use conflicts: there is no CO_2 or other atmospheric contaminant.

In some ways, therefore, solar power looks ideal for decentralized, low-output use and as such a good source for LDCs and remote rural areas: small localized industries, clinics, schools and homes could run off solar collectors and small scale generators, and the distillation of water could also be achieved, and at a medium scale, a research report from Sweden suggests that all of its energy could be generated from the sun by 2015 (Taylor 1978). By contrast, Häfele and Sassin (1978) talk about 'hard' solar energy in which a major proportion of the AD 2030 demand for 35 TW is achieved by solar collection. This, they say, is the outcome of realizing that only solar and nuclear power offer virtually unlimited energy. Their process uses photovoltaic and STEC methods to make hydrogen and then, provided there is still coal available, to synthesize methanol. To get an output of 35 TW, some $1{\cdot}75 \times 10^6$ km^2 of sunny and dry lands would be needed, a figure which we can compare with the 13×10^{16} km^2 currently in arable usage. This kind of 'second agriculture' would require an endowment for construction of at 3–5 times the world's material consumption in 1975, so, on the whole, 'hard' solar energy has the highest capital costs, reckoned in conventional terms, of all the energy strategies; however, to many people, the benefits look commensurately large (Berham 1979, de Winter and Lox 1979).

Stretching the resources

In the face of concern about the supplies of existing fuels and the development of new ones, many people have turned their attention to making better use of what we already have. The most obvious development is energy conservation, by which is generally meant the use of a lower quantity of energy per capita but without any drastic changes in life-style. At its simplest conceptual level, energy conservation may mean the use of waste heat from a power station for the district heating of homes. Insulation of houses by filling cavity walls with a plastic foam, by double glazing, and by lagging pipes and roof spaces is another simple measure. More complex are ideas, for example, of substituting different forms of transport: most methods of public transport for example use energy much more efficiently than the private auto (Table 10.16), especially if the latter is carrying only one person. Conservation, it is claimed, may help to damp down demand in DCs to the point where some of the more controversial technologies such as the FBR are not necessary.

TABLE 10.16 Energy consumption of different passenger transport methods

Mode	*Net energy consumption at average occupancy: kWh/ passenger kilometre*
2-stroke moped	0·20
Small private car in town	0·50
Large private car in town	1·06
Bus (urban)	0·25
Subway (electric)	0·11
Diesel train (<100 km trip)	0·29
Electric train (long distance)	0·09

Source: Foley 1976

New methods of energy storage have also been mentioned, especially in connection with small-scale electrical generation. Currently, the most popular of these are hydrogen and methanol which have the advantage that they can fairly readily be substituted for natural gas and petroleum in combustion devices and in pipelines. New technologies for increasing the efficiency of use of fossil fuels and electricity are also under development, such as magneto-hydrodynamic power (which has a large potential but a difficult technology); the heat pump which uses the principle of the refrigerator in reverse; the extraction of oil from coal and coal gasification; fluidized-bed combustion of coal; and hydrogen-oxygen fuel cells. The role of helium gas in energy-related uses may be so important in future that a conservation programme for it should be started now (Cook 1979). It is important to stress that although these techniques may enable us to squeeze more juice out of the orange, they do not make the fruit any bigger; their role ought not to be underplayed providing they do not blind us to the essentially finite nature of the fossil fuels and burner reactors.

The ecological impact of energy use

Of all resource processes, energy utilization is one of the most manipulative of ecosystems, partly because that is one of its very uses and partly because there are often spin-off effects. If we think first of all of the on-site effects of power generation, then there is common to them all visual intrusion, although to some people the appearance of high technology has its own romance.

Oil extraction offshore may produce gas flares which kill a large number of birds and the structure may interfere with fishing patterns; oil spills and blow-outs are highly biotoxic. On land, petroleum installations produce some local effluent problems even if these are accidental rather than inevitable. Coal production from shaft mines produces tips as well as buildings, and the former are apt to be sources of dust and to harbour vermin. Strip or open-cast mines also create derelict land unless restoration is enforced, they disrupt the hydrological pattern of the region and the run-off is usually acidified from the sulphides found in association with carboniferous rocks. The initial phases of open pit mining may cause eutrophication of streams from 'nutrient dumping' when the topsoil is removed. The hazards to workers, particularly underground, are well known: the most prevalent are chronic respiratory diseases which persist long after the miner has ceased work at that occupation. Health hazards to workers are also found on-site in the nuclear industry: some workers are exposed to radiation and may accidentally get doses beyond the permitted levels. There is the possibility, however small, of a melt-down accident, stored radioactive material may leak into the soil, and even low level wastes like the process waste of uranium refining is now thought to be unsafe when used as a general fill and foundation material as in a few towns in the western United States. Nor should it be thought that the 'alternative' energy sources are without any environmental impacts: in some cases these are not yet properly known and, in many, electricity generation is foreseen which will carry with it the usual ecological changes.

Beyond the immediate sites of extraction and generation, the problems of disposing of wastes are many, and the particular nature of the various wastes are dealt with in chapter 12. At this point, we may note that many environmental contaminants are by-products of power generation, transmission and use (Table 10.17), especially those associated with the lowering of air quality, the calefaction of water, and the additive burdens of heat (for instance a nuclear fission reactor gives off 40 per cent more waste heat for the same amount of electric power as a thermal generating station), and CO_2 in the atmosphere. Overall, the more energy that is available to a society, the greater the environmental impact at a particular time: the control of it is used purposefully to change ecological systems as in the bulldozer and in buildings, and indirectly it causes environmental contamination and allows the outreach of man into all kinds of remote places in search of materials or of recreation (Commoner 1975). Remotest of all, only energy use of a very high intensity has allowed man to visit the Moon and to photograph Saturn.

There is one cultural dimension to a particular energy system which must be discussed, and that is the societal dimensions of nuclear power. Even if the risks from wastes and nuclear reactors themselves are low, fears are still expressed that the nuclear inventory may be a target for terrorism, that more power stations mean more likelihood of the proliferation of nuclear weapons, that storage of wastes for a quarter of a million years means a commitment by human societies of an altogether different dimension from anything undertaken before, and that the centralized, highly technological nature

TABLE 10.17 Some environmental impacts of energy

	Health	*Property*	*Social*	*Quality of life*	*Environmental services*
Exploration					
Oil/gas	—	—	—	Invasion of wilderness	—
Harvesting					
Coal mining	Accidents, black lung	Loss of farmland, subsidence	Use of public lands	Defaced landscape	Acid drainage
Offshore oil	Accidents	—	—	Oil on beaches	Oil as a biocide
Hydroelectric dam	Dam collapse	Loss of farmland	Displacement of residents	Loss of wild rivers	Fish passage, wildlife breeding grounds
Processing					
Oil refining	Air/disease	Air/crops	—	Smells, visibility	Pollution of estuaries
Shale processing	Air/disease	Water consumption	—	Waste piles	Water pollution
Conversion					
Coal power plant	Air/disease	Air/crops-buildings	—	Noise, visibility	Acid rain, CO_2/particles/climate
Fission reactor	Reactor accident that breached containment would produce all classes of impact				
Transportation					
Oil tanker	Fire	Fire, collision	—	Oil on beaches	Oil as biocide
Electrical transmission	Electrocution	Restriction on land use	—	Unsightly towers	—
Plutonium	Leak/cancer	Land contamination/quarantine	Terrorism, nuclear bombs	—	—
Consumption					
Automobile	Air/disease	Air/crops	Suburbanization	Noise, visibility	Paved environment, heat/climate
Waste management					
Radioactive wastes	Leak/mutations	Land use	Terrorism, sabotage	—	Groundwater contamination

Source: Holdren 1975

of nuclear power will mean that its control will be in the hands of a small elite to whom the meaning of power may not be confined to electricity. These essentially cultural matters are controversial but in the outcome of decisions to 'go nuclear' or not, they will affect many ecological systems.

One circle, three layers

William Blake may have said that 'Energy is eternal delight' but to the real world resource manager and to the writers of books it is more like an infernal nuisance. Apart from the complexities of the concepts of energy and power, the multiplicity of units in which it is measured, and the implications of the laws of thermodynamics, there is the complex way in which energy considerations are interwoven with most resource processes, both in terms of supply and of the wastes produced by the energy (Table 10.18). Thus energy becomes a mediator between man and environment (though not in a deterministic sense) and the levels and intensity of use are good indicators of the types of relationship between a human population, its resources and its environment (e.g. Antonini et al 1975, H. T. Odum 1976). At present, the per capita energy use in the United States is 12 kw/yr, in other DCs it is 3·7 kw/yr and in the LDCs 0·5 kw/yr and these figures symbolize much of the problems of resource supply and environmental degradation found in the world, as does the proportion of non-commercial energy (Table 10.19) used in the LDCs and which rarely enters global considerations and planning. But both show the great gap between rich and poor and both point to the difficulties of securing future supplies at an acceptable price.

Energy and life-style

One of the distinctions which has to be made in considerations of future energy use, is that between demand and need. Demand can apparently be bottomless: the LDCs wish to attain the standards exemplified by DC levels of consumption and the DCs see further material gains to be made by using more power. Thus forecasting energy 'demand' becomes subject not only to the usual difficulties of foreseeing the future, but also to the difficult cultural problems of differentiating between what levels of consumption provide a dignified life at a sustainable level and what levels are indicative of living off capital rather than income and are the proximate cause of environmental insults. What is beyond certain doubt is that the rates of growth of energy use to which the world (especially the DCs) has become accustomed cannot be sustained for ever. Quite apart from problems of supply, the laws of thermodynamics ensure that all the energy ends up as heat in the atmosphere and with non-equilibrium sources, this is in addition to the natural flux, thus raising the temperature of the atmosphere. Extrapolations of historical rates of energy use show how quickly impossible conditions can be reached and it is reasonable to infer that severe interference with climatic patterns would occur well before intolerable levels are reached (Fig. 10.9).

Within this overall envelope of restraint imposed by ineluctable laws of physics, there is considerable room for change. The national efficiencies of energy use, for

TABLE 10.18 Energy relations of some resource processes

Resource process (as in chapters 2–9)	*Energy sources*	*Energy intensity (i.e. added energy from human activity) GJ/ha/yr*	*Heat loss characteristics*	*Site of initial heat disposal*			*Other energy-related wastes (not including ecosystem manipulation)*
				A	*FW*	*O*	
Unused land	Solar	0	Dispersed, equilibrium	×			
Wilderness	Solar	~0	Dispersed, equilibrium	×			
Proctected landscapes and ecosystems	Solar with some fossil fuels used for management and by visitors	<0·5	Mostly dispersed and equilibrium. Some concentrations from fossil fuels	×			
Outdoor recreation	Solar, with (usually) seasonally pulsed FF use by recreationists. Low year round FF use for management	<1·0	Mostly dispersed and equilibrium; some FF concentrations along roads and popular places	×			Hydrocarbon combustion products in air; PCS in places with sun and inversions
Water catchment	Basically solar, via falling water. Energy from FF's invested in construction; own or FF energy used in lifting water	n.a.	Equilibrium; but concentration at dams if generators installed	×	×		
Forestry	Mostly solar but FF's used in management and harvesting	Depends on forest type <1·0	Dispersed and equilibrium; small concentration of FF's related to machinery	×			Noise of machinery may disturb wildlife

Agriculture subsistence	Solar energy; non-commerical energy (domestic animals, wind and water, dung, wood) plus some small amounts FF use	0·2–0·5	Dispersed and equilibrium	×			
Grazing	Solar energy; some FF's in managed rangelands of DC's	0·6	Dispersed and equilibrium	×			
Intensive	Solar energy plus FF energy on farm, upstream and downstream	10–150	Mostly dispersed except for very intensive animal-rearing operations. Some added heat	×			
Oceans	Solar, small inputs of fossil fuels	very low	Dispersed			×	
Minerals: industrial processing (and other materials)	Fossil fuels and nuclear power	20,000 (Manhattan Island)	Concentrated in cities and other industrial areas. Non-equilibrium (added) heat	×	×	×	Hydrocarbon combustion products; PCS from petrol engines. Noise may be very high
Power generation	Oil, coal, natural gas etc (FF's); nuclear materials like uranium	1,000,000 (oil refinery)	Concentrated at generating plant and at end use site, some transmission losses	×	×	×	Possible health effects from high voltage power lines. Hydrocarbon combustion products. Radiation from nuclear power generation

Energy intensities mostly from Slesser (1978). A = Atmosphere; FW = Freshwater; O = Oceans; FF = Fossil Fuels; PCS = Photochemical Smog

TABLE 10.19 Non-commercial energy as percentage of total energy consumption

Region	(in terms of heat content) %
Africa	51
C. America	35
S. America	45
Asia	58
N. America	3
Europe	7
Oceania	13

Non-commercial energy is chiefly wood and dung.
Source: Pearson and Pryor 1978

example, are variable: advanced countries like Sweden achieve as high a material standard as the US with a lower per capita consumption (Schipper and Lichtenburg 1976) and an analysis of energy consumption and life-style indicators in DCs showed significant correlations only in terms of economic parameters which were, in any case, capable of various interpretations other than the obvious causal ones (Table 10.20; Mazur and Rosa 1974). It seems as if some nations are having to consume more energy each year to stay at the same level of GNP (probably because of moves to centralized electrical power on grids and larger production units, according to Lindberg (1977)) whereas in others the first moves are being made to reduce the amount of energy lost through inefficient use: conservation measures and new technology are to bring to an end a period of profligate use (Hayes 1979).

This brings us to the idea of low-energy scenarios developed for several DCs in which it is postulated that current material standards can be maintained without recourse to higher levels of consumption and without radical changes in life-style. For the UK one study (Leach et al 1979, Leach 1979) suggests even with a threefold increase of GDP by AD 2025, a series of already known technical fixes could keep energy demand

TABLE 10.20 GNP and energy consumption, mid 1970s

Country	GNP/cap $US	Energy consumption per capita (mtce)
USA	7,060	10,999
Sweden	7,880	6,178
Switzerland	8,050	3,642

Source: various UN statistics

constant at the 1970s level; the chief elements in the strategy would be conservation measures, the renewal of plant, transport and buildings with more energetically efficient replacements and the adoption of heat pumps. Little 'alternative' power enters the scenario but nuclear power retains its peripheral role and could be abandoned if desired. A comparable scenario for the USA projects a real GNP in AD 2000 of twice that of 1973 but with zero energy growth, and includes a provision for more jobs (Freeman 1974).

Low demand is, of course, the present position in the LDCs and in general they are unhappy with it unless they belong to the OPEC group, particularly since firewood is becoming increasingly scarce and expansive (Eckholm 1975). Elsewhere, most of the conventional sources of energy are proving very expensive to buy (e.g. oil) or to develop (e.g. HEP) and, even in the latter case, electricity is not always relevant to the needs of

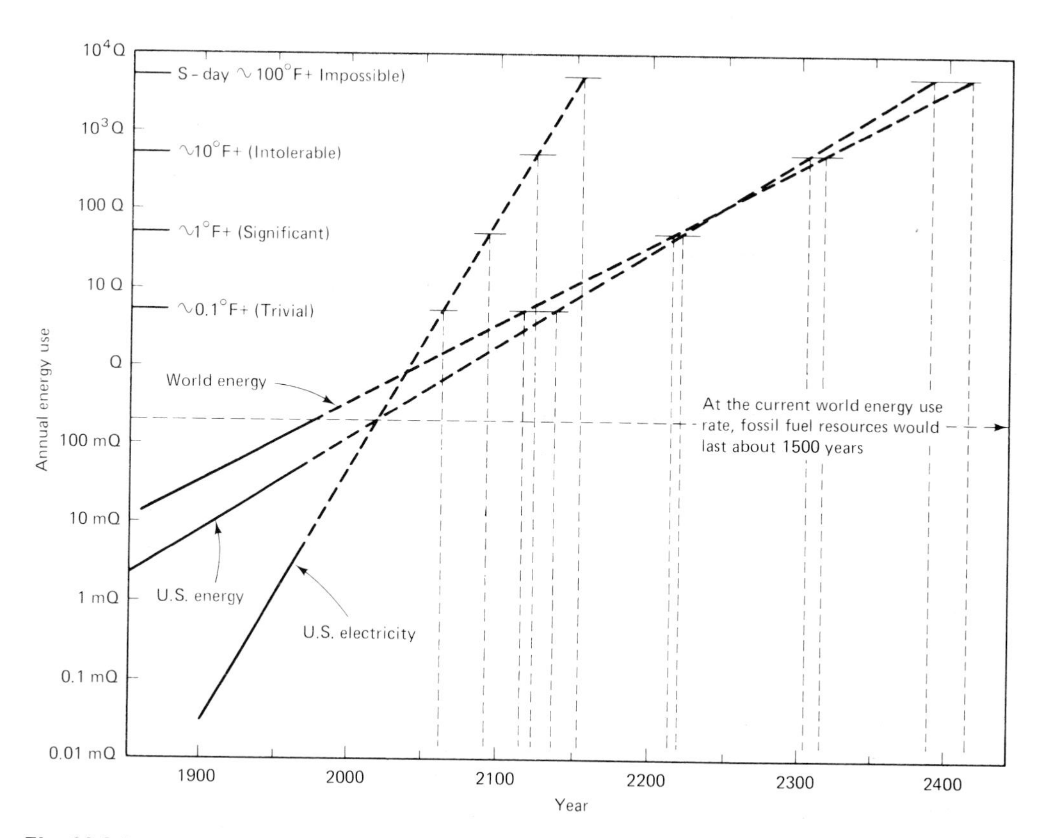

Fig. 10.9 Some extrapolations of energy growth in the USA and the world.
Note: At "S Day" the world would be about 100 F hotter than today and would be uninhabitable; accordingly, the condition is unrealizable. At one tenth the S-day energy use, the earth would still be 10 F hotter and much of it uninhabitable, an intolerable condition. Note how long it is before even trivial heating occurs and how quickly it worsens.
Source: Luten 1974

the rural poor, even though it has been suggested that rural electrification (preferably accompanied by television) is one of the most effective contraceptives available. In the immediate future, there seems a good case for basing development on low-cost, decentralized energy sources such as wood (where this can be 'farmed' rather than depleting the capital of forests) and solar sources (Hayes 1977, Brown 1978). At present, small-scale solar-based technologies are as expensive as diesel power or electricity from a national grid but seem likely to become economically attractive before long (Brown and Howe 1978). Thus small (ca 40 kW) hydroelectric generators, wind, methane from organic wastes, photovoltaic cells and solar flat-plate collectors may be a viable pattern for the foreseeable future and may have the additional advantage of not increasing migration to the cities. It is however unlikely that this kind of development will raise the LDCs to the 4·4 kW consumption envisaged by Hafele and Sassin (1978), compared with their present level of 0·5–2·0 kW; if 4·4 kW proves to be necessary for the proper development of the LDCs, then it will have to come from high-technology solar collection or nuclear power. Conventional economics will not benefit the LDCs in their decision-making, for high prices to them mean simply poverty and are not likely to be a force turning them to substitutes (Revelle 1976). But what help they are likely to receive, other than technology transfer, is contentious: Lindberg (1977) maintains that the international dimensions of the DC energy policies are generally nationalist and neo-mercantilist in tone and content; the dominant mentality is *sauve qui peut*.

The realization of the linkages of energy in resource processes and the need to reduce waste has led to a significant development in energy studies, namely energy accounting. In this, the full cycle of fuel involvement in the use of a material is counted rather than just the obvious. In the example of a motor car, the energy of its running is only the start: 'downstream' there are the energy uses of cleaning up the contamination it creates, for example, and of providing pedestrian foot-bridges or crossing lights, as well as that of junking the car at the end of its life, and reclaiming some of the materials. 'Upstream' there are the energy costs not only of building the car, but of making and transporting the machinery to the factory and of transporting the workers as well. It is difficult to know where to stop in the upstream analysis, unless it is where the relevant materials leave the ground.

Energy analysis has various applications, for example, in the spatial patterns and built form of settlements (Hirst 1976, Jackson 1978), and may reveal surprises: for example that Britain's North Sea oil programme will take several years to pay off the energy 'debts' incurred in developing it, and that a rolling programme of construction of breeder reactors is always in energy debt so long as building continues: only when fabrication stops does a positive energy balance emerge (Chapman 1975). The energy costs, then, of any existing or proposed process are a vital part of any reckoning beyond the directly monetary.

Nor can energy, of all resource processes, be considered outside of the political subset of culture. This book is not the place to summarize even part of the literature on the politics of energy but we need to remember the effects of the spatial inequalities of fossil fuel reserves and the dominant role of Saudi Arabia, the key nature and interests of the large oil companies, the suspicions about the nuclear industry, and the counter-suspicions that led to one book on nuclear power being withdrawn for amendment because of the hysterical accusations levelled against those not in favour of nuclear energy. All these types of consideration necessarily affect not only the development of energy supplies of various kinds but also the kind of society that results along with its environmental relations; furthermore, the 'right' solution for one nation or group of nations may well be different from that adopted by others. The 'best mix' for Europe may not be the same as that for North America, and Japan may have no choice.

So that the fundamental questions about energy in the future, difficult as they are anyway, are further complicated by spatial distributions, by nationalism and by long-standing conflicts. These questions centre around the foci of 'how much energy?' 'from what sources?' and 'what will be the environmental impacts?' Generalized answers are difficult to find: most authorities have their own pet solution. The question of environmental impact, especially of waste heat, is central, for it leads to the notion that some kind of decoupling between ecosphere and econosphere is desirable. To a partial extent this could be achieved by offshore and underground power stations, for example, and in their grand designs for hugely increased global energy productions, most writers suggest that the generation takes place on small islands, or in places where the waste heat will be of little climatic significance, or in 'nuclear parks' where the whole fuel cycle can be kept within one boundary fence. One of the difficulties, though, of grand designs (whether solar or nuclear) is the capital needed, both monetary and energetic (H. T. Odum 1977). All of which suggests that an upper asymptote to energy production may well be reached below the levels at which climate would be seriously affected beyond the scale of the small region, and as a corollary that the LDCs' position will continue to be depressed (Holdren 1975).

To conclude, we have to agree that many of the fundamental questions about energy are 'trans-scientific' or, in other words, moral: not for nothing has nuclear power been described as a 'Faustian bargain'; not for negligible reasons have responsible scientists said that the price of plentiful energy may be authoritarian government. Inside the envelope of these universal considerations, there are many steps which appear to Lovins (1975) as rational, and a considerable though by no means universal body of opinion would agree: firstly that we must adjust to the reality of much slower and even zero growth rates in energy consumption in the DCs; next, that oil and natural gas should be conserved and used sparingly, and that oil must be thought of as a valuable petrochemical feedstock rather than a fuel (Hayes 1976). Coal could be a much more valuable fuel if the environmental and health costs of its use could be minimized. 'Alternative' energy supplies may probably be worth more capital investment than nuclear power (a highly contentious judgement) because of the risks of the latter. Countries not yet industrialized should not get 'hooked' on fossil fuels; LDCs dependent upon them should receive special compensatory aid from the DCs. And we must distinguish need from demand.

Few now would dissent from the idea that the important issues of energy strategy are not technical and economic but social and ethical. Among other things Illich (1974) has pointed out that those with enhanced access to energy are those who will exercise power: the penetration of the linkages of energy are as fundamental as that. We must be careful who takes on the role of Brahman whose one circle or wheel of power determines our three layers of light, fire and darkness.

Further reading

BERHAM, D. 1979: *Solar energy: the awakening science.*
COOK, E. 1976: *Man, energy, society.*
Dialogue 1978: The energy dilemma.
GRAINGER, L. 1978: *Energy resources: availability and rational use.*
HÄFELE, W. 1980: A global and long-range picture of energy developments.
HAYES, D. 1976: *Energy: the case for conservation.*
HAYES, D. 1977: *Energy for development: third world options.*
HOARE, A. 1979: Alternative energies: alternative geographies.
HOLDREN, J. P. 1975: Energy resources.
LINDBERG, L. N. 1977: The energy syndrome.
LOVINS, A. B. 1975: *World energy strategies.*
LOVINS, A. B. 1977: *Soft energy paths.*
McMULLAN, J. T., MORGAN, R. and MURRAY, R. B. 1977: *Energy resources.*
ODUM, H. T. 1976: Macroscopic minimodels of man and nature.
PATTERSON, W. 1976: *Nuclear power.*
SLESSER, M. 1978: Energy in the economy.
SMIL, V. 1979: Energy flows in the developing world.

11

Wastes and pollution

Everything touched by King Midas turned to gold. By a sort of inversion process, pretty well everything modern men touch, including themselves, turns to a waste product sooner or later. Exceptions are rare, and centre around such things as valued landscape views, and objects of 'high culture' such as Old Masters, priceless buildings and symphonies. Even Michelangelos will crumble some day, presumably, and fed through hi-fi technology Brahms may become noise pollution. In the sphere of material use, the creation of waste products is generally accepted as inevitable: food becomes sewage, automobiles become junked cars, while quarries and mines become derelict land. The ways in which some of the wastes arise in an industrial society are shown in Fig. 11.1.

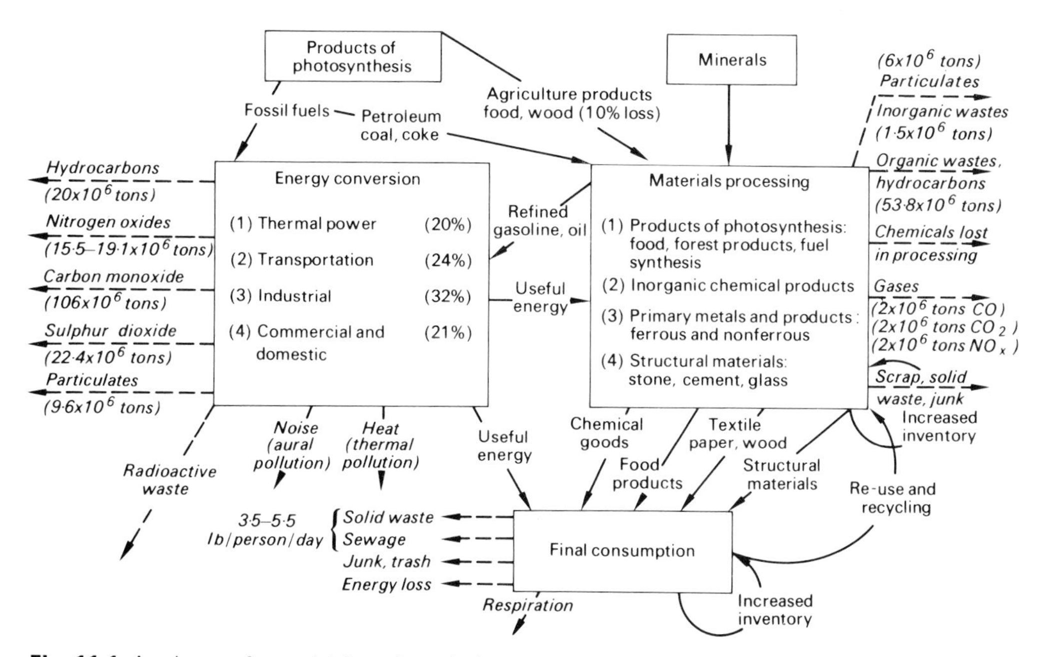

Fig. 11.1 A scheme of material flow through the resource processes of the USA, showing how inputs of energy and materials become converted to various kinds of wastes. Those from energy conversion in Note: Imperial units are used in this diagram. 1 ton = 1,016 kg; 1 lb = 0·45 kg.

Wastes of all kinds may follow one of two paths when they become identified as such. They may be regarded as sources of raw materials (the resource process of re-use or recycling) or ignored as too costly or technically unsuitable to undergo such processing. In the first type of perception, sewage becomes a source of manure, junked cars a veritable treasure-heap of steel, and derelict land an opportunity to create new parkland or housing space. The second perception sees bagged fertilizer as cheaper than dried sludge, new steel as more convenient than reprocessing old cars, and agricultural land as easier to convert to housing than levelled tip heaps; and there is as yet no way of filtering out the organochlorine pesticides from the waters of the oceans. The strategy of re-use is thus heavily dependent upon the relative cost of re-used and new materials and hence of the type of technology used in each process, assuming that supplies *de novo* of the material are still plentiful. Global scarcity would change the relative costs quite strikingly.

Materials and waste heat from energy consumption which currently are not or cannot be re-used are led off into the environment as a means of disposal. Usually the wastes are not put back at the point of first extraction but as close as possible to the processing plant or the sites of use: wastes from coal extraction accumulate at the pit head, garbage in the dustbins of domestic consumers and then at the municipal rubbish dump; heat and gases are dispersed into the atmosphere (Plate 15, page 276). Food wastes are usually discarded into water, with or without processing. In fact, the use of flowing water or the ocean margins as sinks for all kinds of agricultural, urban and industrial wastes is the outstanding feature of the world's resource processes: the hydrological cycle ensures the steady movement of all kinds of wastes into the oceans, and the depths are deliberately used for the dumping of radioactive material and spare barrels of chemical and biological warfare (CBW) agents.

At some point, however, these substances may begin to cause changes which allow them to be identified as pollutants. Pollution may be defined as:

> The introduction by man into the environment of substances or energy liable to cause hazards to human health, harm to living resources and ecological systems, damage to stuctures or amenity, or interference with legitimate uses of the environment.

Thus the major characteristics of pollution are its creation by man of increments to the biogeochemical cycles of natural substances, and its traceability through particular pathways in the environment away from the place of discharge. Its significance is related to its effects on a range of receptors, including man himself and the resources and ecological systems on which we depend, and it is also judged in the social context of damage to structures and amenity so that noise and smells, for example, may depend on their cultural context as to whether they are determined to be pollutants or not. Thus while pollution has ecological dimensions which can be objectively measured, many value judgements have to be made about the acceptibility of various levels of substances and energy; these features can be encapsulated in the simple definition of pollution as something in the wrong place at the wrong time in the wrong quantity (Holdgate 1979).

The two kinds of dimension interact. The death of songbirds from pesticide poisoning may not 'matter', since they are not an 'economic' resource and their niche in the ecosystem can be filled by another group, but their loss is culturally unacceptable to many people. There can be no doubt that acceptability of the pollutant by-products of economic activity varies with affluence: it is the already rich nations which are trying to grapple with pollution. Poor countries and the poorer regions of the industrialized nations are prepared to tolerate the effects of pollution if their living standard is raised by the operation of the earlier parts of the resource process: loss of wildlife through use of DDT brings little sorrow to Asians freed from malaria, and the cessation of pollution from pulp-mill waste is little compensation to the workers of the shut-down plant in an isolated part of Newfoundland. Cultural thresholds are very difficult to quantify, since man possesses the ability to adapt to changing external circumstances and relatively little is known about the way in which levels of tolerance are felt.

TABLE 11.1 Major waste products and their receiving environments

	Environments into which wastes are discharged: X *Environments into which wastes get transferred:* O				
Wastes	*Air*	*Fresh water*	*Oceans*	*Land*	*Clinical effects of residues on humans?*
Gases and associated particulate matter (e.g. SO_2, CO_2, CO, smoke, soot)	X	O	O	O	Yes
Photochemical compounds of exhaust gases	X	O	?	O	Yes
Urban/industrial solid wastes			X	X	No
Persistent inorganic residues, e.g. lead (Pb), mercury (Hg), cadmium (Cd)	X	O	OX	X	Pb—disputed Hg—definitely Cd—definitely
Persistent organic compounds:					
Oil			X	O	No
Organochlorine residues	O	XO	X	X	Disputed
Pharmaceutical wastes		X	O		Unknown
Organic wastes:					
Sewage		X	X		Possible bacteria carrier
Fertilizer residues with N_2, P		O	O	X	Yes, especially N_2
Detergent with P		X	O		No
Radioactivity	X	O	X	O	Yes
Land dereliction				X	No
Heat	X	X	XO		No
Noise	X				Yes
Deliberate wasting e.g. defoliation in CBW		O	O	X	Yes

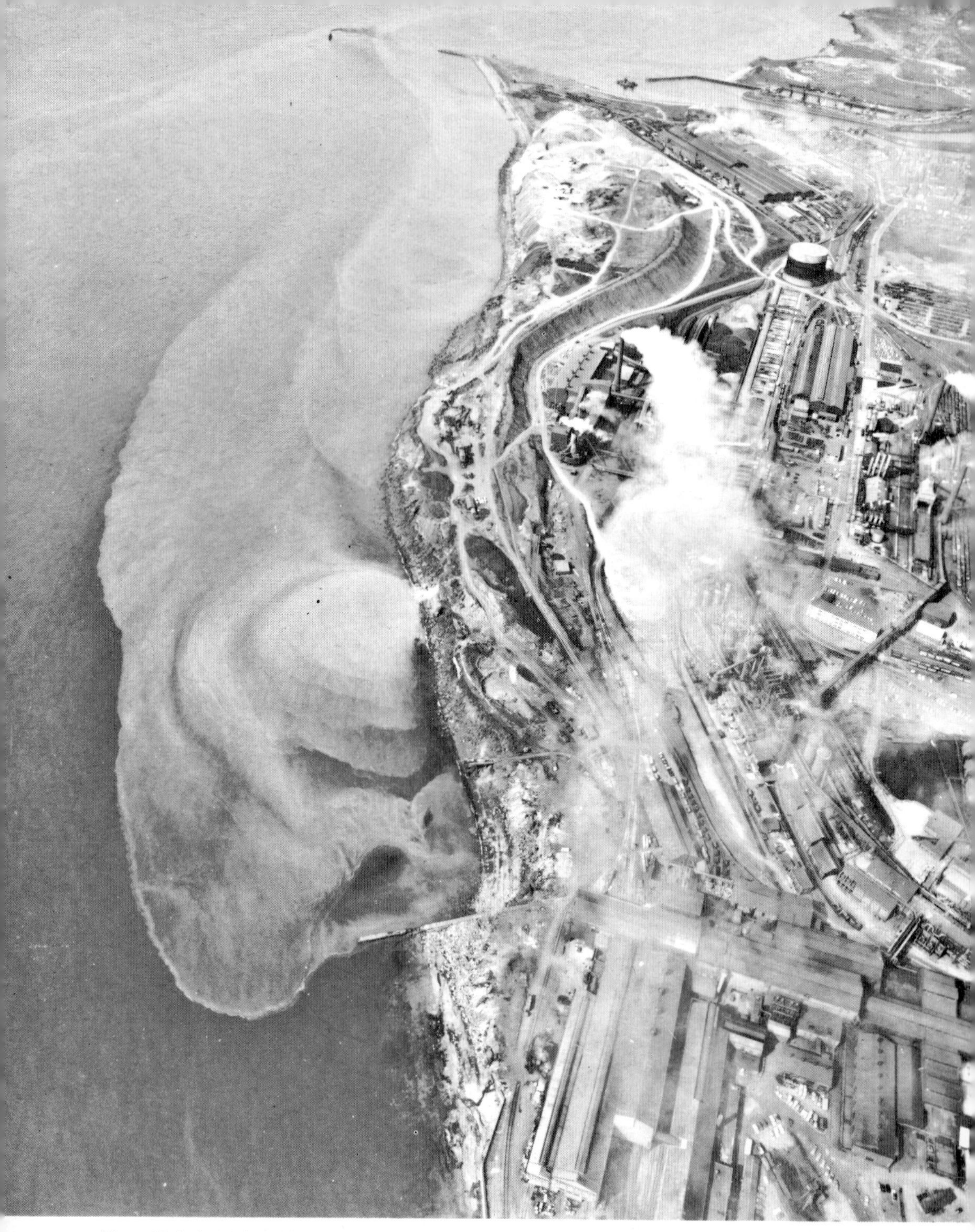

Plate 16 Industrial and urban areas gather together materials and hence dispose of the wastes in a concentrated form. This steel works at Workington, Cumbria, England, can be seen to be emitting wastes into both the sea and the atmosphere. *(Aerofilms Ltd, London)*

Classification of wastes and pollutants

A division of the various types of wastes for purposes of discussion is attempted in this chapter, on the basis of the type of material and its ecological effect when it reaches a concentration sufficient for it to be called a pollutant. This method is chosen in preference to alternative ways of classification by type of environment (air, land, sea, etc.) or by type of receptor, since it involves less repetition. A summary chart of waste material by environment type is presented in Table 11.1.

Gaseous and particulate wastes emitted into the atmosphere

Fossil fuels and water vapour

The gaseous composition of the atmosphere is more or less constant except for the amount of water vapour. By-products of a gaseous and finely divided particulate nature are emitted into it: some are gases already present in the 'natural' air, others are foreign (Schaefer 1970), and some particles such as dust from exposed soils are also present in uncontaminated air. Notably, most of the emissions into the atmosphere are by-products of the combustion of fossil fuels for energy use. (Table 11.2 shows the relative values of common pollutants for one year in the USA. Radioactive wastes are also produced during power generation by nuclear fission, but are dealt with separately below.) The biggest product of such oxidation reactions is water, but this is not usually regarded as a contaminant, possibly since the atmospheric content is normally variable between 1–3 per cent by volume and emissions do not radically change this figure at ground level.

TABLE 11.2 US Gaseous emissions 1970

	$\times 10^6$ *tons/yr*	%
Sulphur oxides	33·9	12·8
Particulate matter	25·6	9·7
Nitrogen oxides	22·7	8·6
Hydrocarbons	34·7	13·2
Carbon monoxide	147·0	55·7
Total	263·9	100

Source: Lynn 1975

CO, SO_2 and H_2S

These three gases comprise the biggest volume of wastes emitted into the atmosphere (along with carbon dioxide, dealt with below). Carbon monoxide, mainly from the internal combustion engine, is the largest of the three and in the USA its volume equals the sum of all the other industrial contaminants (about 150 million tons/yr), although the global production due to human activities ($0{\cdot}02–0{\cdot}2 \times 10^9$ t/yr) is small compared with natural emissions (5–10 × 10^9 t/yr). In cities its concentration is usually between 1 and 55 ppm with an average 10 ppm, but detectable clinical symptoms appear at 100 ppm and humans are killed at 1,000 ppm. Los Angeles has established alert levels for CO at 100, 200 and 300 ppm. Its effect often spreads into buildings, especially during the heating season when it is drawn into apartment blocks and offices by upward currents of warm air: in one office building in New York, permissible CO levels were exceeded for 47 per cent of the time during which the block was being heated. Although deleterious to humans because of its affinity for haemoglobin, thus depriving body tissues of their oxygen supply, CO is not cumulative, and removal of the pollutant source allows a rapid return to a normal level. At high concentrations, however, its effect appears to be synergistic. Its major effect therefore is as a human poison, rather than a factor of ecological change.

Although produced in smaller quantities by resource use than the previous gas sulphur dioxide is much more toxic. It comes from the combustion of coal and fuel oils, at sulphuric acid plants, and in the processing of metal ores containing sulphur. In the atmosphere it lasts an average of 43 days, being converted to SO_3 and reacting with water to produce an aerosol form of sulphuric acid which is toxic to plants at 0·2 ppm and is highly corrosive of iron, steel, copper and nickel, while building materials containing carbonates find these compounds replaced by soluble sulphates. Sulphur dioxide is produced in very large quantities in urban-industrial areas: New York City emits nearly 2 million t/yr from coal, and Great Britain 5·9 million t/yr. A concentration of 1–5 ppm usually produces a detectable physiological response in man, and in London's great smog of December 1952, 1·34 ppm was recorded; elsewhere, urban concentrations up to 3·2 ppm have been noted. Plants are injured at such levels, and animals including man suffer from inflammation of the upper respiratory tract. Like CO, therefore, SO_2 is particularly a pollutant of urban areas, although downwind transport of H_2SO_4 from industrial Britain is claimed to be acidifying the fresh-water bodies of Scandinavia, and acid rain has been similarly identified in the north-east USA (Likens and Bormann 1974). Some sulphates have been claimed to be damaging to health at concentrations 30–40 times lower than, for example, the SO_2 levels set for the USA by the Clean Air Act of 1970. The WHO has now set a long term objective of a level so low (ca 0·3 ppm SO_2/m^3) that even plants are unaffected.

Hydrogen sulphide and its related organic compounds, mercaptans, are sometimes by-products of petroleum processing, coking, rayon manufacture and the Kraft process for making paper pulp. Hydrogen sulphide is also present when anaerobic bacteria are found in considerable quantity, as in inadequately treated sewage. Both these sources are nuisances rather than dangers, their odour being detectable at concentrations of 0·03 ppb (mercaptans) and 0·035–0·10 ppm (H_2S).

Particulate matter

Industrial processes, auto exhausts, bare soil and backyard barbecues alike produce particulate matter, a conspicuous component of air pollution for many centuries. Dust comes from many sources including the gradual comminution of the debris which accumulates in cities and open spaces without vegetation. Soot consists of finely divided carbon and heavy hydrocarbons from combustion processes, and to this may be added finely particulate material from almost every urban-based resource process: ash, flour, rubber, glass, newspaper, lead and fluorides. The last-mentioned are produced in the manufacture of ceramics, bricks and phosphatic fertilizers and, especially in the form of hydrogen fluoride, can cause damage to plants, animals and man. Fluorosis, the mottling of teeth, occurs at concentrations of 20 ppm, but levels are usually less than 0·02 ppm except around poorly designed emission sites. Many other chemicals are present as particulates, but concentrations above 50 mg/1,000 m^3 air seem to be unusual.

The particulate materials have well-known effects. The reduction of visibility is obvious in most cities, soiling of paintwork, buildings and Monday's wash is commonplace, metals are corroded, and immense social and economic costs are incurred because of the aggravation of bronchial illnesses which takes place. Some writers also attribute carcinogenesis in people to one or more of the substances of this class of wastes. Their order of magnitude can be gauged by the fallout: in Great Britain large cities experience depositions of the order of 130–525 $\times 10^6$ kg/ha whereas small towns and rural areas receive 2·6–26·0 $\times 10^6$ kg/ha. But from 1958 to 1974, power stations in Britain burned 35 per cent more coal but emitted 82 per cent less grit and dust, and a 60 per cent rise in cement production was accompanied by a 64 per cent decline in dust emission. On a world scale, Mitchell (1975) estimates that about 4×10^7 t of particulates are present in the atmosphere, of which *ca* 1×10^7 t is derived from human activity. A growth rate in the latter of 4 per cent per annum would produce 60 per cent above present levels by AD 2000 but the rate of the anthropogenic emissions is now greater than the average volcanic dust loading in the past century (Fig. 11.2).

It is difficult to forecast the effect of the loading upon climate, even if there were not additional variables such as CO_2 (see below); Mitchell (1975) suggests that some global cooling would be expected although a net warming is also possible, depending on the layering of the particles and their efficiency as absorbers rather than scatterers of solar radiation in the troposphere. Regional changes of temperature different from planetary average trends may be expected if the particle loading is, as we would expect, spatially non-uniform.

Smog

The combustion of coal, oil or natural gas in power plants and the internal combustion engine both result in the emission of nitrous oxides into the atmosphere (Haagen-Smit and Wayne 1968). Of these sources, the incomplete combustion of the auto engine is by far the most important contributor by a factor of about 6. Because of its motorized life-style and inversion frequency, Los Angeles provides many of the data on nitrous oxides and their atmospheric fate. LA County produces about 0·15 kg of nitrous oxides/cap/day and concentrations of 0·02–0·9 ppm are typical, with an alert level at 3·00 ppm. The major constituent is NO_2 (nitrogen dioxide), which is a respiratory irritant. It also absorbs sunlight, especially in the blue wavelengths, and so appears as a

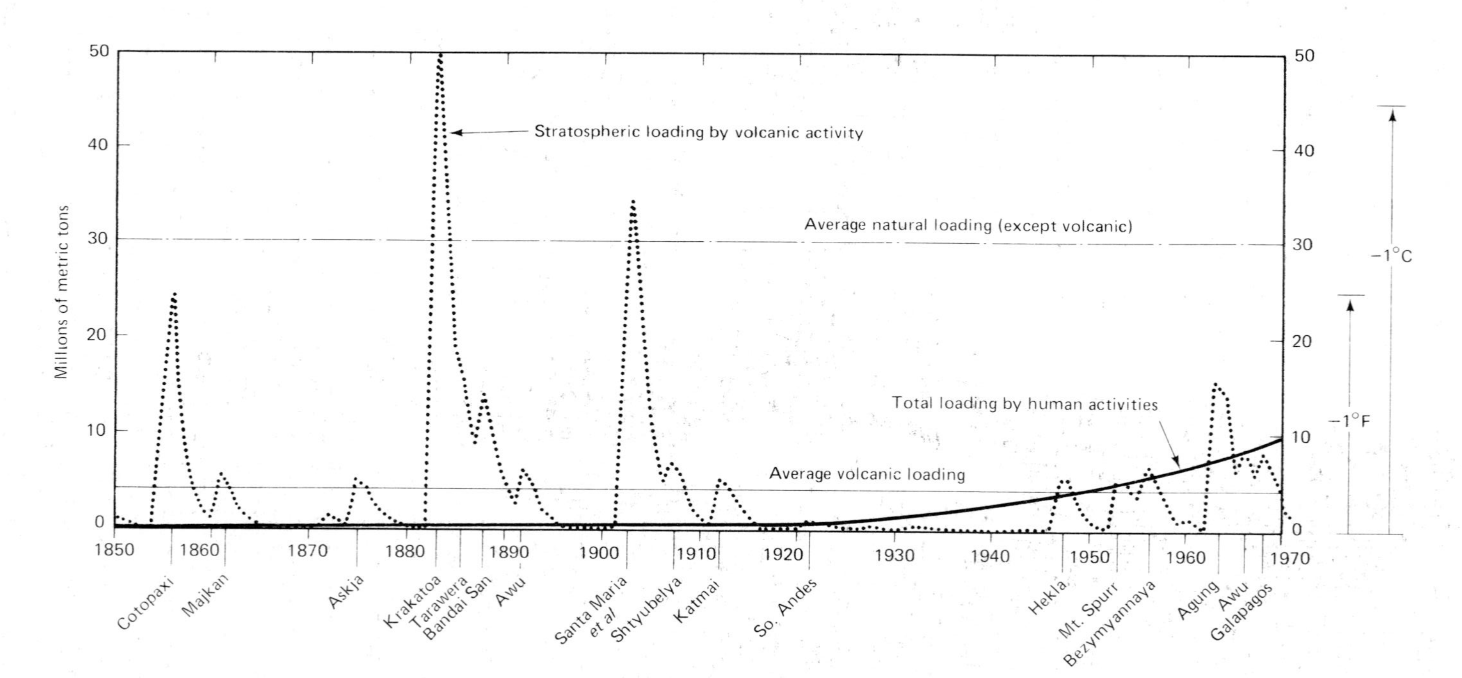

Fig. 11.2 Estimated chronology of annual average global atmospheric particle loading, 1850 to 1972, by volcanic activity (stratospheric loading only, dotted curve), by human activities (heavy solid curve), and by all natural sources other than volcanic (assumed constant at 30 million tons, dash-dotted line). Estimated calibration of volcanic loading curve in terms of planetary temperature influence is shown at right.
Source: Mitchell 1975

yellow-brown gas in which a concentration of 8–10 ppm reduces visibility to one mile. Since it absorbs sunlight, it can undergo photolysis, at a rate of 2 ppm/hr. The chemical reactions are complex but the products include ozone, formaldehyde and nitrous oxide (NO). This latter appears to be oxidized to NO_2, providing a further source of reactants. Numerous other substances are present in the ensuing 'photochemical smog' resulting either from atmospheric reactions or incomplete combustion in auto engines. They include other aldehydes, olefins, ethylene and peroxyacetyl nitrate (PAN): on a smoggy day in Los Angeles city, ozone levels may be at 0·65 ppm (of 0·30 on a non-smoggy day), and oxides of nitrogen at 2·0 ppm (if 0·05–1·30). This somewhat heady brew bears little relation to laughing gas, and indeed lachrymation is one of the usual effects. Ozone concentrations of 1·25 ppm cause respiratory difficulties, and a series of smog-alert levels in Los Angeles and other North American and Japanese cities tell citizens, especially those with chronic respiratory ailments, of the potential dangers of smog levels.

The effects of smog are not confined to people: damage to crops during the 1960s in California exceeded $8 million/yr, and on the eastern seaboard of the USA $18 million. The forests around the Los Angeles basin, a vital reserve of recreation space, suffer from various forms of die-off due to atmospheric pollution and it is thought that ozone is probably the chief agent of biotic damage. Damage to materials is also prevalent, stretched rubber being one of the most vulnerable substances. Los Angeles, due to its geographical peculiarities, is the most quoted example of a smoggy city, and as such often used as a whipping-boy (whipping-girl would be more accurate, in view of its full name) by all opponents of the motor car. But practically all industrial cities suffer to some degree from photochemical smog, and any which are liable to air stagnation, usually in the form of inversions, are vulnerable to smog accumulation; in large Japanese cities such as Kyoto and Tokyo, CO level readout panels are installed at some intersections and sometimes the traffic police wear masks as confirmation of the high levels of exhaust gases. High ozone levels have even been detected in London and the rural areas west of it, suggesting that even in a cyclonic climate, periods of anticyclonic stagnation in summer are sufficient to cause the creation of smog.

The transmission of radiation: chlorofluorocarbons, oxides of nitrogen and carbon dioxide

All the effects so far discussed have been at a local or regional scale and apparent in a short period of time. But the dynamic nature of the atmosphere, its ability to provide reactants for chemical change, and its exchange processes with land and sea all ensure that many of the emissions are translocated far from their source, sometimes in a chemically altered form. Of the substances enumerated, most appear to have no global implications, although since monitoring is so recent and so fragmentary this should be regarded as an interim conclusion. Most of the sulphur compounds end up in the oceans as soluble sulphates with no known effects, for example; photochemical pollutants stabilize either as nitrates or $CO_2 + H_2O$; a global emission of 33,528 t/day of nitrous oxides means a yearly atmospheric increases of 2×10^{-6} ppm.

Some other substances, however, have given rise for concern on a global scale, since their effects seem to be diffused widely within the atmosphere. These effects include a greater efficiency of the transmission of solar radiation, thus leading to fears of rapid global warming or the increase of skin cancer from UV radiation. One such set of substances are the chlorofluorocarbons (CFCs or CFMs) which are used in refrigerator cooling systems and as aerosol propellants. Since 1931 about 8 million tonnes have been made, and most of this has been released to the atmosphere, where it breaks down very slowly and the current production of 700,000 t/yr is bringing about a 1·5–2—fold increase in concentration in the atmosphere since 1970. At one time in the 1970's, models predicted that there would be a rapid depletion in the ozone layer of the atmosphere since the O_3 would be dissociated by reaction with chlorofluorocarbons as well as NO_x from aircraft exhausts, especially those of SSTs. The production of CFCs has been declining in the years after 1974 and any problems caused by them are now thought to be much lower in magnitude than those attributable to carbon dioxide, though by no means insignificant in depleting the ozone layer (Schiff 1979). The same seems to be true of nitrogen oxides emitted by fuel consumption at high altitudes, and diffused further by the manufacture and use of nitrogenous fertilizers. Given the worst case, the number of extra melanomas produced is seen by some authorities to be much lower than those produced by the growing addition of light-skinned people to sunbathing. Effects on global temperature are uncertain and not usually estimated, since by comparison with carbon dioxide they seem relatively small (Holdgate 1979).

Carbon dioxide, CO_2, has been released to the atmosphere at an accelerating rate since the late 19th century. The pre-industrial concentration of CO_2 is thought to have been 265–270 ppm, and the 1976 level 325 ppm; the annual increment is ca 0·7 ppm but this would be at least twice that level if all the CO_2 released from burning fossil fuels remained in the atmosphere. It is estimated that the sea takes up 50–60 per cent of the incremental carbon dioxide, and the rest goes into terrestrial ecosystems via photosynthesis. It is clear that man can affect this latter rate by deforestation, for example, or planting trees on the other hand. A plot of the increase in amount as excess above a nineteenth-century base level, together with predicted increases based on assumptions about fossil fuel consumption, is given as Fig. 11.3.

Why should there be concern over this gas? It seems to be effective in letting through solar radiation from space, and in reducing its outward flux. Thus a global warming is to be expected: a doubling of atmospheric CO_2 might mean an increase in global temperatures of 1·5–2·5°C, although it is by no means certain that the relationship is linear; in any case, the rate of use of fossil fuels seems highly likely to slow down. But put together, it is conceivable that CO_2, CFCs and NO_x could cause changes in regional and global climate and allow the penetration of harmful radiation; they are at present the only types of contaminant for which such widespread effects can currently be suggested. Even slight changes of mean temperature could cause shifts in the boundaries of deserts and the northern limits of cultivation, and could increase the melting rate of polar ice-caps to bring about rises in sea-level which at the very least would be inconvenient and costly. Even though any effects might, objectively, seem minor, they are unlikely to be so in their effects in a world where drought and crop failure cause hunger and where the benefits to one group of economies which generate pollution bring about damage to innocent people elsewhere (Bach et al 1979, Wigley et al 1980).

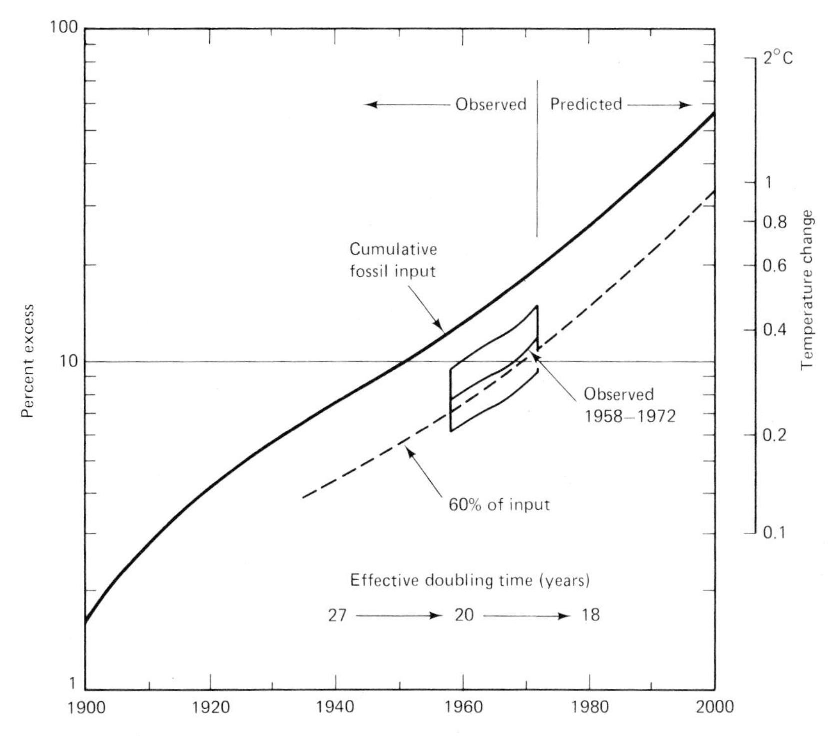

Fig. 11.3 Growth of atmospheric carbon dioxide, AD 1900 to 2000, expressed as percent excess above 19th century base level. Cumulative fossil input curve represents potential atmospheric growth if all fossil CO_2 were to remain in the atmosphere without loss to the oceans or bios7here. Wavy line segments, labelled 'observed 1958–1972', are observed CO_2 growth at Mauna Loa for three assumed 19th century base levels between 285 and 295 ppm. Dashed curve indicates tentative long-term increase of actual atmospheric CO_2 content. Projections after 1970 are based on assumption that fossil CO_2 input will increase 4%/yr to 1980 and 3·5%/yr thereafter.
Source: Mitchell 1975

The deleterious effects to many people of wastes are therefore seen at three scales. There is the point-source danger, as when a noxious gas is imperfectly controlled: an H_2S escape in Mexico in 1950 caused 22 deaths. Regionally, various pollutants doubtless exacerbate existing illnesses (especially respiratory troubles) and maybe induce them. In unusually severe episodes of smog or industrial pollution, premature death is caused. Many other substances are emitted in particulate form into the atmosphere, cause problems in terms of health, damage to materials and living matter, and are aesthetically unacceptable. A newspaper quoted a tentative estimate of the health costs of air pollution in the USA in 1968 as $6·1 × 10^9. They also cause changes in some parameters of regional climate. Measures are now being adopted in most industrial nations to reduce air pollution in cities, but less attention is generally given to substances which do not directly affect the human population. At a global scale, problems become more nebulous. The probability of direct changes in climate resulting from CO_2 emissions seem small this century, given the possible natural cooling trend and the effect of scattering of radiation by particles, but the longer-term possibilities are so serious that efforts to learn about trends of climatic change and the fate of industrially produced CO_2 must claim a high priority (Niehaus 1979, Marchetti 1979).

Economic poisons

The elimination of competitors

In the course of resource processes such as agriculture, many poisons are applied to ecosystems in order to eliminate or diminish the effect of man's competitors for the crop. We also apply poisons to ourselves in order to kill viruses and bacteria which are inimical to our health, to our immediate surroundings to kill the vectors of disease such as the malaria-carrying mosquito, or to kill competitors for stored food such as the rat. All the substances used as poisons interfere with the metabolism of organisms and are effective in small doses. Ideally, the toxic substance breaks down into biologically insignificant compounds as soon as it has performed its work, so that dead organisms contain no residue of poison to be passed to scavengers, and soils would contain no active toxin to be washed off into streams or be available to organisms other than the targets. Some substances used as poisons do not possess this quality, and some are converted to an even more toxic form when they remain in the environment: collectively they are known as persistent toxins (Moriarty 1975, Khan 1977, Perring and Mellanby 1978). Both short-lived and persistent poisons are rarely totally effective or completely harmless: application to any population of plants or animals elicits a spectrum of response from the organisms since all but the most extreme poisons leave a few resistant individuals and if their immunity is genetically transmissible, and poisoning continues at a constant level, then resistant populations may develop. Examples are the bacterium *Staphylococcus* which is often unaffected by penicillin, and the mosquito *Anopheles* (a vector of malaria), some strains of which are immune to the effects of DDT. The toxin is unlikely to affect all the species in a habitat, so that another species may experience a population surge and itself become a weed or pest. Most important of all the unwanted effects of economic poisons is that caused by persistent residues when they undergo ecological amplification. Typically their concentration increases as they pass from one trophic level to another through a food web. The concentrations in successive organisms are cumulative to the point where lethal or sub-lethal effects upon an organism, often far removed in time and space from the target, are sometimes observed: precautions against sheep warble fly end up preventing the reproduction of golden eagles, for example, or anti-gnat measures cause the death of grebes. The longer the food chain the higher the probability of physiologically significant concentrations in the highest predator levels.

What follows is an examination of some of the most commonly used toxic substances which are applied mainly to terrestrial habitats, sometimes to fresh waters. In 1945, less than 20 active ingredients were available, by 1975 about 200 were being marketed in at least 800 formulations in the UK. (Sly 1977). Residues not immobilized in the soils or in organic matter pass through the runoff phase of the hydrological cycle, and if sufficiently long-lived they accumulate in the oceans. Most of them are poisonous to humans if large quantities are inhaled, ingested or absorbed cutaneously as a consequence of escapes during application; few of the toxins discussed here have not been the direct cause of human death. However, it is with the unintentional effects of residues that we are mainly concerned.

Biocides

Chemicals such as copper sulphate (as Bordeaux mixture) have been used for hundreds of years against pests in orchards. Sodium chlorate is a well-known weedkiller, highly explosive and moderately persistent; sulphuric acid is still used for the destruction of potato haulms. These simple substances have largely been replaced by complex organic compounds, of which DNOC is a much-used member. It is very poisonous (but not persistent) and protective clothing must be worn for its application, typically by spray at 10 lb/ac (1·8 kg/ha). Accidents from spray drift have had unpleasant results. More popular still have been the contact hormone weedkillers which affect broadleaved plants only (thus leaving cereals and grasses, for example) and are said not to affect man. Their effect is to stimulate growth hormones to the point where the plant grows so fast it virtually dies of exhaustion. Two popular examples are 2,4-D (non-persistent) and 2,4,5-T (highly persistent). Both have been criticized because spray drift is an indiscriminate killer of vegetation, and there is coming to light an uncomfortable amount of evidence suggesting that teratogenicity of human foetuses can result from certain levels of exposure to any 2,4,5-T which contains the contaminant dioxin. It was dioxin (TCDD) which has contaminated the soil around Seveso in Lombardy, Italy, after an explosion at a chemical factory there in 1976; the area around had to be evacuated and much of the vegetation and top soil scraped up and buried (Cattabeni et al 1978, Bonaccorsi et al 1978). In 1978 the US Environmental Protection Agency introduced an emergency suspension on the use of 2,4,5-T because of the alleged risk that it produces miscarriages. Two types of systemic total herbicides have also been developed. The first group, typified by simazine and monuron, are persistent and prevent regrowth for periods of up to one year; the second group are short-lived: the effect of dalapon lasts for 6–8 weeks, paraquat has a life of a few hours only once it reaches the soil. Dalapon is especially suitable for aquatic weeds since there is little apparent effect upon animal life.

The success and adaptability of the insects has provoked a formidable battery of poisons. There are three main kinds: inorganic compounds, botanicals based on natural substances, and synthetic organic compounds of two major subtypes, the organophosphorus compounds and the chlorinated hydrocarbons. Outstanding among the inorganics is lead arsenite, deployed against caterpillars on fruit trees. Since the combination of lead and arsenic appears particularly frightening, it is comforting that in Britain the Food and Drugs Act 1955 lays down that no food may have > 1 ppm arsenic or > 2 ppm lead content. Lead arsenite is not applied later than 6 weeks before marketing. The botanicals are dominated by powders derived from two plants: rotenone, from the roots of *Derris elliptica*, is often called derris and is non-persistent, though deadly to fish if it drifts or is spilled into water-courses. The flowers of *Chrysanthemum cinerariaefolium* yield pyrethrum, which paralyses insects but is non-persistent and non-toxic to most other groups. In some ways it is an ideal insecticide but is expensive, so that ways of synthesizing it cheaply by industrial methods are being sought, and by the 1970's were showing signs of success.

From 1944 onwards organophosphorus pesticides, most of which are aimed at insects, became commonly used. Most of them are cholinesterase inhibitors which impede impulse transmission in the nervous system so that respiratory failure ensues. Two of the commonest are parathion and TEPP, both of which are very toxic beyond the target group (ca 30 g of TEPP will kill 500 people) but break down very rapidly. Both are suspected of having killed non-target organisms locally, but proof is difficult because of their rapid breakdown. Their close relative malathion is safer, being much less toxic and having also a rapid disintegration rate into non-toxic substances. In many instances it could replace the persistent DDT but is much more costly to manufacture.

The most widespread and heavily used group of pesticides are the organochlorines or chlorinated hydrocarbons. The most famous, DDT, was first synthesized in the nineteenth century but did not come into widespread use until the war of 1939–45, when its efficacy against malarial mosquitoes became known; after 1945 its role in agriculture quickly came to exceed that of medical applications. DDT shares some characteristics with several other commercially available organochlorines. It is highly toxic to insects, though resistant strains can breed, but is poisonous to many other groups as well, of which fresh-water fish are one of the most vulnerable. The whole group is long-lived, with lives of at least 10–15 years (Table 11.3). Thus they persist in food chains, causing both sub-lethal and lethal effects at various trophic levels.

TABLE 11.3 Longevity of chlorinated hydrocarbons in soil

Years elapsed	*Perticide*	*% remaining*
14	Aldrin	40
	Chlordane	40
	Endrin	41
	Heptachlor	16
	BHC	10
	Toxaphene	45
15	Aldrin	28
	Dieldrin	31
17	DDT	39

Note: The experimental design meant that maximum longevity is indicated, so that survival time may be longer than average in nature.
Source: Rudd 1975

A great deal of investigation has been done on the effects of DDT residues, usually metabolites such as DDE and DDD which share its characteristics, although DDE is considerably less toxic to most organisms. DDT and its metabolites appear to be distributed over the whole surface of the globe, including the far oceans and the lower layers of the atmosphere: for example, African pesticides appear in Caribbean winds (Wurster 1969). Analyses of DDT concentration in water are perhaps misleading, since it is scarcely soluble in water (1·2 g/litre) but highly soluble in organic substances (e.g. 100 g/litre in lipids) so that it will always 'flow' from the inorganic world to the organic (Woodwell 1967a). Distribution as airborne droplets may account for its presence in the fat and viscera of fish and penguins in Antarctica, along with other chlorinated hydrocarbons such as BHC, dieldrin and heptachlor epoxide (Tatton and Ruzicka 1967). Its ability to build up in food chains is well documented in an example from estuaries on the east coast of the USA, where a concentration in water of 0·0005 ppm is accompanied by concentrations of 0·33 ppm in *Spartina* grass, 0·4 ppm in zooplankton, 2·07 ppm in needlefish, 3·57 ppm in herons, 22·8 ppm in fish-eating mergansers, and 75·5 ppm in gulls. The DDT present (about 70 per cent as DDE) is all residual (Woodwell *et al.* 1967). Sub-lethal effects are also common: DDT appears to inhibit calcium carbonate deposition in the oviducts of certain birds (e.g. pelicans and the peregrine falcon) and thin-shelled eggs which rarely come to term are the result (Ratchiffe 1970). Very low dosages (of the order of 5 ppb of DDT or 500 ppb of dieldrin) will affect the behaviour and reproduction of fish and amphibians. Some work suggests that the levels of residual pesticides in organisms are due more to the ability of a particular species to excrete the compounds than to biological magnification but this does not alter the significant effects upon biota which have been detected. The time at which the bioride is ingested may be critical: bats for example may be killed by pesticides after hibernation, either during a phase of high feeding intensity upon a sprayed insect population, or during migration when fat reserves containing high levels of toxin are mobilized.

Particularly because of its effect on wild animals, DDT has attracted a good deal of attention in the DCs. Less well known to the public are other chlorinated hydrocarbons which are equally or more toxic in residual form than DDT and its derivatives: BHC, dieldrin, adlrin, endosulphan and heptachlor are examples. In the soil, aldrin, for instance, is transformed into heptachlor epoxide which is more toxic and more persistent than its parent substance. Gradually, most of these compounds have been either banned or used only in restricted circumstances in the DC's: persistent chlorinated hydrocarbons used in England and Wales fell from 460 t/yr in 1963 to 250 t/yr in 1972; in the USA the equivalent figures are 60,000 and 44,000 t/yr. (Holdgate 1979). Aldrin and dieldrin have been phased out.

Substitutes have of course been developed: these are often designed to be more target-specific and always to be non-persistent, breaking up rapidly once they have affected the target organism or group of organisms, and they do not rapidly produce resistant populations. Sevin, the foramidines and the carbamates are examples, as are analogues of parathion such as fenthion, bromophos and iodophenos. But even one of the 'third-generation' pesticides, trithion, killed wild geese in Lincolnshire, England, in 1974–75. The cost of testing the new biocides makes them more expensive than DDT and so their introduction, especially in LDCs, is slow: in the early 1970s the cost of DDT for anti-malarial usage by WHO was $US 0.40/kg whereas malathion cost $1.43/kg, propoxur $3.52/kg and fenetrothion $1.60/kg. Presumably large-scale production will cause the relative costs to fall.

The pathways of the DDT and other persistant toxins have been studied with a view to their eventual fate. Woodwell et al (1971) conclude that most of the DDT produced has either been degraded into an innocuous form or partitioned into places where it will not be available to plants, animals or man. The total biota probably contains less than 1/30 of one year's production of DDT during the 1960's although to this must be added the other persistent toxins. Although able to be transported world-wide, most of the deleterious effects of DDT seem to be regional and so terrestial wild bird populations, for instance, seem to have recovered soon after applications of the biocide have ceased. The fauna which pick up DDT from oceanic food webs seem to lag further behind diminutions in DDT usage.

The persistent organochlorines have attracted a good deal of criticism in recent years, though stoutly defended by those who point to the increased productivity and stability of agriculture gained by their use (Moore 1977, Mellanby 1977). Levels in human body fat are typically 12–16 ppm in DCs, so that in the USA most people are now legally unfit for human consumption. No adverse effects have as yet been noted, but since 1945 marks a zero datum line, no individual can yet have carried the current body burden through a normal life-span. In ecosystems, synergistic effects may exert long-run changes in ecosystems which are impossible to predict (Sassi 1970).

Accidental contamination by organic substances

Oil

The dominant member of this category is crude oil. On one scale its loss during transport may be acutely dangerous to human life, the more so because of the inflammable nature of some of its constituents. More concern is generally expressed with its toxicity to plants and animals and with the aesthetic effects caused by spilt oil being washed up on beaches and shorelines. Fig. 11.4 sets out the toxicities of various fractions of crude oil over a broad spectrum of marine animals, from which it can be seen that the low-boiling aromatic hydrocarbons and non-hydrocarbons provide the greatest source of toxic materials: these are, however, the first to evaporate after a spill.

Some data on the interaction of petroleum and the marine environment are shown in Table 11.4. There is a background seepage from the earth's crust (mostly at the junctions of tectonic plates), which is exceeded by industrial extraction. Much of the crude oil is then transported by sea and there are two main kinds of injection into marine ecosystems. The first of these are chronic losses from refining, transhipments, tanker cleansing and other accidental spills. There were 595 such incidents around the shores of Britain in 1976 and 642 in 1977; in 75 per cent of cases the source could not be identified. The second are the acute incidents where large quantities are lost after an accident, usually a blow-out as at Santa Barbara and the Ekofisk B (North Sea) or the wrecking of a large tanker like the Torrey Canyon, Tampico Maru, Arrow, and the Amoco Cadiz, to name but a few. Although such incidents are taken very seriously and generate a great deal of public concern and scientific work, we must remember that incidents of the size of the Torrey Canyon spill occur about a week in the world as a whole.

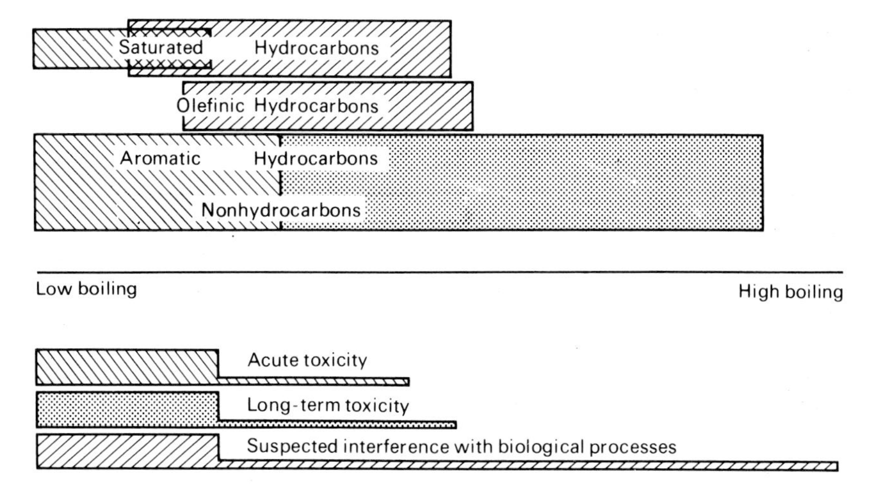

Fig. 11.4 A diagram of the toxicity to marine organisms with different fractions of crude oil. The low-boiling saturated hydrocarbons are the most poisonous but are the first to evaporate.
Source: Blumer 1969

TABLE 11.4 Petroleum and the marine environment

Source	$\times 10^6$ *t/yr*
Natural seepage into ocean	0·6
World oil production (1975)	2·70
Oil transport by tanker (1975)	1·75 (est)
Injections through routine activities	2·25–3·40
Torrey Canyon discharge	0·119
Santa Barbara blowout	0·003–0·011
Amoco Cadiz	0·20

Source: Goldberg and Bertine 1975 with later additions

The scale of the problem has led to a considerable accumulation of knowledge about the effects of oil on marine life. It can kill organisms directly through coating and asphyxiation (e.g. barnacles and other intertidal life), and can kill by poisoning via contact or ingestion (e.g. vacuolar plants and preening birds). Water-soluble toxic compounds can be lethal to fish and invertebrates and result also in the destruction of the sensitive eggs and larval forms of fish. Disruption of the body insulation of birds will also result in their death. Oil causes harmful effects on ecosystems as well: the disruption of food sources is the obvious consequence of death or reduced productivity anywhere in the system. There may be synergistic effects which reduce the resistance to other stresses and it has been suggested that carcinogenic chemicals may enter food chains from oil. Reproductive success may be reduced and the chemical clues vital to survival, reproduction or feeding of some organisms may be masked by the petroleum. A number of factors influence the severity of contamination: the dosage, the physical and chemical nature of the oil, the location of a spill, the time of year and the weather, and the clean-up techniques are all important. Much of the information on the effects of spills comes from episodes of acute poisoning, but many areas in the world have a chronic low-level contamination which causes environmental stress. Little is known about the effects of this in the open oceans but in shallow inshore areas the results include lower species diversity, fluctuations in the dissolved oxygen concentration and near-anerobic conditions on the bottom. At Fawley in England, the effluent from a refinery killed a salt-marsh in the years before 1970 but this has recovered since better waste treatment was introduced.

The effects of oil injections are best recorded in the case of birds. It is estimated that only 5–15 per cent of all those killed by this cause are actually washed on shore: in the case of the *Torrey Canyon* this figure was $40–100 \times 10^3$ and some groups seem to suffer differentially, particularly the auks, loons and grebes. The jackass penguin, endemic to South Africa, is gradually being extirpated from oil lost by tankers rounding the Cape of Good Hope. There are few reports of direct death in adult fish but tainting is common; in the case of shellfish the taint remained up to three years after the incident in the worst cases. Petroleum inhibits photosynthesis by plankton; depending on the grade, at concentrations of 60–200 ppb. But the most obvious visual effects are in the intertidal zone. Non-toxic but lethal effects may be worst here: oil may cause loss of purchase by animals which are then washed away, or alternatively may be cemented to their rocks. By contrast, the benthos is usually ignored in post-spill studies, and studies are complicated by the effects of dispersants and sinkers such as chalk. Oil on the sea-bed certainly kills many of the animals and plants, and the recovery time in the worst affected places is of the order of 5–10 yr. Sunken oil also fouls fishing gear.

Oil spills may also be more severe in their effects in particular parts of the world. The outstanding example is the polar regions where a spill is likely to have very long-lasting effects because the temperatures do not permit the evaporation of the light fractions, because bacterial degradation of the oil is very slow, and because many of the marine organisms are very long-lived and have slow reproduction rates. Coral reefs seem to be most at risk in their living portions which are exposed at low tides; the recovery period is very slow and in the interim the reef may be eroded away. In estuaries, oil may persist for a long time in anaerobic sediments but in the aerobic environments the high density of bacteria means that recovery can be fast.

The effects of a spill must include those of the clean-up techniques. The oil may be skimmed, burnt, absorbed onto peat or straw, or dispersed with emulsifiers, or sunk with chalk or sand. Many of these techniques are directed as saving the intertidal zone but ignore other parts of the ecosystem such as the benthos; thus the short-term tourist interests may be helped by emulsifiers or by chalk but the fisheries are likely to suffer much longer. Scraping of beaches causes only short-term biological damage in contrast to steam cleaning which kills everything. There is the danger therefore that aesthetics and tourism may dictate clean-up methods which are detrimental to longer-term considerations (Nelson-Smith 1970, 1977).

Out of the mass of data now available, scientific opinion (Boesch 1974, GESAMP 1977) seems to consider that the greatest hazards in the open sea are to sea-birds but that light and refined oils can damage the planktonic stages (eggs and larvae) of fish.

Because oil is becoming a very expensive substance, the motivations to lose as little as possible are stronger than with unwanted residues of industrial or agricultural resource processes. But the costs are widely ramified, and include, besides the loss of natural resources, the loss of revenue to resort areas, and the costs of cleaning up by local governement when the source of the spill cannot be traced. There are losses to the polluters if they are fined and even others incurred in investigating responsibility and in prosecution through the courts. The total of these can never be accurately assessed (Hawkes 1961). It will take a great deal of tough action by nation states along their shores to produce a noticeable decline in contaminated beaches (which although undesirable are probably in the long term the least important of the effects of oil pollution) and inshore pollution. The stopping of pollution on the open oceans requires international action of a scale not so far envisaged as practicable, partly because even if Conventions and Treaties are agreed their enforcement requires costs which no user nation is ready to countenance.

PCBs

These are a class of organic substances (polychlorinated biphenyls) which are used in condensers, as heat transfer agents, in transformers and paints, in inks and carbonless copy paper. They are long-lasting, insoluble in water but soluble in fats, and most of them are used in closed-cycle conditions which should make their release to the environment almost impossible. Before the manufacturers change their disposal policy however a number of toxicity episodes in which PCBs were implicated were discovered. It appears that like some residual pesticides, PCBs can undergo biological magnification along a food chain, and Table 11.5 shows the PCB levels found in the Irish sea in 1969 when 12–15,000 seabirds died during a period in the autumn. In birds, PCBs can affect the metabolic rate and the structure and waterproofing of feathers; it is hypothesized that the guillemots which were the main victims were caught at a time of storms and so metabolized PCBs in their fat as well as any which was ingested from their food, leading to the very high levels in their livers.

Japan has notably suffered from PCB contamination, where it has been found in every environmental compartment, to the point where it is possible to chart its pathways (Fig. 11.5). It is usually found at high levels in fish but the highest levels given to humans were in rice-bran oil in 1968; a total of 1200 patients resulted, with various skin lesions being the most common pathology (Tatsukawa 1976). Fish have also exhibited high PCB levels in the Great Lakes (e.g. coho salmon 5 ppm. eels 17 ppm) which led Canada at one point to propose a limit of 2 ppm on fish for human consumption.

There is not yet time perhaps for a response to lower levels of emission, although it seems as if environmental concentrations are not dropping as fast as would be expected. This has led to hypotheses, still under investigation, that there is a recycling pool of PCBs in marine sediments or that photo-oxidation of residual DDT is producing PCBs. Industrial disposal of PCBs means their incineration at very high temperatures

TABLE 11.5 Biological magnification of PCB's, Irish Sea 1969

Sample	*µg/kg wet weight*
Seawater	<0·01
Zooplankton	<10–30
Mussels	10–800
Whiting liver	4,500–27,000
Dead guillemots	
Liver	10,000–200,000
Rest of body	1,000–10,000
Healthy guillemots	
Liver	0–2,000
Rest of body	1,000–7,000

Source: Holdgate 1979

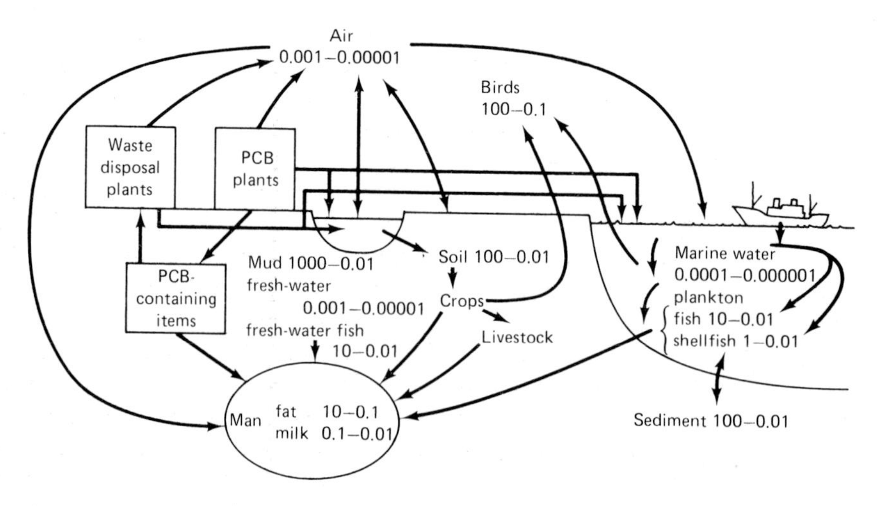

Fig. 11.5 Fate of PCB's in the environment (unit: µg/g).
Source: Tatsukawa 1976

since there is no completely effective method of removing them from water; one more example of how energy costs and availability have very wide linkages.

Derelict land

Creation of waste land

Many resource processes of an industrial kind produce land which is no longer valuable for any other purpose; the UK definition of derelict land, for example, is 'land so damaged by industrial or other development that it is incapable of beneficial use without treatment'. Mining of various kinds is the chief contributor of derelict land: open-pit mining produces a large hold together with dumped overburden and spoil, and the hole may fill with water. Machinery and buildings may with everything else be abandoned when extraction ceases after the average life of pits and quarries of 50 years. Shaft mining produces few holes, but tips of waste material are usual and subsidence may render unusable further areas of land. Surface mining is relatively cheap and over 50 minerals are produced in the USA by this means; costs are held down by the technological capacity now available to highly capitalized operators. Industrial processes such as smelting also create considerable quantities of waste, as do the armed forces, especially at war-training areas, and railways and waterways in periods of decline.

The actual amounts of derelict land in some industrial countries are very high. Measurement is difficult, since official definitions often include only land which is derelict and abandoned. Thus a tip heap still in use is not classified in the UK as derelict land, neither is land covered by any form of planning permission or restoration conditions, however impossible these may be to enforce. Thus in the West Riding of Yorkshire, England, there were officially 2,542 ha of derelict land in 1966, yet functionally derelict land was estimated from an aerial survey at 9,754 ha. Even using the official figures, England and Wales had 45,460 ha of derelict land in 1967, increasing at 1,417 ha/yr. The true amount is probably 101,250 ha; and most of it is a result of the extraction of some 406 million mt/yr of minerals which occurs in Great Britian: for instance, about one-fifth of the National Coal Board's land holdings are derelict land (Barr 1969). In the USA, an estimated 1,295 km^2 of land were disturbed by surface mining at January 1965, with an annual increment of 61,956 ha: an amount equivalent to a rectangle 16 × 40 km. An additional 129,600 ha were or had been in use for roads and exploration activities. Coal dominated the picture, being responsible for 41 per cent of the disturbance, with sand and gravel 26 per cent (US Department of the Interior 1967, US President's Council on Recreation and Natural Beauty 1968, Weisz 1970).

The effects of dereliction extend beyond the immediate sterilization of tracts of land. Open mining may contribute dust to the local atmosphere, and waste material, especially if unvegetated, is a considerable source of silt. On 40 per cent of derelict sites in the USA, soil erosion was taking place on the tips, and strip-mined lands in Kentucky yielded $7{\cdot}1 \times 10^9$ kg/ha of silt whereas undisturbed forest nearby yielded $6{\cdot}5 \times 10^6$ kg/ha. Thus in the USA, 9,280 km of streams and 1,175 ha of impoundments are affected by the operations of open-cast mining of coal (US Department of the Interior 1967). Chemical effects may also occur when water leaches through waste materials with soluble elements in them: sulphur-bearing tailings may well yield streams contaminated with sulphuric acid.

Treatment and reclamation of derelict land are not technically impossible but currently appear expensive. In Lancashire, England, 15·5 ha of derelict land cost the local authority £3,700 to buy, but reclamation and landscape cost £15,000, and the usual cost of reclamation was £32–£284/ha. Nevertheless, authorities in England like Lancashire and Durham County Councils have Treasury-aided reclamation programmes of an ambitious nature. Since 1965 Durham has launched over 80 reclamation schemes on land made derelict by coal extraction; ongoing programmes plan to reclaim 200–250 ha/yr, but even this is not keeping pace with colliery closures. England and Wales have been reclaiming a total of 800 ha/yr in recent years and restoration conditions are usually placed on planning permission to extend mineral extraction, but these are often difficult to enforce (Baker 1970, Lenihan and Fletcher 1976). The attitudes of the industrial revolution are clearly changing with regard to derelict land: where there's muck there's money but there is also dirt, dust, rats and even large-scale death as at Aberfan in south Wales. The psychological effects are also important: authorities of such areas which seek to attract new industry have great difficulty in persuading managers to locate amid mountains of waste or lunar landscapes of dust and water. The use of derelict land as a resource for filling in holes, for making new sites for housing and industry and for creating marinas and nature reserves is a necessity in small countries with large populations, and a desirable aim for others (Karsch 1970).

Organic wastes

Concentration in runoff

Most of the substances considered here are not persistent as are, for instance, chlorinated hydrocarbon pesticides, but contain certain elements present and important in the biosphere (of which nitrogen and phosphorus are the most studied), which are concentrated by man in the course of their use and are then released back into the biosphere. The effects are those of providing a large quantity of an element which may well have been limiting, and of accelerating the natural cycle of that substance in the biosphere.

Sewage

Contamination of fresh waters and shallow offshore seas by sewage is a common occurrence. Sewage is about 99·9 per cent water and 0·02–0·04 per cent solids of which proteins and carbohydrates each comprise 40–50 per cent and fats 5–10 per cent. Water carriage of sewage from urban areas dates only from the 1840s, before which the contamination of water supplies led to epidemics of cholera, typhoid and dysentery. Such outbreaks are still common in LDCs and in pockets of poverty in the DCs such as in migrant labour camps (e.g. in Florida early in 1973) and in regions with many native people (e.g. the Mackenzie Delta of Canada in October 1972). Treatment of sewage to kill the agents of these diseases is an important part of sewage processing where this exists: chlorine is generally used. The wastes of any human population will also contain bacteria and viruses from other ill people and from healthy carriers, which reinforces the need for treatment of the water component of sewage. Contamination of water by untreated sewage is denoted by the presence of the bacterium *Escherichia coli* which is not itself infectious but is an indicator of the presence of the agents of typhoid and dysentery.

Further sewage treatment consists of the separation of the organic matter from the

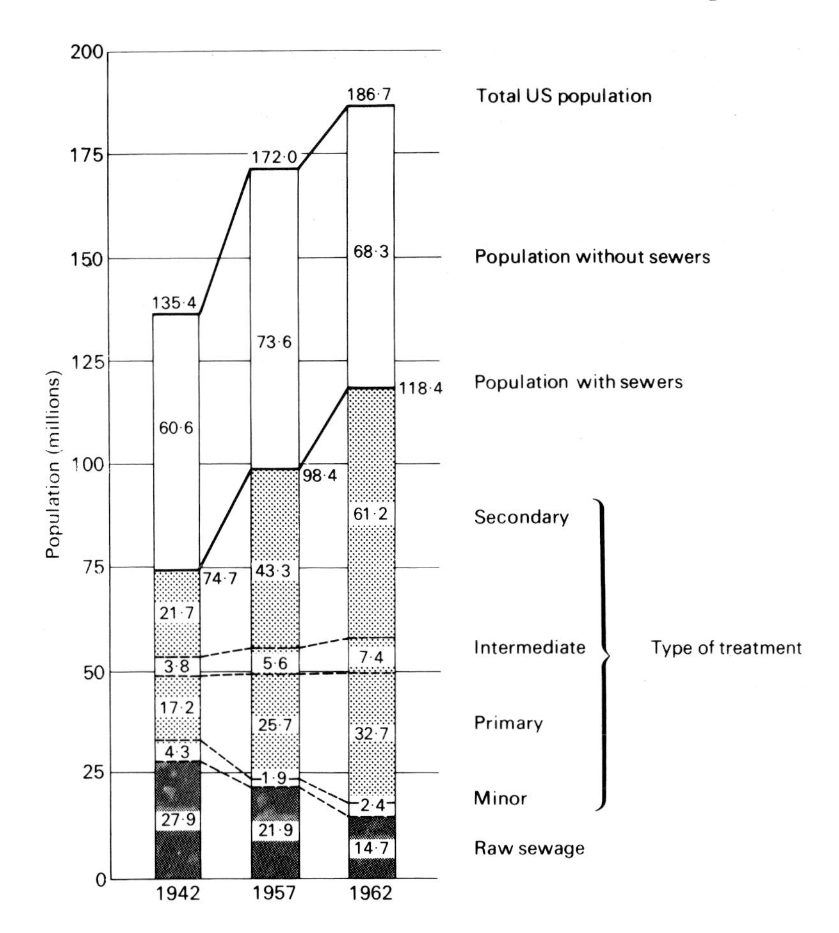

Fig. 11.6 Even in rich countries like the USA, sewerage may scarcely keep pace with the growth in population, although in 1957–62 it began to gain. Lack of sewers does not necessarily mean pollution if effective septic tanks and cesspools are used, but in the case of many large cities raw sewage is dumped into nearby water.
Source: A. Wolman 1965

water and its conversion to a biologically inactive and aesthetically inoffensive state. A well-fed human excretes 14 g N_2 and 1·5 g P per day, of which 20–50 per cent is removed by waste treatment. The end-product is sludge, a valuable fertilizer but one which is currently more expensive than factory-produced material and may contain undesirable concentrations of metals. However, many large cities in the West do not treat their sewage at all: in metropolitan Canada, for instance, 100 per cent of the sewage of Toronto gets a two stage treatment, whereas practically all the production of the population of Montreal and Quebec is poured untreated into the St Lawrence River. In the USA the mixture of treatment and non-treatment means that an estimated 450–680 million kg of nitrogen and 91–250 million kg of phosphorus (Bartsch 1970) reach the surface waters of the continent every year (Fig. 11.6). Even tertiary treatment of municipal sewage will not eliminate the eutrophication of fresh water which calls for the even more expensive process of the removal of inorganic and organic compounds from a stabilized sewage effluent.

Detergents

More phosphorus is added from 'hard' synthetic detergents whose foaming reduces photosynthesis and inhibits oxygenation of the water. Concentrations of < 1 ppm may also inhibit oxygen uptake by organisms. 'Soft' or biodegradable detergents are now available and have mostly replaced their forerunners, but both they and the organic cold-water detergents contain phosphorus; indeed 'enzyme' detergents are in fact phosphate pre-soaks (US Congress 1970).

Agricultural wastes

Fertilizers in which the chief ingredients are nitrogen and phosphorus have been applied in increasing amounts in modern agriculture, especially in the DCs and in intensification programmes in the LDCs. This process is reflected not only in agricultural output but also in the increased run off of N and P not taken up by the plants. The main difficulty associated with these wastes is the eutrophication of fresh waters (see below) but there is also the problem of nitrates entering ground water where they may cause metabolic disorders in domestic animals and in children if used for drinking, including the human condition known as methemoglobinemia, where the binding action of the contaminant prevents the haemoglobin from taking up oxygen; the blue-baby syndrome. In Illinois in the 1970s, the use of fertilizer led to the level of nitrates in drinking water exceeding the FDA tolerance level of 40 mg/l but pressure from the farmers prevented any action to diminish the use of fertilizers. There is indeed some controversy whether farm wastes are the worst contributor to the fresh waters since as Table 11.6 shows, domestic wastes are much higher in nitrogen and sometimes in phosphorus as well. The increasing price of fertilizers (which is linked closely to that of the energy used in manufacture and application) may help to combat this problem (Loehr 1978).

TABLE 11.6 Nutrient concentrations in discharges, USA, 1960's

Source	*Nitrogen (mg/l)*	*Phosphorus*
Domestic waste	18–20	3·5–9·0
Industrial waste	0–10,000	n.a.
Agricultural land	1–70	0·05–1·1
Rural non-agricultural land	0·1–0·5	0·04–0·2
Urban run-off	1–10	0·1–1·5
Rainfall	0·1–2·0	0·01–0·03

Source: Friday and Allee 1976

Agriculture also produces a lot of other residues. Crop residues in the USA amount to about 8 t/family/yr and each year sees 58 million dead birds unfit for eating whose carcasses have to be dispersed. Intensive animal farming produces wastes analogous to a town: the volume of daily wastes from such operations are approximately 10 times those of humans. Each year the human population of the USA produces 153 million m^3 of sludge and its domesticated animals 765 million m^3 (Taiganides 1967), and the biochemical oxygen demand may be 100 times that of normal sewage. In New England dairy farms, each cow is responsible for a net loss to the environment of 119 kg/yr of nitrogen, 34 kg/yr of phosphorus and 53 kg/yr of potassium (Ashton 1970). It is theoretically possible to turn much of this excreta into animal feed, or to ferment it for methane gas, but the economics of the processes have so far precluded their introduction on a large scale. Energy costs could well force changes in the next 20 years.

Eutrophication

The introduction of large quantities of nitrogen and phosphorus into water bodies is one of the major problems of wastes today (Sawyer 1966). Phosphorus is a scarce element in the lithosphere and many ecosystems are adjusted to its scarcity: it is a limiting element in Liebig's sense. The large quantities made available by human-induced concentrations lift such limits and 'blooms' result even at concentrations as low as 50 ppm of phosphorus; nitrogen may then become the limiting factor and organisms such as blue-green algae take over from the plankton because they escape the

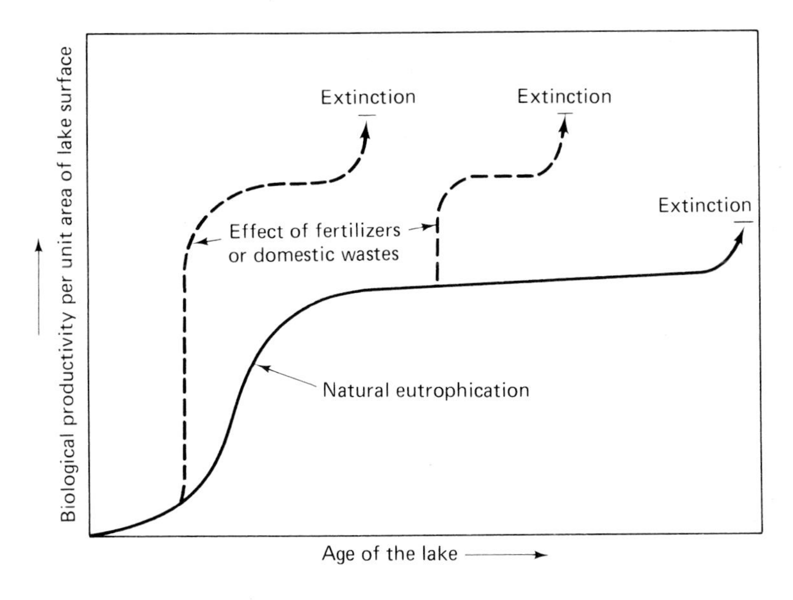

Fig. 11.7

Source: Friday and Allec 1976

nitrogen limits. 'Blooms' of such algae then follow. The additional nitrogen (and nitrogen fixation by human-induced activity is now about 50 per cent of the level in nature) and phosphorus accelerate the ageing of lakes, themselves short-lived in geological terms, by nutrient enrichment or eutrophication (Fig. 11.7). Such a process happens naturally but is speeded up many times by human activities which result in large inputs of phosphorus and nitrogen (OECD 1970). Fig. 11.8 shows the build-up of

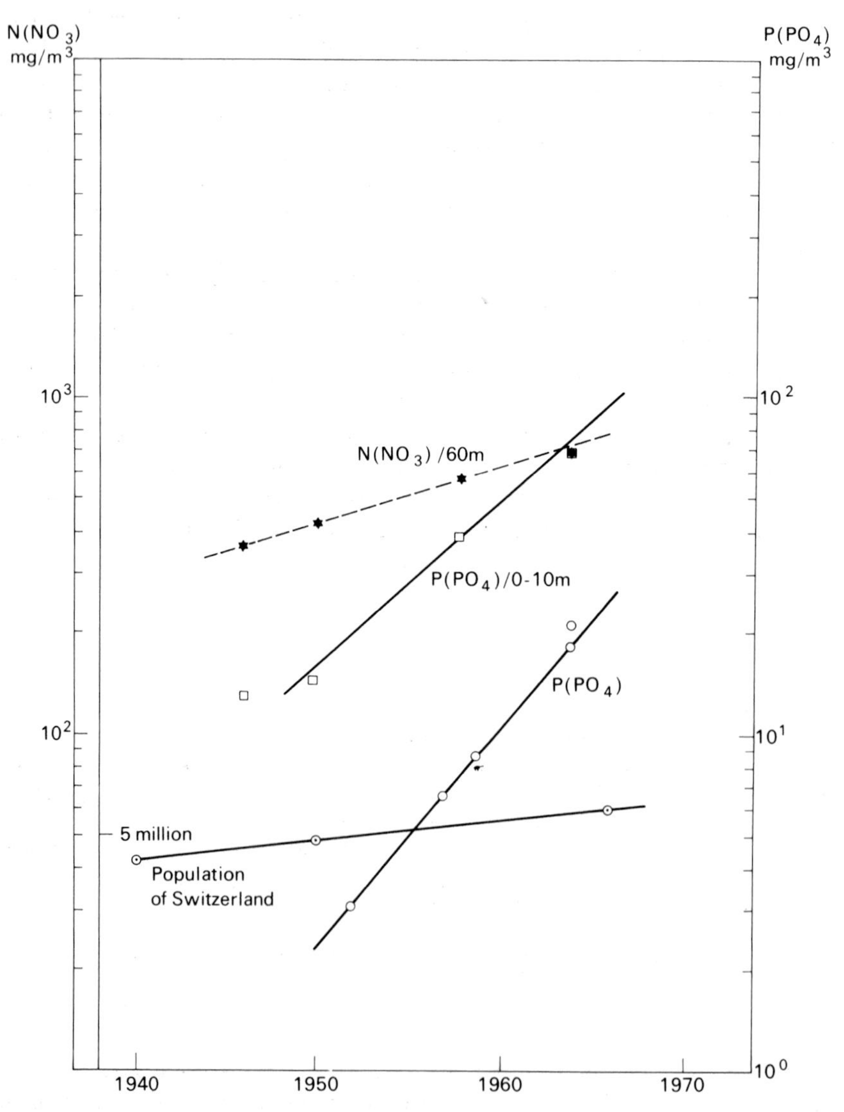

Fig. 11.8 Changes in nutrient concentrations as measured in the Zurichsee and Bodensee in Switzerland, compared with the rate of population growth. The eutrophication is clearly due to factors other than simply population growth.
Source: OECD 1970

nutrients in two European lakes which have undergone enrichment. Many other examples are known, of which the Great Lakes of North America are probably the best documented, especially Lakes Erie and Ontario (Charlier 1970, International Joint Commission 1970). The total dissolved solids in these two lakes have risen by 50 ppm in the last 50 years, for example, and the fish populations have changed almost completely (a situation complicated by the effects after 1945 of the parasitic sea-lamprey, which, however, cannot spawn in the tributaries of Lake Erie). The bottom fauna too has changed sweepingly, especially through increases in midge larvae and tubificid worms. In summer a serious depletion of dissolved oxygen occurs, *Cladophora* piles up in offensive heaps on the shores and blue-green algal scums are formed. Eutrophication is not confined to water: input–output studies of forested watersheds in New Hampshire showed that the aerially derived input of some mineral elements exceeded the loss by runoff. This suggests a gradual enrichment of this habitat, but the long-term effects are difficult to predict because of the considerable powers of nutrient absorption possessed by the forest flora.

The cessation of eutrophication of water bodies is clearly an immense task both technically and institutionally; but failure to achieve it will result in the loss not only of biologically significant elements of diversity but also in the diminution of an important element of environmental quality for humans. However, experiments in suitable lakes, especially in Sweden and the USA, have shown that eutrophication is to some extent reversible, provided that the relevant inputs are curtailed and that remedial treatment is undertaken: Lake Washington near Seattle and Lake Tahoe in California are examples, the latter being described in its pristine state as 'gin-clear'. In Europe, lakes in the French and Bavarian Alps have similarly been cleared up (Hasler 1975).

Radioactive wastes

Effects of radiation

TABLE 11.7 Glossary of terms used in radiation studies

roentgen (r)	a unit of X or gamma radiation exposure producing a charge of 258×10^{-6} coulomb per kg of air.
Curie (Ci)	One curie is the amount of any radioactive nuclide that undergoes 37×10^{9} disintegrations per second.
Half-life	The average time required for half the atoms of an unstable nuclide to transform.
Dose	A measure of the energy actually absorbed in tissue by interactions with ionizing radiation.

Natural radiation is emitted from a wide variety of sources such as X-rays and cosmic rays, but its level is such that a normal dose of 5 roentgens (5r) is accumulated over the first 30 years of life (a dental X-ray is 1r). In addition to this are certain man-made emissions which result from nuclear explosions in the atmosphere and underground and by the use of nuclear fission processes to generate energy. Nuclear testing in the atmosphere has been much reduced since the Test-Ban Treaty of 1967, although China and France have openly set off devices since then. Underground weapons testing has continued, especially by the USA and USSR, and there have been trials of nuclear explosions for non-military purposes as with the US Project Gasbuggy in 1967, when an attempt to improve natural gas yields was made by shattering a 'tight' reservoir rock beneath New Mexico. There has been a considerable rise since 1950 in the number of atomic power plants, and many writers foresee a much greater reliance upon atomic power in the future.

Whenever energy is released by splitting atoms, so are potential contaminators of the ecosphere by the radioactive particles which are the products of the fission reaction. Atmospheric explosions yield numerous isotopes, some very short-lived, others like Sr-90 and Cs-137 having half-lives measured in decades. These are caught up in the atmospheric circulation and gradually come to the earth's surface as 'fallout'. Underground explosions release fewer particles but the chance of contaminating groundwater is always present, especially when the area used in highly faulted geologically.

Radioactive waste from nuclear power

Most phases of the fuel cycle of a fission reactor produce radioactive isotopes. Some are very short-lived and can be released in carefully controlled quantities to the atmosphere or water bodies, others are long-lived and high emitters ('high-level' wastes) and so must be kept away from the environment and from people for a long time. At the end of its 25–40 year life the power plant itself becomes a radioactive waste that must be treated though, because of the novelty of atomic power, there is as yet little experience of that process.

The most abundant wastes from a fission plant are tritium, krypton and argon; these are gases and are released into the atmosphere around the plant and it is calculated that dilution by the atmosphere will prevent them doing any harm. Some other low-level wastes are dumped into deep oceans. More concern is expressed about the high level liquid wastes, especially those which result from fuel reprocessing. Some of the isotopes have relatively long half-lives: for example a PWR may produce 100,000 Ci/yr of strontium and cesium. ^{90}Sr has a half-life of 28 years which means that it takes 300 yr for 1 Ci to drop to 1 μCi and 500 yr to drop to one-millionth of its original activity and the recommended level of safety. Plutonium has a half-life of 24,400 yr. These liquid wastes are not however produced in very large quantities: 1 t of spent nuclear fuel when reprocessed gives 500 litres of liquid wastes. After 25 yr a burner reactor has yielded 25 t of liquid wastes and a breeder reactor 44 t. After extraction from the reactor these wastes are too hot to do anything other than be kept in tanks for some years, after which the preferred technique at present is to solidify them either in glass or a manufactured product called synroc (Ringwood et al 1979). Then 25 t of liquid wastes becomes 2·5 m^3 of solid and 44 t liquid gives 4·4 m^3 of solid material.

The disposal of these wastes causes much concern, for it is imperative that they be isolated from the biosphere for about 250,000 yr, a societal commitment of a high order.

Numerous methods have been suggested for disposing of what will amount to about 2,000 cm^3 of waste per person per year served by the nuclear plant: allowing the wastes in bomb-shaped containers to sink into the Antarctic ice-cap, shooting them into the sun, and placing them on inter-plate subduction zones to be carried into the earth's core, are some suggestions; the most popular however is geological disposal in rock formations without any ground-water circulation and so salt deposits (and in particular abandoned salt-mines) are currently highly favoured; some writers even have misgivings about these and there is still the potential for accidents during transportation.

As far as lay people are concerned, one of their greatest fears with atomic power is of an accident at the plant. Even though experts in risk analysis calculate the chances of a major accident such as an uncontrollable meltdown of the reactor core (the so-called 'China syndrome') as being very small, there have been a large number of minor accidents resulting in the uncontrolled release of radioactive materials into water, air and soil. No full meltdown has yet occurred, though the Fermi breeder in the USA had a meltdown of 3 out of 100 fuel assemblies; according to Medvedev (1977), a large nuclear explosion involving nuclear waste occurred in 1957 or 1958 in the Southern Urals, contaminating an area 100 × 50 km northeast of Chelyabinsk. One of the calculations made for a nuclear reactor is the MCA or Maximum Credible Accident and its consequences: it is expressed in terms of effects upon humans and their activities; estimates of the ecosystemic effects of large scale nuclear war have been made by SIPRI (1977). The MCA calculations reproduced in Table 11.8 are most frequently quoted but some writers believe they are too low by a factor between 2 and 10, especially for cancer deaths.

TABLE 11.8 MCA consequences for a LWR accident, USA

Effect	*Rate or number*	*Duration*	*Best estimate*	*Low estimate*	*High estimate*
Prompt deaths	3,300	—	3,300	825	13,200
Cancer deaths	1,500/yr	30–40 yr	$45–65 \times 10^3$	7,500	180×10^3
Prompt illnesses	49,500	—	49,500	12,375	198×10^3
Thyroid illnesses	8,000/yr	30–40yr	$240–320 \times 10^3$	80,000	960×10^3
Genetic effects	190/yr	many generations	28,500	4,750	171×10^3
Property damage	$\$14 \times 10^9$	—	$\$14 \times 10^9$	$\$2.8 \times 10^9$	$\$28 \times 10^9$

Source: Ehrlich, Ehrlich and Holdren 1977, quoting US Nuclear Regulatory Commission *Reactor Safety Study* (1975), the 'Rasmussen Report'.

The probabilities of an MCA were calculated to be 1 chance in 20 per year with an inventory of 1,000 LWR plants in operation. The safety of breeder reactors is still a matter of controversy because so few have been built, and fission reactors are thought to be safer still in terms of their radioactive products. Of all the environment concerns of recent years, the fears over atomic power may well have the most substance even allowing for the immense care which is taken for even if engineering accidents, weapons proliferation, terrorism, and human error are dismissed as unreal worries. The concentration of power into such a centralized network is a form of societal development which many people reject (Holdren 1974, Pigford 1974, Cohen 1977, Hohenemser et al 1977, Rochlin 1977, La Porte 1978).

Radioactive material in food chains

Such concerns are exacerbated by the biological magnification of radioactive particles. Just as residues of persistent chemicals may accumulate in the biosphere far beyond the initial dose, so may long-lived isotopes. Clay minerals may selectively absorb and concentrate particles, for example, as with Sr-90, Cs-137, Co-60 and Ru-106. As might be expected, however, one of the most frequent processes is accumulation along food chains. Retention at the various stages depends upon many variables such as the differential uptake of various isotopes and their retention in different organs. Ruthenium-106, in edible seaweeds for example, is the critical path for discharges into the Irish Sea from the Windscale plant. In the total biomass, a major fraction of most of the radioactive isotopes is held in the primary production level, but this does not prevent large absolute amounts from reaching high trophic levels (Woodwell 1963, 1967b).

Atmospheric nuclear testing produced very high rates of fallout in arctic regions, especially of Sr-90 and Cs-137. From the atmosphere the particles settle onto, and become incorporated in, the lichens of the tundra and taiga. Cs-137 is effectively

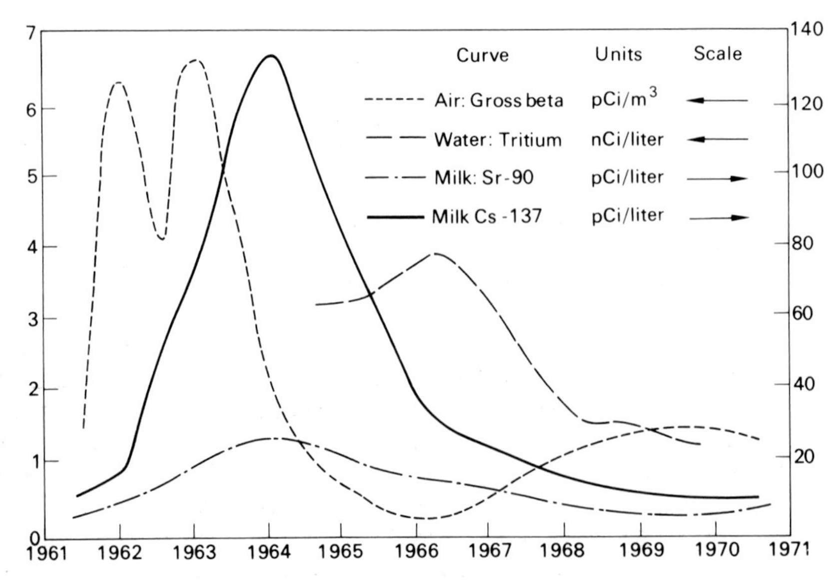

Fig. 11.9 Trends of selected radio-isotopes in the atmosphere, water and milk, as measured in the USA. The effects of the cessation of atmospheric nuclear weapons testing (except by France and China) is apparent.
Source: US President's Council on Environmental Quality 1971

retained in the upper parts of these organisms whereas Sr-90 is translocated through the whole plant, which grows very slowly, incorporating approximately 95 per cent of the fallout (half-lives 3–13 years) which settles on it. The lichens are important winter food for caribou which graze off them, especially the too parts, thus leading to high winter uptake levels of isotopes. Some native peoples depend largely upon caribou meat for food, though not usually in the winter, and adult males may eat 5–6 kg/week. Thus certain northern peoples of Alaska, Canada, Lapland and the USSR have body burdens of Cs-137 and Sr-90 which average 50–100 times higher than those of temperate-zone inhabitants (Hanson 1967a, 1967b). Cessation of atmospheric nuclear testing except by France and China has led to a diminution in fallout rates (Fig. 11.9), although the build-up of Sr-90 in water, breadfruit and land crabs on Bikini atoll has meant that the islanders, allowed back in the early 1970s, were moved out again in 1978.

Metals

Man's concentrations of natural elements

TABLE 11.9 Rates of mobilization of selected materials, 1960's

Element	*Geological flow rate of rivers*	$\times 10^3$ *t/yr*	*Mining rates*
Iron	25,000		319,000
Manganese	440		1,600
Copper	375		4,460
Zinc	370		3,930
Lead	180		2,330
Mercury	5		5
Tin	1·5		166

Source: SCEP 1971. The geological figure for mercury does not include gaseous losses from coastal rocks of the order of 150,000 t/yr.

These elements are often essential to the metabolism of plants and animals but usually at very low concentrations; the effect of man is often to mobilize metals at far higher rates than those found in nature (Table 11.9). At high dosage levels they are frequently toxic to plants, animals and men and several of them are thought to be cumulative. They are of course part of the 'natural' environment but little is known about the development of tolerances, with the exception of the evolution of strains of certain plants, mainly grasses, able to withstand very high levels of, for example, lead and zinc, and thus grow on industrial waste tips formed at the sites of the smelting of these metals. The uses of metals are extremely widespread and, apart from the obvious and visible, include substances such as pesticides where 141 metals can be found in 112 compounds, and in additives like lead in petrol. Material concentrations in industrial areas are likely to be big pools of metals: sewage and urban solid wastes, for example, are high in their concentration of them. While in use, elements like lead, mercury, zinc, selenium, manganese, chromium, copper, cadmium and nickel are likely to be closely observed for toxicity, especially among workers handling them or consumers of products: hence the development of lead-free paints. However, their residual effects, as wastes and in unobservable forms like aerosols, are likely to escape notice until levels somewhere build up to a measurably toxic concentration (Garrels et al 1975).

Lead

The annual industrial production of lead increased rapidly in the 1940s (Fig. 11.10) when lead alkyls (mostly tetraethyl lead) were first used as additives to petrol. Now, about 10 per cent of lead is used this way and so ca 4×10^8 kg/yr of lead is emitted to the atmosphere where it has a residence time of about a month before falling out (Boggess and Wixson 1979). The density of fallout is highest near sources such as highways and smelters but the aerosol form ensures wide distribution: soils near Reading, England, had an input from the air of 38 mg/m^2/yr of lead and lost only 1·4 mg/m^2/yr so that here is a route for lead to enter food chains; analyses from the Greenland ice-cap show an continuing accumulation of residual lead (Fig. 11.10) but in the southern hemisphere the picture is not so simple and most of the lead there is probably from volcanic eruptions (Boultron and Lorius 1979).

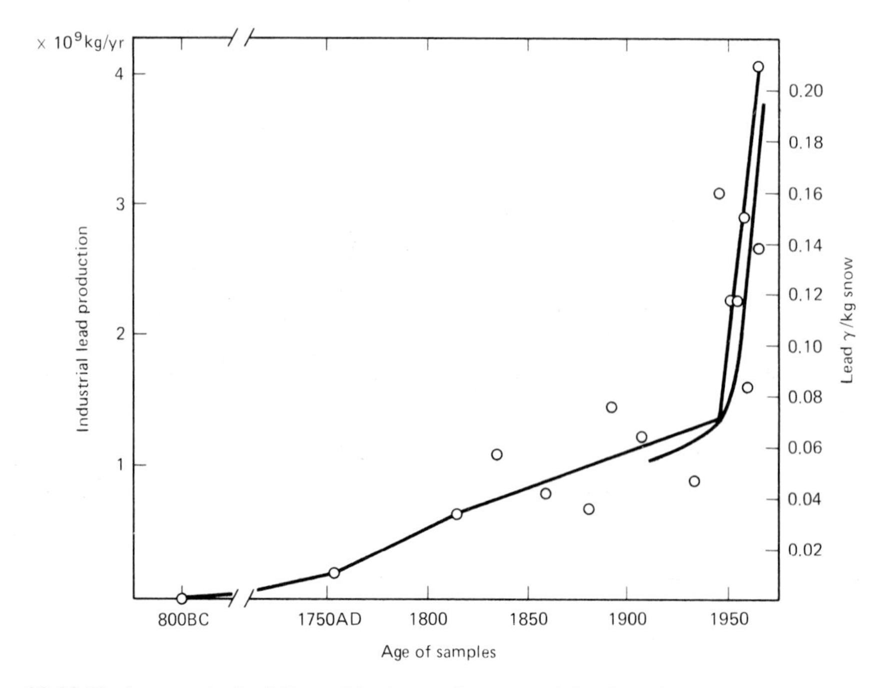

Fig. 11.10 The increase in the fallout of lead onto the snow of the Greenland ice-cap since B.C. 800
Source: Murozumi *et al.* 1969

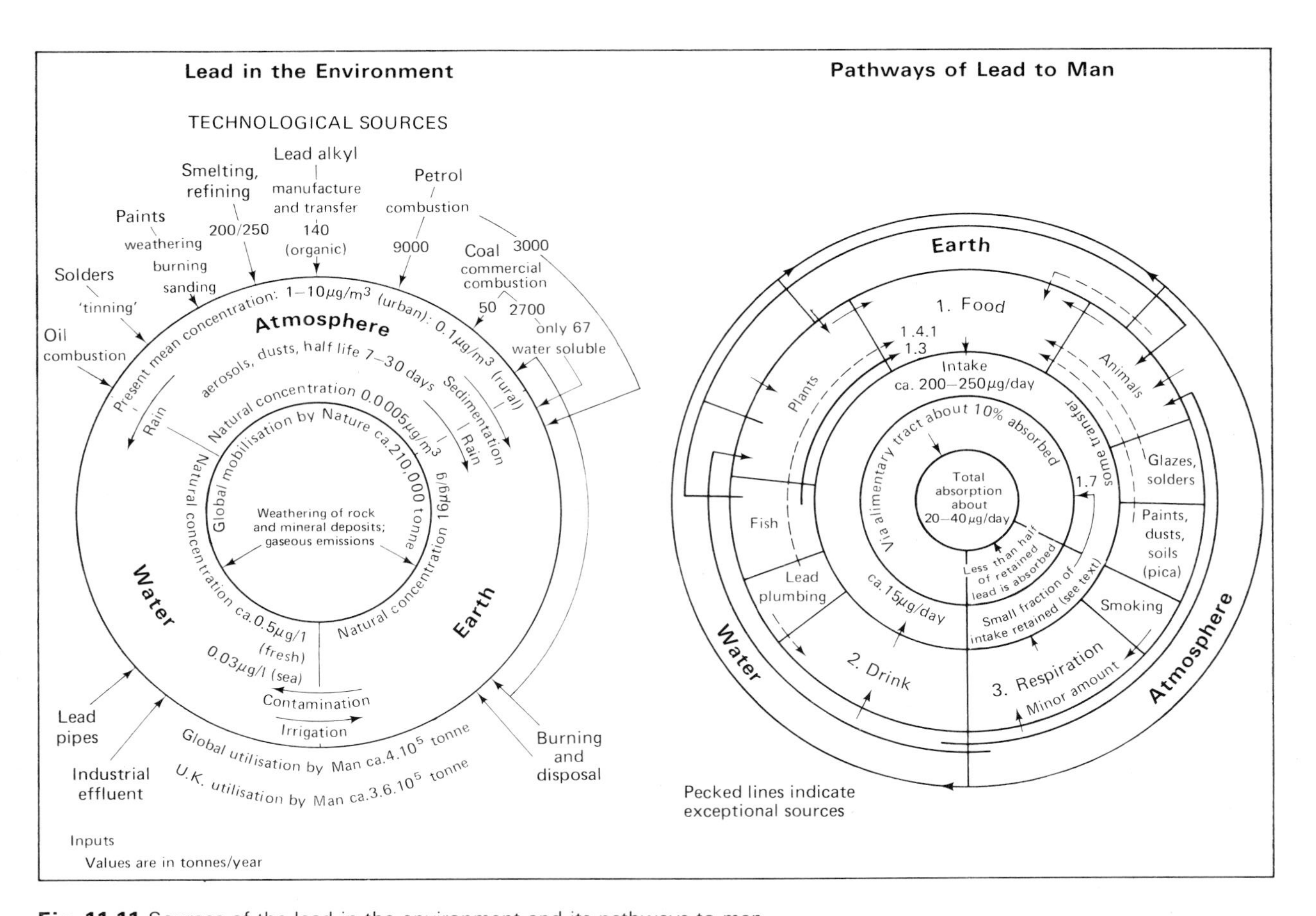

Fig. 11.11 Sources of the lead in the environment and its pathways to man.
Source: DoE 1974

There is a long history of direct poisoning of all kinds of organisms by lead (as Goyer and Chisholm (1972) point out, consumers of moonshine whiskey are at risk for lead poisoning, among other things) but concern at present centres around human intakes and retention of residual lead from wastes (Fig. 11.11). The actual effects of ingested and inhaled lead and their importance spatially with regard to sources such as motorway interchanges are still controversial but it seems that young animals and children are especially at risk from sub-clinical lead poisoning which may impair their mental faculties (Bryce-Smith et al 1978). Most countries are now setting maximum limits for the lead content of air and food (DoE 1974, Moore et al 1977) although some have been deterred by the argument that more oil will be used since consumption by auto engines is higher when lead is omitted. However if the energy costs of producing and adding the lead are reckoned then this argument loses much of its force.

Mercury

If there is some doubt about the status of lead as a toxic agent, then the case of mercury is much clearer (Nriagu 1979). Mercury is used in industrial processes (its main use is in the production of chlorine and caustic soda in which there can be a loss of 0·2 kg Hg per tonne of the chlorine produced), and as an agricultural fungicide, so that it appears in industrial effluent and in runoff, both discharging frequently into lakes or at the coast (Table 11.10).

TABLE 11.10 Mercury concentrations reported in environmental samples

Sample	*Estimated natural levels*	*Concentrations measured in contaminated samples*
Air	2 $\mu g/m^3$	2–20 $\mu g/m^3$
Water:		
Sea water	0·00006–0·0003 ppm	0·0005–0·030 ppm
Fresh water	0·00006 ppm	0·0001–0·040 ppm
Soils[1]	0·04 ppm	0·08–40 ppm
Lake sediments[1]	0·06 ppm	0·08–1,800 ppm
Biological materials:		
Fish	0·02 ppm	0·5–17 ppm
Human blood	0·0008 ppm	0·001–0·013 ppm

[1] Mercury concentration is dependent on organic content.
Source: Harriss 1971

The 'natural' level of mercury is of the order of a release of 5,000 t/yr by chemical weathering, and human use adds about the same quantity (Harriss 1971). Thus while mercury is present in small amounts in all environments, there seem to be two pathways in which it can be carried to receptor organisms at a level sufficient to be toxic. The first of these is the contamination of food, mostly grain, which kills animals and man. A load of seed corn dressed with a mercury fungicide, for example, was made into bread in Iraq in 1972 and deaths were in the hundreds. The second pathway is the discharge of mercury into water where there is an uptake and concentration by fish. This route is the way in which the poisoning incidents at Niigata and Minamata occurred in the late 1950s and early 1960s (the certified number of victims at Minamata is over 3000 and 616 at Niigata), and in which the levels of mercury in some Inuit and Indians in Canada is above 200 ppb, which is the level associated with symptoms of acute exposure at Minamata (Charlebois 1978). complications in the study of mercury toxicity include such variables as the role of methylated mercury: at one stage it was thought that inorganic mercury was usually converted to the more toxic organic methyl form by bacterial action in bottom sediments. It appears now that this reaction is quite restricted in its occurrence and that demethylation processes exist as well (Goldwater and Stopford 1977). The effects of poisoning at Minamata are thought to have been exacerbated by such heavy reliance on a fish diet that other nutritional deficiencies made the toxicity more effectual.

Despite the complexities, most governments have moved to try and limit the exposure of its citizens to mercury: WHO have suggested standards of 1·0 g/l in water and 0·3 mg/week in food.

Cadmium

This is a relatively rare element, often found in association with zinc, and sometimes concentrated in nature along with sulphur, e.g. by bacteria. The amount in river transfer is *ca* $1{\cdot}1 \times 10^6$ kg/yr, and the industrial production has risen from 5×10^9 kg/yr in the 1920s to a current level of about 17×10^9 kg/yr. Approximately 60 per cent of its use is in electroplating but it is also found in pigments, as a stabilizer in plastics, in solder and in Ni-Cd batteries. Its pathways seem to be rather like those of lead, with the exception of automobiles, and it seems to be present in most foods at a low level. The hazards and symptoms of occupational exposure are well known but the effects of chronic exposure to low levels are less easily agreed; kidney malfunction seems to be common, as does emphysema but the response is very variable. The well-known *itai-itai* disease incident in Toyama Prefecture in Japan seems strongly linked to cadmium levels but other dietary factors may have complicated the etiology; high levels in soil and food at Shippham, England, from a mine waste tip, do not seem to have caused any measurable ill effects. Nevertheless, ingestion in food of levels like 20–50 μg/day (smokers twice as much), given a long residence time in the body of the small amount retained, clearly needs monitoring for long-term effects (Fassett 1972, Hiatt and Huff 1975, Webb 1979).

Solid wastes

Products of city metabolism

Urban areas concentrate materials greatly and there is a good deal of waste which has to be removed from the cities. Solid wastes are characterized by a great mixture of substances, including fine dust, cinder, metal and glass, paper and cardboard, textiles, putrescible vegetable material, and plastics. In addition, there is bulky waste such as old refrigerators, washing machines and autos (UK 1970, US President's Council 1971); in Britain these latter comprise 3–4 per cent of the total weight of refuse. The quantities of these materials are high: in the USA a year's solid wastes amount to 168 million t (1965) or an average of 2 kg/cap/day or 907 kg/cap/yr. (The equivalent figure in 1920 was 1·25 kg/cap/day.) The composition of the garbage is changing too: the last 10 years has seen a considerable rise in the proportion of plastics, paper and packaging materials, to the point where Chicago refuse is about 56 per cent paper. The proportion of fuel wastes has, however, fallen: in Britain, the average weekly weight of garbage fell from 16·8 to 12·7 kg between 1935–67, reflecting a fall from 9·70 to 5·46 kg of dust and cinder (Fig. 11.12).

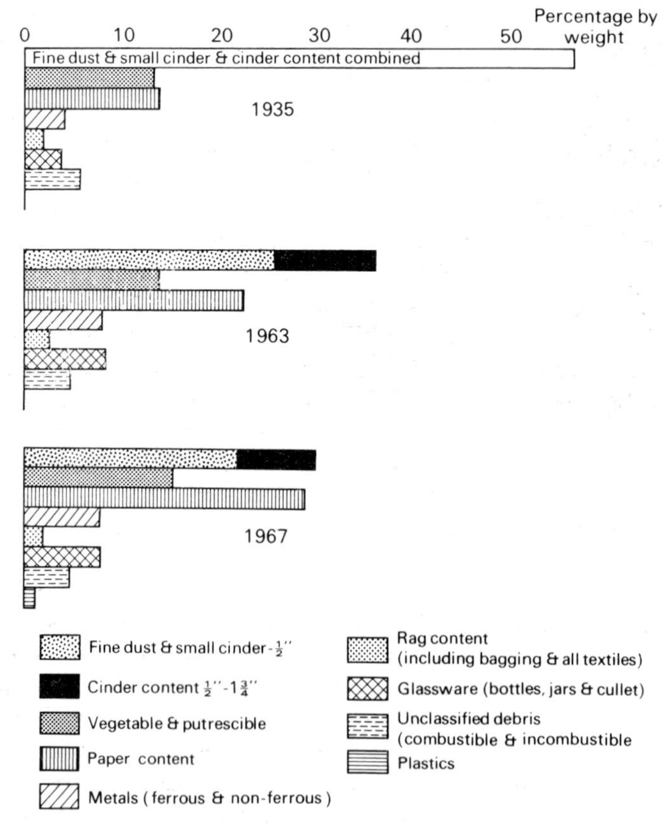

Fig. 11.12 The changing content of urban solid wastes in Britain 1935–67. The reduction of coal wastes, the late appearance of plastics and the increase in paper are the major features.
Source: United Kingdom 1970

Under the carpet?

Such quantities of unwanted material can cause serious disposal problems. The simplest method is crude tipping or open dumping: in 1965 the USA had 17,000–22,000 such places, and in the UK (1963) 115 local authorities still used this method. More satisfactory is controlled tipping or the sanitary landfill. A layer of about 2 m of refuse is covered by at least 25 cm of earth, ash or other inert material, up to the level of the hole chosen. The surface can then be used for housing or sports fields, for example. Before such filling, the wastes can be pulverized by machines to a uniform particle size: by this means the volume is reduced and thus the life of the tip extended, and some of the refuse is more quickly biodegraded. Again, pulverized material can be subjected to fermentation before dumping: the heat generated (65°C) helps to destroy bacteria and insect larvae. Incineration can be carried on inside cities, the residues used for civil engineering purposes, and power generated from combustible material. Temperatures of 870–1,040°C are commonly used, but unless electrostatic precipitators are used, air pollution is substituted for garbage problems; particulate matter from incinerators is estimated to constitute 10–15 per cent of New York City's airborne load. None of these methods is without its side-effects. Uncontrolled tipping contributes smells, windborne litter, dust, flies, rats and complete loss of amenity, and more seriously still, fire (used to consume the combustible components, it may smoulder inside the tip, breaking out sporadically), and contamination of ground water and streams by rainfall which has percolated through the tip. Controlled tipping is not immune to all these difficulties, and even wire fences will not control the airborne litter in a strong wind. Pulverization can create bad dust problems unless it is covered quickly; fermented material, like sewage sludge, could be sold, but it is expensive and is also very high in certain metals. One overall problem is the effect upon amenity, for although everybody wants their garbage taken away, nobody wants to live near the tip or the incinerator. There may also be a shortage of suitable sites within some jurisdictions and thus agreements with neighbours have to be made, or more often large cities have to exert pressures to gain space in outlying rural areas. While the use of worked-out quarries and pits for such purposes is to some extent acceptable, the covering of wildlands such as mud flats and swamps is a much more serious loss in terms both of amenity and biological productivity.

Clearly, refuse is an immense potential source of raw materials and energy, but steps towards re-use are very slow, with the exception of metals. Cars especially can now be virtually pelletized and the scrap sold to steel works. Paper can be recycled and many other materials re-used, but labour costs are very high and few people wish to do the work. Thus successful paper reclamation schemes like those of certain Canadian cities rely on subsidies from local taxes or upon the free labour of volunteer groups. In 1970 the Greater London Council forecast that ferrous metal recovery was the only permanent salvage activity it could foresee in the future. However, power generation is now a reality in some DC cities.

Solid wastes also include a variety of industrial materials, some of which are very toxic, like cyanide wastes or biocides like kepone. These can be neutralized, but unscrupulous industrialists pay truck drivers to dump the wastes by night onto open tips. A series of revelations of this practice in the English midlands in 1972 led to emergency legislation designed to prevent 'fly-tipping' of these materials. Probably only the tip of this particular iceberg has emerged; reports from the USA (Maugh 1979) have shown that many firms tried to dispose of toxic wastes by simply dumping them on waste ground or on open tips; thus health and environmental problems may exist at 1,200–34,000 sites across the country.

Waste heat

Whenever energy performs work, heat is released which is radiated back into the atmosphere. Man concentrates this process spatially, and the resultant heating of air and water is called calefaction. In practical terms the commonest sources of concentrations of waste are power-generating plants. 1 kWh of energy generated from fossil fuels creates 6,326 kJ of heat; the same energy from nuclear fuels gives off 8,962 kJ. The demand for power has meant that in Canada, for example, waste heat from thermal power generation is predicted to rise 14 times in the period 1966–90, and by AD 2000 the amount of waste heat reaching Lake Ontario in January will be equivalent to 8 per cent of the solar energy input in that month (Cole 1969). Most of the heat is carried away as hot water and this produces distinct changes in biota. A body of water at 30–35°C is essentially a biological desert and many game fish require temperatures of <10°C for successful reproduction although they will survive above that temperature. A temperature rise of 10°C will double the rate of many chemical reactions and so the decay of organic matter, the rusting iron and the solution rate of salts are all accelerated by calefaction. The metabolic rate of animals is increased and so they require more oxygen which, however, is lower in concentration at these higher temperatures. Since the rate of exchange of salts and organisms increases, any toxins are liable to exert greater effects, and temperature fluctuations are also likely to affect organisms.

In the USA, the Environmental Protection Agency has recommended that heat discharges, should not raise temperatures by more than 5°F in streams, 3°F in lakes, 4°F in coastal marine environments in winter, and 1·5F in summer; the use of cooling towers can also improve oxygenation, but evaporative loss of water is high. Calefaction is therefore likely to exert a disruptive effect upon aquatic ecosystems, although some ways of putting the waste heat to beneficial use have been suggested. Where power plants are near deep ocean waters, the very cold water at depth could be brought up for efficient cooling and then pumped at its raised temperature into tanks used for raising fish or algae. Experiments in Hawaii and in St Croix in the Virgin Islands have suggested that a 100 mW plant could service a pond volume of 10^5 m^3 and produce 70 t/yr weight of carnivorous (third trophic level) fish. Such a plan would also obviate the disturbance of the aquatic littoral community (Bienfang 1971). Such local developments do not countermand the fact that energy production and use all contribute heat to the atmosphere. The possible global effects are discussed in chapter 10.

Noise

Noise is primarily a feature of cities, as exemplified by J. Caesar's action in banning chariots from the streets of Rome by day, thus producing insomnia by night. Defined as 'sound without value' or 'any noise that is undesired by the recipient' (and measured in a variety of purpose-specific units), noise levels in many urban-industrialized situations are known to be deleterious to human health and efficiency, with effects on the sense organs, cardiovascular system, glandular and nervous systems, while physical pain results at a level of 140 perceived noise decibels (pNdB) (Table 11.11; Molitor 1968, US Federal Council 1968).

TABLE 11.11 Noise levels (pNdB)

Silence	0	Rock band	100
Average residence	40–50	Compressors and hammers on construction work at 10 ft	110
Dishwasher	65	Four-engine jet at 500 ft	118
Auto at 20 ft	70–80	Hydraulic press at 3 ft	130
Light truck	75	Threshold of pain	140
Light truck accelerating	85		
Subway train at 20 ft	95		

Source: US Federal Council 1968

Localized noise derived from sources such as traffic or pan-throwing neighbours may affect personal health but does not have any obvious ecological repercussions. Recreation areas may be affected by noise from motorboats, but environmental noise is particularly associated with aircraft, especially along flight-paths in the vicinity of

Plate 17 The use of fossil fuel usually means noise. At this moment the Indian 'Palace in the Sky' is probably preferable to the Englishman's castle on the ground. *(Guardian, London)*

airfields (Plate 17, page 311), where the frequency of noise is as worrying as its actual intensity. Settlement location may need to be adjusted in such places with consequent ecological disruptions elsewhere. The sonic boom path associated with SST projects such as Concorde will produce noise of a very different order, in the form of sudden but repeated shock waves. These will cause disturbance to wild birds as well as domestic stock and buildings; subsonic flight over land seems to be a necessary condition. In terms of actual damage to hearing, various industrial groups are most at risk, though they can wear appropriate protection. No such avenue seems to be open to the devotees of the discotheque.

Warfare

Many wars are about access to resources and conflict invariably speeds up the use of materials as well as pre-empting the use of space at the scenes of the struggle. Less considered is the impact of war upon human environments, and especially upon the living components of these, including both 'wild' and 'tame' ecosystems (McClintock et al 1974, Huisken 1975).

In pre-industrial times, the deliberate use of environmental damage as a technique of war was known but limited in its application and effects. The transmission of disease to the enemy was one method, used for example in North America when blankets from smallpox hospitals were sent as gifts to the Indians; the poisoning of wells and the use of toxic smokes are recorded from Classical times; the burning of forests to flush out the enemy is mentioned for places as far apart as Italy in 55 BC and Scotland in the 17th century. Industrial age war has of course increased the impact on ecological systems just as it has heightened other resource-environment interactions. The steamship, the motor vehicle and aircraft have enabled the technological equipment of modern war to be taken virtually anywhere; and immense damage can be caused by bombardment from heavy artillery and from the air. The battle-grounds of Flanders are examples of the first case: there are still areas of pock-marked terrain from those years. By contrast, World War II probably had less environmental impact. The battles were more mobile than World War I and the targets of air bombardment were urban and industrial, not biophysical, systems. Nevertheless, considerable soil erosion took place after the tank and artillery battles in the USSR (in the Ukraine for example). There are, too, a great number of instances of forest destruction either by the war itself, or in order to provide resources to pursue it, few of which have been recorded in any systematic fashion; and at such times nobody is too fussy about waste disposal and pollution control (Lumsden 1975).

The potential of modern warfare for environmental alteration has been nowhere more apparent than in the second Indochina war (Westing 1976). United States General William Westmoreland's words, 'Technologically the Vietnam war has been a great success' is a clue to the fact that this war employed the full armory of modern science and technology and employed it in the context of biophysical systems just as much as against the urban-industrial infrastructure, of which there was not a great deal. The bombing of South Vietnam, for example, was aimed at the countryside and no habitat was spared either bombing or artillery fire. In South Vietnam, 587 kg/ha (577 kg/cap) of munitions were delivered; the average zone of devastation from a B-52 mission was 296 ha and $8{\cdot}1 \times 10^6$ ha or 11 per cent of the total area of Indo-China (equivalent to Connecticut or 66 per cent of the area of Wales) was hit, producing 350×10^6 craters, unevenly spread. The blast effects destroyed large areas of trees and crops, the craters often breached hardpans which held up perched water tables and thus made soils more arid; the flying metal killed wildlife and made trees difficult to use in sawmills as well as opening the bark to fungal rot. Some 3×10^9 m^3 of soil was displaced and bared, resulting in soil erosion and nutrient dumping.

Biocides were used on a large scale for the first time in war: to disrupt the wild ecosystems in order to deny cover and to destroy the tame ecosystems in order to deny sustenance. Herbicides (mostly 2,4,5-T, 2,4-D, Picloram, Sodium dimethyl arsenate and Dimethyl arsinic acid) were used; 86 per cent of the missions were against forest and mangrove and 14 per cent against crops. About 10 per cent of South Vietnam (equivalent to 14 per cent of South Vietnam's woody vegetation) was sprayed one or more times, and about 30 per cent of War Zone III (from Saigon to the Cambodian border) was thus treated. Apart from the effects on people (2,4,5-T contains the teratogenic contaminant, dioxin) this programme caused disforestation over large areas, followed by nutrient dumping and eutrophication of fresh waters and by the invasion of pioneer grasses. Although little laterization has occurred in Vietnam, the forest is slow to recolonize for, in some areas, the lack of evaporation and transpiration has led to high water tables which inhibit tree regeneration. The bamboo seral stage seems to last 50 years, and the *Imperata* grass stage 15 years, more if there is fire. Climax forest may take 500 years to come to maturity. Defoliated mangroves appear to be showing no recolonization and so coast erosion has accelerated since there is no protection from storms. In total, some 14 per cent of the standing crop of timber in South Vietnam was destroyed and at least three endangered species were hit hard, especially the kouprey (*Bos sauveti*). Animals that benefitted included malarial mosquitoes able to breed in the craters, and tigers who scavenged on human corpses.

Work on the potential effects of future wars has been carried out to a high degree of apparent accuracy of prediction in the case of nuclear weapons and to a much more general level in the case of Chemical and Biological Warfare (CBW) and other environmental modifiers. The environmental impact of nuclear weapons (Westing 1977) varies not only with their energy yield but with the site of explosion, whether at ground level or in the atmosphere. For example, a medium-sized atomic bomb of 0·91 Mt yield (Hiroshima and Nagasaki were 18 Kt) exploded at ground level, would kill by blast most trees over an area 13×10^3 ha, most vertebrates over 36×10^3 ha and, depending on the weather and site, start wildfires over 33×10^3 ha which would burn most of the vertebrates on 42×10^3 ha. Something like 19×10^3 m^3 of fine dust would be injected into the atmosphere (cf Krakatoa, estimated at 30×10^3 m^3, which affected climates over a wide area), leaving a crater 396 m in diameter with a displaced mass of 9960×10^6 kg. Aerial bursts generate the $NO_x \rightarrow 0_3 \rightarrow 0_2$ reactions and so increase UV radiation which might lead to the damage of the DNA of living creatures, especially small diurnal organisms.

Radiation would lead to the firestorms mentioned above and then to intense irradiation of living organisms. From the bomb of 0·91 Mt 312 ha would get 70 kilorads (kR) in the first 60s (troposphere burst) and 2830 ha would get the same level in the first 24 hours (surface burst). It requires 2 kR to destroy coniferous forest, 10 kR to destroy temperate deciduous forest, and 70 kR to kill prairie grassland. The sub-lethal effects include reduced fertility, shortened life span, higher incidence of neoplasms and an increased mutational frequency, as well as all the other environmental stresses like soil erosion, eutrophication and the eruption of the populations of tolerant species.

Deliberate environmental modification is also becoming more feasible although not yet perhaps a very dire threat to ecosystems. Rainmaking is conceded to be unreliable, although clearly its effects could be very great, given the fine turning of some ecosystems to precipitation input: the density of even jackrabbits is said to reflect rainfall. Fog creation, the control of tsunamis, hurricanes and tornadoes are all speculative as, probably, is the use of electromagnetic radiation to 'burn' holes in the 0_3 layer and fry those beneath. It is also said that low frequency radiation in the ionosphere affects brain rhythms: not only of humans, presumably (Jasani 1975).

By contrast, CBW is well developed and well documented, even to the point where the American patents for nerve gases have, in 1978, been freely available. CS gas harasses a lot of animals via inhalation and has been reported to cause damage to trees. The lethal synthetics ('V' agents) would kill not only humans (1 × 4,000 kg plane load would be lethal to humans over 5,000 ha) but would kill the other exposed vertebrates and a lot of arthropods, since they are cholinesterase inhibitors like organophosphorous insecticides. Vegetation would be unaffected but would remain a source of contaminated food, although the V gases are not persistent or cumulative in food chains. Botulin toxin (1200 ha of lethal area per average plane load) would kill many birds and some mammals, but not domesticated bovids and pigs, nor vultures which is perhaps just as well. Micro-organisms such as anthrax would debilitate most mammal populations and be very difficult or even impossible to eradicate; yellow fever virus would provide a pool of infection for subhuman primates and might become established in places where it is now absent, such as Asia.

Let Tacitus have the last word on the consequences of warfare:

'Ubi solitudinem faciunt: pacem appellant'
'They made a wasteland: they called it peace'

The invasion of the biosphere

As Table 11.1 (page 275) has shown, the various components of the biosphere act as temporary or permanent resting places for man-induced wastes. Many contaminants pass through the atmosphere, for example, but gases such as CO_2 which remain there, and finely divided particulate matter, provide the most likely agents of widespread change. The chemical balance of gases in the atmosphere is apparently the product of biological processes and can be changed by man with possible effects upon the global climate. By contrast, contamination of moving fresh water is primarily a local or regional problem, and if the offending inputs cease, then the biological processes and the hydrological cycle restore the water to a purer condition (Wolman 1971). This may of course take some time, as in the case of the River Rhine, where *inter alia* a German factory discharges wastes with a daily BOD equivalent to the sewage of 4.7 million people, and Alsatian potash mines discharged enough salt to raise the chlorine content of the river from 150 to 350 mg/litre in 30 years. As Coleridge remarked, 'What power divine / Shall henceforth wash the river Rhine?' : the EEC's attempts at sainthood are notorious for their failures. Relatively rapid cleansing is not possible in large lakes where poisoning and eutrophication can bring about apparently irreversible biological death if they proceed too far; small lakes at an early stage of enrichment can be saved. Soils also can accumulate nutrients toxins with as yet unknown long-term effects. The oceans inevitably form the main repository for contaminants from land, fresh water and atmosphere, although they lose some volatile substances to the last of these. Even if accumulation ceased now, the great water bodies would harbour residual pesticides for many years to come, along with numerous long-lived industrial, chemical and pharmaceutical compounds and oil effluent. Apart from their role as a food provider, the oceans play a significant part in the CO_2/O_2 balance of the planet, and the inhibition of photosynthesis by contaminants may seriously affect both (Goldberg and Bertine 1975, Ketchum 1975). In 1972 an international convention to reduce the dumping of waste materials into the sea was signed by 80 countries whose fleets account for about 90 per cent of ocean pollution. It prohibits the dumping of radioactive wastes, CBW agents, oils, cadmium and mercury, and organohalogen compounds. A special clause allows some of these to be dumped if they are immediately hazardous to human health. A second list contains substances which may only be off-loaded in specific locations and quantities with prior permission from a secretariat: arsenic, lead, copper, cyanides and fluorides are included. The accord closely follows the terms of the Oslo Agreement for the North Sea (also of 1972) but includes large maritime nations such as the USA, USSR, Liberia, Japan and Greece. It is impossible so far to see what effect such conventions may have.

On the land, poisoning of animals and plants by various air- and water-borne effluents is a daily occurrence in industrialized countries and cumulative toxins build up steadily, although radioactive fallout has fortunately continued to decline. Mortality among predators at the tops of food chains has been accompanied by sub-lethal effects in these and other animals at lower trophic levels.

Inevitably the possible effects of all these wastes upon man has been the subject of most concern. Apart from direct poisoning during toxin application, mortality from waste products is not apparently very high, though deaths from contaminated drinking water or sewage-laden sea-water off resorts are doubtless not publicized by the communities in which they occur. Directly traceable incidents like Minamata disease are generally rare, and features like smog perhaps exacerbate existing ailments rather than induce new ones. But the increased volume and incidence of pollutants is so recent that no adult has yet gone through a life-span carrying for example the body-burden of DDT now common. Not until those born after 1950 have gone through a normal twentieth-century Western urban existence without showing significant damage can we say that the present levels of contamination are harmless. Many organisms certainly have shown genetic responses to pollutants (Cook and Wood 1976).

More important than considerations of the personal health of individuals is the 'health' of the systems of the biosphere, for man is inextricably bound up in the webs of these systems and cannot exist apart from them, and his effect has often been to break down barriers between natural systems, to allow energy and matter to flow between the new subsystems. It is notable therefore that the effects of the various forms of pollution upon ecosystems can be generalized and indeed predicted. Putting together the evidence from ecological change caused by radioactivity, eutrophication, toxins, defoliation and deforestation, Woodwell (1970b) summarizes his findings that

> pollution operates on the time scale of succession, not of evolution, and we cannot look to evolution to cure this set of problems. The loss of structure involves a shift away from complex arrangements of specialized species toward the generalists; away from forest, toward hardy shrubs and herbs; away from those phytoplankton of the open ocean that Wurster proved so very sensitive to DDT, towards those algae of the sewage plants that are unaffected by almost everything including DDT and most fish; away from diversity in birds, plants and fish toward monotony; away from tight nutrient cycles toward very loose ones with terrestrial systems becoming overloaded; away from stability toward instability especially with regard to sizes of populations of small rapidly producing organisms such as insects and rodents that compete with man; away from a world that runs itself through a self-augmentive, slowly moving evolution, to one that requires constant tinkering to patch it up, a tinkering that is malignant in that each act of repair generates a need for further repairs to avert problems generated at compound interest.

To which may be added that all these shifts represent a movement away from systems which are highly valued in aesthetic and other non-economic terms to those of lower acceptability and value: in other words, a lowering of what we choose to call environmental quality. So far, however, the worst cases of inquination have been regional: the loss in O_2 levels in the lower Great Lakes and the Baltic, the industrial effluents in the Rhine, Minamata, Seveso, the levels of PCBs and organochlorines in coastal birds, the air pollution in the London of the 1950's and the smogs of North American and Japanese cities (Holdgate 1979). The potential global problems are CO_2 and particulate matter, and perhaps the dissociation of ozone, but no definite effects can as yet be traced to them.

The solution to the ecological downgrading caused by contamination can only be multidimensional. A suitable technology is an obvious starting place and may indeed be the easiest phase to achieve. Efficacious methods of material 'sieving' to prevent contamination of biospheric systems are available for many resource processes, although in some cases they merely transfer the site of the disposal problem. But nevertheless, sewage can be treated and smokes can be scrubbed, mine wastes can be reburied along with atomic wastes, and non-residual third-generation pesticides will replace the chlorinated hydrocarbons. But some contaminations are beyond the reach of technology: as long as industry persists, carbon dioxide and heat will be generated and led off into the atmosphere, and no technique for sieving out the DDT and related substances now in the oceans has been, or is likely to be, invented. So while recovery of many wastes is technically possible, there are some contaminant-caused problems to which no technological solution is at all likely. Movement towards greater sifting of the by-products of resource processes is likely to be accelerated by a shortage of the initial supply of the material and by the increased costs of energy since recycling often uses less than the processing of virgin materials.

Current economic and social values dicate that the proper response to a materials shortage is to develop a new process which will render hitherto inaccessible sources usable or to achieve substitution by another material. The complexity of contemporary technology means however that long 'lead times' are inevitable for new processes, so that the time-scale of problem accumulation is much slower than the time-scale of finding solutions. Substitution may not always be possible because of scale factors: water is the obvious example here. Recycling of materials is, therefore, not popular in the DCs at present but is likely to become more applicable because of cost factors and because of public concern, for example over disposable but non-returnable articles like plastic bottles and paper products (Kok 1976). This movement may gain impetus if ways of accounting are devised that include the full social costs of environmental contamination in its various forms; 'making the polluter pay' is unpopular in DCs because of the fear that industrial firms will become less competitive in world markets On the other hand, pollution control seems to be a creator of employment and hence a distributor of wealth.The loss of environmental quality due to insidious degradation is scarcely likely, however, to be susceptible to such analysis and will likely be ignored. A basic change in human values that puts healthy ecosystems before an unending supply of new materials would be ideal but seems destined to remain a low priority, although nearly all nations can point to an impressive array of legislation and regulation designed to clear up gross contamination; Japan for example has been very successful in its crash programme to improve air and water quality (McKnight et al 1974, Kelley et al 1976).

The atmosphere and the deep oceans are also a commons in Hardin's (1968) sense, and so international agreement is necessary for any of the clean-up processes to be effective. The scale and magnitude of the agreements necessary, for example to lower nitrogen and phosphorus levels in estuaries or to prevent untreated sewage reaching the seas, are immense and less capable of solution than the development of new technology. The guide-lines issuing from the UN 1972 Stockholm conference make general references to pollution but offer no suggestions for implementing technological solutions on an international basis. 'The just struggle of the people of all countries against pollution should be supported' is not a blueprint for environmental cleanliness. Solutions can be found to multi-national problems, like the Rhine, the North Sea, and the Mediterranean but they involve much political bargaining and nationalistic righteousness which may seem to make polluted waters seem positively pristine by comparison.

Finally, we must remember that pollution and contamination exist as the end-parts of resource processes and cannot sensibly be viewed outside this context; attempts to 'cure' pollution without considering the whole of the relevant resource process are as useful as trying to cure lung cancer with aspirins. Hence the primary importance of recycling wastes as sources of raw materials. The magnitude of the processes is inevitably linked to the ever-increasing rates of production of material goods and hence to population levels. People are not pollution, as some slogans aver, but there is little doubt that they are the cause of it, particularly the affluent ones; the example of many LDCs in reclaiming every possible material from every waste output is both a precept and a probable portent. Pessimists tend to take as symbolic the story (only slightly embroidered from LaMore 1971) of the rich Texan who was buried in a king-size grave propped up in the front seat of a Cadillac convertible with the hi-fi playing and the air-conditioning full on. One mourner was heard to say to another, 'Man, that's livin'.'

Further reading

BUTLER, G. C. (ed.) 1978: *Principles of ecotoxicology*.

CHADWICK, M. J. and GOODMAN. G. T. (eds.) 1975: *The ecology of resource degradation and renewal*.

EHRLICH, P., EHRLICH, A. and HOLDREN, J. P 1977: *Ecoscience: population, resources, environment*.

HENSTOCK, M. E. 1976: The scope for materials recycling.

HODGES, L. 1973: *Environmental pollution*.

HOLDGATE, M. W. 1979: *A perspective of environmental pollution*.

KELLEY, D. R., STUNKEL, K. R. and WESCOTT, R. R. 1976: *The economic superpowers and the environment*.

McKNIGHT, A. D., MARSTRAND, P. K. and SINCLAIR, T. C. (eds.) 1974: *Environmental pollution control. Technical, economic and legal aspects*.

MURDOCH, W. W. (ed.) 1975: *Environment. Resources, pollution and society*.

PERRING, F. H. and MELLANBY, K. 1977: *Ecological effects of pesticides*.

SCHULTZ, V. and KLEMENT, A. W. (eds.) 1963: *Radioecology*.

SHEETS, T. J. and PIMENTEL, D. (eds.) 1979: *Pesticides: contemporary roles in agriculture, energy and the environment*.

SINGER, S. F. (ed.) 1975: *The changing global environment*.

STERN, A. C. (ed.) 1976: *Air pollution*.

STUDY OF CRITICAL ENVIRONMENTAL PROBLEMS (SCEP) 1970: *Man's impact on the global environment: assessment and recommendations for action*.

WOODWELL, G. M. 1970: Effects of pollution on the structure and physiology of ecosystems.

Part III

The perception of limits

12

Resources and population

This section opens with a discussion of the growth of an future prospects for man's numbers. This leads into a consideration of some of the spatial and social consequences of the interaction of population and resources, an evaluation of the developing concepts of environmental management especially as enunciated in the DCs, and a brief discussion of 'development'.

Population

Historical perspectives

Many general references have been made to pupulation growth and its relationship to resource use. Since it is people who use materials and environments, both as metabolic requirements and cultural accessories, an examination of the past, present and probable future numbers of people and their distribution is considered at this point. There are numerous specialized works on demography and population geography (e.g. J. I. Clarke 1965, Zelinsky 1966, Bogue 1969, Petersen 1969, Trewartha 1969, Zelinsky, Kosinki and Prothero 1970, McKeown 1976), and only their general conclusions are presented here.

Estimates of the world's population before about AD 1650 are, as a UN publication sharply puts it, 'vague reconstructions', although perhaps the stage when the totals were one and two is reasonably well documented. Estimates for prehistoric times and for the first 1600 years AD are generally based on calculations of culture area by population density, where the latter is inferred from values for present-day examples of such economies as hunting and gathering or shifting agriculture. The taking of censuses began in 1655 and is now common, although many of them are probably not very reliable; on a world basis we may, however, expect a degree of cancellation of errors. The estimated numbers up to the present can be seen in Fig. 12.1, and Table 12.1 gives Deevey's (1960) estimates of population total and density for various periods.

Though the data are scanty before AD 1650 they give a consistent picture of a population with a very slow rate of growth (Fig. 12.1). High infant mortality rates and low longevity meant that rates of increase were small; in Roman times the average life expectation for men was about 30 years, and this was not altered significantly until the coming of scientific medicine in Europe. The replotting of the curve for total world population upon a logarithmic basis reveals a number of surges in population (Fig. 12.2), the first of which was coincident with the Neolithic revolution and the advent of agriculture, when population increased 25-fold due to increase in the means of

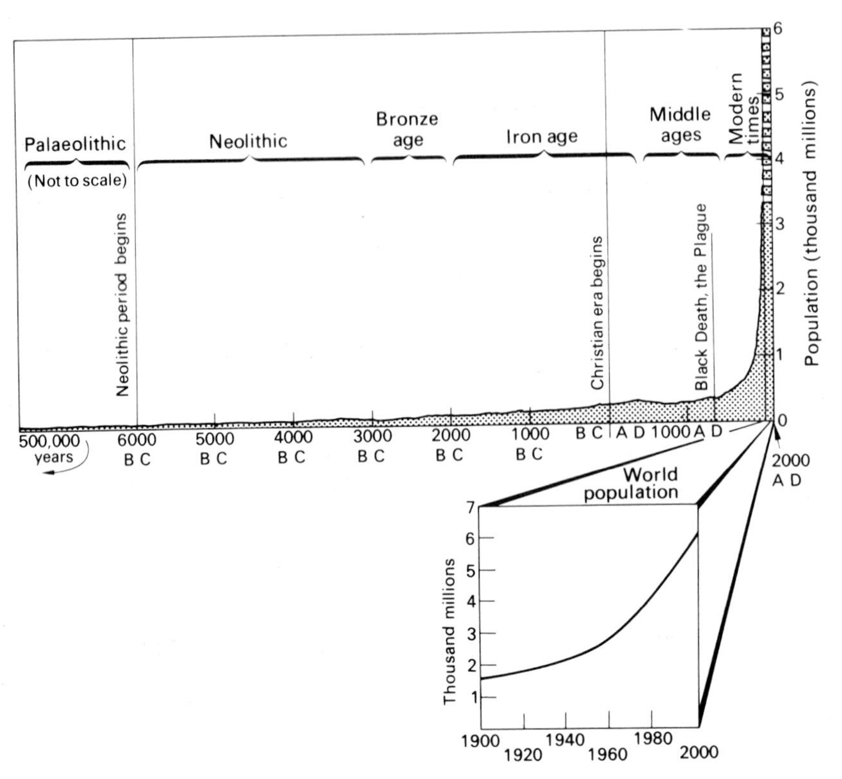

Fig. 12.1 The growth of the world's human population since the Palaeolithic, and projected at current rates of increase to AD 2000. Catastrophes like the Black Death had remarkably little long-term effect.
Source: Trewartha 1969

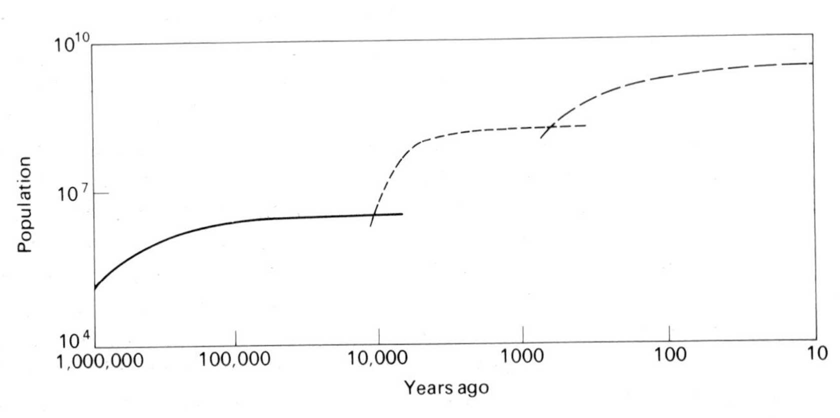

Fig. 12.2 A logarithmic plot of the population curve: surges in population are inferentially connected by Deevey (1960) with the invention of tool making (solid line), the agricultural revolution of the Neolithic (short dashes) and the scientific-industrial revolution (long dashes). Other writers (e.g. Durand 1967) interpret the latest surge as originating before the scientific-industrial revolution.
Source: Deevey 1960

TABLE 12.1 Population growth

Years ago (base-date = 1960)	*Cultural stage*	*Area populated*	*Assumed density per km²*	*Total population (millions)*
1,000,000	Lower Palaeolithic (hunting–gathering)	Africa	0·00425	0·125
300,000	Middle Palaeolithic (hunting-gathering)	Africa and Eurasia	0·012	1·0
25,000	Upper Palaeolithic (hunting–gathering)	Africa and Eurasia	0·04	3·34
10,000	Mesolithic (hunting–gathering)	All continents	0·04	5·32
6,000	Village farming and early urban	Old World New World	1·0 0·04	86·5
2,000	Village farming and urban	All continents	1·0	133
310 (1650)	Farming and industrial	All continents	3·7	545
210 (1750)	Farming and industrial	All continents	4·9	728
160 (1800)	Farming and industrial	All continents	6·2	906
60 (1900)	Farming and industrial	All continents	11·0	1,610
10 (1950)	Farming and industrial	All continents	16·4	2,400

Source: Deevey 1960

subsistence. Once agriculture had become firmly established (4000–3000 BC) three main zones of occupance existed: high-density areas of agriculture and gathering around the agricultural hearths and diffusion areas; areas with thinly spread gathering and hunting cultures; and unoccupied areas. By about 2000 BC the agricultural occupance of much of Eurasia and the northern half of Africa is postulated. Beyond this core lay the food gatherers and hunters, and the uninhabited areas were still considerable, though reduced in size. Within the agricultural areas densities were higher in favoured places and highest of all where cities had developed. At the time of the Classical world, 200 BC–AD 200, the populations of the world are dominated by empires. The Indian subcontinent held 100–140 million and mainland China 71 million; Imperial Rome under Augustus had some 54 million people who constituted a quarter to a fifth of the world's population. Between AD 1000–1600 a steady but slow growth is seen on most continents (Table 12.2) in spite of intermittent checks imposed by famine and plague. The exception is the Americas, where the colonists and conquistadors were responsible for reducing the aboriginal populations to fractional levels through warfare and disease, although it has been argued that a group like the Maya were on the verge of collapse, having outrun their resource base.

A population of 470–545 million in AD 1650 seems very large by comparison with the figures for BC/AD 0 or 4000 BC, but becomes small in the context of the accelerated growth after AD 1650. The major components of the growth 1750–1979, continent by continent, are shown in Table 12.3, using Durand's (1967) work, to which reference

TABLE 12.2 Approximate population (millions) of the world and its subdivisions, AD 1000–1600

Year	World	Europe	Asiatic Russia	South-east Asia	India	China Major[1]	Japan	South-east Asia, Oceania	Africa	The Americas
1000	275	42	5	32	48	70	4	11	50	13
1100	306	48	6	33	50	79	5	12	55	17
1200	348	61	7	34	51	89	8	14	61	23
1300	384	73	8	33	50	99	11	15	67	28
1400	373	45	9	27	46	112	14	16	74	30
1500	446	69	11	29	54	125	16	19	82	41
1600	486	89	13	30	68	140	20	21	90	15

[1] China proper, plus Manchuria and Korea, Outer Mongolia, Sinkiang and Formosa.
Source: Desmond 1965

TABLE 12.3 Estimates of population growth 1750–1979 (millions)

Areas	*1750*	*1800*	*1850*	*1900*	*1950*	*1979*
World	*791*	*978*	*1,262*	*1,650*	2,515	4,321
Asia (exc. USSR)	*498*	*630*	*801*	*925*	1,381	2,498
Africa	*106*	*107*	*111*	*133*	222	457
Europe (exc. USSR)	*125*	152	208	296	327	483
USSR	*42*	56	76	134	180	264
North America	*2*	*7*	26	82	166	244
South and Central America	*16*	*24*	38	74	162	352
Oceania	*2*	*2*	2	6	13	23

Figures in italics are those which lack a firm foundation.
Source: Durand 1967; Population Reference Bureau 1979

should be made for discussion of the reasons for the growth in each individual continent. The major point of interest is the phase of accelerated population growth which has lasted to the present. Its inception varies from place to place and is prior to 1750 in Europe, Russia and America and probably before 1650 in China. The absolute numbers would of course rise quite steeply even with a constant rate of population growth, but as Table 12.4 shows, the actual rate itself has undergone acceleration, with the period 1950–70 having the highest rates.

Close inspection of estimates of growth for individual nations has revealed little relationship between the development of industrialization and the expansion of population. Why there should then have been a simultaneous upturn in rates of growth in the eighteenth and early nineteenth centuries is not known. No simple explanation of causality is acceptable and Durand's (1967) hypothesis seems more plausible: he suggests that the stimulus of agricultural improvement in the sixteenth and seventeenth

TABLE 12.4 Annual rate of increase (per cent)

Areas	*1750–1800*	*1800–1850*	*1850–1900*	*1900–1950*	*1969*	*1979*
World	*0·4*	*0·5*	*0·5*	*0·8*	1·9	1·7
Asia (exc. USSR)	*0·5*	*0·5*	*0·3*	*0·8*	2·0	1·8
Africa	*0·0*	*0·1*	*0·4*	*0·7*	2·4	2·9
Europe (exc. USSR)	*0·4*	*0·6*	*0·7*	*0·6*	0·8	0·4
USSR	*0·6*	0·6	1·1	0·6	1·0	0·8
North America	—	2·7	2·3	1·4	1·1	0·7
South and Central America	*0·8*	*0·9*	1·3	1·6	2·9	2·7
Oceania	—	—	—	1·6	1·8	1·3

Figures in italics are those which lack a firm foundation.
Source: Durand 1967; Population Reference Bureau 1969, 1979

centuries provided the potential for considerable population expansion but that this was held back by the transmission of diseases which followed exploration and trade in the same period. By the eighteenth century sufficient resistance to imported infections had built up for the population potential to be realized. McKeown (1976), in contrast, emphasizes the improvement in nutrition in the 18th and 19th centuries as the chief element in the decline of infectious diseases. At any rate, by the end of the nineteenth century the dichotomy of demographic process between industrial and less developed countries was apparent: the former had death control (giving lower rates of infant mortality and greater longevity) and a measure of birth control, but the latter had experienced only the initial impact of the death control techniques which are dominated by modern medical practices such as antisepsis, immunization, disease vector control and pharmaceutical improvements, together with a knowledge of nutrition science.

Present numbers, distribution and density

Estimates of the populations of the continents are given in Table 12.5 and the percentage contributions of the various areas shown in Table 12.6. These are summarized in the map of population density, Fig. 12.3. Such statements of distribution reveal that Europe and Asia together contain over three-quarters of mankind, and Asia has more than half, whereas less than 10 per cent live in the southern hemisphere. Between 20°N–60°N in the Old World are found four-fifths of the population, in a zone which includes also most of the great deserts of the Old World, along with the Alpine and Himalayan mountain systems. Within this zone are two great concentrations of man: south-east Asia, where approximately one-half of the world's people live on one-tenth of the habitable area, and Europe (including European Russia) where one-fifth of the gobal population occupy less than one-twentieth of its ecumene.

There are in addition secondary concentrations in the eastern half of North America, California, coastal Brazil, the Plate estuary, the valley of the Nile, west Africa, south-east Africa and south-east Australia. Looked at politically, six states (China, India,

TABLE 12.5 Estimated world population, mid-1979 (millions)

World	4,321	Latin America	352
Africa	457	Europe	483
Asia	2,498	USSR	264
North America	244	Oceania	23

Source: Population Reference Bureau 1979

TABLE 12.6 Continental percentages of world population, 1920–73

	1920	*1960*	*1973*	*1979*
Africa	7·9	8·5	9·7	10·6
Asia	53·3	56·1	57·2	57·8
North America	6·5	6·6	6·0	5·6
Latin America	5·0	6·9	7·9	8·1
Europe	18·1	14·2	12·2	11·1
USSR	8·7	7·1	6·4	6·1
Oceania	0·5	0·5	0·5	0·5

Source: Population Reference Bureau 1969, 1973, 1979

Pakistan, Bangladesh, Japan and Indonesia) have nearly half the world's population and, contrarywise, some pocket populations like Gambia (0·6 million in 1979) and Iceland (0·2 million in 1979) continue to exist. Considerable variations in density are revealed by Fig. 12.3. The environmentally inhospitable parts of the globe are principally those too cold, too dry or too high for mass settlement and at concentrations of $< 1/km^2$ contrast with the favoured agricultural areas like the valley of the Ganges and maritime east Asia, and with intensely used industrial and agricultural zones, such as west and central Europe, both at $100/km^2$. The inner cores of Western cities have traditionally held the densest agglomerations of people, but migration to suburbs is lessening the populousness of such places.

Future growth and numbers

Interest in the growth rates of populations and the consequent future numbers centres around the availability of resources for them and also their effects upon the biosphere, especially that of the wastes they produce. But just as the reasons for population growth in the past can never all be known, so future growth cannot be predicted without making certain assumptions about demographic variables and the way in which these are affected by social and economic conditions. Some projections therefore assume constant fertility while others may assume that a form of demographic transition to

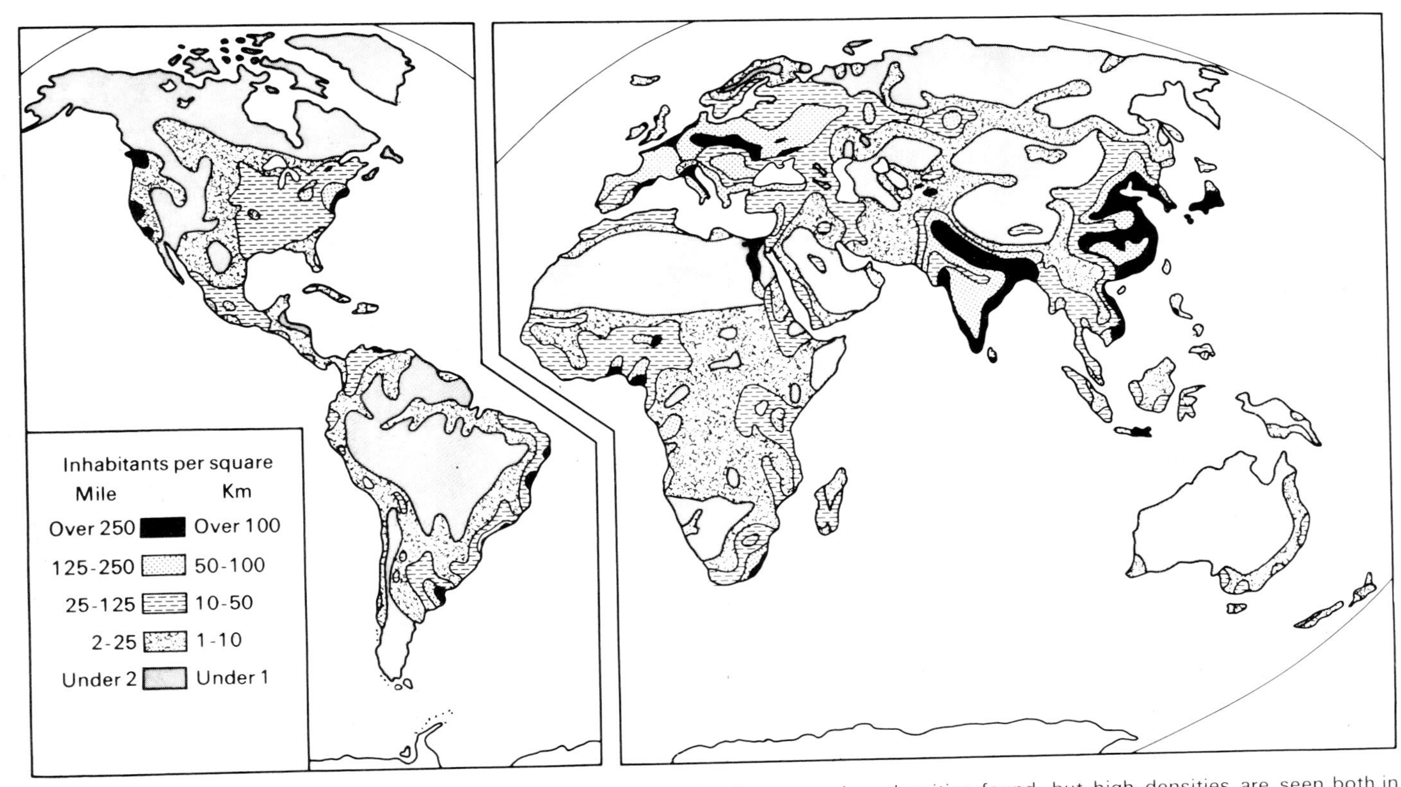

Fig. 12.3 World population density: only in regions of unfavourable climate are low densities found, but high densities are seen both in urban-industrial regions of the world and in rural-agricultural regions, especially in Asia.
Source: Trewartha 1969

TABLE 12.7 Growth rates and doubling times

Increase % p.a.	*Years to double population*
0·1	693
0·5	139
1·0	70
1·5	47
2·0	35
2·5	28
3·0	23
3·5	20
4·0	18

Source: Hardin 1969

lower birth rates will occur as nations become industrial, as happened in nineteenth-century Europe. Both of these assumptions may turn out to be untrue.

The simplest measure of population growth is the annual percentage increase which relates easily to doubling time (Table 12.7). The best estimate of the present world rate (1979) is 1·7 per cent p.a., which gives a doubling time of about 41 years. This average rate is composed of a multitude of national rates of greater or lesser accuracy of estimation, most of which are within the range 1–2·5 per cent but with extreme variants (Table 12·8). By continents, Europe appears to have the lowest rate of growth and Latin America the highest; West Germany has the lowest national rate of growth and Kenya the highest. (For the purposes of this discussion migration has been ignored; although significant nationally, as in Lebanon, it is not relevant at the global scale.) Demographically two main groups exist, those with a birth-rate (BR) over 35/1,000 and those whose BR is < 35. The first group comprises most of the LDCs and has two

TABLE 12.8 Population growth rates 1969 and 1979 (per cent per annum)

	Mean		*Individual countries 1979*	
	1969	*1979*	*Highest*	*Lowest*
World	1·9	1·7		
Africa	2·4	2·9	3·5	1·1
Asia	2·0	1·8	3·7	0·9
N. America	1·1	0·7	0·8	0·6[1]
Latin America	2·9	2·7	3·5	0·9
Europe	0·8	0·4	2·2	−0·2
USSR	1·0	0·8		
Oceania	1·8	1·3	3·1	0·8[2]

[1]Canada and USA only. [2]Data for 6 countries only.
Source: Population Reference Bureau 1969 and 1979

sub-groups, the first of which has BR > 35 and death rate (DR) > 15/1,000. This is best represented in Africa and Asia, where DRs remain high at present but are falling under the impact of modern medicine. The second sub-group has DRs < 15 and is responsible for the very high growth-rates of Latin America, since BR remains high but DRs have fallen substantially. A BR of < 35 is characteristic of the DCs and a few LDCs which have low absolute populations. A subdivision can be made on the basis of BR = 20; above this level are a few DCs whose birth rates seem bound to fall and thus bring them into a typical DC position where BR < 20. These countries have gone through a demographic transition to low birth rate and low death rate.

The different demographic types can be linked to variations in family characteristics, as detailed by Petersen (1969). These types are conceptualized in terms of the demographic transition in Europe but may sometimes have a relevance elsewhere to populations whose cultural characteristics are changing. In medieval times there was a traditional pre-industrial family where late marriage and non-marriage helped to depress birth rates. Guilds permitted marriage of their apprentices only when their service was complete, and younger children who would not inherit their father's farm, for example, might well enter the Church, whose clergy were supposed to be celibate. This family type was succeeded during the industrial revolution by the proletarian family in which early marriage was favoured by the availability of work within the factory system. There were few institutional barriers to sexual relations and illegitimacy was high, since in some countries the mother could force support from the putative father. Additionally, enough children put out to work at an early age could ensure a comfortable living for the mother. The third type may be called the rational family and was coincident with the rise of a large middle class. A sense of parental responsibility for limiting family size arose which was helped by the availability of contraceptive methods, and the age of marriage also began to rise.

All such generalizations are of course abstractions from a multitude of cultures, and as Zelinsky (1966) points out, it is easy to forget that all demographic characteristics are a result of cultural practices: the adoption of modern medicine is merely one of these. Any real understanding of population growth must therefore start with an assessment of the cultural variables involved, which naturally includes the peoples' perception of themselves and their physical and social environments.

Population projections

The pitfalls inherent in making forecasts of population levels beyond the immediate future are considerable (Dorn 1965). For example, it is often assumed that as the LDCs industrialize they will undergo the same type of demographic transition to low BR and low DR as did Europe in the nineteenth century. Yet there is no certainty of this since very different cultures are involved, apart from the assumption of the inevitability of industrialization in the poorer countries.

The assumptions underlying projections must therefore be clear if the estimates of future numbers are to have any usefulness. Two types of assumptions are most often used:

TABLE 12.9 World and continental population projections to 2000

	UN medium variant: base years 1969 and 1979 (Millions)			
	1969	*Base year*	*1979*	*Variation (%)*
World	6,130		6,168	+ 0·6
Africa	768		831	+ 7·6
Asia	3,458		3,588	+ 3·6
North America	354		289	−18·4
Latin America	638		597	− 6·4
Europe	527		521	− 1·1
Oceania	32		30	− 6·2
USSR	353		311	−11·8

Source: Population Reference Bureau 1969 and 1979

(1) constant fertility projections, in which the recent growth rate is projected into the future, i.e. a simple extrapolation of present trends; and

(2) changed fertility projections, in which assumptions are made about the socio-economic conditions affecting natural increase and migration, principally the former.

The uncertainties are so great that the UN projections, for example, employ three variants (high, medium and low) based on differing assumptions about the impact of medical and contraceptive programmes and of nutrition standards in various parts of the world.

The medium variant is the most often quoted: it assumes that fertility will decline during the 1970s and 1980s at rates previously observed in some areas. The outcome of projections to e.g. AD 2000 depends on the base year as well as refinements in technique and data availability such as new censuses. Table 12.9 therefore shows not only the medium variant projection for continents based on 1979 but the equivalent figure from a 1969 base year, and the difference between these estimates. The latest data exhibit two salient features: first that even the medium variant involves a 42 per cent increase in absolute numbers between 1979 and the end of the century. The second is that a large number of the additional people will be Asians: they above all will be the most numerous, perhaps nearly 60 per cent of the world total. Thus it seems that as Freyka (1973) suggested, the present ratio of about 30 : 70 between the rich and the poor will become more like 20 : 80 and perhaps not even stop there. Stasis populations for the world seem unlikely to be achieved before the middle of next century and before a level of 8,500 million, although a really sharp decline in fertility might stabilize numbers at *ca* 6,000 million by AD 2020.

Recent rises in population at the rate of nearly 2 per cent per annum do not seem very large unless the characteristics of an exponential curve with this increment are considered carefully. At present the doubling time for the world population is 41 years and the projection of this kind of doubling time and the absolute numbers already achieved soon bring about very high levels of people. In AD 2600 a number of the order of 1×10^{15} could be reached: each person would then have about 0·5 m^2 of room and it would be Standing Room Only; in about AD 4000 the earth would be a mass of humanity expanding outwards at the speed of light. Fremlin (1964) explored the life-style of a population about 875 years hence which at a growth rate of 2 per cent per annum had become $60{,}000 \times 10^{12}$. To cope with such numbers the whole planet would be covered with a 2,000 storey building, each person having about 7·5 m^2 of living space in the 1,000 storeys devoted to the people rather than plant for food producing and refrigeration; the temperature of the outer skin would be at 1,000°C. Even the shipping off of people to other planets has been put forward from time to time, an idea scathingly reviewed by Hardin (1969).

These apparent fantasies of huge numbers and SF-type technologies have one serious purpose: to make us aware that absolute limits exist. If there were not the nutrition limit set by photosynthesis or even photosynthesis plus 'unconventional' foods, then there are the space limits and the heat limits. Therefore let nobody be persuaded that the earth has some infinite capacity for people.

Population control

Attempts to vary rates of increase have a long history (Benedict 1970). Most of our knowledge (and a great deal of folklore) has been concerned with efforts to increase fecundity, particularly in pre-industrial societies with high rates of infant mortality. Modern medicine is also a death-control mechanism, particularly as it affects longevity and infant mortality. Death control is usually more acceptable than birth control, although there is considerable evidence for the latter in pre-industrial groups, especially if we include abortion and infanticide as methods of reducing increase.

Modern concern to limit birth stems from two types of motivation. The first is concern with the physical and economic health of an individual family, or a particular woman, where hardship occurs if too many children are born too quickly. However, the spacing of children does not necessarily mean a small number of births and medicine in many places enhances their chance of survival. A second reason for advocating birth control is the economic, social and ecological health of an entire community or nation, especially where resources are scarce or where the rate of growth is placing great strains on the ability of the community services to cope with it. Observation of the areas of fastest growth will often reveal a certain coincidence of rapid increase with poorly developed social infrastructures. The response to a desire for a slower growth rate comes in two phases. The first of these may be called the family planning phase, in which the efforts of government or private programmes are directed at individuals and families; a world perspective is maintained by the highly effective International Planned Parenthood Federation (IPPF). Emphasis is placed on the advantages both to parents and to children of smaller or well-spaced families. Sterilization and contraceptive techniques are widely offered. However, abortion, usually illegal, is probably of more significance than any of the physical and hormonal contraceptives. Likewise, late

Plate 18 Jumbo-sized population increases bring forth commensurate communications media. How many loops would an Air India Boeing 747 (see Plate 17) have bought? *(Camera Press Ltd, London)*

marriage is a very effective factor in limiting births, as has been shown in China. The targets for family planning programmes usually aim at slowing a growth rate (e.g. by 1 per cent p.a.) in LDCs (Plate 17, above) or at eliminating 'unplanned' births in the DCs. The UN's World Population Plan of Action (adopted at Bucharest in 1974) incorporates those types of aims.

The very features that make family planning acceptable render it ineffective for the second phase, which is population control. This is linked to the circumstances of the community rather than the individual and envisages either a specific target population to be achieved by a slower rate of growth or a stabilization level with zero growth or even a diminution in absolute size. The persuaders aim at replacement level only, i.e. a maximum of two children per family. Again, sterilization and contraception are the most widely offered means, although the popularity of legal abortion is growing. No country has yet adopted a distinct control policy with announced targets in terms of absolute numbers, but some of the governments of the richer DCs are under strong pressure from neo-Malthusian pressure groups to do so. Opposition to both family planning and population control has often been strong, and the number of countries which have official and semi-official programmes is perhaps surprisingly high (IPPF 1972). An emphasis in Asian countries is noteworthy, especially where ethnic Chinese are concerned, and it is interesting that Singapore has perhaps the most stringent relationships between family size, taxation and welfare benefits. Too zealous an attitude to population control, especially male sterilization, led to considerable upheaval in India in the mid-1970's (Mitra 1978), but in Nepal and the Phillipines, for example, tax deductions have been reduced or eliminated for children.

A common theme of most opponents of population control is the right of the individual family to decide on its size: their procreative activities are no concern of the community. This is becoming less acceptable as it is more widely realized that every new individual requires resources which have significance to a wider group: the community, the nation or even the whole world. In hunting societies, where the limits of the environment are closely perceived, population control appeared to be adjusted to the needs of the community, but the coming of agriculture seems to have put the onus on the family. First agriculture and then industrialization have allowed environmental limits to be perceived as virtually infinite. The advent of modern ecology has brought back into focus the limits to the carrying capacity of the planet, and the realization of an upper limit to the number of people that the earth can support. In the face of this fact the options appear to be twofold. If the population exceeds the carrying capacity then the natural checks of famine, disease and war will probably operate: the latter can now produce some very large-scale oscillations indeed; or we can bring to bear all the knowledge of science, values and communication that we possess in an effort to level off the exponential curve. The fact that in 1979 the world increase had dropped back to 1·7 per cent per annum from its earlier high of 2·0 per cent shows not only the drop in the birth rates in DCs but declines in fertility in some LDCs as well (Table 12.10, Salas 1978). Very few of them have yet come near the 20·0 level characteristic of the transition to a DC demographic pattern.

On the other hand, some countries are actively concerned that their population is not

TABLE 12.10 Recent birth rates in selected DCs

	Rate per 1000 population				
Sri Lanka	1953	39·4	Taiwan	1959	41·0
	1963	34·6		1963	35·6
	1968	32·1		1970	27·2
	1979	26·0		1979	24·0
Hong Kong	1961	35·5	Singapore	1957	42·7
	1969	20·7		1966	28·6
	1979	18·0		1970	22·0
				1979	17·0
South Korea	1958	44·8	Tunisia	1961	49·6
	1963	38·2		1968	43·0
	1966	32·0		1979	36·0
	1970	30·0			
	1979	24·0			

Source: Various UN data

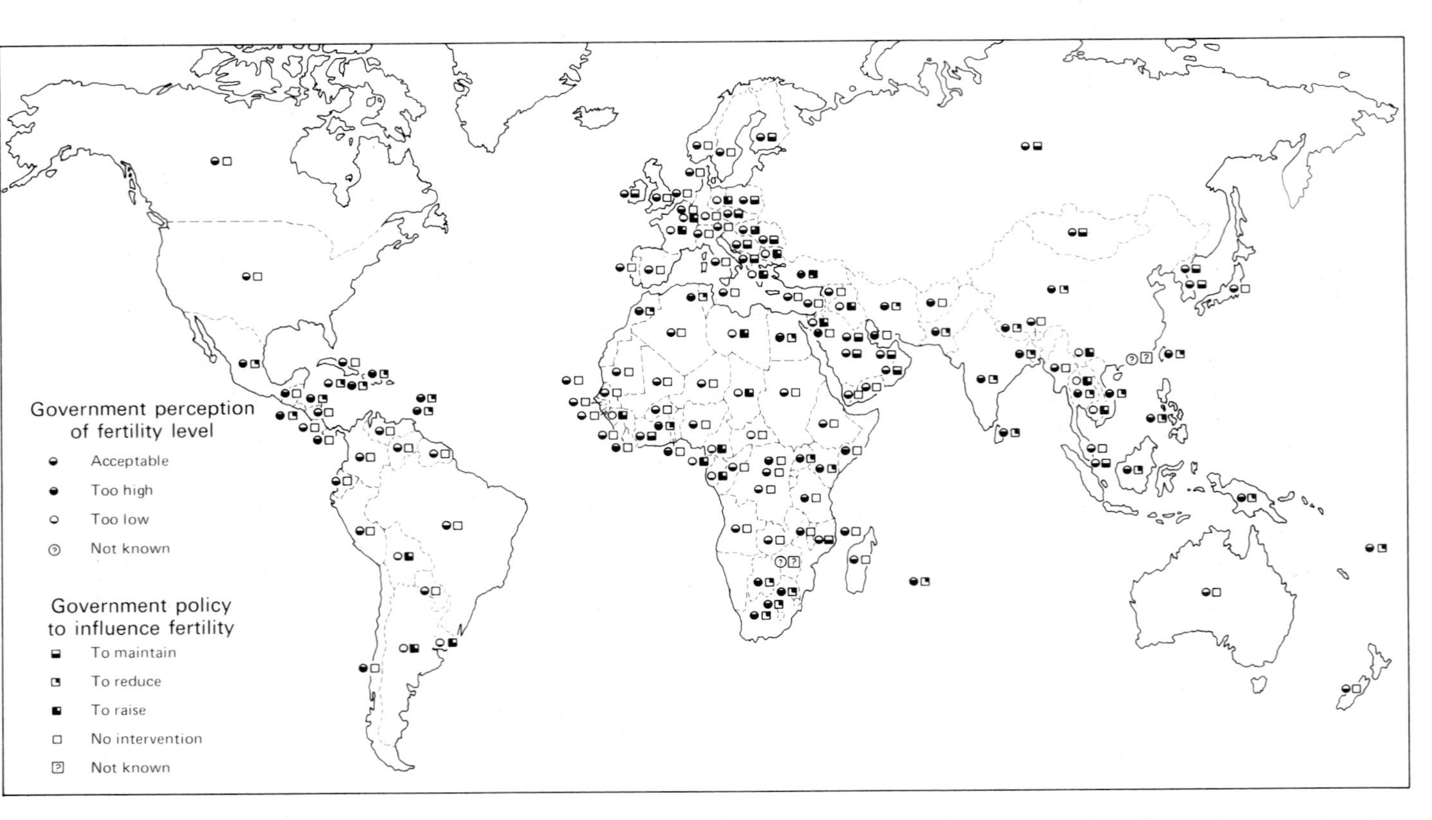

Fig. 12.4 An overview of the attitudes of the governments of individual nations to the fertility of their own nations. For each country there is a symbol for perception of fertility level and for policy to influence fertility.
Source: IPPF 1979

high enough and give pronatal financial inducements; notable among this group are the USSR and other Socialist republics of eastern Europe. Government inducements to breed in France, for example, having been less than successful, it may be wondered if this may not be the cheapest way to bring about declines in fertility.

A major problem is that although the problems caused by a rapdily increasing population can be identified on a world scale, their solution must inevitably be national. This produces a growing fear in the LDCs and among minority groups that 'they' are trying to subjugate 'us' by controlling 'our' numbers. As far as the LDCs are concerned, no power on earth can now stop them far exceeding the DCs numerically, but since concern over the global population–resources–environment balance comes mainly from the DCs it would seem prudent for them to practise what they preach in terms of population control.

Summary

The complex factors of demography boil down to one or two salient features as far as we are concerned here:

(1) Demographic history shows a rapid expansion of population after the seventeenth century, following thousands of years of very slow growth.

(2) The world population, currently (1979) at 4321 million, has very uneven distribution, with a particular concentration in Asia.

(3) A world growth rate of approximately 1·7 per cent p.a. gives a doubting time of 41 years, irrespective of the absolute numbers.

(4) There is general discrepancy between the high rates of increase in the LDCs and a low rate (nearing stability in Sweden, Norway, the UK, West Germany, for example) in the DCs.

(5) A rapid rate of urbanization prevails in LDCs.

(6) Birth control programmes in LDCs are beginning to reduce fertility levels though in most of these nations annual increases of 2·0 per cent and above are still common.

(7) Fertility control is a complete cultural matter and is not simply a matter of providing a technology.

At the United Nations world population conference in 1974, considerable emphasis was put on the role of economic development ('the best contraceptive') and it was clear that population control cannot be considered just as a medical or paramedical process: it is embedded in a social and economic fabric (Wilber 1978). A Plan of Action was agreed but in 1979 it seemed that not a great deal had been done although 80 per cent of the population of the LDCs now reside in nations whose governments wish to reduce their rate of population growth (Tabah 1979).

Population–resource relationships

As populations have grown, so have the magnitudes of those resource processes which supply materials, so that the graph of population growth also describes in general terms the use of energy or the consumption of food. The important exception is that, world-wide, some resource processes are increasing in volume even faster than the population, i.e. per capita use is also going up: energy and water use are examples. This situation is most striking in the DCs, where access to technology creates a different kind of access to resources beyond those which are necessary to sustain life. In the LDCs, cultural consumption is by no means absent (particularly in those with a well-developed social stratification), but there is much greater emphasis on the provision of necessities such as food, shelter and employment. The contrast between the population–resource relationships of different types of countries allows the construction of a regional classification of the type put forward by Zelinsky (1966) and which appears as Fig. 12.5.

A diagnostic feature of these regions is the role of technology as part of the culture. The degree of application of mechanical and electronic technology, backed by expert knowledge, has been seen to alter the perception of available resources and hence lead to prosperity. Inherent in the role of technology as a creator of resources is its power to destroy them, either by misapplication or by means of the waste products which it creates; in addition it may have little to do with the long-term ecological stability of a resource process. Thus, while cognizant of the vital role of technology as a means of gaining access to environmental resources, we must not be so blinded as to fail to recognize its spin-off effects, not the least of which are its use for its own sake and the economic constraints created by the expensive machinery which is now widespread (Galbraith 1967). On the other hand, technology increases the chances of averting or minimizing economic and ecological disasters through the ingenuity it confers.

Type A regions constitute technological source areas, where invention, research and development are at a high level. This is combined with a large land area which is well stocked with available and potential resources and has a low or moderately sized population relative to its size and to other nations. The high level of technology means that the people have access to the resources and also create the effective demand to ensure their use; prosperity allows the purchase from other countries of what they lack. The high levels of material wealth have been gained at the price of widespread environmental damage during phases of very rapid development and exploitation, and it is not surprising that the strongest public pressures for high-quality environments and 'environmentalist' crusading have come from the epitome of this group, the USA. In these countries, economic growth (i.e. the expansion of the magnitude of material-using resource processes) is in general a publicly espoused aim and its measures are seen as measures of the general welfare of the people. In the wealthiest, there may be some questioning of the purpose of such growth, particularly since the desired rates of expansion are very difficult to achieve in energy- and machine-intensive economies especially at a time when energy costs are rising so quickly. This type of population–resource situation may be labelled 'the US type', but it also includes Canada, Australia, New Zealand, the USSR, part of Argentina and, with qualifications, Uruguay, south Brazil, South Africa and Zimbabwe.

Type B may be categorized as 'the European type'. This is also a technology source area but differs from the US type because the population is high in both absolute numbers and density, the countries smaller and their heritage of resources less

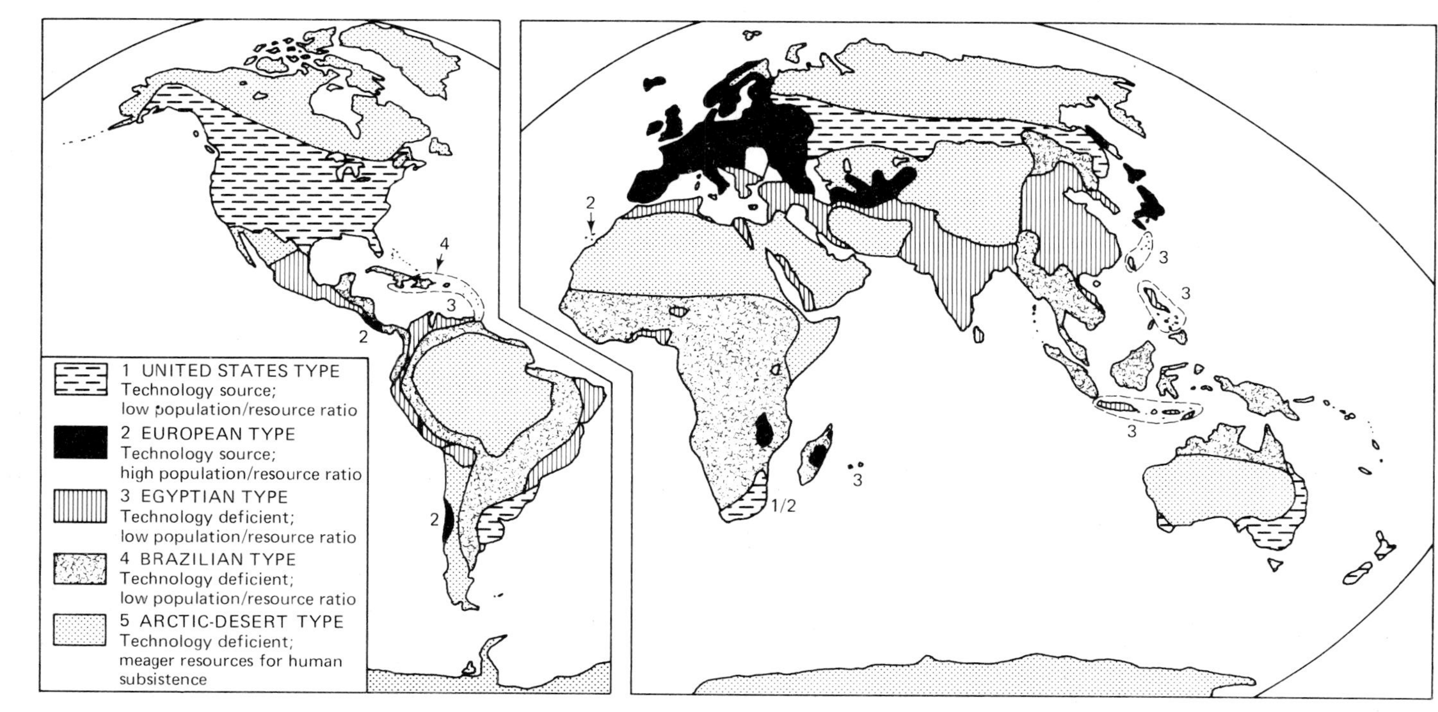

Fig. 12.5 Generalized population/resource regions of the world. (Most of Egypt falls outside the category that is named after it.)
Source: Zelinsky 1966

Plate 19 Immigration and natural increase, along with a small land base, produce squatter settlements like these in Hong Kong. Although its economy is perhaps closest to the European type, this colony's problems are exacerbated by its isolated location. *(World Council of Churches, Geneva)*

abundant and more fragmentary than those of larger nation-states. There is less room for trial and error than with type A and so a more 'conservative' attitude towards land and water resources has developed which is less profligate and damaging than the US type, although the number of people makes considerable environmental manipulation inevitable. Many of the man-made ecosystems, such as European agriculture, are essentially very stable, but the margin of stability is not very wide and most such countries rely on a 'ghost acreage' of food-producing land and water outside their national boundaries. Economically, an elaborate system of international exchange has to develop in order to ensure material supplies for the urban-industrial base. Thus trade with LDCs is widespread, partly as a historical result of colonialism, and international blocs (of which the EEC is the latest and largest) are commonplace. Again, economic growth is an explicitly stated goal of public policy, often being said to be necessary in order to clear up the legacy of a long history of industrial impact upon the environment and upon people. However, the value of growth is being questioned to some extent, and we may expect nations with a very slowly growing population and a culture capable of accepting changes in social aims, such as Sweden, to revise their attitudes towards growth in the near future. As would be expected, the European countries themselves provide type-specimens of this category, along with nations like Japan and Israel. In transit to this group are nations like Chile, together with Hong Kong (Plate 18, opposite) and the island of Puerto Rico; aspiring candidates who have passed some of the examinations would appear to be Mexico and Libya; the People's Republic of China in now adopting a 'modernism' would seem to be aiming for this category.

Type C envisages countries with large resource bases but which are deficient in technology. Their populations are relatively low, so that there is very little sense of pressure upon resources, except perhaps in the urban areas, but material standards are low as well. Their status is not permanent: increased technological development will allow them to move into the European category, but absence of such development if combined with rapid population growth may well depress them into type D countries. Such nations much desire economic growth and may even want population growth in order to try and achieve it faster. But they are hindered by economic factors such as the control of many of their exportable resources by outside interests or dependence upon world prices which may not move in their favour. This group may be labelled 'the Brazilian type' and is concentrated in three main areas: Indochina-Malaya (but excluding Vietnam, Java and the north and central Philippines), tropical Africa, and South America. There is an exceptional group in the shape of the OPEC nations which are, in pockets, extremely rich but whose terrain and overall state of development suggests that class C is appropriate. Standards of resource management vary widely and are critical in determining the movement of nations out of this category.

Type D is the most unfortunate group. Not only is there a deficiency of appropriate technology but the population presses hard upon the resources and is generally growing at 2 per cent p.a. or more. The capacity to deal with population–resource imbalances is frequently lacking, so that both the means of subsistence and the means of employment are deficient. Institutional factors may exacerbate the troubles of such nations. Small size of territory may impinge harshly upon a growing population, as may for example the activities of a landowning class who occupy all the productive land and grow cash crops for export upon it. This forces the peasants onto marginal lands which may erode quickly, depriving them of basic nutrition at the same time as their population levels far outrun any animal protein supply. Development programmes are most likely to misfire is such areas, not only because the limitations of the environment may be misunderstood and prediction of ecological side-effects inaccurate, but because of institutional factors centring around resistance to change on the part of the entrenched rich as well as the bewildered poor. This type can be labelled 'the Egyptian type' and includes part of southern Europe as well as classic LDCs in Africa like Ruanda-Burundi, Latin American and Caribbean areas like Haiti, Jamaica, Central America and the Andean highlands, India, Pakistan, Bangladesh, Sri Lanka, Syria and the islands of the Pacific and Indian Oceans. The UN has even designated a group within a group: the poorest nations of all—the Most Seriously Affected (MSAs). Escape from this category has been limited: Japan and China are outstanding examples; Iran looked as if it might be moving out of this category but the 1979 Revolution with its emphasis on the rejection of Western ideas has at any rate retarded this process.

Type E territories possess low resource bases because of the constraints of physical geography, and their populations are low, consisting either of people living on subsistence economies or those engaged in developing resources like minerals and oil. This 'arctic-desert' type also encompasses the oceans, only the fringes of which are subject to any form of inhabitation.

Management

The achievement of the economic growth espoused by most Western countries has usually been the result of private entrepreneurial ventures either aided or resisted by government. In centrally planned economies, the state has made all the running. But during the period since 1920 resource development has been increasingly subject to attempts to achieve an orderly, programmed rate of development, rather than submit to the irregular pyrotechnics of unfettered free enterprise.

The first stage is usually termed *resource management*. Attempts are made to optimize yields from a particular resource process by exerting governmental control or influence upon part of the resource process. Such influence may be used to bring about regional development, as in the New Deal period of the USA, or to realize the maximum use of a potential resource, as with water-impoundment schemes which supply water for irrigation, urban-industrial purposes, recreation, and for the generation of hydro-electric power. A feature of resource-management philosophies in recent years has been to try to minimize the impact of economic development upon people and ecosystems not directly parts of the particular resource process. At a simple level this may mean building fish ladders alongside dams to enable migratory salmonids to reach their spawning grounds, or screening a new electricity-generating plant with mounds of earth and tree-plantings. More complex, and usually expensive, ameliorative measures include the whole battery of technological devices for removing toxic or harmful contaminants from industrial wastes, or the construction of new sewage-treatment schemes for large urban areas. The underlying values behind the abstraction labelled resource management are clear. They are devoted to development and change rather than preservation of the natural state, and to expansive growth of the scale of economic activity. Since exponential growth has for so long been the norm, it accepts the values of such a phase as correct, even when admitting that there may be deleterious side-effects. Above all, the concept acknowledges no upper asymptote to development, either because it cannot contemplate the reality of such a feature or because it believes that technological and social change will make the idea redundant. This set of values leads inevitably to problems in the balance of population and resources.

The heightened awareness of the inter-relationships of biota and their inanimate surroundings has made possible the further step of identifying the wider concept of *environmental management* (Bennett and Chorley 1978, Holling 1978.) It is recognized that it is the bioenvironmental systems of the planet which provide resources and that any resource process must be rationally managed in order to ensure a sustained yield—

TABLE 12.11 Attitudes in environmental management

Purpose of environmental management	*Resource-population type*			
	USA	*European*	*Brazilian*	*Egyptian*
Useful, i.e. for materials	Dominant aim but decreasing emphasis	Dominant but decreasing; imports always significant	Dominant as industrialization proceeds	Basic development often for export to USA and European types
Life-supporting, i.e. food and ecological stability, gaseous exchange	Increasing realization of wider implications	Importance only recently realized; too densely populated to achieve 'natural' stability	Food important, otherwise environment not much valued, although knowledge of implications sometimes present	Struggle for food often dominant, knowledge of other processes discounted
Beauty	Strong motivation for preservation as reaction to over-use	Always a feature of culture; now getting stronger in face of increased impact	Residual or marginal land may thus be employed	Wildlife on residual land may hold interest, otherwise little valued

preferably one which is capable of due increase, but in which the existence of limits is recognized. In addition the deployment of an individual resource process must not be inimical to the operation of others with which it may share space or identical ecosystems. It is realized that the environment has simultaneously to be useful as a provider of materials, to be beautiful as a provider of recreation, wildlife and valued landscapes, and to be life-supporting as a provider of space, food and essential biological systems such as those which produce oxygen and carbon dioxide (O'Riordan 1971a). All things to all men, in fact, in which varying types of society exhibit different attitudes (Table 12.11) and determine their own orders of priority. It would be unreal to pretend that environmental management is at present much more than a concept except in the relatively simple situation of Antarctica, but the formation of Ministries and Departments of the Environment in some countries may encourage integrative planning. If, however, we consider process-response situations such as the avoidance of flood hazard (O'Riordan 1971b) as environmental management, then the number of instances may be increased.

The purpose of environmental management is to produce resources but simultaneously to retain a sanative, life-supporting environment. There is, therefore, an attempt to reconcile the demands of socio-economic systems with the constraints of the biosphere. To this end, long-term strategies are essential and are based on two aims:

the first is to reduce stress on ecosystems from contamination or over-use; the second is to pursue short-term strategies that are sufficiently flexible to preserve long-term options: no resource processes should be developed which bring about irreversible environmental changes. As a set of values, environmental management is ambivalent towards economic growth. It recognizes that there is an absolute limit to the materials and surface of the planet but sees no reason to prevent the use of the resources up to that limit provided that some ecological stability can be maintained, whether by preserving the natural systems or by increasing man-directed inputs of energy, matter and information.

Frameworks of resource management in developed countries

Within the United States and European types of resource–population relationships, the purposes of resource management differ between the free-market economies and the planned economies, i.e. between capitalist and socialist nations. In the free-market countries resource management comprises multiple aims: the provision of the material needs of the community by encouraging individuals to invest money and time and hence realize personal profits; the provision of environments for recreation and wildlife by protecting land from certain kinds of resource process; and the permitting of certain places to be used for waste disposal. Woven throughout the pattern are the roles of various levels of government: encouraging here, discouraging there and preventing everywhere.

In principle, governments aim at maximizing the general good of the community, (although they are often criticized for adopting the values of only one section of the population), and hence often will provide resources which individuals cannot profitably develop, such as nature conservation and countryside recreation. But the overall aim is economic growth, usually as measured by increases in GNP per capita. In this aim, there is little significant difference between free-market economies and Socialist centrally planned ones. In the latter, the government acts as sole entrepreneur and 'profits' are to be distributed over the entire population. Theoretically, integrated resource and even environmental management should occur, but the evidence seems to indicate that the USSR, for example, experiences problems of environmental contamination just as severely as the West (Gerasimov *et al.* 1971, Goldman 1971, Pryde 1972). More thorough Communism, as practised in the People's Republic of China, may have diminished contamination problems (Unger 1971); it will be interesting to see how China deals with the by-products of its industrialization programme. She adopted an Environmental Protection Law in September 1979.

Along with citizen groups, most governments have tried to mitigate the impacts of resource development (especially industry) by a variety of measures, while retaining the general aim of an expanding economy. Such activities are often referred to as 'conservation', although they fall well short of the integrative approach to man–environmental relationships which perhaps better deserves that label. Another name for tidying-up activities is 'the search for environmental quality' and a general attention to the landscape effects of industrialization and transportation is often a key element. Thus highway signs are regulated or eliminated, factories landscaped, and auto junkyards screened with trees.

A wider concern with the protection of valued landscapes and open space may also be part of the same movement, so that green belts around cities may be created, as around the major cities of Britain. Alternatively, urban growth may be directed along particular axes, with 'green wedges' in between them; ideally this brings undeveloped land close into the city centre. In many European countries, the motivation to set aside valued landscapes is part of the 'conservation' movement: the 'Naturparken' of Denmark and West Germany and the National Parks of England and Wales are examples. Along with such movements often goes the desire to provide more open space and access to resources for outdoor recreation, which can frequently be combined with landscape protection. Here again, the protagonists are often called 'conservationists'. Many of the State Parks of the USA, the Provincial Parks of Canada, and local arrangements such as the East Bay Regional Park District in California or the parks of the Metropolitan Toronto and Region Conservation Authority, are part of the same resource process. Much of the current effort in nature protection is also carried on in order to save biota and habitats from encroachment because of their national or regional significance in aesthetic or educational terms, rather than for their part in the stability of the biosphere on a global scale. The encouragement of a rare predator or wader to breed successfully on the margins of its range will elicit devotion and money that no unspectacular alga suffering from biocide toxicity can hope to receive.

The strongest of all these drives in the DCs is for pollution control. Many environmental contaminants are not only aesthetically undesirable but can be shown to be directly damaging to humans, and so in the industrial nations considerable legislation is being passed (and sometimes enforced) to control wastes. Water Quality Acts, such as those of the USA, set standards for the concentration of various effluents of a toxic or unaesthetic character. In the UK a standing Royal Commission on Environmental Pollution has been set up and all the water authorities are expected to deal with offences against the pollution laws. Noise is another area in which attempts to control levels are being made, in particular from aircraft and heavy vehicles. Few would deny that anti-pollution moves bring about improvements in the environmental quality for many people. However, the clamour is directed against one phase only of the resource process (the last) and does not consider the magnitude of and necessity for the earlier parts of the process.

To this extent pollution abatement, like the other operations discussed in this section, is truly cosmetic. It improves the surface, but if the ecology of the resource process is unstable then it will not be corrective. Indeed, it may hide the need for more radical re-orientations of resource processes and can thus be used as camouflage by entrepreneurs and politicians. An example of the cosmetic approach can be seen in the declaration by the European Conservation Conference, meeting in 1970 (Council of Europe 1971). The human rights proposed are those of having air and water 'reasonably free from pollution', freedom from noise, and access to coast and countryside. Member nations are urged to combat pollution and, somewhat vaguely, to ensure the conservation of the European environment. Individuals, states article 30, should be ready to pay the costs of conservation. The other provisions largely concern planning for 'rational' use of resources and the reduction of pollution, especially the unwanted effects of the internal combustion engine, jet aircraft and chemicals such as pesticides, fertilizers and detergents. The aims and desirability of population and industrial growth are nowhere mentioned and the whole document is clearly and acceptance of current trends, subject to some improvements in national and regional planning and cleaning up of the more serious pollutants.

The next step beyond cosmetic procedures is to try to assess the ecological impact of a proposed development before it is put into effect. If the percussive effect of the change is deemed unacceptable (by standards that are not usually defined objectively but emerge empirically from public quasi-judicial proceedings) then another site is sought. Indeed, any enquiry may be specifically directed to choose between limited possibilities, as was the Roskill Commission of 1971 on the site of London's third airport. Symbolically, however, this Commission was not allowed to investigate whether London ought to have another major airfield, merely where it should be, so that only one phase of the resource process was considered. In the same way, the enquiry in 1977 into the extension of the nuclear fuel reprocessing plant at Windscale in England, was not supposed to cover the need of the UK either to produce more electricity or to reprocess wastes from, for example, Japan.

In the USA the 1969 Environmental Protection Act sets forth environmental policies for the nation which go beyond the loosely formulated suggestions of the European Declaration quoted above. For example, one of the objectives for the USA is to 'achieve a balance between population and resource use which will permit high standards of living and a wide sharing of life's amenities'. As far as actions by the Federal Government are concerned, it must 'utilize a systematic, interdisciplinary approach which will insure the integrated use of the natural and social sciences and the environmental design acts in planning and decision-making which may have an effect upon man's environment'. Any legislation or other Federal action which affects the quality of the human environment has to be investigated for its environmental impact and any adverse effects pointed out, along with alternatives to the proposed action; the relations between local and short-term uses of the environment and the maintenance of long-term productivity must be stated too. This Act tries therefore to impose limitations on the magnitude of resource processes if there are environmentally detrimental effects (Caldwell 1971). Such Environmental Impact Statements are now required at State and lower levels, and the legal process has spread to other nations in the developed world (e.g. SCOPE 1979, Golden et al 1978).

Nations with well-established Socialist governments such as those of the USSR and eastern Europe have theoretically an institutional structure which will enable them to avoid the environmentally stressful features of free-market economics. Yet in performance they rarely seem any better and sometimes manifestly worse. A variety of reasons seems to account for such a position. Firstly, although integrated control of a resource process should exist, *de facto* it often does not. Secondly, resource managers are as keen as any capitalist to dispose of their wastes as cheaply as possible and thus externalize the costs. Because there is an ideological attachment to industrial and general economic growth, this cannot be slowed down because of difficulties over aesthetics or contamination. Lastly, because Marxism is held to be so superior to capitalism, it is hard to admit that the end-product may be the same and that Marxist-Leninist fish are as dead as capitalist fish. Socialist governments may well outshine the Western nations at tasks like pollution control when they have decided to undertake the job; in fields like nature conservation their record is excellent. We may surmise nevertheless that Socialist nations will find it harder than most to deviate from goals of perpetual economic growth, particularly while they feel militarily threatened.

Resource management in Brazilian and Egyptian categories

The purposes of the development of resources in these two groups are twofold. Firstly, they are used for regional and national purposes in order to produce metabolic and cultural materials such as food, housing and roads, and also to provide jobs and create wealth to be used as capital for further development. Secondly, many bioenvironmental resources are developed for export to European and US type countries. Oil is an obvious instance, but many metal ores and crops like cocoa, coffee and rubber also enter this category. Such use brings income to the supplier, but it also brings dependence upon demand in the DCs; where the product is dispensable or subject to fashion or is overproduced, then the fortunes of the producers (as happens with cocoa and coffee) are fickle indeed; on the other hand where the product is so important, as with oil, that the industrial nations are virtually dependent upon the supplies, then the LDCs force up the price and nationalize expensive plants virtually with impunity.

The desire of the countries of the Third World to be masters of their own destinies leads them into the process of 'development', undertaken with substantial assistance from the richer nations. This is usually undertaken in response to particular exigencies such as under- or malnutrition or a chronic lack of employment or shelter, or to ensure the occupational survival of a political leader. As Caldwell (1971) points out, the process of development proceeds from a set of assumptions about the relationship of man and nature which were developed in the West during the phase of the industrial revolution and few if any of which (particularly those about the long-term effects on the stability of ecosystems of constantly increasing impacts of a technology based on fossil fuels) are verified by scientific evidence. Indeed, some writers aver that development is primarily for the benefit of the industrial nations: Woodis (1971) says that in one of its operations in an African country, a rubber company takes home three times as much profit as the nation's entire revenue; that 28 per cent of UK overseas aid goes to pay back interest on former aid; and that the drop of 15 per cent in export prices of raw materials from tropical Africa in 1955–9 entailed a loss of twice the annual amount of foreign aid. Some large multi-national companies have annual economies many times the size of some small newly independent nations; it has even been rumoured that a corporation thought about buying one such country and running it as a tourist enterprise, having renamed it Tarzania.

Given that even the most exploitive resource-development programme will leave a residue of investment capital, or that disinterested aid from an international agency has been granted, development is not without hazard. The assumptions referred to above ensure the dominance of technical and economic factors in evaluating priorities at the expense of behavioural and ecological considerations (e.g. Day and Singh 1979). In the desire to transform simple agrarian societies into more complex industrially based economies, it may be forgotten that both traditional systems and natural ecosystems have passed the evolutionary tests of selection for survival but that the new ecosystems and new cultural values are not guaranteed to share such properties. The more interference with the old order, the greater the chance of unpredictable synergistic effects, and the success of a practice in one place may encourage unwarranted optimism about its potential in another. The hazards of the imposition of cultural complexes based on high-energy societies upon those accustomed to the flow of only solar energy, or for example the development on small islands of mining by techniques usually employed on large continents, may be disastrous: Taghi Farvar and Milton (1972) bring together a large number of unhappy case-histories; and in the same volume Caldwell considers that there are six barriers to success in 'development', only one of which is ecological, the others being derived from various parts of the cultural milieu, including the political. Development thus provides us with an example of a situation where, within the ecological envelope of facts, the cultural world of values is all-important and cannot in the least be ignored, a conclusion of significance when considering any alterations in the nineteenth-century assumptions which underlie so much of our thinking about man and nature. In the meantime many millions of individual people live far below their full potential as humans (Jequier 1976, Gonzalez 1978, Pearson and Pryor 1978, MacAndews and Sen 1979, Mayur 1979, Bassow 1979).

The area in which outside ideas can claim a little success is nature conservation, especially where this is a source of revenue from tourism. More wide-ranging environmental protection is virtually absent, with the possible exception of forest reserves established to prevent the denudation of steep hillsides or to safeguard a supply of wood for construction and fuel. Long-run ecological stability with no visible benefits is therefore, and understandably so, subordinate to economic growth in the short-term.

The 'radical alternative' to this view of development is based on a 'small is beautiful' rational. It stresses local self-sufficiency on a labour-intensive basis, the use of 'soft' or 'appropriate' technology which will not cause environmental stress and is non-capital-intensive but which will employ a lot of people, and a re-valuation of traditional cultures and ways (Dworkin 1972, Dunn 1978, Norman 1978). The whole is sometimes summed up as 'eco-development'. Though there are many pockets of it in LDCs few governments have adopted as national policy a programme which eschews Western ideas of development; perhaps the nearest are Papua-New Guinea (Winslow 1977) and the rural self-sufficiency programme in Tanzania. It is noteable, too, that in the late 1970's India reversed its priorities and put the development of agricultural self-reliance first. In general, the value of self-sufficient communities of a more or less permanent kind is being re-assessed and found to be much higher than Western culture has placed it in recent times (Clarke 1977).

Really renewable resources

We need to remind ourselves at this point that although stocks of many materials are very large and it is possible to recycle them, that the only really renewable resources are those created by living organisms (and especially as Net Primary Productivity), and that out most basic need is to eat, and to drink. Putting these two together in terms of NPP/capita and the food-producing or purchasing ability of a nation, as has been done by S. R. Eyre in his *The Real Wealth of Nations* (1978), produces a quite different regionalization from those discussed so far. His Group I consists of nations with high potential NPP/capita and a good mineral base: of 31 constituents, most are in the tropics (e.g. Brazil, Venezuela, Zambia, Guyana) with a few temperate representatives like Canada and Australia. A second group includes the USA, USSR, Sweden, South Africa, Malaysia and Chile; a third group is mostly dry or cold countries but with low populations (Botswana, Namibia, Saudi Arabia and Iceland for example). Group IV (23 nations) have rather low NPP/cap figures and in general low mineral production/capita: Portugal, Nepal, France, Denmark, China and Czechoslovakia are included. The last group have a small mineral income and a high density of population and so are the worst off. Jamaica, Japan, most of western and central Europe, Egypt, and Hong Kong are included. This regionalization excludes the effects of trade and the present economic order but is without doubt significant in terms of what basic renewable resources and food-purchasing power are, in the last analysis, available to nation-states.

Growth and progress

It seems pertinent to remind ourselves that growth of economies and populations on the scale to which we are accustomed is a feature of the last 150 years only. But expansion rates of 5 per cent p.a. and above have come to be the normal situation for resource managers, their mentors, and of those who laid down the ground rules for both of them. Because growth has been a normal situation and because it appears to offer solutions to most problems, including those it has itself created, it has become almost everywhere a desirable goal and indeed equated to a large extent with the concept of 'progress'. Earlier in this century some questioning of the equivalence was heard, albeit faintly: one character in Aldous Huxley's *Point Counter Point* (1928) explodes, 'Progress! You politicians are always talking about it. As though it were going to last. Indefinitely. More motors, more babies, more food, more advertising, more everything, for ever. . . .' and Huxley's ideas have a contemporary air. Whether, where and how far the expansion of the magnitudes of resource processes can follow their present trajectories into the future is the subject of the last chapter of this book.

Further reading

ALLISON, A. (ed.) 1970: *Population control.*
ALLISON, L. 1974: *Environmental planning.*
BENJAMIN, B., COX, P. R. and PEEL, J. (eds.) 1973: *Resources and population.*
BISWAS, A. K. and M. R. 1976: State of the environment and its implication to resource policy development.
BROOKFIELD, H. C. 1975: *Interdependent development.*
BROWN, L. R. 1976: *World population trends: signs of hope, signs of stress.*
——— 1979: *Resource trends and population policy: a time for reassessment.*
DASMANN, R. F., MILTON, J. P. and FREEMAN, P. H. 1973: *Ecological principles for economic development.*
DICKSON, D. 1974: *Alternative technology.*
DUNN, P. D. 1978: *Appropriate technology.*
DWORKIN, D. M. (ed.) 1974: *Environment and development.*
GÜRKAYNAK, M. R. and AYHAN LECOMPTE, W. (eds.) 1979: *Human consequences of crowding.*
HOLLING, C. S. (ed.) 1978: *Adaptive environmental assessment and control.*
KELLEY, D. R., STUNKEL, K. R. and WESCOTT, R. R. 1976: *The economic superpowers and the environment.*
MACANDREWS, C. and CHIA LIN SIEN (eds.) 1977: *Developing economies and the environment. The Southeast Asian experience.*
MAULDIN, W. P. 1980: *Population trends and prospects.*
O'RIORDAN, T. 1971: *Perspectives on resource management.*
PEACH, W. N. and CONSTANTIN, J. A. 1972: *Zimmermann's World Resources and Industries.* 3rd edn.
PEARSON, C. and PRYOR, A. 1978: *Environment: north and south. An economic interpretation.*
PRYDE, P. R. 1972: *Conservation in the Soviet Union.*
REINING, P. and TINKER, I. (eds.) 1975: *Population: dynamics, ethics and policy.*
RIDKER, R. G. 1972: Population and pollution in the United States.
SAI, F. T. 1977: *Population and national development—the dilemma of developing countries.*
STRONG, M. F. (ed.) 1972: *Environment and development.*
STYCOS, J. M. 1974: The population debate: colours to the mast.
TAGHI FARVAR, M. and MILTON, J. P. (eds.) 1972: *The careless technology: ecology and international development.*
TEITELBAUM, M. S. 1974: Population and development: is a consensus possible?
VANN, A. and ROGERS, P. (eds.) 1972: *Human ecology and world development.*
ZELINSKY, W. 1966: *A prologue to population geography.*

13

An environmental revolution?

The last chapter of this book first examines the extension of the present patterns of resource processes, in particular the environmental and social problems which are created, and then moves to consider two main sets of reactions to the disharmonies which are evident. The first of these is the argument that technological development will eventually provide solutions; the second, by contrast, advocates a radically different approach to the relations of man and nature. The book ends with a consideration of alternative models of the future based largely upon these two types of reaction.

Impacts upon the environment

There is no doubt that an increasing disharmony between man and nature is becoming apparent, especially in phenomena such as malnutrition, soil erosion, gross pollution, and the attrition of the aesthetic qualities of parts of the environment which are valued in several cultures. One of the major concerns is that the increasing magnitude of resource processes is creating a set of environmental problems which in turn may impair not only the usefulness of the environment but also its life-supporting capability, its ability to absorb wastes and its beauty. The environmental problems created by more people using more materials can be divided into those with an environmental linkage, and those with a largely social linkage (Russell and Landsberg 1971). The former group in turn comprises regional problems such as sewage, sulphur dioxide fallout, the habitat requirements of migratory birds, or particular geographical entities such as the Rhine or the Nile and the uses made of them.

Much stronger anxiety, however, has been expressed about global problems such as food supply and the consequences of agricultural intensification, residual pesticides, the effects of the contamination of the oceans by oil, and the alteration of atmospheric processes by increased loads of carbon dioxide and particulates. The most vivid statement of this view is Ehrlich's (1971) scenario for 'ecocatastrophe' in which poisoning of the oceans by organochlorine pesticides reduces food supplies to Asia at the same time as biocide-resistant strains of pest virtually eliminate the land-based food supplies. The demand for food of a very high and rapidly growing population is the trigger-factor of a worldwide nuclear war which he predicted would take place by 1979, but which happily had not come about at the time of going to press. A more penetrating analysis which emphasizes the role of population growth in creating environmental disharmony is given by Ehrlich and Holdren (1971). They suggest that the increases in human numbers have caused a totally disproportionate impact on environment. For example, the provision of minerals and fossil fuels to an expanding population even at fixed levels of consumption requires that as the nearest and richest ores are worked out, then the use of lower-grade ores, deeper drilling and extended supply networks all increase the per capita use of energy and hence the per capita impact on the environment which is symbolized by the increase in the use of oil (Table 13.1) per person even during the years of such rapid population growth.

TABLE 13.1 World per capital use of oil

	Barrels
1910	0·19
1930	0·70
1950	1·52
1960	2·58
1978	5·23

Source: Brown 1979; McHale 1972

Similarly, the environmental impact of supplying water needs rises dramatically when the local supply is outrun: ecological, aesthetic and economic costs are incurred in diverting supplies to the growing region. Increase of food production likewise requires energy and material uses disproportionate to the population fed because of the need to obtain and distribute water, fertilizer and pesticides (see pp. 285–6). Some indication of the environmental consequences of all the additional inputs required to bring about a modest increase in per capita food supply (Table 13.2) can be inferred from previous chapters.

TABLE 13.2 World increases in agricultural activity

Increase, %	*1955–65*		*1965–1976*		
Tractors in use	63		48		
Nitrogenous fertilizers	146		300		
Phosphate fertilizers	75		210		
Food production index	1967	1970	1975	1976	$\frac{1976}{1967}$%
(1961–65 = 100)	113	120	136	140	23
Total agricultural production	112	119	134	137	22
Food production per capita	1966	1970	1975	1977	
(1970 = 100)	97	100	103	103	6

Source: Various UN data

The role of population density is also relevant, since some proponents of growth argue that countries with a low density can afford high rates of growth. Density is not necessarily a good criterion for the effect of populations upon the biosphere: industrial nations, for example, gather in resources from a very wide area. The Netherlands is often quoted as an example of a very dense but wealthy population, but it is the second largest importer of protein in the world and also imports the equivalent of 27 million t/yr of coal. In addition, many environmental problems are independent of the distribution of population, especially those involving contamination of the oceans and atmosphere.

A cooler, more detailed view is taken by the authors of the SCEP (Study of Critical Environmental Problems) group (1970), who conclude that the current 'ecological demand' (i.e. the stress put upon planetary bioenvironmental systems by man's demands upon them) is not yet sufficiently great to cause a breakdown. Nevertheless, they point out the rapid rises in demands for materials, energy and space, and by considering Gross Domestic Product (GDP) minus services as an index of 'ecological demand' (IED), they calculated that there was an increase of 5–6 per cent p.a. in the period 1950–1963, a level which as Table 13.3 shows has been sustained into the late 1970's, with perhaps a hint of a slow-down in more recent years. But at 5 per cent p.a., the next doubling of population (in 41 years at 1·7 per cent p.a.) would increase the IED by about six-fold.

TABLE 13.3 Index of ecological demand

	GDP minus services, 1970 = 100	
	Overall	*Per capita*
1960	59	71
1965	77	85
1970	100	100
1975	121	110
1976	128	115

Source: *UN Statistical Yearbook* 1977

These rates explain why environmental problems appear to have erupted so suddenly and why many students fear that the future will bring more problems than exist at present to the point where a few more doublings of population will bring about the likelihood of the breakdown of the systems of the biosphere: a lily which grows from insignificance to complete cover of the pond in thirty days has only covered half the water surface on the twenty-ninth day.

The differences between Ehrlich and SCEP, therefore, is only one of time: they agree that the planet cannot for much longer go on supplying the demands made on it for the supply of materials and the provision of valued environments, while absorbing the impact of the disposal of wastes, all within the framework of a human population doubling every 35 years.

The role of population growth in creating environmental problems has been closely examined by Commoner (1972a, 1972b), who suggests that the misapplication of technology is the most important factor in the growth of environmental disharmony. In the United States, for example, crop yield in Illinois increased 10–15 per cent in 1962–8 while the quantity of nitrogenous fertilizer doubled; similarly for non-degradable detergents and non-returnable bottles Commoner calculates that it is the technology that creates the difficulties, not the rise in the numbers using it. While there is no doubt that 'ecologically faulty' technology exists, it seems that none of these arguments meet those of Ehrlich and Holdren (1971) quoted above, and they manifestly do not apply to those parts of the world where technology is distinctly lacking; to some extent Commoner appears to be lacking a wider perspective beyond the particular situation of the USA. More 'appropriate technology' for DC's would certainly be a help, however.

Thus while not every environmental ill can immediately be laid at the door of population growth, there emerges a strong suggestion that nowhere does the latter bring any benefits. To LDCs it brings medical and social problems; to DCs it may bring environmental problems beyond those of aesthetics; and in both groups, high densities of population add their own special difficulties.

Social linkages

The environmental disharmonies caused by rises in the scale of resource processes are paralleled by some negations of the social benefits which the growth has brought to most people. Economic growth expressed as GDP and Gross National Product (GNP) appears not to solve social problems, even of privileged areas. In the USA, evidence of primary malnutrition may be found among the poorer people; in less specific terms economic growth does little to narrow the gap between rich and poor in technologically based societies. If real net income rises by 10 per cent for the entire population, then the differences between the well-off and the deprived remain exactly the same as before; furthermore, the hard core of the poor do not share in economic growth and are dependent upon handouts from those whose benefit is direct. In DCs, economic growth is not really necessary to remove the very poor from their predicament: a minor redistribution of wealth (e.g. the reduction of the military budget by 25 per cent would achieve the desired end (Mishan 1971)). Indeed, economic growth which goes into military expenditure might well be employed in a more useful fashion: one atomic submarine and its missiles would pay for $150 million in technical aid and one aircraft carrier could be replaced by 12,000 secondary schools (McHale 1972).

Another social argument advanced in favour of economic growth is that it produces more goods so that individuals and societies have more choice, which contributes to greater economic welfare. Such enhanced freedom tends to be illusory because there is constant replacement of models: one cannot choose a car from all the types ever made, only from those currently on the market, and the difference between alternative makes at the same price is more or less negligible. The proliferation of choice may bring forward harmful products which commercial interests have not properly evaluated, and much of it is not necessary: a choice between 50 makes of transistor radios at roughly the same cost is scarcely essential. Lastly, satisfaction with earnings may depend more upon an individual's place within the income structure than with his absolute level of wealth. Trade unions seem often to be most concerned with 'preserving the differentials'. A penetrating analysis of the social limits to growth in industrialized societies is made by Hirsch (1977).

The greatest externality of economic growth as measured now is its failure to take account of its environmental impact, largely because a money value can not be placed on environmental values lost. This book has chronicled numerous side-effects of economic growth, none of which will feature in the balance-sheets of those who caused them. Some measures of total goods and services such as GDP actually count deleterious impacts as components of the growth: if because of industrial emissions more people become ill, then the costs of extra hospital building and medical equipment adds to the GDP; if they die as a result of the same factor then presumably the undertaker's fees are added too. Such methods of accounting also omit the hidden costs of loss of earnings due to environmentally related illness. As was said in Part I, Boulding (1971) points out that GNP is simply a measure of decay (of food, clothing, gadgets and gasoline, as they are used), and the bigger the economic system the more it decays and the more that has to be produced simply to maintain it.

Since economic growth depends upon the extension of the use of machines, many humanitarian writers (of which the most well-known is probably J. Ellul (1964)) have stressed that technology provides its own imperatives, some of which lead to attrition of the quality of human relations; electronic means of communication are probably the most sinister of these since, for example, 'the boss' can be isolated entirely from 'the worker' when audiovisual links are effected. Similarly the social worker may never have actually to see or to smell the derelict, and the doctor will be able to diagnose his patient from a safe distance. Paradoxically, the same media could bring about a sense of closeness on a world scale, the 'global village' to which McLuhan refers. The large size of organizations needed to control resource processes inevitably means impersonality and alienation, and the flexing of corporate power (of the 50 largest economic entities in the world 9 are corporations; in a recent year General Motors grossed more money than the total GNP of Switzerland, and Standard Oil more than that of Denmark) as in the case of a company which allegedly offered large rewards to anybody who could prevent the election of a particular politician in a LDC. One very pessimistic writer (Wesley 1974) thinks that machines will make all carbon-based life virtually extinct to the point where man's ecological role will be reduced to one of 'vanishing importance.'

More strictly relevant to our present theme is the rapid obsolescence which is associated with the products of machine technology. The social consequences are obvious since people will spend time and energy trying to possess the latest model, either willingly or at the behest of advertising. Probably more important is the heavy demand upon resources which such changes make, especially in an open-flow resource process without any recycling. Planned obsolescence is therefore a development which, although perhaps making sense (and a good deal of money for somebody) in terms of contemporary economics, is inimical to rational resource and environmental management.

The environmental changes wrought by industrial growth have brought with them effects on human health. Not only are there the occupational illnesses and hazards of high-risk groups but also a more general suite of diseases which are associated with the developed countries: coronary heart disease, cancer of the large intestine and lung, diabetes, obesity and dental caries are among the list which has replaced the infectious illnesses so common in the LDC's. Multiple causation is possible, but lack of exercise and a diet high in animal fats and processed, fibre-free, foods seem obvious factors in many of these illnesses (Burkitt 1973).

Social consequences of urbanization

Economic growth is very largely industrial growth, since advanced agriculture is an outgrowth of industry, dependent upon it for chemical inputs and for power supply, and industrial activity is largely carried on in cities and conurbations. The role of the city and its life-styles cannot therefore be overlooked in any consideration of the future of resource processes.

The major externality of industrial growth is its environmental effect, already discussed. But there are other effects which, assessed in social terms, are part of the unrest which causes the seeking of alternative ways of living. Behind the anxieties lies the thought that the transition to an urban-industrial life-style, after spending 90 per cent of his evolutionary history as a hunter, may set up psychic strains in man. Many of the traits which conferred a selective advantage in the hunting stage are now of distinctly less value, especially since population densities are so high in cities: aggressiveness is one such characteristic. Many of the biological mechanisms of adaptation to environmental struggles and stresses are now useless, and this is posing its own threat because there are insufficient challenges in bland surroundings which require little effort and thus provide few outlets for traits like aggression except in ritualized forms such as spectator sports (Pugh 1978). On the face of it, man appears to be a very adaptable species, for the overall health and survival rates of the inhabitants of Western cities are very good even though the people are isolated from nature. The price of such successful biological adaptability is, according to Dubos (1967), paid in social terms: people no longer mind ugliness, exhaust fumes and other contaminants, and even regard such conditions as normal. Existence of these ideas suggests that because so many industrial and urban conditions do not immediately threaten biological survival, then we can give up perceiving the objective reality of their true effects, both in terms of environmental impact and in living conditions for man which are, to say the least, sub-optimal. Fortunately, our adaptability is cultural as well as biological and so it seeks to encompass the future as well as acknowledge the past.

A good deal of research and argument has been conducted over whether or not the modern city is a good habitat for man. (Plate 20, page 357). On the face of it, crowding and sensory overload (Milgram 1970, Calhoun 1971) might be expected to produce social pathologies (as happened with Calhoun's (1962) rats) like stress, mental health breakdown, and crime, but the problem is more complex than that (Gürkaynak et al 1979) since some cultures produce very large cities with low crime rates, such as Japan. Apparently the characteristics of the culture can be intensified by crowding, one way or the other. Nevertheless it appears that certain types of building attract more crime than

TABLE 13.4 Magnitude of city metabolism (daily flows in tonnes of a city of 1 million population, USA)

Input					*Output*		
Water			615,157	BECOMES	Sewage		492,125
	Coal	2,952			Refuse		1,968
	Oil	2,755					
Fuels			9,349		Air pollutants	CO	443
	Nat. gas	2,657				SO_2	148
	Motor	985				Particulates	148
						NO_x	98
Food			2,000			Hydrocarbons	98

Source: McHale 1972

others, especially where there is public space not overlooked by any inhabitants, and very large cities (>500,000 population) seem to have higher crime rates than smaller places. But other factors may be at work here: at a larger scale, for example, it looks as if homicide rates are highest in the poor and the very rich countries but with an area of very low rates in the moderately affluent nations (Fig. 13.1). It can be said in favour of the city, too, that it has provided a centre for innovative behaviour of kinds which would be too conspicuous for adoption in most rural environments: many reform groups, for example, start in cities.

One incontrovertible drawback of cities is their ecological instability, dependent as they are upon inputs of food, water and power, and upon having their wastes removed (Table 13.4). Weiner (1950) points out that disconnection of the water supply to New York City for six hours is reflected in the death rate, while power disconnections cause hypothermia among the old and enhanced pregnancy rates in the fertile. In such a sense cities appear parastic, but many dwellers in rural areas are similarly dependent upon supplies bought by money: the farmers and smallholders who might survive a real famine would be overrun by urban hordes long before harvest time, whatever the crop, so that in reality the two are interdependent. The city is therefore ecologically difficult (and, on the whole, anathema to ecologists) because it masses demands for all types of resources, and generates others external to it, such as countryside recreation. It concentrates wastes, the disposal of which may create high levels of contamination of ecosystems (Table 13.4)

Probably none of these problems is intractable, and a fruitful field for technological development exists in the neutralization of the deleterious processes. Socially the city is ambivalent, with some evidence that very large agglomerations may enhance socially undesirable behaviour. The city is, however, a centre of innovative thought and hence is in the vanguard of the study of both its own deficiencies and those of the economic growth which caused it to arise; many of the new approaches to man-environment relations have their origins in the cultural and intellectual ferment of urban conglomerations.

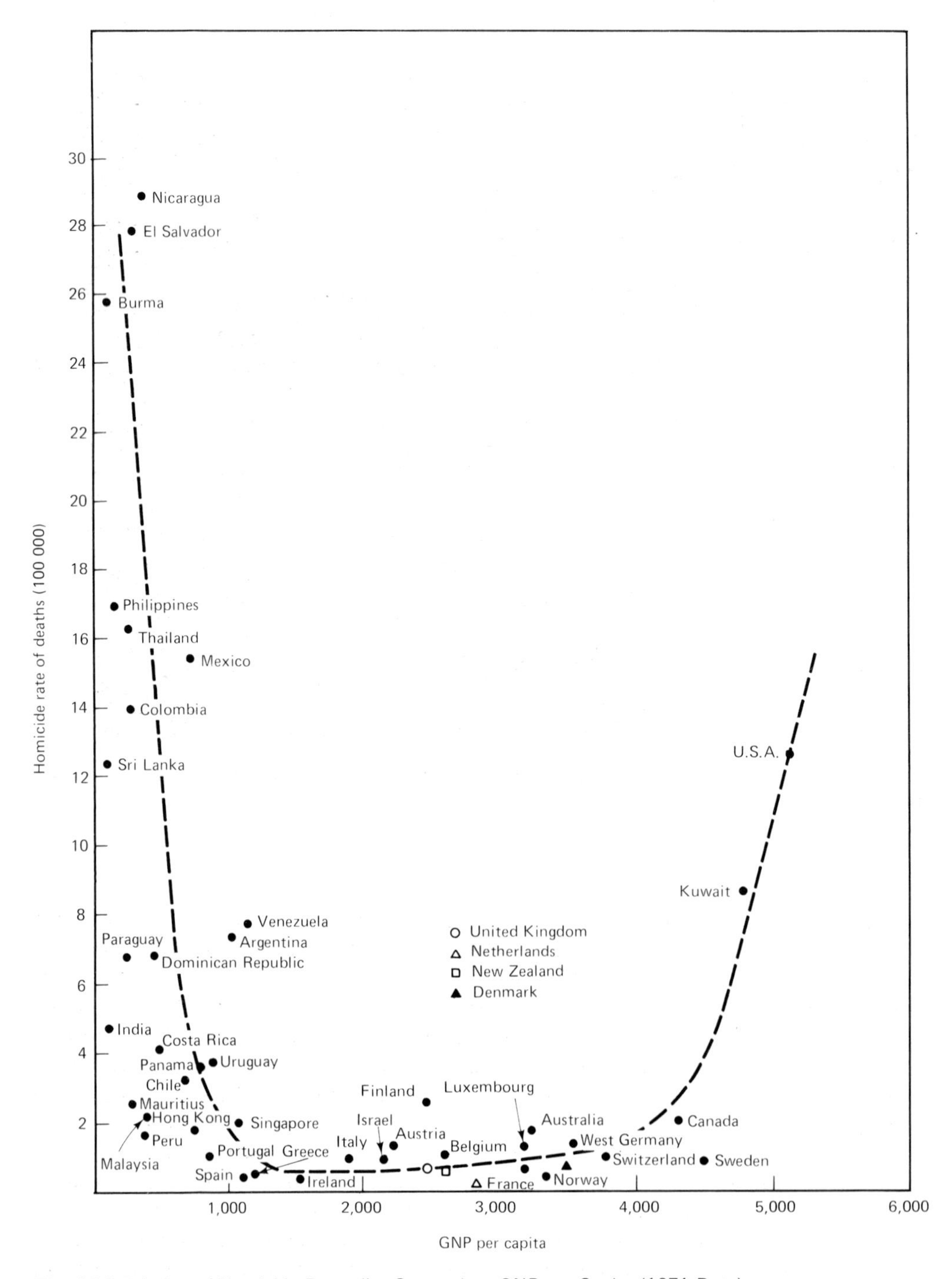

Fig. 13.1 Relation of Homicide Rates (by Country) to GNP per Capita (1971 Data)
Note: The line on this figure is the original author's, and is not statistically fitted.
Source: Watt et al 1977

Plate 20 The negative qualities of city life can be exemplified by the famous *Oshiya* (pushing boys) of the Tokyo subway during rush-hours. *(Keystone Ltd, London)*

Ethical views

A further aspect of the feedback from our social and environmental situation is the failure of man's use of the earth to measure up to certain ethical principles which have been principally formulated, in this context, in the West. The first of these may be summarized as the 'duty to posterity' argument. Its basis is the exhortation to pass on the inheritance of one generation in an unimpaired condition to the next. In terms of resource processes this means the avoidance of foreclosing options for the future by making irreversible changes in the present. Obedience to such a dictum would appear to be ecologically sound as well, since it would have the effect of retaining maximum biological diversity. The difficulty comes with acknowledgement of the effects of increasing technological capability, which often removes the closure imposed at an earlier time or, by creating substitutes, throws the whole resource process into irrelevance. So it has been argued that since we cannot know what the future will hold, we automatically do the best for future generations by maximizing present benefits since this will build up capital in terms of wealth and knowledge which will enable future problems to be tackled successfully. Both versions are perhaps rather too exclusively economic in their orientation to find much favour with either ecologists or the promoters of ethics, who tend to be rather puritan by nature (Boulding 1970).

A second ethical theme is the idea of each generation of men as stewards of the earth who hold it for only a limited period and who are obliged to account for their tenure. This is basically a theistic idea, although a demythologized version seems to have a fairly wide secular appeal. One of its statements is in Aldo Leopold's (1949) 'land ethic' in which men are adjured to respect the qualities of the earth and to gain their living from it without damaging the chances of fruitful yield in the future. Although the formulation of his exposition derives rather strongly from its origin in erosion and flooding problems in north-central North America, the general idea could be applied to resource processes as a whole, except that the criteria for what constitutes acceptable environmental manipulation as distinct from damage are left undefined.

Christian writers such as Black (1970) and Montefiore (1971) emphasize the idea of man as a temporary steward of the earth and the necessity of avoiding 'over-use' (Taylor 1975), and also suggest that nature must be one of the objects of disinterested love which they are enjoined to profess. A concise and sensitive discussion of most of the Western arguments, their relations with science, and everyday realities, is provided by Passmore (1980). Toynbee (1972) even went so far as to suggest that monotheistic religions encourage environmental exploitation and that only a reversion to pantheism and the religions of the East will suffice to alter present values. Unless instant conversions are made, we may harbour a suspicion that, even if efficacious, the time-span involved would be too long to alleviate some of the most pressing difficulties. In so far as the social costs of industrialization have been ignored in both capitalist and socialist countries, the so-called 'work ethic' has been a contributory factor, and as Bruhn (1972) suggests, it is time for ethics and science to unite to produce a guide to individual and collective behaviour which will stress the quality of life rather than the mere production of goods in the DCs; inevitably, the importance of 'work' (which many might suspect to have been strongly implanted by Protestant capitalism in the nineteenth century) would require radical re-evaluation. Ironically, this might come about through technological developments such as the microprocessor rather than from exhortations in the quality Sunday newspapers or the cultural TV channel where there is one.

To consider a last creed, Marxists have not been strong protagonists of a unique body of environmental ideas and indeed have been largely defensive since many other writers have suggested that Marxism and free-market capitalism are both basically nineteenth-century materialist world-views which will equally result in environmental degradation (Daly 1971, Fry 1976).

Summary

The social difficulties created by continued expansion of present-day processes in the DCs are concisely summarized by Wagar (1970). In the first place we cannot be certain that present arrangements will go on, since neither free-market nor centrally planned economies have as yet developed systems capable of rapid effective response to environmental attrition; rather, resource management by crisis has been the rule. Technological unemployment threatens domestic tranquility and there is a reliance upon defence industries for growth which might not be sustained. Again, the levels of organization are so complex that chain reactions of failure are very easy; and lastly growth can be sustained only by disproportionate inputs of energy and materials whose environmental impact has become clear (Brown 1978a).

The most vigorous statement of this type of view was made by Meadows et al (1972) as the result to their computer-based modelling of world population-resource-environment futures. Their results suggested that curtailment of present trajectories needed to be undertaken quite soon if resource depletion and environmental instability were to be minimized. They were strongly attacked (e.g. Cole et al 1973, Curnow 1975) by those who saw either various flaws in the modelling (e.g. aggregated parameters) or who employed the technique of the intricate defensive. The general conclusion, that economic growth of the type with which we are familiar, cannot go on for ever, seems uncontroversial enough. In a longer perspective perhaps the main value of this work has been to sound the trumpet and at the very least recruit a whole new generation of thinkers and modellers who will be able to point out alternative futures with greater accuracy (Fowles 1978).

A great number of commentators are agreed that all is not well with the man–environment relationship and that without some changes, troubles are likely. There are, however, considerable differences in views as to what should be the first priorities and, more significantly, the long-term purpose of different directions. For the purposes of discussion we shall classify the responses of our problems into two groups. The first allows that many problems exist but suggests that continual economic growth and the creation and widespread distribution of more wealth and technology will provide sufficient amelioration. The second requires a reorientation of material use and a greater concern for the functioning of the natural systems of the biosphere. The first approach will be labelled 'the technological fix' and the second, the 'environmentalist viewpoint'; here they are necessarily presented as black-and-white alternatives without the intermediate shades which inevitably occur in real life.

The technological fix

This set of ideas admits that there are significant problems in the relations of population, resources and environment, but suggests that the difficulties are correctable by only relatively minor alterations in technology and institutions. Prognostications of ecological instability are discounted, and to Maddox (1972b), the state of mind of the catastrophists is more dangerous than what they predict.

The importance of population growth as a factor in problems of imbalance is subject to close scrutiny. An eventual levelling-off is regarded as desirable (and often as a by-product of other processes such as economic development), but the components of growth may be viewed differentially: for some regions a drastic retardation is required, for others no particular action seems necessary. Notably, the argument that the DCs should come quickly to zero growth because they dominate such a large proportion of the world's resource processes is not admitted as valid, since it is their growth which brings benefits for everybody else in its wake. In the DCs it is pointed out that many environmental problems are caused less by population growth than by the misapplication of technology by those in search of quick and high profits, such as the manufacturers of detergents, pesticides, chemical fertilizers, large overpowered automobiles, flip-top beer cans and electric carving knives. Commoner's (1972a) argument, that the ecologically unsound developments of the past two decades would have created severe trouble even if the population had been stable, is usually quoted, although it is apparent that, once launched, such developments can be fuelled by higher levels of population.

Reliance upon economic development to alleviate resource-related difficulties is especially strong in the LDCs, with obvious cause. Some also see higher populations as an eventual strength rather than a problem and may rely on political change to bring about better material conditions. In a wider perspective, they are not willing to be the environmental saviours of the DCs by acting as reservoirs of oxygen-producing plants or picturesquely backward tourist spots (Castro 1972). Many see reduced rates of population and economic growth in the DCs as deleterious (Wolfers 1971, Sills 1973), since demand for their raw materials is thereby lessened; it seems, however, a curious world in which the rich must consume ever more to support the poor, and the implications of the political thralldom such linkages bring are not lost on many LDCs.

Economists who are also optimists point out that scarcity of any material resource will enforce substitution for it, but they note too that alleviation of shortages may be slow, and the ability of the price mechanism to give adequate forewarnings of the need for the development of substitutes in the face of the rapidity and magnitude of the doublings of demand now taking place has not yet been tested. Once a shortage is evident and there is an expectation that prices will rise faster than the rate of interest, then it pays a resource manager to stop production altogether, bringing about an unstable situation. Another improvement which can be made possible by economics is the internalizing of costs, particularly in the case of pollution control, and free-market economists incline to the view that if the costs are fully accounted then pollution becomes uneconomic; more intervention-minded writers (Brubaker 1972) ask if the contaminator can then purchase an unbounded right to pollute, and thus end up in favour of the concurrent imposition of governmental regulations.

The most vigorous polemic in favour of continued growth of a conventional kind is associated with W. Beckerman (1974), encapsulated in his statement that '... in the sober and considered opinion of the latest occupant of the second oldest Chair in Political Economy in this country [the UK], although life on this Earth is very far from perfect there is no reason to think that continued economic growth will make it any worse.'

Most important of all to those who advocate increased development, albeit directed to different places and sometimes to new ends, is the faith that technology will eventually provide answers. At a simple level of appeal, an energy-resource supplier will say at his annual shareholders' meeting, 'Why can't the environmentalists go away and leave us alone: something will turn up'; in a much more elaborate version, Eastlund and Gough (1969) envision the day when an ultra-high-temperature plasma acts as a fusion torch to reduce any material to its elements for separation so that on a large scale urban sewage could be processed, all soild waste recycled, electricity produced through fuel cells, and water desalted, all using only heavy hydrogen from sea water as a basic material and producing as a 'waste' only helium which will itself be quite valuable. (Waste heat will of course be produced too.) As always, science is a two-edged weapon, and believers in technology share with environmentalists a concern over what criteria would be used to decide what was to be produced from such a horn of plenty, and who would decide which buttons to press. In the longer run it seems indisputable that the growth-oriented view of the world's future depends upon abundant energy supplies and in foreseeable terms this means not only thermal nuclear reactors but breeders as well and presumably the eventual certainty of fusion power.

Another kind of world: the environmentalist viewpoint.

As a reaction both to those who foresee imminent and inevitable doom and to those who see solutions mainly in terms of continued technological development, a set of views which emphasize the biological and physical limitations of the planet as a home for man have been put forward. Their answer lies in the adoption of a totally different strategy of man–environment relations, which is characterized largely by steady states of everything pertaining to man, be it population levels or economic activity.

The envelope of the alternative is clear. This is the objective knowledge about the ecosphere and its functioning which modern science is beginning to provide. As ecology, for example, becomes more sophisticated, its level of contribution will increase: Boulding (1966a) did not get into much trouble when he described ecologists as 'a lot of bird-watchers'. Indeed, ecology's major contribution so far has been to provide a holistic conceptual framework for thinking about the ecosphere rather than detailed, accurate and predicative data about the way it will behave in any given circumstances, although improvements will no doubt be brought about as a result of current research.

One major general contribution of ecology has been the knowledge of limits. Since ecosystems function within the wider limits of finite solar radiation and finite cycling times of scarce mineral elements, we become aware of a set of finite limits to food production and of a finite boundary even to a population supported by continuously recycled materials. Terrestrial space is clearly finite and cannot be created, give or take a polder or two. Wastes may also produce limits, especially heat where there are probably not only safe limits beyond which atmospheric systems may be unpredictably altered, but absolute limits beyond which the transport out into space cannot take place at a sufficiently fast rate. The thresholds at which these limits might be expected to operate, other than at local scales, and the rates of activity which might bring them down upon us are uncertain, but no amount of technology or wishful thinking can make them go away, since they are inescapable parts of the physical constitution of the planet. An additional set of limits may be produced by the breakdown of the behaviour patterns of humans under conditions of high density and environmental stress (Catton 1976).

The evidence, as discussed above, is not conclusive, but since there is clearly a cultural dimension to tolerance of crowding, then expressions of aversion to particular conditions are probably significant. Thus if people feel that Britain, for example, is overcrowded, then perhaps it is overcrowded and action needs to be taken to stabilize and possibly reduce the population.

Within the envelope of the limits just discussed, a conception of an alternative man–environment relationship has grown up. Although deriving its fundamental thesis from the observations of biology, it has been given added impetus by inputs such as the NASA pictures of this fragile but fertile planet spinning alone in an infinity of hostile space. Thus the outstanding articulator of the idea, K. E. Boulding (1966c) has called it the 'spaceship earth economy'. He contrasts it with our present economy, which he calls the 'cowboy economy', being characterized by flamboyance, waste and a taste for burning candles at both ends: if we book a passage on the *Titanic*, there is no point in going steerage.

Basic principles of the 'spaceship earth' concept

The purpose of this alternative model is to establish a dynamic equilibrium. Populations are stable, or oscillate with only a small amplitude around a constant level. All materials are conserved by undergoing recycling. There is thus an end to exponential growth on a world scale, although individual parts may exhibit growth within the overall limits. The idea is thus analogous to preferred limits of population in hunting societies, as distinct from populations which necessitate the use of all the food resources of the groups' territory.

Although the basic concept is simple, the specific conditions for its possible success need more detailed examination, and we must begin with the search for ecological understanding. At the traditional scales of ecological research, work is still essential to secure the fundamental understanding of how ecosystems work and, indeed, whether the ecological model of the functioning of nature is a correct one. In particular, more knowledge of the predictability of ecosystem response to various natural and anthropogenic stimuli is required, as is a deeper understanding of the nature of ecological stability and the conditions under which it can be expected (Woodwell and Smith 1969). Knowledge of the complete system of 'man and his total environment' is required, and in particular the continuous adaptation of environment by living organisms (Lovelock 1979).

Although the working of component parts of this ecosphere need to be beter understood, a feeling for the whole is essential, especially for the repercussions of unwanted changes in unexpected places. In general, a determination to regard the phenomena of both physical and cultural systems holistically is necessary, unpopular though this is with 'pure' scientists who see an invasion of their carefully guarded swimming pools by the unwashed hordes from across the tracks. The integrated view is characteristic of geography as an intellectual discipline, and geography deals with the cultural dimension as well as the ecological, but geographers have shown few signs of wishing to put their feet into the turbulent waters of environmental debate, with certain honourable exceptions (Burton and Kates 1964, Zelinsky 1970, Eyre 1971, 1978).

In practical terms, the 'spaceship' economy is predicated upon the attainment of a stable population at a world scale. Growth of world population must clearly stop at some time, and the equilibrium approach prefers an all-out effort to bring this about by cultural and technical means rather than leave it to the alternatives. These latter are likely to be either psychological, with stress causing spontaneous abortion, inadequate parental care of infants or possibly enhanced rates of foetal re-absorption, or directly Malthusian, through famine, war and disease.

In bringing about population stability, priority needs to be given to the DCs and the Egyptian type of LDC. The importance of the European-US-type nations is twofold. Firstly, they are the dominant users of resources and producers of contaminants. Statistics for almost any part of a global resource process will show the USA itself in the lead for per capita use, followed closely by other nations of the US type, and hotly pursued by the European-type countries. A further reduction of the already slow rate of increase of population would help to remove the fundamental cause of their need for resources, although cultural demands would doubtless keep consumption levels high and the actual reduction of usage rates would have to be a cultural decision of a different kind from agreeing to attempt to stabilize population. Secondly, Western material culture is the object of aspirations in many LDCs who think that population control is basically an imperialist gambit to keep the Third World in its 'proper' place. It is essential therefore to practise what is preached, and to make it clear that population stabilization by the rich ought ultimately to be for the greater benefit of the poor.

The need for population control in resource-poor and technology-poor nations is self-evident. It is probably their only way of moving away from the precarious balance between adequacy and insufficiency, whether of food or employment. Intensified and more productive agriculture may improve living standards, but it inevitably causes a wholesale drift to the cities, where industrialization rarely proceeds fast enough to provide jobs for the influx, partly because capital accumulation per capita is low at times of rapid population growth. Furthermore, the basic resources for the industrial development of many of the populous LDCs are lacking: cheap energy, in particular, is likely to be difficult to acquire. Additionally, their rapid rates of population growth (often in excess of 2 per cent p.a.) impose strains upon their social structures, especially in the case of newly independent nations lacking political stability. Not the least problem is the high proportion of young people in a rapidly growing population, many of whom will have material and educational aspirations on a scale undreamed of by their parents. Urgency is less marked in nations of the Brazilian character, since they can for a time contain their people by developing hitherto unattractive areas, witness the Brazilian plans for the Amazon Basin. But there can be no question of their freedom from eventual limits, and prudence would seem to dictate a halt to growth before their population-resources imbalances become national rather than local or regional.

The fragility of many of the biomes in these countries, especially those in the tropics, serves to underline the virtues of leaving a wide margin of safety which is not dependent upon technology, an implement which is still generally lacking in such places and sometimes unwisely used where present.

An equilibrium economy requires not only the attainment of a stable population but also its maintenance. If this is to be achieved, it will fundamentally require the acceptance of the notion that the maximum size of an individual's family is for society to decide, and not for the parents alone. In effect the situation will have been brought back to the condition of 'primitive' hunting groups which limited their populations to the number sustainable on a preferred food source rather than let it expand to the boundary of the entire supply of nutrition. A fairly drastic social adjustment consisting of the yielding up of an individual choice is therefore indicated. Those who argue the unacceptability of such a change must be asked to consider the other choices which will become unavailable in a condition of a serious imbalance of population, resources and environment. Even if the these of Parsons (1971) are not all accepted, few of the developments of technology which some believe will avert any 'environmental crisis' seem to propose increased personal liberty and choice as their concomitants, let alone their immediate purposes.

The cycling of materials

This necessary condition is based on the premise that some materials, e.g. metals, will be in short supply because of the demands of increased populations wishing to share in the benefits of an industrial way of life. In addition, the environmental alterations inevitably associated with the extraction of lower qualities of materials will be diminished if supplies can be re-used (US National Commission on Materials Policy 1973. Thomas 1974, Barton 1979). Relatively few materials are actually destroyed in the course of their use, so that given the appropriate technology and cost structure, recycling becomes a feasibility for many items: those containing metals and wood products (Plate 21, opposite) are obvious examples, but the idea can be applied to many diverse materials, including lubricating oil in autos, water, sewage and textiles. Some losses are unavoidable, as for example the escape of metals through friction; new sources would be used only to make up such losses, but additions to the total stock would presumably be made only after satisfaction of the strictest criteria. Probably the critical feature of the recycling economy would be the availability of energy to power the recycling processes, and in view of present attitudes towards the future costs of energy, Berry (1972) has suggested that the production and recycling of goods be evaluated from the point of view of energy economies (Table 13.5).

TABLE 13.5 Energy consumption in bottling

kWh/litre beer bottle, inc. production and packaging	
Returnable glass bottle (19 trips)	0·9375
Returnable glass bottle (8 trips)	1·2298
Non-returnable PVC bottle	1·3288
Non-returnable steel can	2·2188
Non-returnable glass bottle	3·2341

Source: *The Guardian* (London), 13 Oct 1977, p. 14

Plate 21 Things to come? A recycling centre run by voluntary labour in Berkeley, California. Here, bottles are sorted into various kinds and the truck behind serves as a temporary repository for old newspapers. *(I. G. Simmons)*

Using as measures the loss of thermodynamic potential involved in the manufacture of a new car and its subsequent recycling, he observes that, given increased demand for cars and present technology, recycling is a questionable process compared with extending the life of the machine, which would diminish by two to three times the energy used, since fewer new cars would be needed. Even bigger savings could be achieved by improvements in the technology of the basic recovery and fabrication processes of metals, where thermodynamic savings of five to ten times the present expenditure could be made as well as using less water and causing less environmental contamination. So, economies of energy use brought out by recycling are small compared with extended lifetimes of goods and smaller still compared with a technology which loses less energy; yet since recycling is more economically feasible it would be a useful start (a saving of 1,000 kWh/vehicle in the USA is equivalent to eight to ten generating stations) towards a policy of thrift. Industrialization of the LDCs, adds Berry, may only be achievable if policies of thermodynamic thrift are brought into effect.

Re-use has the further advantage that all wastes become important sources of raw materials and thus environmental contamination is reduced, except for the non-re-usable substances, to which pollution-control measures will have to be applied. Pollution control *per se*, although useful and necessary at an early stage of moving to a closed-cycle economy, now becomes only a last-ditch treatment of intractable wastes.

Moving towards a recycling of materials will be difficult economically. Theoretically the price of recycled materials should make them more attractive as supplies from fresh sources become more expensive; but because of the nature of exponential curves, the limits of supply may be suddenly upon us before the price mechanism has had time to react so as to show the need for research and development into recycling technology and the accompanying changes in social patterns. Effective legislation and taxation to internalize pollution costs and governmental inputs into recycling technology to prime the pump are therefore urgent priorities now.

Ecosystem balance

The ideas of overall economic and ecological stability embodied in the spaceship earth economy have to be translated into quantities of various types of ecosystem. Because the world cannot return to a pre-agricultural economy, the ecosystems of man's creation must also included. A valuable start on the analysis needed to provide the criteria for balance between different types of ecosystems has been provide by E. P. Odum (1969). He emphasizes that nature is a mosaic of ecosystems at different levels of succession, some of them 'young' with low inherent stability and diversity but high productivity; by contrast, mature ecosystems have high stability and diversity but their productivity may be lower. Globally, the major ecosystems of the natural world exhibit a developed state of internal symbiosis, good nutrient conservation, high resistance to external perturbation, and low entropy. On the other hand, as Odum puts it,

> Man has generally been preoccupied with obtaining as much 'production' from the land as possible, by developing and maintaining early successional types of ecosystems, usually monocultures. But, of course, man does not live by food and fibre alone; he also needs a balanced CO_2–O_2 atmosphere, the climatic buffer provided by oceans and masses of vegetation, and clean (that is, unproductive) water for cultural and industrial uses. Many essential life-cycle resources, not to mention recreational and a esthetic needs, are best provided man by the less 'productive' landscapes.

Different types of ecosystems must therefore form a balanced pattern. As in nature there is a mix of mature, stable systems with immature, unstable ones, so in the man-mainpulated world there must be an intercalation of man-dominated simple systems alongside complex natural ecosystems. Four types of ecosystems may be distinguished:

(1) Non-vital systems, or the 'built environment' of urban and industrial areas. These are in fact dependent upon imports from outside, e.g. of oxygen and water, for their continued existence; in the natural state the nearest analogy is probably a volcano,

for it too emits sulphur dioxide, carbon dioxide and particulate matter into the atmosphere. Other non-vital systems of the natural world might include ice-caps (which, however, act as reservoirs of water and affect climate) and the most barren deserts, to which cities are not infrequently compared by writers with bucolic yearnings, although perhaps cities are morphologically more akin to karst terrain.

(2) Intensively used biotic systems with high productivity and capacity for high yield to man. Like natural systems at an early stage of succession, man-directed systems such as agriculture have a low diversity of species and are unstable. However, they are easily exploitable and hence highly valued in economic terms. As in comparable natural situations, the growth qualities of the plants and animals selected for rapidity, and quantity of production is valued above quality.

(3) Compromise areas, such as those devoted to multiple-use management of forests, or intermixtures of forest, agriculture and grazing or other wild land, are analogous to natural ecosystems approaching, but not yet at, maturity. They are characterized by higher diversity than the earlier stages of succession and by the development of complex webs of interdependence between organisms, in contradistinction to the linear chains, dominated by grazing, of earlier stages.

(4) The mature wild systems, i.e. areas of 'climax' vegetation which are basically unaltered by man. In addition to being important reservoirs of biotic diversity, they are vital agents in gaseous interchange, and provide a source of recreational and aesthetic pleasure. Their metabolism is diametrically different from that of early successional phases since it is a highly bound network of food webs with high internal conservation of nutrients by cycling, and has very high resistance to external perturbation. The characteristics of such ecosystems, compared with those in early stages of succession, are set out in Table 13.6

TABLE 13.6 Trends in the development of ecosystems

Ecosystem attributes	*Developmental stages*	*Mature stages*
Food chains	Linear, predominantly grazing	Weblike, predominantly detritus
Biomass supported/unit energy flow	Low	High
Total organic matter	Small	Large
Species diversity-variety component	Low	High
Size of organism	Small	Large
Life cycles	Short, simple	Long, complex
Mineral cycles	Open	Closed
Nutrient exchange rate between organisms and environment	Rapid	Slow
Internal symbiosis	Undeveloped	Developed
Nutrient conservation	Poor	Good
Stability	Poor	Good
Entropy	High	Low

Source: E. P. Odum 1969

These natural systems play a role far beyond the value which conventional economics accords them. They are the 'anchormen' of total global stability, and perform a function which is 'protective' rather than 'productive' but no less vital (Watt et al 1977). If the oceans are included as mature systems, we may note the suggestions quoted by Odum that the oceans are the governor (in the mechanical sense) of the biosphere, slowing down and controlling the rate of decomposition and nutrient regeneration, thereby creating and maintaining the highly aerobic terrestrial atmosphere to which the higher forms of life, ourselves included, are adapted.

Perhaps, too, we should remind ourselves at this point of Georgescu-Roegen's (1971) view that most of economic processes (and hence our transformations of ecosystems) are entropy-creating and that we have to dig deep into stocks of negative entropy like fossil fuels to rescue from such trends enough comfort for our lives. 'Every time we produce a Cadillac we irrevocably destroy an amount of low entropy that could otherwise be used for producing a plough or a spade ... every time we produce a Cadillac we do it at the cost of decreasing the number of human lives in the future ... it is definitely against the interest of the human species as a whole, if its interest is to have a lifespan as long as is compatible with its dowry of low entropy.' In the very long run, reliance on negative entropy in the shape of solar energy capture seems essential and we ought to look at ecosystem transformation from that viewpoint, which again emphasizes the key nature of energy flow as a guide to systems and their stability (Lugo 1974).

Social and economic adjustments

The primary and ineluctable requirement for the implementation of the closed-cycle economy is population control, which must be aimed at producing a stable world population, with the initial priorities of achieving stable or even declining populations in the US and European types of resource–population situation, and stability in the UAR group. The Brazilian type can possibly be allowed to grow a little more, but not for long.

Alongside this, the measures described by Georgescu-Roegen (1976) although developed in a different context, seem to apply to any attempts to formulate a programme for transition to a steady state. First, the production of the instruments of war: it seems logical to divert the resources used in their manufacture to a more life-enhancing end. Thereafter (in priorities but parallel in time), the LDCs must be aided to reach a good, though not necessarily luxurious life; both they and the DCs may have to alter current outlooks to some extent. In view of the discussions on energy, it seems sensible (a) to avoid all waste of it until the giant amounts providable by solar or fusion sources become both omnipresent and cheap; (b) to achieve levels of population that can be fed by an organic agriculture not based on fossil fuels; (c) to reduce energy levels in industry by making goods last longer, i.e. designed so that they are repairable and marketed in a way that the demand for a new fashion does not arise at frequent intervals. In a less tangible fashion we all need to rid ourselves of the idea that it is worthwhile shaving faster in order to have time to design a better razor which shaves faster so as to save time to ... and so on. (All authors of books on environmental problems are of course bearded). Similar prescriptions, with more emphasis on the role of toxins in the biosphere are given by Woodwell (1974).

The basis of value of environmental resources will also undergo change in a recycling economy. Whereas present economic systems value throughput and high turnover allows low prices, the opposite is likely to occur in a revised system. Because of the need to conserve materials, articles which have a long life, preferably free of losses through wear, are likely to be relatively less expensive than those requiring frequent trips through the recycling process. Thus the Chippendale chair ought to cost less than the cardboard one, and the tough dress that lasts their whole adult lives will be easier on the ladies' fashion budget than the frilly number lasting only a few parties. (Such an equation ought to bring home the truly revolutionary nature of the spaceship earth concept.) The same argument, *mutatis mutandis*, will apply to more costly material possessions like vehicles, and the current idea of building disposable towns will be seen as an evanescent efflorescence of an age that denied its own limitations.

The final disappearance of the goal of unlimited growth may also be a signal for a major redistribution of incomes. The gap between rich and poor should diminish, partly because less growth will provide fewer opportunities for rapidly gained entrepreneurial profits, and partly because a more coherently planned economy may well value more highly the present poor. A lessened emphasis on the quantity of industrial production will reverse the present trend to energy-intensive production in favour of labour-intensive outputs. This should encourage production of quality goods and services and perhaps restore to the manual workers some of the dignity and job-satisfaction which mass-production has taken away from them. Since labour is likely to be in shorter supply in a stable population, unemployment might reasonably be expected to diminish, and re-evaluation of those currently rejected as too old (rather than incapable) for particular tasks is likely to occur.

The role of industry must change. Nobody advocates a return to some pastoral condition where factories are unheard of and the cross-stitching on the shepherds' smocks is sullied only by the spots of blood from their inevitable tuberculosis, but certain shifts in industrial policy are advocated. Most important is the abandonment of the 'technological fix' as an article of faith, acknowledging that in many respects natural systems are much more efficient than those which are man-made. Next in priority is a realization that for some of the difficult problems caused by population growth, there are no simple (or even complex) technological panaceas worth having (Ehrlich and Holdren 1969). The disparate time-scales upon which population growth and technological development operate, and the complexity of man-environment relations outside the narrowly conceived context of food production, are all taken to suggest that present technology will not suffice to maintain the expected populations of the world, even given optimistic assumptions. Rather is it the case that technology will only become effective and life-enhancing in a stable economy, when the possibility of applying several technological schemes along a broad front (instead of the current piecemeal approach which usually creates several difficulties for each one it removes) becomes more feasible.

Given the abandonment of the ideology of growth, the redirection of technological effort becomes possible and its contribution to a harmonious relation of man and nature immense. New criteria of success will be essential, especially with regard to the environmental consequences and in the discarding of those material uses which are totally ephemeral and add nothing to the quality of anybody's life. Beyond such a stage it is possible that the kind of 'de-development' proposed by Ehrlich and Harriman (1971) may be essential in the European and US groups of nations. Here the control of 'needs' induced by advertising, the freeing of industrial policy from the hands of a few individuals unaccountable to the society which supports them, the enhancement of private affluence at the expense of public squalor, and all the excrescences of the 'disposable society' will have to be firmly controlled without undue delay, as requisites of the closed-cycle economy. Adaptation to a new type of economy, together with the many unsolved problems of the LDCs, will ensure a series of challenges to industry of an unprecedented nature. There is no desire to abandon technology, but neither should there be any intention to allow the big corporations (and their Socialist equivalents) to dominate the resource processes of the biosphere; control of the harmful technology described by Commoner (1972a, b) is necessary but is only a first step.

Brief, if insufficient, mention must be made of some of the institutional adjustments which would be necessary to the success of an equilibrium economy. At a global scale, international agreement on certain movements towards stability will be necessary, especially where the prohibition of poisoning or over-use of commons like the oceans is concerned. Moves towards lessening nuclear testing underground and towards the elimination of nuclear weapons altogether would also be a useful adjunct. Co-ordination of the programmes of nation states and economic groups like the EEC as they alter their goals would be a function of the UN, but a tighter global control is probably an unrealistic prognosis. The main effort would have to come from independent states and the blocs to which they have given over some of their sovereignty: it is they who would lead their people through the difficult task of reorientating their way of life.

At the level of national and local institutions, and even that of individual people, the way in which a transition to a different life-style might be achieved is difficult to see, and not a major concern at this point. Several writers have however addressed themselves to the socio-economic consequences of the steady state; in general they seem to feel that it is a feasible aim provided that the necessary alterations in institutions and goals are achieved in an orchestrated manner and in an evolutionary rather than revolutionary spirit (e.g. Ayres and Kneese 1971, Eversley 1975, Ophuls 1977, Pirages 1977, Watt et al 1977, Daly 1978).

It is of course possible that a different type of process might occur: that the changed consciousness of individuals may itself work towards a life-style which is consonant with long-term ecological stability, in the framework of what E. F. Schumacher (1973) called 'Buddhist economics' (Pirsig 1974, Capra 1976). Those members of affluent societies who seek alternative life-styles which involve self-sufficiency and low-impact living generally regard themselves as being in the vanguard of this new wave, just as did the 'hippy' people of the 1960s. All such movements are in danger of creating and sustaining their own mythologies (John 1979) and it is difficult to avoid the thought that in Type B nations at least, the need for conversion to economic Buddhism is greatest in the cities and that the flight of the 'alternatives' to the countryside may represent a reluctance to face the major challenges that are now apparent. What might occur, in an almost unnoticed fashion, is a satiation of Western people with goods and services and a turning to lower growth rates especially in energy use and a concern that extends beyond the immediately material, or even a 'satisficing' viewpoint that declines to exchange leisure for more money: why work 5 days when 3 will suffice for your needs? Anathema though it must be to the inheritors of the 'work ethic', there is perhaps some evidence that in one or two nations of the West, such attitudes are emerging (Nossiter 1978). Perhaps this is one result of the years of high profile exposure of 'the environmental crisis' which culminated in 1972 at Stockholm. (It may be said that the problems are still with us even though the media have moved on and talk of 'the energy crisis' as if it were unrelated. But even if the high water-mark of concern about the environment has passed, there is now the need for a steady and sustained campaign along the same lines, until environmental linkages are automatically considered in any decision at any level).

Adjudication: limits of all kinds

If the human species is to continue to depend upon the natural systems of the biosphere for life-support, for materials, for valued surroundings and for waste disposal, then it must acknowledge the existence of limits.

The most obvious of these are the ecological limits of which numerous examples have been given in this work, outstanding among which is the SCEP (1970) calculation that every doubling of population increases the 'ecological demand' by six times. Within these limits, a new linkage is needed in which the socio-economic systems realize the importance of the protective systems and reflect this value in their economies. This will contrast with the present situation where wild-lands are usually regarded as raw materials which acquire value only if transformed to something else. And in this connection we ought to remember Lovelock's (1979) suggestion that the key ecosystems of the planet which require protection are firstly the tropical forests and secondly the estuaries, wetlands, and muds on the continental shelves.

Politicians must also realize that these limits prevent them from promising unlimited 'growth' to their constituents. More realistically, those leaders might well work towards a condition in which all people had a 'right' to life-support from the planet. A fundamental flaw in the world's legal systems allows individuals to appropriate portions of the complex cycles upon which we all depend. They may contaminate part of the water cycle or the air, or may clear protective forest in order to enhance short-term gain from exploitive agriculture. The ownership of land and water must not therefore confer the right to remove it from the life-support systems, and those who wilfully release wastes which disorder those systems must be made to desist: a process which is beginning to take effect.

If limits are to be culturally accepted, then the poor have to be reassured that the rich are not merely pulling up the ladder behind them, having achieved their desired level of material prosperity. This applies both to rich and poor in industrial nations, and to rich and poor countries collectively. In industrial countries with clearly differentiated social strata most of the pressures both for cosmetic conservation and 'environmentalism' come from the 'middle class' and are often denounced on that account. There is often truth in the idea that cosmetic acts merely impose middle-class values upon the whole of the society and use up wealth which might be better spent upon necessities for the poor. Wider concepts of the closed-cycle equilibrium type, however, are basically advantageous to the less privileged groups in society.

At present, the poor have a smaller share in the proceeds of economic growth, for example; the industrial workers suffer most from pollution both at work and in their homes, for the better-off can always move house to the cleaner suburbs and handle nothing more contaminating than the latest issue of *Playboy*. The much vaunted example of London's clearer air following smoke abatement programmes has merely transferred the pollution from London to the areas around the coking plants, and to Scandinavia. Some LDCs may forgo the energy-intensive economies of the West and opt for intermediate technology aimed at improving rural life, but those where industry has been made the basis of economic development will wish to attract more capital. Such LDCs can scarcely be expected to stabilize their economies at the present stage, but many will try to be more selective in their accession to industrial status, and some may reject this path altogether; the relationship between the two groups at a time when DCs are stabilizing economies but LDCs still expanding is a difficult and complex issue which needs a great deal of study.

Admission of limits to the carrying capacity of the planet need not carry with it the automatic connotation of doom if all expansion of flows through resource processes is not stopped immediately. It is possible that the choice of chaos or survival currently postulated is a false one. There may be two kinds of limits in the capacity of the earth. The first of these is the *absolute limit* in which a man-made stability replaces that of natural systems. A first stage might be the conversion of the earth largely to a food–man monoculture, but later an immense technological contribution would be mandatory. The human populations that could be supported would be huge, but there would be little room for other than technological life-supported systems: the machines would indeed dominate the earth. The alternative might be a *preferred limit*, in which the natural systems continue to be crucial for our survival and in which they are valued accordingly. A diversity of habitat and culture would remain (unlikely to happen in the absolute-limit alternative) but the supportable population would be lower than in the technology-dependent world; this latter alternative is of course identical with the spaceship earth concept. The two limits are depicted in Fig. 13.2.

There may be a third alternative in which the intensive use of cheap energy means that all man's needs can be produced in industrial plant which is as much as possible decoupled from the biosphere. Man's material needs would be supplied by an economic system parallel to but in minimal interaction with natural systems. Plans for underground and offshore nuclear power stations could presumably be extended to all kinds of manufacturing and processing installations so that, given adequate technology and favourable geology, even food could be produced from atomic and molecular building blocks in underground and submersible factories, or even in space, leaving the

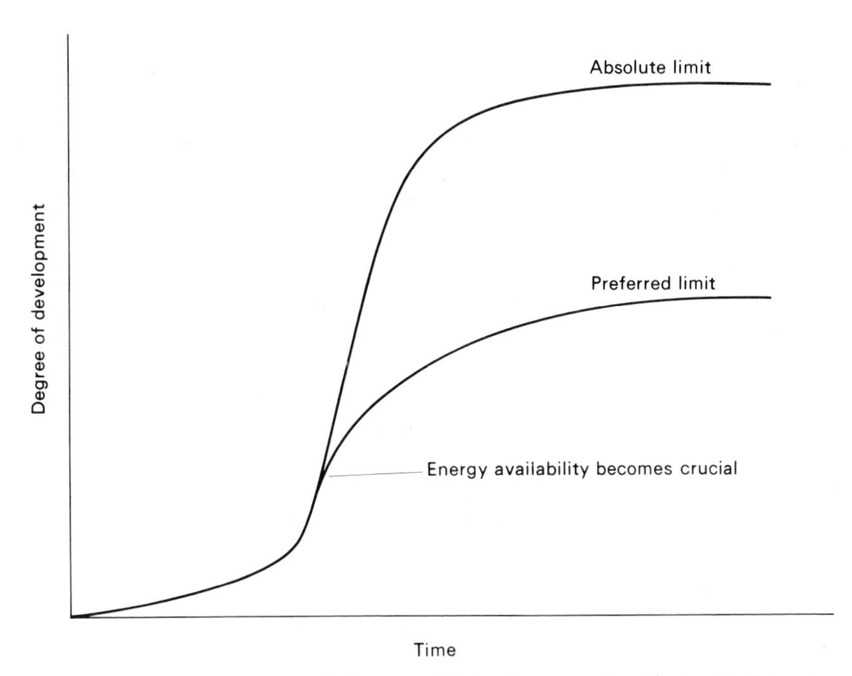

Fig. 13.2 A scheme of two alternative limits to world development: the absolute limit involves replacement of natural life-support systems by man-made processes, the preferred limit does not. The availability of cheap and safe energy is crucial to the pursuit of the absolute limit.

biosphere relatively free of human impact, although many people would still want to live on the earth's surface.

In spite of its apparent reasonableness, this alternative has an air of science fiction about it. But both it and the system which depends upon the supersession of natural systems depend upon the provision of a cheap, clean and ubiquitous energy supply whose waste heat can be safely radiated into space without a major disturbance of atmospheric patterns. And so attention to developments such as controlled fusion reactions, alternative methods of energy conversion, more locally autonomous power sources of a non-fossil and non-nuclear type, more efficient technology which will effectively utilize energy inputs (few steam plants today operate beyond an efficiency of 45 per cent), and energy-conservative buildings, domestic heating systems and transport networks, are all vital and pressing needs, as must be the study of the disposal of waste heat at high levels of fossil or nuclear power generation. Without them only the environmentalist model offers hope of any quality of life.

If there is a choice between these alternatives, what criteria are to be used in making decisions? There is no doubt that the preferred limit is safer: it gives much wider margins for error, and forecloses fewer options than the absolute limit. It also offers a greater diversity of all kinds to individual people: of different employments, different places to visit and different cultures to exist alongside. But it is a world in which growth is no longer equated with progress; it is analogous to the condition of men in hunting societies who limited their populations to those who could be fed by a preferred food supply such as meat, rather than an absolute food supply which included rodents and roots.

The outlook of those who advocate the preferred limit is essentially optimistic. Those who propose totally technological solutions are seeking to avoid dangers, but the spaceship earth proponents see a different kind of world. They do not wish to return to some ill-remembered Eden without machines (indeed they want more technology, not less, but applied in a different manner), but to create a world of stability in which the resources are used equitably by an adequately nourished population which does not poison its own habitat with the wastes. They see it as the most hopeful way of coming to grips with man's most fearsome problems: the containment of population growth, the management of energy flows, the development of stable ecological and political orders, and the formulation of a coherent political and ethical doctrine for human behaviour in relation to natural systems, and furthermore one which neither attempts the unforeseeable nor commits the irrevocable. Bringing this about is a task of immense difficulty, because it can only be made meaningful to diverse groups of men in terms of the lineaments of their own culture: success is unlikely if new ways are imposed internationally or by foreign-educated governors. Because local or regional instabilities are at present countered by importing resources from elsewhere, the world becomes more intensely coupled and interdependent. But because of the accelerating likelihood of instability brought about by the exponential growth of numbers of people, consuming materials and energy, and subsequently discharging waste, all men should recognize that ecological instability (which may well manifest itself initially in the form of climatic fluctuations) is their common enemy, one which will not distinguish between rich and poor, black, white or brown, PhDs and peasant farmers.

The alternatives are clear: to try to develop to an absolute limit in one of its forms, to aim for a preferred limit, or to adopt no overall strategy and allow present-day institutions to respond to individual problems as they arise, which is likely to mean management by crisis. As was said at the beginning of this book, men are material-using animals. We must also make moral choices.

Further reading

BECKERMAN, W. 1974: *In defence of economic growth.*
BOULDING, K. E. 1966: The economics of the coming spaceship earth.
BRANDT, W. 1980: *North-south: a programme for survival.*
BROWN, L. R. 1978b: *The Twenty-Ninth Day.*
BURCH, W. R. and BORMANN, F. H. (eds.) 1975: *Beyond growth: essays on alternative futures.*
DALY, H. E. 1978: *Steady state economics.*
EHRLICH, P. and HOLDREN, J. P. 1969: Population and panceas: a technological perspective.
EHRLICH, P. and HOLDREN, J. P. 1971: Impact of population growth.
FREEMAN, C. and JAHODA, M. (eds.) 1978: *World futures. The great debate.*
GOLDSMITH, E. *et al.* 1973: *Blueprint for survival.*
HIRSCH, F. 1977: *The social limits to growth.*
IUCN 1980: *World Conservation Strategy.*
LOVELOCK, J. E. 1979: *Gaia: A new look at life on earth.*
MESAROVIC, M. and PESTEL, E. 1975: *Mankind at the turning point.*
MISHAN, E. J. 1977: *The economic growth debate.*
OPEN UNIVERSITY 1974: Course S26 (The Earth's Physical Resources) Block 6, *Implications: limits to growth?*
OPHULS, W. 1977: *Ecology and the politics of scarcity.*
O'RIORDAN, T. 1976: *Environmentalism.*
PASSMORE, J. 1980: *Man's responsibility for nature*, 2nd edn.
PIRAGES, D. C. (ed.) 1977: *The sustainable society.*
WARD, B. 1979: *Progress for a small planet.*
WATT, K. E. F. *et al.* 1977: *The unsteady state.*
WHITE, G. F. 1980: Environment.
WOODWELL, G. M. 1974: Success, succession and Adam Smith.

Bibliography

ABERG, B. and HUNGATE, F. P. (eds.) 1967: *Radioecological concentration processes*. New York and London: Academic Press.

ACKERMANN, W., WHITE, G. F. and WORTHINGTON, E. G. (eds.) 1973: *Man-made lakes: their problems and environmental effects*. Washington DC: American Geophysical Union Monograph, No. **17**.

ADAMS, A. B. (ed.) 1964: *First World Conference on National Parks*. Washington, DC: US Government Printing Office.

ADAMS, W. P. and HELLEINER, F. M. (eds.) 1972: *International geography*. Toronto: University of Toronto Press.

AITKEN, P. L. 1963: Hydroelectric power generation. In Institute of Civil Engineers, 34–42.

ALEXANDER, M. 1974: Environmental consequences of rapidly rising food output. *Agroecosystems* **1**, 249–64.

ALLISON, A. (ed.) 1970: *Population control*. Harmondsworth: Pelican.

ALLISON, L. 1974: *Environmental planning: a political and philophical analysis*. London: Allen and Unwin. New York: Rowan and Little.

AMBIO **6** (1), 1977, Special issue.

AMIDON, E. and GOULD, E. M. 1962: *The possible impact of recreation development on timber production in three California National Forests*. Berkeley: US Forest Service Pacific SW Experimental Station Technical Paper **68**.

ANDREWARTHA, H. A. and BIRCH, L. C. 1954: *The distribution and abundance of animals*. Chicago: University of Chicago Press.

ANTONINI, G. A., EWEL, K. C. and TUPPER, H. M. 1975: *Population and energy. A systems analysis of resource utilization in the Dominican Republic*. Gainesville, Fla.: University of Florida Press.

ARKCOL, D. B. 1971: Agronomic aspects of leaf protein production. In N. W. Pirie (ed.), 9–18.

ARMSTRONG, P. H. 1974: Some examples of the use of current ecosystem models as frameworks for land use studies. *GeoWest* **3**. Perth: Univ of W. Australia.

ASHTON, M. D. 1970: *The relationship of agriculture to soil and water pollution*. Report on the 1970 Cornell Agricultural Waste Management Conference. Washington, DC: Reports of the UK Scientific Mission in North America, UKSM 70/12.

AYRES, R. U. and KNEESE, A. V. 1971: Economic and ecological effects of a stationary economy. *Ann. Rev. Ecol. Syst.* **2**, 1–22.

BACH, W., PANKRATH, J. and KELLOGG, W. (eds.) 1979: *Man's impact on climate*. Amsterdam and New York: Elsevier.

BAINES, G. B. K. 1977: The environmental demands of tourism in coastal Fiji. In J. Winslow (ed.) *The Melanesian environment*. Canberra: ANU Press, 448–57.

BAKER, F. C. 1970: *Derelict land*. 'The Countryside in 1970', 3rd Conference, Report 18. London: HMSO.

BAKUZIS, E. V. 1969: Forestry viewed in an ecosystem concept. In G. M. Van Dyne (ed.), 189–258.

BALIKCI, A. 1968: The Netsilik Eskimos: adaptive processes. In R. B. Lee and I. De Vore (eds.), 1968a, 78–82.

BARAM, M. S., RICE, D. and LEE, W. 1978: *Marine mining of the continental shelf. Legal, technical and environmental considerations*. Cambridge, Mass.: Ballinger.

BARBOUR, I. G. (ed.) 1973: *Western man and environmental ethics*. Reading, Mass and London: Addison-Wesley.

BARKHAM, J. P. 1973: Recreational carrying capacity: a problem of perception. *Area* **5**, 218–22.

BARNARD, C. (ed.) 1964: *Grasses and grasslands*. London and Melbourne: Macmillan.

BARNETT, H. J. and MORSE, C. 1963: *Scarcity and growth*. Resources for the future series. Baltimore and London: Johns Hopkins Press for RFF.

BARRETT, E. C. and CURTIS, L. F. 1981: *Introduction to remote sensing of the environment*. London: Chapman and Hall. New York: Halsted Press. 2nd edn.

BARR, H. M., CHADWICK, B. A. and THOMAS, D. L. (eds.) 1972: *Population resources and the future: non-Malthusian perspectives*. Provo, Utah: Brigham Young University Press.

BARR, J. 1969: *Derelict Britain*. Harmondsworth: Pelican.

— (ed.) 1971: *The environmental handbook*. New York: Ballantine. London: Friends of the Earth.

BARRY, R. G. 1969: The world hydrological cycle. In R. J. Chorley (ed.), 11–29.

BARTON, A. F. M. 1979: *Resource recovery and recycling*. Chichester: Wiley.

BARTSCH, A. F. 1970: Accelerated eutrophication of lakes in the United States: ecological response to human activities. *Environmental Pollution* **1**, 133–40.

BASSOW, W. 1979: The third world: changing attitudes towards environmental protection. *Ann. Amer. Acad. Pol. Soc. Sci.* **444**, 112–120.

BAUMHOFF, M. A. 1963: *Ecological determinants of aboriginal California populations*. University of California Publications in Archaeology and Ethnology **49** (2). Berkeley and Los Angeles: University of California Press.

BAYFIELD, N. 1971: Some effects of walking and skiing on vegetation at Cairngorm. In E. Duffey and A. S. Watt (eds.) 469–84.

BEAUMONT, P. 1968: Quanats on the Varamin Plain, Iran. *Trans. Inst. Brit. Geogr.* **45**, 169–79.

— 1977a: Water in Kuwait. *Geography* **62**, 187–97.

— 1977b: Water and development in Saudi Arabia. *Geogr. J.* **143**, 42–60.

BECKERMAN, W. 1973: Growthmania revisited. *New Statesman*, 19 Oct. 550–2.

— 1974: *In defence of economic growth*. London: Cape.

BECKINSALE, R. P. 1969a: Human use of open channels. In R. J. Chorley (ed.), 331–43.

— 1969b: Human responses to river regimes. In R. J. Chorley (ed.), 487–509.

BELL, F. W. 1978: *Food from the sea. The economics and politics of ocean fisheries*. Boulder, Colo.: Westview Press.

BENEDICT, B. 1970: Population control in primitive societies. In A. Allison (ed.), 165–80.

BENNETT, C. F. 1975: *Man and earth's ecosystems*. New York and Chichester: Wiley.

BENNETT, E. 1978: Threats to crop plant genetic resources, in J. G. Hawkes (ed.), *Conservation and agriculture*. London: Duckworth, 113–122.

BENNETT, R. J. and CHORLEY, R. J. 1978: *Environmental systems. Philosophy, analysis and control*. London: Methuen.

BENJAMIN, B., COX, P. R., and PEEL, J. (eds.), 1973: *Resources and population*. London and New York: Academic Press.

BERHAM, D. 1979: *Solar energy: the awakening science*. London: Routledge and Kegan Paul.

BERKOWITZ, D. A. and SQUIRES, A. M. (eds.) 1971: *Power generation and environmental change*. Cambridge, Mass.: MIT Press.

BERRY, R. 1972: Recycling, thermodynamics and environmental thrift. *Bull. Atom. Sci.* **28** (5), 8–15.

BEZDEK, R. H., HIRSHBERG, A. S. and BABCOCK, W. H. 1979: Economic feasibility of solar water and space heating. *Science* **203**, 1214–20.

BICCHIERI, M. G. (ed.), 1972: *Hunters and gatherers today*. New York: Holt, Rinehart and Winston.

BIENFANG, P. 1971: Taking the pollution out of waste heat. *New Sci.* **51**, 456–7.

BIRDSELL, J. B. 1968: Some predictions for the Pleistocene based upon equilibrium systems among recent hunter-gatherers. In R. B. Lee and I. De Vore (eds.), 1968a, 229–40.

BLACK, J. 1970: *The dominion of man: the search for ecological responsibility*. Edinburgh and New York: Edinburgh University Press.

BISWAS, A. K. 1974: Water. In F. S. Sargeant (ed.) *Human ecology*. Amsterdam: North Holland, 206–33.

— (ed.) 1978: *United Nations Water Conference. Summary and Main Documents*. Oxford and New York: Pergamon Press. Water Development Supply and Management, vol. 2.

BISWAS, A. K. and M. R. 1976: State of the environment and its implications to resource policy development. *BioScience* **26**, 19–25.

BLUMER, M. 1969: Oil pollution of the ocean. In D. P. Hoult (ed.), 5–13.

BOESCH, D. F., HERSHNER, C. H. and MILGRAM, J. H. 1974: *Oil spills and the marine environment*. Cambridge, Mass.: Ballinger for the Ford Foundation.

BOGGESS, W. R. and WIXSON, B. G. 1979: *Lead in the environment*. Tunbridge Wells, England: Castle House Publications. New York: International Publications Services.

BOGUE, D. J. 1969: *Principles of demography*. New York: Wiley.

BOLTON, J. R. 1978: Solar fuels. *Science* **202**, 705–711.

BONACCORSI, A., FANELLI, R. and TOGNONI, 1978: In the wake of Seveso. *Ambio*, **7**, 234–39.

BORGSTROM, G. 1965: *The hungry planet*. New York: Macmillan. London: Collier-Macmillan.

BORMANN, F. H. and LIKENS, G. S. 1969: The watershed-ecosystem concept and studies of nutrient cycles. In G. M. Van Dyne (ed.), 49–76.

BORMANN, F. H., LIKENS, G. E. and EATON, J. S. 1969: Biotic regulation of particulate and solution losses from a forest ecosystem. *BioScience* **19**. 600–610.

BORMANN, F. H., LIKENS, G. E., FISHER, D. W. and PIERCE, R. S. 1968: Nutrient loss accelerated by clear-cutting of a forest ecosystem. *Science* **159**, 882–4.

BOUGHEY, A. S. 1968: *Ecology of populations*. New York: Macmillan.

BOULDING, K. E. 1962: *A reconstruction of economics*. New York: Science Editions.

— 1966a: Discussions in: F. Fraser Darling and J. P. Milton (eds.), 291–2.

— 1966b: Ecology and economics. In F. Fraser Darling and J. P. Milton (eds.), 225–34.
— 1966c: The economics of the coming spaceship earth. In H. Jarrett (ed.), 3–14.
— 1970: Fun and games with the Gross National Product: the role of misleading indicators in social policy. In H. W. Helfrich (ed.), 157–70.
— 1971: Environment and economics. In W. W. Murdoch (ed.), 359–67.
BOULTRON, C. and LORIUS, C. 1979: Trace metals in Antarctic snows since 1914. *Nature, Lond.* **277**, 551–54.
BOX, T. W. and PERRY, R. A. 1971: Rangeland management in Australia. *J. Range Management* **24**, 167–71.
BOYD, R. 1972: World dynamics: a note. *Science* **177**, 516–19.
BRACEY, H. C. 1970: *People and the countryside.* London and Boston: Routledge and Kegan Paul.
BRADY, N. C. (ed.) 1967: *Agriculture and the quality of our environment.* Washington, DC: Publication **85**, AAAS.
BRANDT, W. 1980: *North-South: a programme for survival.* London and Sydney: Pan Books. Cambridge, Mass.: MIT Press.
BRAIDWOOD, R. 1970: The agricultural revolution. In J. Janick (ed.), 4–12.
BRITISH TRAVEL ASSOCIATION-KEELE UNIVERSITY 1967: *The pilot national recreation survey, Report 1.* London: BTA.
BROOKFIELD, H. C. 1975: *Interdependent development.* London: Methuen. Pittsburgh: University of Pittsburgh Press.
BRONN, L. R. 1979: *Resource trends and population policy: a time for reassessment.* Washington DC: Worldwatch Institute, Paper **29**.
BROWN, C. L. 1976: Forests as energy sources in the year 2000. *Journal of Forestry.* **74**, 7–12.
BROWN, H. 1954: *The challenge of man's future.* New York: Viking Press.
— 1970: Human materials production as a process in the biosphere. In *Scientific American* (ed.), 115–24.
BROWN, H., BONNER, J. and WEIR, J. 1963: *The next hundred years.* New York: Viking Press.
BROWN, L. R. 1970: Human food production as a process in the biosphere. In *Scientific American* (ed.), 95–103.
BROWN, L. R. 1975: The multinational corporations: economic colossus of modern times. *Horizons USA*, No. **9**, 8–12.
— 1976: *World population trends: signs of hope, signs of stress.* Washington DC: Worldwatch Paper No. **8**.
— 1978a: *The global economic prospect: new sources of economic stress.* Washington DC: Worldwatch Paper No. **20**.
— 1978b: *The twenty-ninth day.* New York and London: Norton.
— 1979: *Resource trends and population policy: a time for reassessment.* Washington DC: Worldwatch Paper **29**.
BROWN, L. R. and ECKHOLM, E. P. 1975: Man, food and environment, in W. W. Murdoch (ed.), *Environment.* Sunderland, Mass.: Sinauer, Second edition, 67–94.
BROWN, L. R. and FINSTERBUSCH, G. 1972: *Food.* London and New York: Harper and Row.
BROWN, N. L. (ed.) 1978: *Renewable energy resources and rural applications in the developing world.* Boulder, Colo.: Westview Press, AAAS Selected Symposium No. **6**.

BROWN, N. L. and HOWE, J. W. 1978: Solar energy for village development. *Science* **199**, 651–57.

BRUBAKER, S. 1972: *To live on earth: Man and his environment in perspective*. Resources for the future series. Baltimore and London: Johns Hopkins Press for RFF/New York: Mentor Books.

BRUHN, J. G. 1972: The ecological crisis and the work ethic. *Int. J. Environ. Studs.* **3**, 43–7.

BRYAN, R. 1973: *Much is taken, much remains*. Belmont, Calif.: Wadsworth.

BRYCE-SMITH, D., MATHEWS, J. and STEPHENS, R. 1978: Mental health effects of lead on children. *Ambio* **7**, 192–203.

BRYSON, R. A. and WENDLAND, W. M. 1970: Climatic effects of atmospheric pollution. In S. F. Singer (ed.), 1970a, 130–38.

BURCH, W. R. and BORMANN, F. H. (eds.) 1975: *Beyond growth: essays on alternative futures*. New Haven: Yale University School of Forestry and Environmental Studies.

BURKART, A. J. and MEDLIK, S. 1974: *Tourism. Past, present and future*. London: Heinemann. New York: International Publications Service.

BURINGH, P., VAN HEEMST, H. D. J. and STARING, C. J. 1975: *Computation of the absolute maximum food productivity of the world*. Wageningen: Agricultural University Department of Tropical Soil Sciences.

BURKITT, D. P. 1973: Diseases of modern economic development, in G. M. Howe and J. Loraine (eds.), *Environmental medicine*. London: Heinemann, 140–44.

BURLEY, J. and STYLES, B. T. 19776: *Tropical trees. Variation, breeding and conservation*. London: Academic Press, Linnean Society Symposium Series No. **2**.

BURRELL, I. S. 1973: National Parks. The big three—conservation, recreation and education. *Journal of Environmental Management* **1**, 201–5.

BURTON, I. and KATES, R. W. 1964: Slaying the Malthusian dragon: a review. *Econ. Geogr.* **40**, 82–9.

— (eds.) 1965: *Readings in resource management and conservation*. Chicago and London: University of Chicago Press.

BURTON, I., KATES, R. W. and SNEAD, R. E. 1969: *The human ecology of coastal flood hazard in Megalopolis*. Chicago: University of Chicago, Department of Geography Research Paper **115**.

BURTON, I., KATES, R. W. and WHITE, G. F. 1978: *Environment as hazard*. New York and Oxford: OUP.

BURTON, J. A. 1975: The future of small whales. *New Sci.* **66**, 650–51.

BUTLER, G. C. (ed.) 1978: *Principles of ecotoxicology*. Chichester and New York: Wiley, for SCOPE and ICSU. SCOPE vol. 12.

BURWELL, C. C. 1978: Solar biomass energy: an overview of US potential. *Science* **199**, 1041–48.

BUTZER, K. 1971: *Environment and archaeology: an introduction to Pleistocene geography*. Chicago: Aldine Press. Second edition.

CAHN, R. 1968: *Will success spoil the National Parks?* Boston: Christian Science Monitor Reprints.

CALAPRICE, J. R. 1976: Mariculture: ecological and genetical aspects of production. *Journal of Fisheries Research Board Canada* **33**, 1088–93.

CALDER, N. 1967: *The environment game*. London: Secker and Warburg.

— (ed.) 1968: *Unless peace comes*. Harmondsworth: Pelican.

CALDWELL, L. K. 1971: *Environment: a challenge to modern society*. New York: Doubleday Anchor Books.
— 1972: An ecological approach to international development: problems of policy and administration. In M. Taghi Farvar and J. P. Milton (eds.), 927–47.
CALDWELL, M. 1971: World resources and the limits of man. In P. H. G. Hettena and G. N. Syer (eds.), 15–37.
CALHOUN, J. B. 1962: Population density and social pathology. *Sci. Amer.* **206** (2), 139–48.
— 1971: Psycho-ecological aspects of population. In P. Shepard and D. McKinley (eds.), 111–33.
CANTLON, J. E. 1969: The stability of natural populations and their sensitivity to technology. In G. M. Woodwell and H. H. Smith (eds.), 197–205.
CAPRA, F. 1976: *The Tao of physics*. London: Fontana Books.
CARPENTER, K. J. 1969: Man's dietary needs. In J. Hutchinson (ed.), 61–74.
CARSON, R. 1963: *Silent spring*. Boston: Houghton Mifflin. Harmondsworth: Penguin.
CASSIDY, N. G. and PAHALAD, S. D. 1953: The maintenance of soil fertility in Fiji. *Fiji Agric. J.* **24**, 82–6.
CASTRO, J. A. DE A. 1972: Environment and development: the case of the developing countries. In D. A. Kay and E. B. Skolnikoff (eds.), 237–52.
CATTABENI, F., CAVALLARO, A. and GALLI, G. (eds.), 1978: *Dioxin: toxicological and chemical effects*. London and New York: Spectrum.
CATTON, W. R. 1976: Can irrupting man remain human? *BioScience* **26**, 262–7.
CENTRAL INTELLIGENCE AGENCY [of the USA] 1974: *Potential implications of trends in world population, food production, and climate*. Washington, DC: CIA Office of Political Research, Report OPR-401.
CENTRE FOR AGRICULTURAL STRATEGY, 1978: *Phosphorus: A resource for UK agriculture*. Reading University (UK): CAS Report No. **2**.
CHADWICK, M. J. and GOODMAN, G. T. (eds.) 1975: *The ecology of resource degradation and renewal*. Oxford: Blackwell. New York: Halsted Press.
CHAPMAN, P. F. 1970: Energy production—a world limit? *New Sci.* **47**, 634–6.
CHAPMAN, P. 1975: *Fuel's paradise: energy options for Britain*. Harmondsworth: Penguin.
CHARLEBOIS, C. T. 1978: High mercury levels in Indians and Inuits (Eskimos) in Canada. *Ambio* **7**, 204–10.
CHARLIER, R. H. 1970: Crisis year for the Great Lakes. *New Sci.* **44**, 593–6.
CHERFAS, J. 1978: Rationality and the slaughter of seals. *New Sci.* **77**, 724–26.
CHORLEY, R. J. (ed.) 1969: *Water, earth and man*. London: Methuen. New York: Barnes and Noble.
CHOW, B. G. 1977: The economic issues of the fast breeder reactor program. *Science* **195**, 551–56.
CHRISTY, F. T. and SCOTT, A. 1967: *The common wealth in ocean fisheries: some problems of growth and economic allocation*. Resources for the future series. Baltimore and London: Johns Hopkins Press.
CIRIACY WANTRUP, S. V. 1938: Soil conservation in European farm management. *J. Farm Econ.* **20**, 86–101.
CIRIACY WANTRUP, S. V. and PARSONS, J. J. (eds.) 1967: *Natural resources: quality and quantity*. Berkeley and Los Angeles: University of California Press.
CIVIC TRUST, 1964: *A Lee Valley Regional Park*. London: The Civic Trust.
CLARK, J. G. D. 1954: *Excavations at Star Carr*. Cambridge: Cambridge University Press.

CLARKE, C. 1977: *Land use and population growth*. London: Macmillan, 2nd edn.

CLARKE, F. J. P. 1978: Unconventional energy sources. In L. Grainger (ed.). *Energy resources: availability and rational use*. London: IPC/World Energy Conference 1977, 108–118.

CLARKE, J. I. 1972: *Population geography*. Oxford and New York: Pergamon Press. 2nd edn.

CLARKE, W. C. 1977: The structure of permanence: the relevance of self-subsistence communities for world ecosystem management. In T. Bayliss-Smith and R. Feachem (eds.). *Subsistence and survival. Rural ecology in the Pacific*, London and New York: Academic Press, 363–84.

CLAWSON, M. 1963: *Land and water for recreation*. Chicago: Rand McNally.

CLAWSON, M. and KNETSCH, J. L. 1966: *Economics of outdoor recreation*. Baltimore and London: Johns Hopkins Press for RFF.

CLAWSON, M., LANDSBERG, H. H. and ALEXANDER, L. T. 1969: Desalted water for agriculture: is it economic? *Science* **164**, 1141–8.

CLAWSON, M. 1974: *Forest policy for the future*. Washington DC: RFF.

— 1975: *Forests for whom and for what?* Baltimore and London: Johns Hopkins Press for RFF.

CLAYRE, A. (ed.) 1977: *Nature and industrialization*, Oxford and New York: Oxford University Press.

CLOUD, P. 1968: Realities of mineral distribution. *Texas Quarterly* **11**, 103–26.

— 1969: Minerals from the sea. In NAS/NRC, 135–55.

— 1975: Mineral resources today and tomorrow. In W. W. Murdoch (ed.), *Environment. Resources, pollution and society*. Sunderland, Mass: Sinauer, 2nd edition, 97–120.

COHEN, B. L. 1977: The disposal of radioactive wastes from fission reactors. *Sci. Amer.* **236**, 21–31.

COHEN, J. M. 1976: *The food crisis in prehistory*. New Haven and London: Yale University Press.

COLE, G. F. 1974: Management involving grizzly bears and humans in Yellowstone National Park, 1970–73. *BioScience* **24**, 335–38.

COLE, H. S. D., FREEMAN, C., JAHODA, M. and PAVITT, K. L. R. (eds.) 1973: *Thinking about the future*. London: Chatto and Windus for Sussex University Press. New York: Universe Books (as *Models of Doom*).

COLE, L. C. 1969: Thermal pollution. *BioScience* **19**, 989–92.

COLES, J. M. and HIGGS, E. S. 1969: *The archaeology of early man*. London: Faber and Faber. New York: Penguin.

COLLIER, B. D., COX, G. W., JOHNSON, A. W. and MILLER, P. C. 1973: *Dynamic ecology*. Englewood Cliffs, NJ: Prentice Hall.

COMMONER, B. 1972a: *The closing circle*. London: Cape. New York: Knopf/Bantam.

— 1972b: The environmental costs of economic growth. In R. Dorfman and N. S. Dorfman (eds.), 261–83.

— 1975: Energy, environment and economics, in G. D. Eppen (ed.), *Energy: the policy issues*. London and Chicago: University of Chicago Press, 25–40.

CONKLIN, H. 1954: An ethnoecological approach to shifting agriculture. *Trans. N. Y. Acad. Sci.*, Ser. II, **17**, 133–42.

— 1957: *Hanunoo agriculture in the Philippines*. Rome: FAO Forestry Development Paper 12.

COOK, E. 1971: The flow of energy in an industrial society. *Sci. Amer.* **224** (3), 135–44.

— 1975a: Flow of energy through a technological society. In J. Lenihan and W. W. Fletcher (eds.). *Energy resources and the environment.* Glasgow and London: Blackie, 30–62.
— 1976: *Man, energy, society.* San Francisco and Reading: Freeman.
— 1979: The helium question. *Science* **206**, 1141–47.
COOK, L. M. and WOOD, R. J. 1976: Genetic effects of pollutants. *Biologist* **23**, 129–39.
COOPER, J. P. (ed.) 1975: *Photosynthesis and productivity in different environments.* Cambridge: The University Press, IBP Studies No. 3.
COPPOCK, J. T. and DUFFIELD, B. S 1975: *Recreation in the countryside: a spatial analysis.* London: Macmillan. New York: St. Martin's Press.
COSTIN, A. B. and FRITH, H. J. (eds.) 1971: *Conservation.* Ringwood, Victoria: Penguin.
COSTIN, A. B. and GROVES, R. H. (eds.) 1973: *Nature conservation in the Pacific.* Canberra: ANU Press.
COTTRELL, A. 1978: *Environmental economics.* London: Edward Arnold. New York: Halsted Press.
COULSON, J. C. 1972: Grey seals on the Farnes: kindness kills. *New Sci.* **54**, 142–5.
COUNCIL OF EUROPE 1971: *The management of the environment in tomorrow's Europe.* Strasbourg: European Information Centre for Nature Conservation.
COUNTER INFORMATION SERVICES, n.d.: *The Rio Tinto-Zinc Corporation Limited Anti-Report.* London: CIS.
COUNTRYSIDE COMMISSION OF ENGLAND AND WALES: *Annual reports.* London: HMSO.
COUPER, A. 1978: Marine resources and environment. *Progress in Human Geography.* **2**, 296–308. London: Edward Arnold. New York: Cambridge University Press.
COUSENS, J. 1974: *An introduction to woodland ecology.* Edinburgh: Oliver and Boyd. New York: Longman.
COX, G. W. (ed.) 1969: *Readings in conservation ecology.* Ecology Series. New York: Appleton-Century-Crofts.
CRISP, D. J. (ed.) 1964: *Grazing in terrestrial and marine environments.* Oxford: Blackwell.
— 1975: Secondary productivity in the sea, in NAS *Productivity of world ecosystems.* Washington DC: NAS, 71–89.
CROSLAND, C. A. R. 1971: *A social democratic Britain.* London: Fabian Society Tract 404.
CURNOW, R. C. 1975: Population, pollution and natural resources—an inevitable clash. In M. J. Chadwick and G. T. Goodman (eds.): *The ecology of resource degradation and renewal.* Oxford: Blackwell; B.E.S. 15th Symposium, New York: Halsted Press. 407–20.
CUSHING, D. H. and WALSH, J. J. (eds.) 1975: *The ecology of the seas.* Oxford: Blackwell. Philadelphia: Saunders.
CUSHING, D. H. 1978: *Science and the fisheries.* London: Edward Arnold (Studies in Biology, No. 85). Baltimore: University Park Press.
DALY, H. E. 1971: A Marxian-Malthusian view of poverty and development. *Population Studies,* **25**, 25–37.
— 1978: *Steady state economics.* San Francisco and Reading: Freeman.
DARMSTADTER, J. (with TEITELBAUM, P. D. and POLACH, J. G.) 1971; *Energy in the world economy: a statistical review of trends in output, trade and consumption since 1925.* Baltimore and London: Johns Hopkins for RFF.
DASMANN, R. F. 1964b: *Wildlife biology.* New York: Wiley.

— 1968: *A different kind of country*. London and New York: Macmillan.

— 1976: *Environmental conservation*. New York and Chichester: Wiley. 4th edn.

DASMANN, R. F., MILTON, J. P. and FREEMAN, P. H. 1973: *Ecological principles for economic development*. New York and Chichester: Wiley-Interscience.

DAVIDSON, J. and LLOYD, R. 1977: *Conservation and agriculture*. Chichester and New York: Wiley-Interscience.

DAWS, G. 1977: Tourism in Hawaii: benefits and costs. In J. Winslow (ed.): *The Melanesian environment*. Canberra: ANU Press, 429–34.

DAY, R. H. and SINGH, I. 1979: *Economic development as an adaptive process. The green revolution in the Indian Punjab*. London: Cambridge University Press.

DEEVEY, E. S. 1960: The human population. *Sci. Amer*. **203** (9), 195–204.

— 1970: Mineral cycles. In *Scientific American* (ed.), 83–92.

— 1971: The chemistry of wealth. *Bull. Ecol. Soc. America* **52**, 3–8.

DELWICHE, C. C. 1970: The nitrogen cycle. In *Scientific American* (ed.), 71–80.

DOE [Department of the Environment of the UK] 1974: *Report of the National Park Policies Review Committee*. London: HMSO, (Chairman: Lord Sandford).

— 1974: *Lead in the environment and its significance to man*. London: HMSO, Pollution Paper No. **2**.

DESHLER, W. W. 1965: Native cattle keeping in eastern Africa. In A. Leeds and A. P. Vayda (eds.), 153–68.

DESMOND, A. 1965: How many people ever lived on earth? In L. K. Y. Ng and S. Mudd (eds.), 20–38.

DESPRAIRIES, P. 1978: Worldwide petroleum supply limits. *World energy resources 1985–2020*. London: IPC for World Energy Conference 1977, 1–47.

DETWYLER, T. R. (ed.) 1971: *Man's impact on environment*. New York and Maidenhead: McGraw-Hill.

DE VOS, A. 1969: Ecological conditions affecting the production of wild herbivorous mammals on grasslands *Advances in Ecological Research* **3**, 137–83.

DE WINTER, F. and COX, M. (eds.) 1979: *Sun—mankind's future source of energy*. Oxford and New York: Pergamon, 3 vols.

DIALOGUE, **11** (1978), 3–41, *The energy dilemma*.

DICKSON, D. 1974: *Alternative technology*. London: Fontana. New York: Universe.

DIMBLEBY, G. W. 1962: *The development of British heathlands and their soils*. Oxford Forestry Memoir 23.

DOMBEY, N. 1979: Can we afford to make the fast reactor safe? *Nature, Lond*. **280**, 270–72.

DORFMAN, R. and DORFMAN, N. S. (eds.) 1972: *Economics of the environment*. New York and London: Norton.

DORN, H. F. 1965: Pitfalls in population forecasts and projections. In I. Burton and R. W. Kates (eds.), 21–37.

DORST, J. 1970: *Before nature dies*. London/Harmondsworth: Collins/Penguin. Boston/New York: Houghton Mifflin/Penguin.

DOUGHTY, R. 1975: *Feather fashions and bird preservation*. Berkeley, Los Angeles and London: University of California Press.

DOUGLASS, R. W. 1975: *Forest recreation*. Oxford and New York: Pergamon Press. 2nd edn.

DOWNS, J. F. and EKVALL, R. B. 1965: Animal and social types in the exploitation of the Tibetan plateau. In A. Leeds and A. P. Vayda (eds.), 169–84.

DUBOS, R. 1967: *Man adapting*. New Haven and London: Yale University Press.

DUCKHAM, A. N. and MASEFIELD, G. B. 1970: *Farming systems of the world*. London: Chatto and Windus.

DUCKHAM, A. N., JONES, J. G. W., and ROBERTS, E. H. 1976: An approach to the planning and administration of human food chains and nutrient cycles. In A. N. Duckham, J. G. W. Jones and E. H. Roberts (eds.): *Food production and consumption: the efficiency of human food chains and nutrient cycles*. Amsterdam, Oxford and New York: North Holland Publishing Co., 461–517.

DUFFEY, E. 1974: *Nature reserves and wildlife*. London: Heinemann.

DUFFEY, E. (ed.) 1976: *The biotic effects of public pressure on the environment*. Monk's Wood Experimental Station Symposium 3. Great Britain: The Nature Conservancy.

DUFFEY, E. and WATT, A. S. (eds.) 1971: *The scientific management of animal and plant communities for conservation*. Oxford: Blackwell.

DUNN, F. L. 1968: Epidemiological factors: health and disease in hunter-gatherers. In R. B. Lee and I. De Vore (eds.), 1968a, 221–8.

DUNN, P. D. 1978: *Appropriate technology*. London: Macmillan.

DUNNE, T. and LEOPOLD, L. B. 1979: *Water in environmental planning*. San Francisco and Reading: Freeman.

DURAND, J. D. 1967: The modern expansion of world population. *Proc. Amer. Philos. Soc.* **111**, 136–59.

DURET, M. F., PHILLIPS, G. J. VEEDER, J. I. WOLFE, W. A. and WILLIAMS, R. M. 1978: The contribution of nuclear power to world energy supply, 1985 to 2020. *World Energy Resources 1985–2020*. London: IPC for World Energy Conference 1977, 109–34.

DUVIGNEAUD, P. (ed.) 1971: *Productivity of forest ecosystems*. Paris: UNESCO.

DUVIGNEAUD, P. and DENAEYER-DE SMET, S. 1970: Biological cycling of minerals in temperate deciduous forests. In D. Reichle (ed.), 199–229.

DWORKIN, D. M. (ed.) 1974: *Environment and development*. Indianopolis, Ind.: SCOPE, Miscellaneous Publication.

DWYER, D. J. 1958: Utilization of the Irish peat bogs. *Geogr. Rev.* **48**, 572–3.

EARL, D. E. 1975: *Forest energy and economic development*. Oxford: Clarendon Press. New York: Oxford University Press.

EASTLUND, B. J. and GOUGH, W. C. 1969: *The fusion torch. Closing the cycle from use to re-use*. Washington, DC: US Atomic Energy Commission.

ECKHOLM, E. P. 1975: *The other energy crisis: firewood*. Washington DC: Worldwatch Paper **1**.

ECKHOLM, E. P. 1976: *Losing ground*. New York and London: Norton.

EDEN, M. J. 1978: Ecology and land development: the case of Amazonian rainforest. *Trans. Inst. Brit. Geogr.* NS **3**, 444–64.

EGLER, F. E. 1970: *The way of science: a philosophy of ecology for the layman*. New York: Hafner.

EHRENFELD, D. W. 1970: *Biological conservation*. New York: Holt, Rinehart and Winston.

— 1976: The conservation of non-resources. *American Scientist* **64**, 648–56.

EHRLICH, P. 1971: Ecocatastrophe! In J. Barr (ed.), 205–13.

EHRLICH, P. and EHRLICH, A. 1970: *Population resources environment. Issues in human ecology*. San Francisco: Freeman. 2nd edn.

EHRLICH, P., EHRLICH, A. and HOLDREN, J. R. 1977: *Ecoscience: population, resources, environment*. San Francisco and Reading: Freeman, 3rd edn.

EHRLICH, P. and HARRIMAN, R. L. 1971: *How to be a survivor*. New York: Ballantine/Friends of the Earth.

EHRLICH, P. and HOLDREN, J. P. 1969: Population and panaceas: a technological perspective. *BioScience* **19**, 1065–71.

— 1971: Impact of population growth. *Science* **171**, 1212–17.

EHRLICH, P., HOLDREN, J. P. and HOLM, R. W. (eds.) 1971: *Man and the ecosphere: readings from 'Scientific American'*. San Francisco and Reading: Freeman.

EKISTICS **43** (254), 1977: Special issue.

ELLUL, J. 1964: *The technological society*. New York and London: Vintage Books.

ELTON, C. S. 1958: *The ecology of invasions by animals and plants*. London: Methuen. New York: Halsted Press.

— 1966: *The pattern of animal communities*. London: Methuen. New York: Halsted Press (as *Animal Ecology*).

ENGLAND, R. E. and DE VOS, A. 1969: Influence of animals on pristine conditions in the Canadian grasslands. *J. Range Management* **22**, 87–93.

EVERSLEY, D. 1975: *Planning without growth*. London: Fabian Society Research Series 321.

EYRE, S. R. 1978: *The real wealth of nations*. London: Edward Arnold. New York: St. Martins Press.

FAIR, G. M. 1961: Pollution abatement in the Ruhr district. In H. Jarrett (ed.), 171–89.

FALKENMARK, M. and LINDH, G. 1974: How can we cope with the water resources situation by the year 2015? *Ambio* **3**, 114–22.

FARIS, G. T. M. A. 1966: *A contribution to the economic geography of present day forestry and forest products in the Sudan*. University of Durham, MA thesis.

FARMER, B. H. 1969: Available food supplies. In J. Hutchinson (ed.), 75–95.

— (ed.) 1977: *Green revolution? Technology and change in rice-growing areas of Tamil Nadu and Sri Lanka*. Cambridge Commonwealth Series, London: Macmillan. Boulder, Colo.: Westview Press.

FASSETT, D. W. 1972: Cadmium. In D. H. K. Lee (ed.): *Metallic contaminants and human health*. New York and London: Academic Press, 97–124.

FIMREITE, N. 1970: Mercury uses in Canada and their possible hazardous sources of mercury contamination. *Env. Polln.* **1**, 119–31.

FIREY, W. J. 1960: *Man, mind and land: theory of resource use*. Glencoe, Illinois: Free Press. London: Greenwood Press.

FISHER, J., SIMON, N. and VINCENT, J. 1969: *Wildlife in danger*. London: Collins.

FITTER, R. S. R. 1963: *Wildlife in Britain*. Harmondsworth: Pelican.

FITZSIMMONS, A. K. 1976: National Parks: the dilemma of development. *Science* **191**, 440–44.

FLAWN, P. 1966: *Mineral resources: geology, engineering, economics, politics, law*. Chicago: Rand McNally. Chichester: Wiley.

FLÖHN, H. 1973: Der Wasserhauchalt der Erde. *Naturwissenschaften* **60**, 310–48.

FOE (Friends of the Earth) 1978: *Whale manual '78*. London: FoE.

FOELL, W. K. 1978: Long-term policy assessment of energy/environment systems: a conceptual and methodological framework. In D. F. Burkhart and W. H. I. Helson (eds.), *Environmental Assessment of Socioeconomic Systems*. London and New York: Plenum Press, 183–202.

FOLEY, G. 1976: *The energy question*. Harmondsworth and New York: Penguin (Pelican books).

FOOD AND AGRICULTURE ORGANIZATION (annually): *Yearbook of fishery statistics*. Rome: FAO.

— (annually): *Yearbook of forest products*. Rome: FAO.

— (annually): *Production Yearbook*. Rome: FAO.

— 1963: *World forest inventory*. Rome: FAO.

— 1967: *Wood: world trends and prospects*. Rome: FAO.

— 1970: *Indicative world plan for agriculture*, 2 vols. Rome: FAO.

— 1978: *The state of food and agriculture 1977*. Rome: FAO.

FORD, E. D. 1971: The potential production of forest crops. In P. F. Wareing and J. P. Cooper (eds.) 172–85.

FORD FOUNDATION 1974: *A time to choose. America's energy future*. Cambridge, Mass: Ballinger.

FORRESTER, J. 1971: *World dynamics*. Cambridge: Wright-Allen Press. Cambridge, Mass: MIT Press. 2nd edn.

FOWLES, J. 1978: *Handbook of futures research*. Westport, Conn: Greenwood Press.

FRANCHETEAU, J. and 14 others, 1979: Massive deep-sea sulphide ore deposits discovered on the East Pacific Rise. *Nature, Lond.* **277**, 523–28.

FRANKEL, O. H. and HAWKES, J. G. (eds.) 1975: *Crop genetic resources for today and tomorrow*. Cambridge University Press: IBP Studies No. **2**.

FRANKEL, O. 1978: Conservation of crop genetic resources and their wild relatives: an overview. In J. G. Hawkes (ed.): *Conservation and agriculture*. London: Duckworth, 123–150.

FRANKLIN, J. F. 1977: The biosphere reserve program in the United States. *Science* **195**, 262–69.

FRASER DARLING, F. 1956: Man's ecological dominance through domesticated animals on wild lands. In W. L. Thomas (ed.), 778–87.

FRASER DARLING, F. and EICHORN, N. 1967: *Man and nature in the National Parks: reflections on policy*. Washington, DC: Double Dot Press for the Conservation Foundation.

FRASER DARLING, F. and MILTON, J. P. (eds.) 1966: *Future environments of North America*. New York: Natural History Press.

FREEMAN, C. and JAHODA, M. (eds.) 1978: *World futures. The great debate*. London: Martin Robertson. New York: Universe.

FREEMAN, S. D. (ed.) 1974: *A time to choose. America's energy future*. Final report by the energy policy project of the Ford Foundation. Cambridge, Mass.: Ballinger.

FREJKA, T. 1973: The prospects for a stationary world population. *Sci. Amer.* **228** (3), 15–23.

FREMLIN, J. H. 1964: How many people can the world support? *New Sci.* **24**, 285–7.

FRIDAY, R. E. and ALLEE, D. J. 1976: The environmental impact of American agriculture, in J. Wreford Watson and T. O'Riordan (eds.): *The American environment: perceptions and policies*. London and New York: Wiley, 213–39.

FRISSEL, M. J. (ed.) 1978: *Cycling of mineral nutrients in agricultural ecosystems*. Amsterdam, Oxford and New York: Elsevier Developments in Agriculture and Managed Forest Ecology, 3. (Reprinted from *Agro-Ecosystems* **4** (1/2), 1–354.

FRY, C. 1976: Marxism versus ecology. *The Ecologist* **6** (9), 328–32.

FULLER, W. A. and KEVAN, P. G. 1970: *Productivity and conservation in northern circum-*

polar lands. IUCN Pubs. New Series 16. Morges: IUCN.

GABOR, D. 1963: *Inventing the future*. Harmondsworth: Pelican.

GALBRAITH, J. K. 1967: *The new industrial state*. Boston: Houghton Mifflin. Harmondsworth: Penguin.

GAMBELL, R. 1972: Why all the fuss about whales? *New Sci.* **54**, 674–6.

— 1976: Population biology and the management of whales. *Applied Biology*, **1**, 247–343.

GARNISH, J. D. 1978: Progress in geothermal energy. *Endeavour*, NS **2** (2), 66–71.

GARRELS, R. M., MACKENZIE, F. T. and HUNT, C. 1975: *Chemical cycles and the global environment*. Los Altos, Calif.: Kaufmann.

GARVEY, G. 1972: *Energy, ecology, economy*. New York and London: Norton.

GATES, D. M. 1971: The flow of energy in the biosphere. *Sci. Amer.* **224** (3), 88–100.

GEERTZ, C. 1963: *Agricultural involution: the processes of ecological change in Indonesia*. Berkeley and Los Angeles: University of California Press.

GEORGE, C. J. 1972: The role of the Aswan High Dam in changing the fisheries of the southeastern Mediterranean. In M. Taghi Farvar and J. P. Milton (eds.), 159–78.

GEORGESCU-ROEGEN, R. 1971: *The entropy law and the economic process*. Cambridge, Mass and London: Harvard U. P.

— 1975: Energy and economic myths. Part 1. *The Ecologist* **5** (5), 164–74; Part 2. *The Ecologist* **5** (7), 242–52.

— 1976: Energy and economic myths. In N. Georgescu-Rogen (ed.) *Energy and economic myths*. Oxford and New York: Pergamon, 3–36.

GERASIMOV, I. P., ARMAND, D. L. and YEFRON, K. M. (eds.) 1971: *Natural resources of the Soviet Union: their use and renewal*. San Francisco and Reading: Freeman.

GERASIMOV, I. P. and GINDIN, A. M. 1977: The problem of transferring runoff from northern and Siberian rivers to the arid regions of the European USSR, Soviet Central Asia, and Kazakhstan, in G. L. White (ed.) *Environmental effects of complex river development*. Boulder, Colo.: Westview Press, 59–70.

GESAMP (Group of Experts on Scientific Aspects of Marine Pollution) 1977: *Impact of oil on the marine environment*. Rome: FAO.

GIBBENS, R. P. and HEADY, H. F. 1964: *The influence of modern man on the vegetation of Yosemite Valley*. Berkeley: University of California Agricultural Experiment Station Manual 36.

GLACKEN, C. J. 1967: *Traces on the Rhodian shore*. Berkeley and Los Angeles: University of California Press.

GOLDBERG, E. D. and BERTINE, K. K. 1975: Marine pollution in W. W. Murdoch (ed.) *Environment*. Sunderland, Mass.: Sinauer, 2nd edn., 273–95.

GOLDEN, J., OUELLETTE, R. P., SAARI, S. and CHEREMINISOFF, P. N. 1978: *Environmental Impact Data Book*. Chichester and New York: Wiley.

GOLDMAN, C. R., MCEVOY, J. and RICHERSON, P. J. 1973: *Environmental quality and water development*. San Francisco and Reading: Freeman.

GOLDMAN, M. I. 1971: Environmental disruption in the Soviet Union. In T. R. Detwyler (ed.), 61–75.

— 1972: *The spoils of progress: environmental pollution in the Soviet Union*. Cambridge, Mass. and London: MIT Press.

GOLDSMITH, E., ALLEN, D. et al. 1972: Blueprint for survival. *The Ecologist* **2**(1), 2–43. Reprinted 1973, Harmondsworth: Pelican.

GOLDSMITH, F. B., MUNTON, R. J. C. and WARREN, A. 1970: The impact of recreation on the ecology and amenity of semi-natural areas: methods of investigation used in the Isles of Scilly. *Biological Journal of the Linnean Society* **2**, 287–306.

GOLDSMITH, F. B., 1974: Ecological effects of visitors in the countryside. In A. Warren and F. B. Goldsmith (eds.) *Conservation in practice*. Chichester: Wiley., 217–31.

GOLDWATER, L. J. and STOPFORD, W. 1977: Mercury, in J. Lenihan and W. W. Fletcher (eds.): *The chemical environment*. Glasgow and London: Blackie, Environment and Man, Vol. 6, 38–63.

GONZALEZ, N. L. (ed.) 1978: *Social and technological management in dry lands*. Boulder, Colo.: Westview Press.

GOODLAND, R. J. A. and IRWIN, H. S. 1975: *Amazon jungle: green hell to red desert* Development in Landscape Management and Urban Planning No. 1, Amsterdam and New York: Elsevier.

GOODMAN, D. 1975: The theory of diversity—stability relationships in ecology. *Quart. Rev. Biol.* **50**, 237–66.

GOUROU, P. 1966: *The tropical world: its social and economic conditions and its future status*. Chichester: Wiley. New York: Halsted Press. 4th edn.

GOVETT, M. H. and GOVETT, G. J. S. 1976: Deficiency and measuring world mineral supplies. In G. J. S. Govett and M. H. Govett (eds.), *World mineral supplies. Assessment and perspective*. Developments in economic geology no. **3**. Amsterdam: Elsevier.

GOYER, R. A. and CHISHOLM, J. J. 1972: Lead. In D. H. K. Lee (ed.), 57–95.

GRAINGER, A. 1980: The state of the world's tropical forests. *The Ecologist* **10** (1–2), 6–54.

GRAINGER, L. (ed.) 1978: *Energy resources: availability and rational use*—conference proceedings digest, World Energy Conference 1977. London: IPC.

GRAYSON, A. J. (ed.) 1976: Evaluation of the contribution of forestry to economic development. HMSO, London: Forestry Commission Bulletin no. **56**.

GREEN, M. B. 1978: *Eating oil: energy use in food production*. Boulder, Colo. and London: Westview Press.

GREER, C. 1978: *River management in modern China*. Boulder, Colo.: Westview Press.

GREGORY, D. P. 1973: The hydrogen economy. *Sci. Amer.* **228** (1), 13–21.

GREGORY, R. 1971: *The price of amenity*. London: Macmillan.

GRIBBIN, J. 1975: Climatic change and food production. *Food Policy*, **1**, 301–12.

GRIGG, J. 1970: *The harsh lands: a study in agricultural study*. Focal Problems in Geography Series. London: Macmillan. New York: St. Martin's Press.

GULLAND, J. A. 1970: The development of the resources of the Antarctic seas. In M. W. Holdgate (ed.), Vol. 1, 217–23.

— 1975: The harvest of the sea: potential and performance. In W. W. Murdoch (ed.), *Environment*. Sunderland, Mass.: Sinauer, 2nd edition, 167–89.

GULLAND, J. E. (ed.) 1972: *The fish resources of the ocean*. London: Fishing News (Books) Ltd.

GÜRKAYNAK, M. R. and AYHAN LE COMPTE, W. (eds.) 1979: *Human consequences of crowding*. New York and London: Plenum Press. NATO Conference Series III, vol. 10.

GUSTAVSON, M. R. 1979: Limits to wind power utilization. *Science* **204**, 13–17.

GWYNNE, P. 1972: Nuclear power goes to sea. *New Sci.* **55**, 474–6.

HADEN-GUEST, S. (ed.) 1956: *A world geography of forest resources*. American Geographical Society Special Pub. **33**. New York: Ronald Press.

HÄFELE, W. and SASSIN, W. 1978: A future energy scenario. In L. Grainger (ed.). *Energy resources: availability and rational use.* London: IPC for the World Energy Conference 1977, 190–96.

HÄFELE, W. 1980: A global and long-range picture of energy developments. *Science* **209**, 174–82.

HAMMOND, A. L., METZ, W. D. and MAUGH, T. H. 1973: *Energy and the future.* Washington, DC: AAAS.

HANSON, W. C. 1967a: Radioecological concentration processes characterizing Arctic ecosystems. In B. Aberg and F. P. Hungate (eds.), 183–91.

— 1967b: Caesium-137 in Alaskan lichens, caribou and eskimos. *Health Physics* **13**, 383–9.

HARDEN-JONES, F. R. (ed.) 1974: *Sea fisheries research.* London: Elek.

HARDIN, G. 1959: Interstellar migration and the population problem. *J. Hered.* **50**, 68–70.

— 1968: The tragedy of the commons. *Science* **162**, 1243–8.

— (ed.) 1969: *Population, evolution and birth control.* San Francisco and Reading: Freeman. 2nd edn.

— 1973: *Exploring new ethics for survival. The voyage of the spaceship Beagle.* Harmondsworth and New York: Penguin.

HARDY, R. W. F. and HAVELKA, U. D. 1975: Nitrogen fixation research: a key to world food? *Science* **188**, 633–43.

HARRAR, J. G. and WORTMAN, S. 1969: Expanding food production in hungry nations: the promise, the problems. In C. M. Hardin (ed.), *Overcoming world hunger.* The American Assembly, 1969. New York: Prentice-Hall, 89–135.

HARRIS, D. R. 1978: The environmental impact of traditional and modern agricultural systems. In J. G. Hawkes (ed.), *Conservation and agriculture.* London: Duckworth, 61–70.

HARRISS, R. C. 1971: Ecological implications of mercury pollution in aquatic systems. *Biol. Cons.* **3**, 279–83.

HARTE, J. A. and SOCOLOW, R. H. 1971b: The Everglades: wilderness versus rampant land development in south Florida. In J. A. Harte and R. H. Socolow (eds.), 181–202.

HASLER, A. D. 1975: Man-induced eutrophication of lakes, in S. F. Singer (ed.). *The changing global environment.* Dordrecht: Reidel, 383–399.

HAWKES, A. L. 1961: A review of the nature and extent of damage caused by oil pollution at sea. *Trans 26th North American Wildlife Conference*, 343–55.

HAWKES, J. G. (ed.) 1978: *Conservation and agriculture.* London: Duckworth.

HAYES, D. 1976: *Energy: the case for conservation.* Washington, DC: Worldwatch paper No. **4**.

— 1977: *Energy for development: third world options.* Washington DC: Worldwatch Paper No. **15**.

HAYES, E. T. 1979: Energy resources available to the United States, 1985 to 2000. *Science* **203**, 233–39.

HEATHERTON, T. (ed.) 1965: *Antarctica.* London: Methuen for the New Zealand Antarctic Society.

HECKMAN, C. 1979: *Rice field ecology in Northeastern Thailand.* The Hague: Junk.

HEIZER, R. F. 1955: *Primitive man as an ecologic factor.* Kroeber Anthropological Society Papers No. **13**. Berkeley: University of California Press.

HELFRICH, H. W. (ed.) 1970: *The environmental crisis.* New Haven and London: Yale University Press.

HELLIWELL, J. R. 1973: Priorities and values in nature conservation. *Journal of Environmental Management* **1**, 85–127.
HENDEE, J. C. and STANKEY, G. H. 1973: Biocentricity in wilderness management *BioScience* **23**, 535–8.
HENDRICKS, S. B. 1969: Food from the land. In NAS/NRC, 65–85.
HENSTOCK, M. E. 1976: The scope for materials recycling. *Conservation and Recycling* **1**, 3–17.
HETTENA, P. H. G. and SYER, G. N. (eds.) 1971: *Decade of decision*. London: Academic Press for the Conservation Society.
HEWITT, K. and HARE, F. K. 1973: *Man and environment. Conceptual frameworks*. AAG Commission on College Geography Resource Paper No. **20**. Washington DC: AAG.
HIATT, V. and HUFF, J. E. 1975: The environmental impact of cadmium. *Int. J. Env. Studs*. **7**, 277–85.
HICKLING, C. F. 1970: Estuarine fish farming. *Adv. Marine Biol*. **8**, 119–213.
HIJSZELER, C. C. W. J. 1957: Late-glacial human cultures in the Netherlands. *Geol. en Mijnbouw* **19**, 288–302.
HILL, A. R. 1975: Ecosystem stability in relation to stress caused by human activities. *Canadian Geographer* **19**, 206–19.
— 1976: The environmental impacts of agricultural land drainage. *Journal of Environmental Management* **4**, 251–74.
HIRSCH, F. 1977: *Social limits to growth*. (Twentieth Century Fund Study) London: Routledge and Kegan Paul. Cambridge, Mass: Harvard Univ. Press.
HIRST, E. 1976: Residential energy use alternatives: 1976 to 2000. *Science* **194**, 1247–52.
HOARE, A. G. 1979. Alternative energies: alternative geographies. *Progress in Human Geography* **3**, 506–37. London: Edward Arnold. New York: Cambridge University Press.
HOARE, A. G. and HAGGETT, P. 1979: Tidal power and estuary management—a geographical perspective. In Severn, R. T., Dineley, D., Hawker, L. E. (eds.) *Tidal Power and Estuary Management*, Colston Papers No. **30**, Scientechnica, 14–25.
HODGES, L. 1973: *Environmental pollution*. Philadelphia: Holt, Rinehart and Winston. Eastbourne: Holt-Saunders.
HOHENEMSER, C., KASPERSON, R. and KATES, R. 1977: The distrust of nuclear power. *Science* **196**, 25–34.
HOLDGATE, M. W. (ed.) 1970: *Antarctic ecology*. 2 vols. London and New York: Academic Press.
— 1978: The balance between food production and conservation. In J. G. Hawkes (ed.). *Conservation and agriculture*. London, Duckworth, 227–242.
— 1979: *A perspective of environmental pollution*. Cambridge: University Press.
HOLDREN, J. P. 1974: Hazards of the nuclear fuel cycle. *Bull. Atom Sci*. **30** (8), 14–23.
— 1975: Energy and prosperity. *Bull. Atom Sci*. **31**, 26–28.
— 1975: Energy resources. In W. W. Murdoch (ed.), *Environment*, Sunderland, Mass.: Sinauer, 2nd edn. 121–45.
HOLDRIDGE, I. R. 1959: Ecological indications of the need for a new approach to tropical land use. *Econ. Bot*. **13**, 271–80.
HOLLING, C. S. 1969: Stability in ecological and social systems. In G.M. Woodwell and H.H. Smith (eds.), 128–41.

— (ed.) 1978: *Adaptive environmental assessment and management*. Chichester and New York: Wiley/IIASA, International Series on Applied Systems Analysis No. 3.
— 1978: Obergurgl: development in high mountain regions of Austria. In C.S. Holling (ed.), *Adaptive environmental assessment and management*. Chichester and New York: Wiley/IIASA, 215–42.
HOLM, L. G., WELDON, L. W. and BLACKBURN, R. D. 1969: Aquatic weeds. *Science* **166**, 699–709.
HOLMES, G. D. *et al.* (eds.) 1975: A discussion on forests and forestry in Britain. *Phil. Trans. Roy. Soc. B***271**, 45–232.
HOLT, S. J. 1971: The food resources of the ocean. In P. Ehrlich, J. P. Holdren and R. W. Holm (eds.), 84–96.
HOLZ, R. (ed.) 1973: *The surveillant science*: London and Boston: Houghton Mifflin.
HOOD, D. W. (ed.) 1971: *Impingement of man on the oceans*. New York: Wiley.
HOPKINS, W. H. and SINCLAIR, J. D. 1960: Watershed management in action in the Pacific southwest. *Proc. Soc. Amer. For.*, 184–6.
HORSFALL, J. G. 1970: The green revolution: agriculture in the face of the population explosion. In H. W. Helfrich (ed.), 85–98.
HOULT, D. P. (ed.) 1969: *Oil on the sea*. New York and London: Plenum Press.
HOUSE, P. W. and WILLIAMS, E. R. 1971: *The carrying capacity of a nation*. Lexington, Mass.: Lexington Books.
HOWARD, N. J. 1964: Introduced browsing animals and habitat stability in New Zealand. *J. Wildlife Management* **28**, 421–9.
HOWE, G. M. and LORAINE, J. A. (eds.) 1973: *Environmental Medicine*. London: Heinemann.
HUBBERT, M. KING 1962: *Energy resources: a report to the NAS/NRC*. Washington, DC: NAS/NRC Pubn. 1000D.
— 1969: Energy resources. In NAS/NRC, 157–242.
— 1971: The energy resources of the earth. *Sci. Amer.* **224** (3), 61–70.
HUGHES, M. K. 1974: The urban ecosystem. *Biologist* **21**, 117–26.
HUISKEN, R. H 1975: The consumption of raw materials for military purposes. *Ambio* **4**, 229–33.
HUNT, G. M. 1956: The forest products industries of the world. In S. Haden-Guest (ed.), 83–111.
HUNTER, J. M. 1966: Ascertaining population carrying capacity under traditional systems of agriculture in developing countries: notes on a method employed in Ghana. *Prof. Geogr.* **18**, 151–4.
HURD, L. E., MELLINGER, M. V., WOLF, L. L. and MCNAUGHTON, S. J. 1972: Stability and diversity at three trophic levels in terrestrial successional ecosystems. *Science* **173**, 1134–6.
HUTCHINSON, G. E. 1970: The biosphere. In *Scientific American* (ed.), 1–11.
— 1978: *An introduction to population ecology*. New Haven and London: Yale Univ. Press.
HUTCHINSON, J. (ed.) 1969: *Population and food supply*. Cambridge: Cambridge University Press.
HUTCHINSON, J. B. 1978: The Indian achievement, In J. G. Hawkes (ed.) *Conservation and agriculture*. London: Duckworth, 217–22.
HYNES, H. B. N. 1970: The ecology of flowing waters in relation to management. *Journal of Water Pollution Control Federation* **42**, 418–24.

ILLICH, I. D. 1974: *Energy and equity*. London: Calder and Boyars.
INSTITUTE OF CIVIL ENGINEERS 1963: *Conservation of water resources in the United Kingdom*. London: ICE.
INSTITUTE OF ECOLOGY 1972: *Man in his living environment*. Madison: University of Wisconsin Press.
INTERNATIONAL JOINT COMMISSION, CANADA AND THE USA 1970: *Pollution of Lake Erie, Lake Ontario and the international section of the St. Lawrence River*. Ottawa: Information Canada.
INTERNATIONAL PLANNED PARENTHOOD FEDERATION (annually): *Family planning in five continents*. London: IPPF.
IRUKAYAMA, K. 1966: The pollution of Minamata Bay and Minamata disease. *Adv. Wat. Polln. Res.* **3**, 153–80.
ISAACS, J. D. and SCHMIDT, W. R. 1980: Ocean energy: forms and prospects. *Science* **207**, 265–73.
ISAKOV, Y. A. 1978: Scientific bases of the preservation of natural ecosystems in Zapovedniks. In J. G. Nelson et al. (eds.) 527–46.
IUCN 1966–: *Red data book*. Morges: IUCN.
— 1971: *United Nations list of National Parks and equivalent reserves*. Brussels: Hayez, 2nd edn.
— 1980: *World conservation strategy*. Gland, Switzerland.
IVES, J. D. and BARRY, R. G. 1974: *Arctic and Alpine environments*. London: Methuen. New York: Barnes and Noble.
JACKSON, C. I. (ed.) 1978: *Human settlements and engrgy*. Oxford and New York: Pergamon Press for the UN.
JACKSON, I. J. 1977: *Climate, water and agriculture in the tropics*. London and New York: Longman.
JACOBSEN, T. and ADAMS, R. M. 1958: Salt and silt in ancient Mesopotamian agriculture. *Science* **128**, 1251–8.
JANICK J. (ed.) 1970: *Plant Agriculture*. San Francisco and Reading: Freeman/Scientific American.
JANSSON, A-M and ZUCCHETTO. J. 1978: *Energy, economic and ecological relationships for Gotland, Sweden—a regional systems study*. Stockholm: Swedish Natural Science Research Council, Ecological Bulletins no. **28**.
JARRETT, H. (ed.) 1961: *Comparisons in resource management*. Lincoln, Nebraska: Bison Books.
— 1966: *Environmental quality in a growing economy*. Resources for the future series. Baltimore and London: Johns Hopkins Press for RFF.
JASANI, B. M. 1975: Environmental modifications: new weapons of war? *Ambio* **4**, 191–98.
JENNINGS, P. R. 1976: The amplification of agricultural production. In Scientific American, *Food and agriculture*. San Francisco: Freeman, 125–33.
JENNY, H. 1961: Derivation of state factor equations of soils and ecosystems. *Soil Sci. Am. Proc.* **25**, 385–8.
JEQUIER, N. (ed.) 1976: *Appropriate technology: problems and promises*. Paris: OECD.
JOHANNES, R. E. 1978: Traditional marine conservation methods in Oceania and their demise. *Ann. Rev. Ecol. Syst.* **9**, 349–64.
JOHANSON, T. B. and STEEN, P. 1979: Solar Sweden. *Bull. Atom. Sci.* **35** (8) 19–22.
JOHN, B. S. 1979: Alternative mythology. *Resurgence* No. **74**, 4–5.

JOHNSON, P. L. 1971: Remote sensing as a tool for study and management of ecosystems. In E. P. Odum, 468–83.

JOHNSON, W. A., STOLTZFUS, V. and CRAUMER, P. 1977: Energy conservation in Amish agriculture. *Science* **198**, 373–78.

JONES, G. 1979: *Vegetation productivity*. London and New York: Longman.

JORDAN, C. F. and KLEINE, J. R. 1972: Mineral cycling: some basic concepts and their application in a tropical rain forest. *Ann Rev. Ecol. Syst.* **3**, 33–49.

JORGENSEN, J. R., WELLS, C. G. and METZ, L. J. 1975: The nutrient cycle in continuous forest production. *Journal of Forestry* **73**, 400–03.

JUST, E. 1976: World economic development and future mineral consumption. In G. J. S. Govett and M. H. Govett (eds.). *World mineral supplies. Assessment and Perspective.* Amsterdam, Oxford and New York: Elsevier Developments in Economic Geology No. **3**, 175–84.

KALININ, G. P. and BYKOV, V. D. 1969: The world's water resources, present and future. *Impact of Science on Society* **19** (2), 135–50.

KARR, J. R. and SCHLOSSER, I. J. 1978: Water resources and the land-water interface. *Science* **201**, 229–34.

KARSCH, R. F. 1970: The social costs of surface mined coal. In A. J. Van Tassel (ed.), 269–90.

KASSAS, M. 1972: Impact of river control schemes on the shoreline of the Nile delta. In M. Taghi Farvar and J. P. Milton (eds.), 179–88.

KATES, R. W. 1962: *Hazard and choice perception in flood plain management*. Chicago: University of Chicago, Department of Geography Research Paper **89**.

KAY, D. A. and SKOLNIKOFF, E. B. 1972: *World eco-crisis: international organizations in response*. Madison and London: University of Wisconsin Press.

KELLEY, D. R., STUNKEL, K. R. and WESCOTT, R. R. 1976: *The economic superpowers and the environment: the United States, the Soviet Union and Japan*. San Francisco and Reading: Freeman.

KEMP, R. H. and BURLEY, J. 1978: Depletion and conservation of forest genetic resources. In J. G. Hawkes (ed.). *Conservation and agriculture*. London: Duckworth, 171–186.

KEMP, W. B. 1970: The flow of energy in a hunting society. *Sci. Amer.* **224** (3), 104–15.

KETCHUM, B. H. 1975: Biological implications of global marine pollution. In S. F. Singer (ed.). *The changing global environment*. Dordrecht: Reidel, 311–28.

KHAN, M. 1977: *Pesticides in aquatic environments*. London and New York: Plenum Press.

KIMMINS, J. P. 1977: Evaluation of the consequences for future tree production of the loss of nutrients in whole-tree harvesting. *Forest Ecology and Management* **1**, 169–83.

KLEIN, D. R. 1970: Tundra ranges north of the boreal forest. *J. Range Management* **23**, 8–14.

KNEESE, A. V. 1977: *Economics and the environment*. Education series Harmondsworth and New York: Penguin books.

KNETSCH, J. and CLAWSON, M. 1967: *Economics of outdoor recreation*. Baltimore: Johns Hopkins Press for RFF.

KOK, F. 1976: Economic and environmental implications of paper production and recycling. In A. P. Carter (ed.). *Energy and the environment: a structural analysis*. Hanover, N. H.: Brandeis University Press, 101–60.

KOK, L. T. 1972: Toxicity of insecticides used for Asiatic rice borer control to tropical fish in rice paddies. In M. Taghi Farvar and J. P. Milton (eds.), 489–98.

KORZUN, V. I. (ed.) 1978: *World water balance and water resources of the earth*. Paris: UNESCO. Studies and Reports in Hydrology No. 25.

KOVDA, V. 1970: Contemporary scientific concepts relating to the biosphere. In UNESCO, 13–29.

— 1977: Soil loss: an overview. *Agro-Ecosystems* **3**, 205–24.

KRUTILLA, J. V. (ed.) 1972: *Natural environments: studies in theoretical and applied analysis*. Resources for the future series. Baltimore and London: Johns Hopkins Press for RFF.

KUPCHANKO, E. E. 1970: Petrochemical waste disposal. In M. A. Ward (ed.), 53–65.

LACK, D. 1966: *Population studies of birds*. Oxford: Clarendon Press. New York: Oxford University Press.

LADEJINSKY, W. 1970: Ironies of India's green revolution. *Foreign Affairs* **48**, 758–68.

LAGLER, K. F. 1971: Ecological effects of hydroelectric dams. In D. A. Berkowitz and A. M. Squires (eds.), 133–57.

LAMORE, G. E. 1971: At last—a revolution that unites. In M. A. Strobbe (ed.), 126–32.

LANDSBERG, H. H. 1971: Energy consumption and optimum population. In S. F. Singer (ed.), 1971, 62–71.

LA PORTE, T. R. 1978: Nuclear waste: increasing scale and socio-political impacts. *Science* **201**, 22–28.

LARKIN, P. A. 1978: Fisheries management—an essay for ecologists. *Ann Rev. Ecol. Syst.* **9**, 57–73.

LAURIE, I. C. (ed.) 1979: *Nature in cities*. Chichester and New York: Wiley-Interscience.

LAUT, P. 1968: *Agricultural geography*. 2 vols. Sydney: Nelson.

LEACH, G. 1975: Energy and food production. *Food Policy* **1**, 1–12.

— 1976: Industrial energy in human food chains. In A. N. Duckham, J. G. W. Jones and E. H. Roberts (eds.). *Food production and consumption: the efficiency of human food chains and nutrient cycles*. Amsterdam and Oxford: North-Holland Publishing Co., 371–82.

LEACH, G. 1979: A future with less energy. *New Sci.* **81**, 81–83.

LEACH, G., LEWIS, C., ROMIG, F., FOLEY, G. and VAN BUREN, A. 1979: *A low energy strategy for the United Kingdom*. London: Science Reviews and IIED.

LEARMONTH, A. T. A. and SIMMONS, I. G. 1977: *Man-environment relations as complex ecosystems*. Open University course D204 Unit **8**.

LECOMBER, R. 1975: *Economic growth versus the environment*. London: Macmillan. New York: Halsted Press.

LECOMBER, R. 1979: *The economics of natural resources*. London: Macmillan.

LEE, D. H. K. (ed.) 1972: *Metallic contaminants and human health*. London and New York: Academic Press.

LEE, R. B. 1968: What hunters do for a living, or how to make out on scarce resources. In R. B. Lee and I. De Vore (eds.), 1968a, 30–43.

— 1969: !Kung Bushman subsistence: an input:output analysis. In A. P. Vayda (ed.), 47–79.

LEE, R. B. and DE VORE, I. (eds.) 1968a: *Man the hunter*. Chicago: Aldine Press.

— 1968b: Problems in the study of hunters and gatherers. In R. B. Lee and I. De Vore (eds.), 1968a, 3–12.

LEEDS, A. and VAYDA, A. P. (eds.) 1965: *Man, culture and animals*. Washington, DC: AAAS Publication **78**.

LENIHAN, J. and FLETCHER, W. W. (eds.) 1975: *Food, agriculture and the environment.* London and Glasgow: Blackie, Environment and man series.

LENIHAN, J. and FLETCHER, W. W. (eds.), 1976: *Reclamation.* London and Glasgow: Blackie, Environment and man series.

LEOPOLD, A. 1949: *Sand County almanac* and *Sketches here and there.* Galaxy Books. New York and Oxford: Oxford University Press.

LEOPOLD, A. S. 1963: Study of wildlife problems in National Parks. *Trans. 28th North American Wildlife Conference,* 28–45.

LEWIS, C. W. 1977: Fuels from biomass-energy outlay versus energy returns: a critical appraisal. *Energy* **2**, 241–48.

LEWIS, G. M. 1969: Range management viewed in the ecosystem framework. In G. M. Van Dyne (ed.), 97–187.

LEWIS, R. S. and SPINRAD, B. I. (eds.) 1972: *The energy crisis.* Chicago: Educational Foundation for Nuclear Science.

LIETH, H. 1965: Versuch einer kartographischen Darstellung der Productivitat der Pflanzendecke auf die Erde. *Geographisches Taschenbuch 1964/5,* 72–80.

— 1972: Modelling the primary productivity of the world. *Nature and Resources* **8** (2), 5–10.

LIKENS, G. E. and BORMANN, F. H. 1971: Mineral cycling in ecosystems. In J. A. Wiens (ed.), 25–67.

— Acid rain: a serious regional environmental problem. *Science* **184**, 1176–79.

LIKENS, G. E., BORMANN, F. H., PIERCE, R. S., EATON, J. S. and JOHNSON, N. M. 1977: *Biogeochemistry of a forested ecosystem.* New York, Heidelberg and Berlin: Springer-Verlag.

LIKENS, G. E., BORMANN, F. H., PIERCE, R. S. and REINERS, W. A. 1978: Recovery of a deforested ecosystem. *Science* **199**, 492–96.

LINDBERG, L. N. 1977: Comparing energy policies: policy implications. In L. N. Lindberg (ed.). *The energy syndrome: comparing national responses to the energy crisis.* Lexington, Mass.: Lexington Books. Farnborough: Teakfield. 357–82.

LINDEMANN, R. L. 1942: The trophic-dynamic aspect of ecology. *Ecology* **23**, 399–418.

LIVINGSTONE, R. S. and MCNIEL., B. (eds.) 1975: *Beyond petroleum. Vol 1: Biomass energy chains.* Palo Alto, Calif.: Stanford University Institute for Energy Studies.

LOEHR, R. C. (ed.), 1978: *Food, fertilizer and agricultural residues.* Ann Arbor: Ann Arbor Science Publishers. Chichester: Wiley.

LOFTAS, T. 1972: *The last resource.* Harmondsworth: Penguin.

LONGHURST, A., COLEBROOK, M., GULLAND, J., LE BRASSEUR, R., LORENZEN, C. and SMITH, P. 1972: The instability of ocean populations. *New Sci.* **54**, 500–502.

LOVELOCK, J. E. 1979: *Gaia—A new look at life on eath.* Oxford and New York: Oxford University Press.

LOVINS, A. B. 1977: *Soft energy paths: toward a durable peace.* Cambridge, Mass.: Ballinger. Plymouth: Macdonald and Evans.

— 1975: *World energy strategies: facts, issues and options.* Cambridge, Mass.: Ballinger. Plymouth: Macdonald and Evans.

LOOMIS, R. S. and GERAKIS, P. A. 1975: Productivity of agricultural ecosystems. In J. P. Cooper (ed.). *Photosynthesis and productivity in different environments.* Cambridge: Cambridge University Press, IBP Studies No. **3**, 145–72.

LOVERING, T. S. 1968: Non-fuel mineral resources in the next century. *Texas Quarterly* **11**, 127–47.

— 1969: Mineral resources from the land. In NAS/NRC, 109–34.
LOWE, V. P. W. 1971: Some effects of a change in estate management on a deer population. In E. Duffey and A. S. Watt (eds.), 437–56.
LUCAS, R. C. 1964: Wilderness perception and use: the example of the boundary waters canoe area. *Nat. Res. J.* **3**, 384–411.
— 1965: *Recreational capacity of the Quetico-Superior area.* St Paul, Minn.: US Forest Service Lakes States Research Paper LS-15.
— 1978: Impact of human presence on parks, wilderness and other recreational lands. In K. A. Hammond, G. Macinko and W. B. Fairchild (eds.). *Sourcebook on the environment. A guide to the literature.* Chicago and London: University of Chicago Press, 221–39.
LUGO, A. E. 1974: The ecological view. In H. G. T. Van Raay and A. E. Lugo (eds.). *Man and environment.* Rotterdam: The University Press, 293–316.
LUMSDEN, M. 1975: 'Conventional' war and human ecology. *Ambio* **4**, 223–28.
LUTEN, D. B. 1971: The economic geography of energy. *Sci. Amer.* **224** (3), 165–75.
— 1974: United States requirements. In A. J. Finkel (ed.), *Energy, the environment and human health.* Acton, Mass.: Publishing Sciences Group, 17–33.
L'VOVICH, M. I. 1978: Turning the Siberian waters south. *New Scientist* **79**, 834–6.
LYNN, D. A. 1975: Air pollution. In W. W. Murdoch (ed.), *Environment.* Sunderland, Mass.: Sinauer, 2nd edn., 224–49.
MACANDREWS, C. and SEN, C. L. 1979: *Developing economies and the environment. The southeast Asian experience.* Singapore: McGraw-Hill.
MACARTHUR, R. H. 1955: Fluctuations of animal populations, and a measure of community stability. *Ecology* **36**, 533–6.
MACARTHUR, R. H. and WILSON, E. O. 1967: *The theory of island biogeography.* Princeton: Princeton Univ. Press.
MACFADYEN, A. 1964: Energy flow in ecosystems and its exploitation by grazing. In D. J. Crisp (ed.), 3–20.
MACKENZIE, W. J. M. 1978: *Biological ideas in politics.* Harmondsworth: Pelican books.
MCCLINTOCK, M., RUSSELL, R. B., SCOVILL, H., WEISS, E. B., WESTING, A. H. and ZOLLA, S. 1974: *Air, water, earth, fire: the impact of the military on world environmental order.* San Francisco: Sierra Club International Series No. **2**.
MCMULLAN, J. T., MORGAN, R. and MURRAY, R. B. 1977: *Energy resources.* London: Edward Arnold. New York: Halsted Press.
MCCORMICK, W. T., FISH, L. W., KALISCH, R. B. and WANDER, T. J. 1978: The future for world natural gas supply. *World energy resources 1985–2020.* London: IPC for World Energy Conference 1977, 48–56.
MCGAUHEY, P. 1968: Earth's tolerance for wastes. *Texas Quarterly* **11**, 36–42.
MCHALE, J. 1969: *The future of the future.* New York: Ballantine.
MACINTOSH, N. A. 1970: Whales and krill in the twentieth century. In M. W. Holdgate (ed.), Vol. 1, 195–212.
MCKEOWN, T. 1976: *The modern rise of population.* London: Edward Arnold. New York: Academic Press.
MCKNIGHT, A. D., MARSTRAND, P. K. and SINCLAIR, T. C. (eds.) 1974: *Environmental pollution control. Technical, economic and legal aspects.* London and New York: Allen and Unwin.
MACNEILL, J. W. 1971: *Environmental management.* Ottawa: Information Canada.

MCVEAN, D. N. and LOCKIE, J. D. 1969: *Ecology and land use in upland Scotland.* Edinburgh: Edinburgh University Press.

MADDOX, J. 1972a: The case against hysteria. *Nature* **235**, 63–5.

— 1972b: *The doomsday syndrome.* London: Macmillan. New York: McGraw Hill.

MAKHIJANI, A. and POOLE, A. 1976: Energy and agriculture in the third world. Cambridge, Mass.: Ballinger.

MARCHETTI, C. 1979: Constructive solutions to the CO_2 problem. In W. Bach et al. (eds.): *Man's impact on climate.* Amsterdam and New York: Elsevier, 299–311.

MARGALEF, R. 1968: *Perspectives in ecological theory.* Chicago: University of Chicago Press.

MARKS, P. L. and BORMANN, F. H. 1972: Revegetation following forest cutting: mechanisms for return to steady-state nutrient cycling. *Science* **176**, 914–15.

MARSH, G. P. 1864: *Man and nature, or physical geography as modified by human action.* Reprinted 1965, Harvard and London: Belknap Press/Harvard Univ. Press. (D. Lowenthal, ed.).

MARTINELLI, W. 1964: *Watershed management in the Rocky Mountain alpine and subalpine zones.* Fort Collins, Colo.: US Forest Service Rocky Mountain Research Station Research Note RM-36.

MARX, W. 1967: *The frail ocean.* New York: Sierra Club/Ballantine.

MATELES, R. I. and TANNENBAUM, S. R. (eds.) 1968: *Single cell protein.* Cambridge, Mass.: MIT Press.

MAUGH, T. 1978: Tar sands: a new fuels industry takes shape. *Science* **199**, 756–60.

— 1979: Toxic waste disposal—a growing problem. *Science* **204**, 819–23.

MAULDIN, W. P. 1980: Population trends and prospects. *Science* **209**, 148–57.

MAY, R. M. 1975: Stability in ecosystems: some comments. In W. H. Van Dobben and R. H. Lowe-McConnell (eds.). *Unifying concepts in ecology.* The Hague: Junk, 161–68.

MAYUR, R. 1979: Environmental problems of developing countries. *Ann. Amer. Acad. Pol. Soc. Sci.* **444**, 89–101.

MAZUR, A. and ROSA, E. 1974: Energy and life-style. *Science* **186**, 607–610.

MEADOWS, D. H., MEADOWS, D. L., RANDERS, J. and BEHRENS, W. W. 1972: *The limits to growth: a Report for the Club of Rome's Project on the predicament of Mankind.* London: Earth Island Press. New York: Universe.

MEDVEDEV, Z. 1977: Facts behind the Soviet nuclear disaster. *New Sci.* **74**, 761–64.

MEGAW, J. V. S. (ed.) 1977: *Hunters, gatherers and first farmers beyond Europe: an archaelogical survey.* Leicester: University Press. New York: Humanities Press.

MEGGERS, B. J. 1971: *Amazonia: man and nature in a counterfeit paradise.* Chicago: Aldine.

MEGGERS, B. J., AYENSU, E. S. and DUCKWORTH, W. D. (eds.) 1973: *Tropical forest ecosystems in Africa and South America: a comparative review.* Washington, DC: Smithsonian Institution Press.

MEIER, R. L. 1969: The social impact of a nuplex. *Bull. Atom. Sci.* **26**, 16–21.

MELLANBY, K. 1975: *Can Britain feed itself?* London: Merlin Press.

— 1977: The future prospect for man. In F. H. Perring and K. Mellanby, 180–184.

MESAVORIC, M. and PESTEL, E. 1975: *Mankind at the turning point: the Second Report to the Club of Rome.* London: Hutchinson. New York: Dutton/NAL.

METCALF, R. L. 1972: DDT substitutes. *CRC Critical Reviews in Environmental Control*

3 (1), 25–59.

METZ, W. D. 1978: Energy storage and solar power: an exaggerated problem. *Science* **200**, 1471–73.

MICKLIN, P. P. 1969: Soviet plans to reverse the flow of rivers: the Kama-Vychegda-Pechora project. *Canad. Geogr.* **13**, 199–215.

MIKOLA, P. 1970: Forests and forestry in subarctic regions. In UNESCO, *Ecology of the subarctic regions*. Paris: UNESCO, 295–302.

MILGRAM, S. 1970: The experience of living in cities. *Science* **167**, 1461–8.

MILLER, D. H. 1977: *Water at the surface of the earth. An introduction to ecosystem hydrodynamics*. New York, San Francisco and London: Academic Press.

MILLER, R. S. and BOTKIN, D. B. 1974: Endangered species: models and predictions. *American Scientist* **62**, 172–81.

MISHAN, E. J. 1967: *The costs of economic growth*. London: Staples Press.

— 1971: *Twenty-one popular economic fallacies*. Harmondsworth: Pelican. New York: Praeger.

— 1977: *The economic growth debate*. London and New York: Allen and Unwin.

MITCHELL, B. 1979: *Geography and resource analysis*. London and New York: Longman.

MITCHELL, J. M. 1970: A preliminary evaluation of atmospheric pollution as a cause of the global temperature fluctuation of the past century. In S. F. Singer (ed.), 1970a, 139–55.

— 1975: A reassesssment of atmospheric pollution as a cause of longterm changes of global temperature. In S. F. Singer (ed.), *The changing global environment*. Dordercht: Reidel, 149–73.

MITRA, A. 1978: *India's population: aspects of quality and control*. New Delhi: Abhinav Publication, 2 vols.

MOISEEV, P. A. 1970: Some aspects of the commercial use of the krill resources of the antarctic seas. In M. W. Holdgate (ed.), Vol. 1, 213–16.

MOLITOR, L. 1968: *Effects of noise on health*. Council of Europe Public Health Committee Report CESP (68), Strasbourg: Council of Europe.

MONCRIEF, L. W. 1970: The cultural basis for our environmental crisis. *Science*, **170**, 508–12.

MONTAGUE, A. (ed.) 1962: *Culture and the evolution of man*. New York: Oxford University Press.

MONTEFIORE, H. 1970: *Can man survive?* London: Fontana.

MOORE, M. R., CAMPBELL, B. C. and GOLDBERG, A. 1977: Lead. In J. Lenihan and W. W. Fletcher (eds.). *The chemical environment*. Glasgow and London: Blackie. Environment and Man, vol. 6, 64–92.

MOORE, N. W. 1969: The significance of the persistent organochlorine insecticides and the polychlorinated biphenyls. *Biologist* **16**, 157–62.

— 1977: The future prospect for wildlife, In F. H. Perring and K. Mellanby, 175–180.

MORE, R. J. 1969: The basin hydrological cycle. In R. J. Chorley (ed.), 65–76.

MORGAN, W. T. W. (ed.) 1972: *East Africa: its peoples and resources*. Nairobi: Oxford University Press. 2nd edn.

MORIARTY, F. 1975: *Pollutants and animals. A factual perspective*. London: Allen and Unwin.

MORLEY, S. G. 1956: *The ancient Maya*. Stanford: Stanford University Press.

MUNN, R. E. (ed.) 1979. *Environmental impact assessment*. Chichester and New York:

Wiley. SCOPE 5, 2nd edn.

MURDOCH, W. W. (ed.) 1975: *Environment*. Stamford, Conn.: Sinauer. 2nd edn.

MURDOCK, G. P. 1968: The current status of the world's hunting and gathering peoples. In R. B. Lee and I. De Vore (eds.), 1968a, 13–20.

MUROZUMI, M., CHOW, T. J. and PATTERSON, C. 1969: Chemical concentrations of pollutant lead aerosols, terrestrial dusts and sea salts in Greenland and Antarctic snow strata. *Geochimica et Cosmochimica Acta* **33**, 1274–94.

MURRA, J. V. 1965: Herds and herders in the Inca state. In A. Leeds and A. P. Vayda (eds.), 185–215.

MUSGROVE, P. J. and WILSON, A. D. 1970: Power without pollution. *New Sci.* **45**, 457–9.

MYERS, N. 1972a: National parks in savannah Africa. *Science* **178**, 1255–63.

— 1972b: *The long African day*. New York: Macmillan. London: Collier-Macmillan.

— 1975: The whaling controversy. *American Scientist* **63**, 448–55.

— 1976: An expanded approach to the problem of disappearing species. *Science* **193**, 198–202.

— 1978: Forests for people. *New Sci.* **80**, 951–53.

— 1979: *The sinking ark*. Oxford and New York: Pergamon Press.

NACE, R. L. 1969: Human use of ground water. In R. J. Chorley (ed.), 285–94.

NASH, C. E. 1970a: Marine fish farming. Part 1. *Marine Pollution Bull.* **1**, 5–6.

— 1970b: Marine fish farming. Part 2. *Marine Pollution Bull.* **1**, 28–30.

NASH, R. 1967: *Wilderness and the American mind*. New Haven and London: Yale University Press.

NAS/NRC 1969: *Resources and man*. London and San Francisco: Freeman.

NATIONAL ACADEMY OF SCIENCES [of the USA] 1975: *Underexploited tropical plants with promising economic value*. Washington DC: NAS.

— 1976: *Energy for rural development: renewable resources and alternative technology for developing countries*. Washington DC: NAS.

NELSON, J. G. and SCACE, R. C. (eds.) 1969: *The Canadian National Parks: today and tomorrow*. Calgary: University of Calgary Studies in Land Use History and Landscape Change National Park Series 3. 2 vols.

NELSON. J. G., NEEDHAM, R. D. and MANN, D. L. (eds.) 1978: *International experience with National Parks and related reserves*. Waterloo, Ontario: University of Waterloo, Department of Geography, Publication Series no. 12.

NELSON-SMITH, A. 1970: The problem of oil pollution of the sea. *Adv. Mar. Biol.* **8**, 215–306.

— 1977: Biological consequences of oil spills. In J. Lenihan and W. W. Fletcher (eds.). *The marine environment*. Glasgow and London: Blackie. Environment and Man, vol. 5, 46–69.

NEWBOULD, P. J. 1971a: Comparative production of ecosystems. In P. F. Wareing and J. P. Cooper (eds.), 228–38.

NEWCOMB, R. M. 1979: *Planning the past: historical landscape resources and recreation* Folkestone: Dawson. Hamden, Conn.: Shoe String.

NEW ZEALAND FOREST SERVICE 1970: *Conservation policy and practice*. Wellington: New Zealand Forest Service.

NG, L. K. Y. and MUDD, S. (eds.), 1965: *The population crisis*. Bloomington: Indiana University Press.

NIEHAUS, F. 1979: Carbon dioxide as a constraint for global energy scenarios. In W.

Bach et al. (eds.). *Man's impact on climate.* Amsterdam and New York: Elsevier, 285–97.

NILLSON, P. O. 1976: The energy balance in Swedish forestry. In C. O. Tamm (ed.) *Man and the boreal forest.* Stockholm: Swedish Natural Science Research Council Ecological Bulletin 21, 95–101

NORMAN, C. 1978: *Soft technologies, hard choices.* Washington DC: Worldwatch Paper No. **21**.

NOSSITER, B. D. 1978: *Britain: A future that works.* London: Andre Deutsch. Boston: Houghton Mifflin.

NOTESTEIN, F. W. 1970: Zero population growth: what is it? *Family Planning Perspectives* **2** (3), 20–24; reprinted in W. H. Davis (ed.), 1971, 107–11.

NRIAGU, J. O. (ed.) 1979: *The biogeochemistry of mercury in the environment.* Amsterdam: Elsevier-North Holland. Topics in Environmental Health vol. 3.

NUMATA, M. 1974: Conservation of flora and vegetation of Japan. In M. Numata (ed.), *The flora and vegetation of Japan.* Tokyo: Kodansha Ltd. Amsterdam, London and New York: Elsevier, 269–77.

O'CONNOR, F. B. 1964: Energy flow and population metabolism. *Sci. Progr.* **52**, 406–14.

ODUM, E. P. 1969: The strategy of ecosystem development. *Science* **164**, 262–70.

— 1971: *Fundamentals of ecology.* 3rd edn. Philadelphia: Saunders. Eastbourne: Holt-Saunders.

— 1975: *Ecology.* Philadelphia: Holt, Rinehart and Winston. Eastbourne: Holt-Saunders.

ODUM, E. P. and ODUM, H. T. 1972: Natural areas as necessary components of man's total environment. *Trans. 37th North American Wildlife and Natural Resources Conference*, 178–89.

ODUM, H. T. 1957: Trophic structure and productivity of Silver Springs, Florida, *Ecol. Monog.* **27**, 55–112.

— 1971: *Environment, power and society.* London and New York: Wiley.

— 1976: Microscopic minimodels of man and nature. *Systems Analysis and Simulation in Ecology* **4**, 249–280.

— 1977: Net benefits to society from alternative energy investments. *Trans. 41st N. American Wildlife and Natural Resources Conference*, 327–38.

ODUM, H. T., KEMP, W., SELL, M., BOYNTON, W. and LEHMAN, M. 1977: Energy relationships of man and estuaries. *Env. Management* **1**, 297–315.

ODUM, H. T. and ODUM, E. C. 1976: *Energy basis for man and nature.* New York and Maidenhead: McGraw-Hill.

OECD 1970: *Scientific fundamentals of the eutrophication of lakes and flowing waters, with particular reference to nitrogen and phosphorus as factors in eutrophication.* Prepared by A. Vollenweider. Paris: OECD.

OORT, A. H. 1970: The energy cycle of the earth. In *Scientific American* (ed.), 13–23.

OPEN UNIVERSITY, 1974: Course S26 (The Earth's Physical Resources) Block 6, *Implications: limits to growth?*

OPENSHAW, K. 1974: Wood fuels the developing world. *New Sci.* **61**, 271–2.

OPHULS, W. 1977: *Ecology and the politics of scarcity.* San Francisco and Reading: Freeman.

O'RIORDAN, T. 1971a: Environmental management. *Progress in Geography* **3**, 173–231.

— 1971b: *Perspectives on resource management.* London: Pion Press. New York: Academic Press.

— 1976: *Environmentalism*. London: Pion Press.

O'RIORDAN, T. and DAVIS, J. 1976: Outdoor recreation and the American environment. In J. Wreford Watson and T. O'Riordan (eds.). *The American environment: perceptions and policies*. London and New York: Wiley, 259–76.

O'RIORDAN, T. and MORE, R. J. 1969: Choice in water use. In R. J. Chorley (ed.), 547–73.

ORRRC 1962a: *Recreation for America*. Washington, DC: US Government Printing Office.

— 1962b: *Report 23, Projections to the Years 1976 and 2000*. Washington DC: US Government Printing Office.

OTHERMER, D. F. and ROELS, O. A. 1973: Power, fresh water, and food from cold, deep sea water. *Science* **182**, 121–5.

OVINGTON, J. D. 1957: Dry matter production by *Pinus sylvestris L.*. *Ann. Bot.* **21**, 287–314.

— 1962: Quantitative ecology and the woodland ecosystem concept. *Adv. Ecol. Res.* **1**, 103–92.

— 1965: *Woodlands*. London: English Universities Press.

PADDOCK, W. C. 1971: Agriculture as a force in determining the United States' optimum population size. In S. F. Singer (ed.), 1971, 89–95.

PADDOCK, W. C. and PADDOCK, M. 1967: *Famine 1975!* Boston: Little, Brown.

PAGE, J. J. and CREASEY, S. C. 1975: Ore grade, metal production, and energy. *J. Research U.S. Geol. Survey*, **3**, 9–13.

PAINE, R. T. 1969: A note on trophic complexity and community stability. *Amer. Nat.* **103**, 91–3.

PARK, C. C. 1980: *Ecology and environmental management*. Folkestone: Dawson. Boulder, Colo.: Westview Press.

PARKER, B. C. (ed.) 1972: *Conservation problems in Antarctica*. Blacksburg, Va.: Virginia Polytechnic.

PARKINS, W. E. 1978: Engineering limitations of fusion power plants. *Science* **199**, 1403–1408.

PARSONS, J. 1971: *Population versus liberty*. London: Pemberton Books.

PASSMORE, J. 1980: *Man's responsibility for nature: ecological problems and western traditions*. London: Duckworth. New York: Scribners. 2nd edn.

PATMORE, J. A. 1971: *Land and leisure in England and Wales*. Newton Abbott: David and Charles. New Jersey: Fairleigh Dickinson University Press.

— 1975: *People, places and pleasure*. Hull: The University; an inaugural lecture.

PATTERSON, W. 1976: *Nuclear power*. Harmondsworth and New York: Penguin (Pelican books).

PAWLEY, W. H. 1976: World picture—present and future. In Duckham *et al.*, (eds.), 13–24.

PAYNE, P. and WHEELER, E. 1971: What protein gap? *New Sci.* **50**, 148–50.

PEACH, W. N. and CONSTANTIN, J. A. 1972: *Zimmermann's World Resources and Industries*. London and New York: Harper and Row. 3rd end.

PEARCE-BATTEN, A. 1979: The future of renewable energy. *Focus* **29** (3), 1–16.

PEARCE, D. W. and ROSE, J. (eds.) 1975: *The economics of natural resource depletion*. London: Macmillan.

PEARSON, C. and PRYOR, A. 1978: *Environment: north and south. An economic interpretation*. Chichester and New York: Wiley.

PEARSON, F. A. and HARPER, F. A. 1945: *The world's hunger*. Ithaca, N. Y.: Cornell

University Press.

PEEL, R. F., CURTIS, L. F. and BARRETT, E. C. (eds.) 1977: *Remote sensing of the terrestrial environment*. London: Butterworth.

PEREIRA, H. C. 1973: *Land use and water resources*. Cambridge and New York: Cambridge University Press.

PERKINS, D. F. 1978: The distribution and transfer of energy and nutrients in the Agrostis-Festuca grassland ecosystem. In O. W. Heal and D. F. Perkins (eds.). *Production ecology of British moors and montane grasslands*. Berlin, Heidelberg and New York: Springer-Verlag Ecological Studies No. **27**, 375–95.

PERRING, F. H. and MELLANBY, K. (eds.) 1977: *Ecological effects of pesticides*. London, New York and San Francisco: Academic Press, Linnean Society Symposium No. 5.

PESON, P. (ed.) 1974: *Ecologie forestière*. Paris: Gunthier-Villars Editeur.

PETERS, W. and SCHILLING, H.-D. 1978: An appraisal of world coal resources and their future availabilty. In *World energy resources 1985–2020*. London: IPC for the World Energy Conference 1977, 57–86.

PETERSON, W. 1969: *Population*. 2nd edn. London and New York. Macmillan.

PIERROU, U. 1976: The global phosphorus cycle. In B. H. Svensson and R. Söderlund (eds.). *Nitrogen, phosphorus and sulphur—global cycles*. Stockholm: Natural Science Research Council Ecological Bulletins, No. **22**, 75–88.

PIGFORD, T. 1974: Environmental aspects of nuclear energy production. *Ann. Rev. Nuclear Science* **24**, 515–59.

PIGOTT, C. D. 1956: The vegetation of Upper Teesdale in the North Pennines. *J. Ecol.* **44**, 545–86.

PIMENTEL, D. 1976: World food crisis: energy and pests. *Bull. Entomol. Soc. America* **22**, 20–26.

— (ed.) 1978: *World food, pest losses and the environment*. Boulder, Colo. and London: Westview Press.

PIMENTEL, D. and PIMENTEL, M. 1979: *Food, energy and environment*. London: Edward Arnold. New York: Halsted Press.

PIPER, A. M. 1965: *Has the US enough water?* US Geological Survey Water Supply Paper **1797**. Washington, DC: US Government Printing Office.

PIRAGES, D. C. (ed.) 1977: *The sustainable society*. New York and London: Praeger.

PIRIE, N. and CRAGG, J. B. (eds.) 1955: *The numbers of men and animals*. Edinburgh and London: Oliver and Boyd. Inst. Biol. Symp. **4**.

PIRIE, N. W. 1969: *Food resources: conventional and novel*. Harmondsworth: Penguin.

— 1970: Orthodox and unorthodox methods of meeting world food needs. In J. Janick (ed.), 223–31.

— 1971: Leaf protein: its agronomy, preparation, quality and use. *IBP Handbook* **20**. Oxford: Blackwell. New York: Lippincott.

PIRSIG, R. M. 1974: *Zen and the art of motorcycle maintenance: an inquiry into values*. New York: Bantam Books. London: Corgi books.

POLLOCK, N. C. 1974: *Animals, environment and man in Africa*. Farnborough: Saxon House.

POLSTER, H. 1961: *Neuere Ergebnisse auf dem Gebiet der Standortsokologischen Assimilations- und Transpirations Forschung an Forstgewechse*. Berlin: Sitzber. Deut. Akad. Landwirtschaftwiss, **10**, 1.

POMEROY, L. R. 1970: The strategy of mineral cycling. *Ann. Rev. Ecol. Syst.* **1**, 171–90.

— 1974: *The ocean's food web: a changing paradigm*. BioScience **24**, 499–504.

POPULATION REFERENCE BUREAU (annually): *World population data sheet.* New York: PRB.

PRICKETT, C. N. 1963: Use of water in agriculture. In Institute of Civil Engineers, 15–29.

PRIESTLEY, J. B. and HAWKES, J. 1955: *Journey down a rainbow.* London: Cressett Press.

PROBSTEIN, R. F. 1973: Desalination. *American Scientist* **61**, 280–93.

PROVINCE OF ALBERTA 1968: *Water diversion proposals of North America.* Edmonton: Department of Agriculture Water Resources Division.

— 1969: PRIME: *Alberta's blueprint for water development.* Edmonton: Department of Agriculture Water Resources Division.

PRYDE, P. R. 1972: *Conservation in the Soviet Union.* New York and Cambridge: Cambridge University Press.

PUGH, G. E. 1978: *The biological origin of human values.* London: RKP. New York: Basic Books.

PUGH, N. J. 1963: Water supply. In Institute of Civil Engineers, 9–14.

PYKE, M. 1970a: *Man and food.* London: Weidenfeld and Nicolson. New York: McGraw Hill.

— 1970b: *Synthetic food.* London: Murray.

— 1971: Novel sources of energy and protein. In P. F. Wareing and J. P. Cooper (eds.), 202–12.

RANDERSON, P. F. and BURDEN, R. F. 1972: Quantitative studies of the effect of human trampling on vegetation as an aid to the management of semi-natural areas. *J. Appl. Ecol.* **9**, 439–57.

RAPPORT, D. J. and TURNER, J. E. 1977: Economic models in ecology. *Science* **195**, 367–73.

RAPPAPORT, R. A. 1971: The flow of energy in an agricultural society. *Sci. Amer.* **224** (3), 116–32.

RATCLIFFE, D. A. 1977: *A nature conservation review.* Cambridge and New York: Cambridge University Press, 2 vols.

RAVEN, P. H. 1976: Ethics and attitudes. In J. B. Simmons, R. I. Beyer, P. E. Brendham, G. Ll. Lucas and V.T.H. Parry (eds.). *Conservation of threatened plants.* New York and London: Plenum Press.

RAY, A. J. 1975: Some conservation schemes of the Hudson's Bay Company 1821–50: an examination of the problems of management in the fur trade. *J. Historical Geography* **1**, 49–68.

RAY, G. C. 1976: Critical marine habitats. *Proceedings of the 1st International Conference on Marine Parks and Reserves.* Morges: IUCN PubsNS No. **37**, 15–59.

READER'S DIGEST PUBLICATIONS 1970: *The living world of animals.* London: Reader's Digest Assn.

REED, C. A. 1969: The pattern of animal domestication in the prehistoric Near East. In P. J. Ucko and G. W. Dimbleby (eds.), 361–80.

REED, C. B. 1975: *Fuels, minerals and human survival.* Ann Arbor: Ann Arbor Science Publishers Inc. Chichester: Wiley.

REED, T. B. and LERNER, R. M. 1973: Methanol: a versatile fuel for immediate use. *Science* **182**, 1299–1304.

REICH, C. 1971: *The greening of America: how the youth revolution is trying to make America liveable.* Harmondsworth: Penguin. New York: Random House.

REICHLE, D. (ed.) 1970: *Analysis of temperate forest ecosystems*. London: Chapman and Hall.

REINING, P. and TINKER, I. (eds.) 1975: *Population: dynamics, ethics and policy*. Washington DC: AAAS.

RENDEL, J. 1975: The utilization and conservation of the world's animal genetic resources. *Agriculture and Environment* **2**, 101–19.

RENNIE, P. J. 1955: The uptake of nutrients by mature forest growth. *Plant and Soil* **7**, 49–95.

REVELLE, R., KHOSLA, A. and VINOVSKIS, M. (eds.) 1971: *The survival equation: man, resources and his environment*. Boston: Houghton Mifflin.

REVELLE, R. 1976: Energy use in rural India. *Science* **192**, 969–75.

REX, R. W. 1971: Geothermal energy—the neglected energy option. *Bull. Atom. Sci.* **27** (8), 52–6.

RICHARDS, P. 1952: *The tropical rain forest*. Cambridge and New York: Cambridge University Press.

RICHARDSON, S. D. 1971: The end of forestry in Great Britain. *Advmt. Sci.* **27**, 153–63.

RICH, V. 1979: Water quality threatens Polish agriculture. *Nature, Lond.* **280**, 266.

RICKER, W. 1969: Food from the sea. In NAS/NRC, 87–108.

RIDKER, R. G. 1972: Population and pollution in the United States. *Science* **176**, 1085–90.

— 1973: To grow or not to grow: that's not the relevant question. *Science* **182**, 1315–18.

RINGWOOD, A. E., KESSON, S. E., WARE, N. G., HIBBERSON, W. and MAJOR, A. 1979: Immobilisation of high level nuclear reactor wastes in SYNROC. *Nature, Lond.* **278**, 219–23.

ROBERTS, B. 1965: Wildlife conservation in the Antarctic. *Oryx* **8**, 237–44.

ROBERTS, K. 1978: *Contemporary society and the growth of leisure*. London and New York: Longmans.

ROCHLIN, G. I. 1977: Nuclear waste disposal: two social criteria. *Science* **195**, 23–31.

RODIN, L. E. and BAZILEVIC, N. I. 1966: The biological production of the main vegetation types in the northern hemisphere of the old world. *For. Abstr.* **27**, 369–72.

ROGERS, F. C. 1971: Underground power plants. *Bull. Atom. Sci.* **27** (8), 38–41, 51.

ROSSIN, A. D. and RIECK, T. A. 1978: Economics of nuclear power. *Science* **201**, 582–89.

ROTHE, J. P. 1968: Fill a dam, start an earthquake. *New Sci.* **39**, 75–8.

ROUTLEY, R. and ROUTLEY, V. 1977: Destructive forestry in Australia and Melanesia. In J. Winslow (ed.): *The Melanesian environment*. Canberra: ANU Press, 374–97.

ROWNTREE, R. A., HEATH, D. E. and VOILAND, M. 1978: The United States park system. In J. G. Nelson *et al.* (eds.) *op. cit.*, 91–140.

RUBINOFF, I. 1968: Central American sea-level canal: possible biological effects. *Science* **161**, 857–61.

RUDD, R. D. 1974: *Remote sensing: a better view*. North Scituate, Mass.: Duxbury Press.

RUDD, R. L. 1975: Pesticides. In W. W. Murdoch (ed.). *Environment*. Sunderland, Mass: Sinauer, 2nd edn., 325–53.

RUIVO, M. (ed.) 1972: Marine pollution and sea life. London: Fishing News (Books) Ltd.

RUSSELL, C. S. and LANDSBERG, H. H. 1971: International environmental problems—a taxonomy. *Science* **172**, 1307–14.

RYTHER, J. 1969: Photosynthesis and fish production in the sea. *Science* **166**, 72–6.

SAARINEN, T. F. 1966: *Perception of the drought hazard on the Great Plains*. Chicago: University of Chicago Department of Geography Research Papers 106.

SABINS, F. 1978: *Remote sensing. Principles and interpretation*. San Francisco: Freeman.

SABLOFF, J. A. 1971: The collapse of classic Maya civilization. In J. A. Harte and R. H. Socolow (eds.), 16–27.

SAGE, B. L. 1970: Oil and Alaskan ecology. *New Sci.* **46**, 175–7.

SAI, F. T. 1977: *Population and national development—the dilemma of developing countries*. London: IPPF occasional Essay No. **2**.

SALAS, R. M. 1978: Is population growth slowing down? *Populi* **5** (2), 3–5.

SANDBACH, F. 1980: *Environment, ideology and policy*. Oxford: Blackwell.

SAN PIETRO, A., GREER, F. and ARMY. T. J. (eds.) 1967: *Harvesting the sun*. London and New York: Academic Press.

SARGENT, F. 1969: A dangerous game: taming the weather. In G. W. Cox (ed.), 569–82.

SARGEANT, F. S. (ed.) 1974: *Human ecology*. Amsterdam: North-Holland 206–33.

SASKATCHEWAN-NELSON BASIN BOARD 1972: *Water supply for the Saskatchewan-Nelson Basin: a summary report*. Ottawa: Information Canada.

SASSI, T. 1970: The harmful side effects of pesticide use. In A. J. Van Tassel (ed.), 361–95.

SATER, J. E., RONHOVE, A. G. and VAN ALLEN, L. C. 1972: *Arctic environments and resources*. Washington, DC: Arctic Institute of North America.

SAUER, C. O. 1952: *Agricultural origins and dispersals*. New York: American Geographical Society (reprinted 1969, Cambridge, Mass.: MIT Press).

— 1961: Fire and early man. *Paideuma* **7**, 399–407; reprinted in J. Leighley (ed.), 1963, *Land and life*. Berkeley, Los Angeles and London: University of California Press, 288–99.

SAWYER, C. N. 1966: Basic concepts of eutrophication. *J. Wat. Polln. Control Fedn.* **38**, 737–44.

SCHIFF, H. I (ed.) 1979; *Stratospheric ozone depletion by halocarbons: chemistry and transport*. Washington, D.C.: National Academy of Sciences.

SCHIPPER, L. and LICHTENBERG, A. J. 1976: Efficient energy use and well-being: the Swedish example. *Science* **194**, 1001–1013.

SCHNEIDER, S. and MESIROW, L. 1976: *The Genesis strategy*. New York and London: Plenum Press.

SCHODDE, R. 1973: General problems of fauna conservation in relation to the conservation of fauna in New Guinea, In A. B. Costin and R. H. Groves (eds.): *Nature conservation in the Pacific*. Canberra: ANU Press, 123–144.

SCHOFIELD, E. A. (ed.) 1978: *Earthcare: global protection of natural areas*. Boulder, Colo.: Westview Press.

SCHUBERT, J. and LAPP, R. E. 1957: *Radiation*. New York: Viking Press.

SCHULTZ, A. M. 1967: The ecosystem as a conceptual tool in the management of natural resources. In S. V. Ciriacy Wantrup and J. J. Parsons (eds.). 139–61.

SCHULTZ, V. and KLEMENT, A. W. (eds.) 1963: *Radioecology* London: Chapman and Hall.

SCHURR, S. H. (ed.) 1972: *Energy, economic growth and the environment*. Resources for the future series. Baltimore and London: Johns Hopkins Press for RFF.

'SCIENTIFIC AMERICAN' (ed.) 1970: *The biosphere*. San Francisco and Reading Freeman.

— (ed.) 1976: *Food and agriculture*. San Francisco and Reading: Freeman.

SCOTTER, G. W. 1970: Reindeer husbandry as a land use in northern Canada. In W. A. Fuller and P. G. Kevan (eds.), 159–69.

SCRIMSHAW, N. 1970: Food. In J. Janick (ed.), 206–14.

SEARS, P. B. 1956: The importance of forests to man. In S. Haden-Guest (ed.), 3–12.

SEHLIN, H. 1966: The importance of open-air recreation. In *First International Congress on Leisure and Tourism*, Theme 1, Report 1. Rotterdam: Alliance Internationale de Tourisme.

SENGE, T. 1969: The planning of national parks in Japan and other parts of Asia. in J. G. Nelson and R. C. Scace (eds.), 706–21.

SEWELL, W. R. D. and BURTON, I. 1971: *Perceptions and attitudes in resource management*. Ottawa: Information Canada. Canadian Department of Energy, Mines and Resources Resource Paper 2.

SHEAIL, J. 1976: *Nature in trust*. Glasgow and London: Blackie & Son.

SHEETS, T. J. and PIMENTEL. D. (eds.): *Pesticides. Contemporary roles in agriculture, energy, and the environment*. New York and London: Plenum Press.

SHELL OIL COMPANY 1972: *The national energy position*. Houston: Shell Oil Co.

SHEPARD, P. and MCKINLEY, D. 1971: *Environ/mental: essays on the planet as a home*. Boston: Houghton Mifflin.

SIEGLER, D. S. (ed.) 1977: *Crop resources*. New York and London: Academic Press.

SILLS, D. 1975: The environmental movement and its critics. *Human Ecology* **3**, 1–41.

SIMENSTAD, C. A., ESTES, J. A. and KENYON, K. W. 1978: Aleuts, sea otters and alternate stable-state communities. *Science* **200**, 403–11.

SIMMONDS, N. W. (ed.) 1976: *Evolution of crop plants*. London and New York: Longman.

SIMMONS, I. G. 1966: Wilderness in the mid-20th century USA. *Town Planning Rev*. **36**, 249–56.

— 1967: How do we plan for change? *Landscape* **17**, 22–4.

— 1973: The protection of ecosystems and landscapes in Hokkaido, Japan. *Biol. Cons.* **5**, 281–9.

— 1974: National parks in developed countries. In A. Warren and F. B. Goldsmith (eds.) *Conservation in practice*. London: Wiley, 393–407.

— 1975: *Rural recreation in the industrial world*. London: Edward Arnold. New York: Halsted Press.

SIMMONS, I. G. and VALE, T. 1975: Problems of the conservation of the California coast redwood and its environment. *Environmental Conservation* **2**, 29–38.

SIMONS, M. 1969: Long term trends in water use. In R. J. Chorley (ed.), 535–44.

SINGER, S. F. (ed.) 1970a: *Global effects of environmental pollution*. Dordrecht and Hingham, Mass.: Reidel.

— 1970b: Human energy production as a process in the biosphere. In Scientific American (ed.), 105–13.

— (ed.) 1971: *Is there an optimum level of population?* New York: McGraw-Hill.

— (ed.) 1975: *The changing global environment*. Dordrecht and New York: Reidel.

SINGH, R. B. 1978: National parks, game sanctuaries and public reserves of India. In J. G. Nelson, R. D. Needham and D. L. Mann (eds.). *International Experience with National Parks and Related Reserves*. Waterloo, Ont.: Dept. of Geography Publication series No. **12**, 235–69.

SIOLI, H. 1973: Recent human activities in the Brazilian Amazon region and their ecological effects. In B. Meggers *et al*. (eds.) *Tropical forest ecosystems in Africa and South America: a comparative review*. Washington DC: Smithsonian Institution Press, 321–4.

SINHA, R. (ed.) 1978: *The world food problem: consensus and conflict*. Oxford and New York: Pergamon Press.

SIPRI (Stockholm International Peace Research Institute) 1977: see Westing 1977.

SKOLIMOWSKI, H. 1976: Ecological humanism. *Tract* **19** & **20**, 3–41.

SLESSER, M. 1975: Energy requirements of agriculture. In J. Lenihan and W. W. Fletcher (eds.). *Food, agriculture and the environment.* Glasgow and London: Blackie, 1–20.

— 1978: *Energy in the economy*. London: Macmillan. New York: St. Martins Press.

SLY, J. M. A. 1977: Changes in the use of pesticides since 1945. In F. H. Perring and K. Mellanby, 1–6.

SMIL, V. 1979: Energy flows in the developing world. *American Scientist* **67**, 522–31.

SMITH, C. H. and TOOMBS, R. B. 1978: Development of conventional energy resources. In L. Grainger (ed.), *Energy resources: availability and national use*. London: IPC for the World Energy Conference 1977, 17–25.

SMITH, K. 1972: *Water in Britain*. London: Macmillan.

SMITH, V. L. (ed.) 1978: *Hosts and guests. The anthropology of tourism.* Oxford: Blackwell. Philadelphia: Univ. of Pennsylvania Press.

SNAYDON, R. W. and ELSTON, J. 1976: Flows, cycles and yields in agricultural ecosystems. In A. N. Duckham *et al.*, 43–60.

SÖDERLUND, R. and SVENSSON, B. H. 1976: The global nitrogen cycle. In B. H. Svensson and R. Söderlund (eds.): *Nitrogen, phosphorus and sulphur—global cycles.* Stockholm: Swedish Natural Science Council Ecological Bulletins No. **22**, 23–73.

SOLOMON, M. E. 1976: *Population dynamics.* Studies in Biology 18. London: Edward Arnold. Baltimore: University Park Press. 2nd edn.

SONNENFELD, J. 1966: Variable value in space landscape: an enquiry into the nature of environmental necessity. *J. Social Issues* **22**, 71–82.

SPEDDING, C. R. W. 1975: Grazing systems. In R. L. Reid (ed.), *Proc III World Conference on Animal Production.* Sydney: Sydney University Press, 145–57.

SPEIGHT, M. C. D. 1973: *Outdoor recreation and its ecological effects.* University College London: Discussion Papers in Conservation no. **4**.

SPENCER, J. E. 1966: *Shifting cultivation in S. E. Asia.* Library reprint series. Berkeley, Los Angeles and London: University of California Press.

SPIEGLER, K. S. and BERGMAN, J. I. 1974: *Optimal expansion of a water resources system.* New York: Academic Press.

SPILHAUS, A. 1972: Ecolibrium. *Science* 175, 711–15.

SQUIRES, R. H. 1978: Conservation in Upper Teesdale: contributions from the palaeo-ecological record. *Trans. Inst. Brit. Geogr.* NS **3**, 129–50.

STAMP, L. D. 1969: *Nature conservation in Britain.* New Naturalist Library series. London: Collins. New York: Collins-World.

STANHILL, G. 1974: Energy and agriculture: a national case study. *Agro-Ecosystems* **1**, 207–15.

STANKEY, G. H. 1976: Forest management policy: its evolution and response to public values. In J. Wreford Watson and T. O'Riordan (eds.). *The American environment: perceptions and policies.* Chichester: Wiley, 241–58.

STARK, N. 1972: Nutrient cycling pathways and litter fungi. *BioScience* **22**, 355–60.

STEELE, F. and BOURNE, A. (eds.) 1975: *The man/food equation.* New York and London: Academic Press.

STEINER, D. and CLARKE, J. F. 1978: The Tokamak: model T fusion reactor. *Science* **199**, 1395–1403.

STEINHART, C. and STEINHART, J. 1974: *The fires of culture.* Belmont, Calif.: Wadsworth

Publishing Co.
STERN, A. C. (ed.) 1976: *Air pollution.* London and New York: Academic Press, 3rd edn., 5 vols.
STERNBERG, H. O'R. 1975: The Amazon river of Brazil. *Geogr. Zeitschrift* Heft **40**.
STEWART, C. M. 1970: Family limitation programmes in various countries. In A. Allison (ed.), 204–21.
STODDART, D. R. 1968: The Aldabra affair. *Biol. Cons.* **1**, 63–70.
STODDART, D. R. and WRIGHT, C. A. 1967: Ecology of Aldabra atoll. *Nature, Lond.* 213, 1173–7.
STONE, E. C. 1965: Preserving vegetation in parks and wilderness. Science **150**, 1261–7.
STOTT, D. H. 1962: Checks on population growth. In A. Montague (ed.), 355–76.
STOTT, P. A. 1978: Tropical rain forest in recent ecological thought: the reassessment of a non-renewable resource. *Progr. in Phys. Geogr.* **2**, (1), 80–98. London: Edward Arnold. New York: Cambridge University Press.
STREETER, D. C. 1971: The effects of public pressure on the vegetation of chalk downland at Box Hill, Surrey. In E. Duffey and A. S. Watt (eds.), 459–68.
STROBBE, M. A. (ed.), 1971: *Understanding environmental pollution.* St Louis: Mosby. London: Year Book Medical Books.
STRONG, M. F. (ed.) 1972: *Environment and development.* The Hague: Mouton.
STUDY OF CRITICAL ENVIRONMENTAL PROBLEMS 1970: *Man's impact on the global environment: assessment and recommendation for action.* Cambridge, Mass.: MIT Press.
STUDY OF MAN'S IMPACT ON CLIMATE 1971: *Inadvertent climate modification.* Cambridge, Mass.: MIT Press.
STYCOS, J. M. 1974: The population debate: colours to the mast. *New Internationalist*, no. **20**, 21–4.
SUMMERS, C. 1971: The conversion of energy. *Sci. Amer.* **224** (3), 149–60.
— 1978: Grey seals: the 'con' in conservation. *New Scientist* **80**, 694–95.
SWADLING, P. 1977: Depletion of shellfish in the traditional gathering beds of Pari. In J. Winslow (ed.), *The Melanesian environment.* Canberra: ANU Press, 182–87.
SWANK, W. G. 1972: Wildlife management in Masailand, East Africa. *Transactions 37th North American wildlife and natural resources conference*, 278–87.
SWEET, L. E. 1965: Camel pastoralism in N. Arabia and the minimal camping unit. In A. Leeds and A. P. Vayda (eds.), 129–52.
TABAH, L. 1979: A deeper understanding since Bucharest. *People* **6** (2), 5–7.
TAGHI FARVAR, M. and MILTON, J. P. (eds.) 1972: *The careless technology: ecology and international development.* New York: Natural History Press.
TAIGANIDES, E. P. 1967: The animal waste disposal problem. In N. C. Brady (ed.), 385–394.
TALBOT, L. M. and TALBOT, M. H. (eds.) 1968: *Conservation in tropical south east Asia.* IUCN Publications New Series **10**. Morges: IUCN.
TAMURA, T. 1970: *Marine aquaculture.* Washington DC: NTIS, 2 vols.
TANSLEY, A. G. 1935: The use and abuse of vegetational concepts and terms. *Ecology* **16**, 284–307.
TATSUKAWA, R. 1976: PCB pollution of the Japanese environment. In K. Higuchi (ed.): *PCB poisoning and pollution.* Tokyo: Kodansha Ltd. London and New York: Academic Press, 147–79.
TATTON, J. O'G. and RUZICKA, J. H. A. 1967: Organochlorine pesticides in Antarctica. *Nature, Lond.* **215**, 346–8.

TAYLOR, G. 1978: Sweden strides towards a solar society. *New Sci.* **79**, 550–52.
TAYLOR, J. V. 1975: *Enough is enough: a biblical call for moderation in a consumer oriented society*. London: SCM Press. Minneapolis, Minn.: Augsburg.
TAYLOR, I. J. and SENIOR, P. J. 1978: Single cell proteins: a new source of animal feeds. *Endeavour* NS **2** (1) 31–4.
TEITELBAUM, M. S. 1974: Population and development: is a concensus developing? *Foreign Affairs* **52**, 742–60.
— 1975: Relevance of demographic transition theory for developing countries. *Science* **188**, 420–25.
TERBORGH, J. 1975: Faunal equilibria and the design of wildlife preserves. In F. B. Golley and E. Medina (eds.): *Tropical Ecological Systems. Trends in Terrestrial and Aquatic Research*. Berlin, Heidelberg and New York: Springer-Verlag Ecological Studies No. **11**, 369–80.
— 1976: Island biogeography and conservation: strategy and limitations. *Science* **193**, 1027–29.
THOMAS, C. 1974: *Material gains*. London: Friends of the Earth/Earth Resources Research.
THOMAS, T. M. 1973: World energy sources: survey and review. *Geogr. Rev.* **63**, 246–58.
THOMAS, W. L. (ed.) 1956: *Man's role in changing the face of the earth* Chicago and London: University of Chicago Press.
THOMPSON, L. M. 1975: Weather variability, climatic change and grain production. *Science* **188**, 535–41.
TOYNBEE, A. 1972: The religious background of the present environmental crisis. *Int. J. Environ. Studs.* **3**, 141–6.
TREWARTHA, G. T. 1969: *A geography of population: world patterns*. New York and Chichester: Wiley.
TUAN, YI-FU 1968: Discrepancies between environmental attitude and behaviour: examples from Europe and China. *Canad. Geogr.* **13**, 176–91.
— 1970: Our treatment of the environment in ideal and actuality. *American Scientist* **58**, 244–49.
TURNER, J. 1962: The *Tilia* decline: an anthropogenic interpretation. *New Phytol.* **61**, 328–41.
UCKO, P. J. and DIMBLEBY, G. W. (eds.) 1969: *The domestication and exploitation of plants and animals*. London: Duckworth. New York: Aldine.
UNESCO 1970: *Use and conservation of the biosphere*. Paris: UNESCO Natural Resources Research **X**.
— 1972: *Ecological effects of increasing human activities in tropical and subtropical forest systems*. Paris: UNESCO Man and the Biosphere Programme (MAB) report series No. **3**.
— 1972: *Ecological effects of human activities on the value and resources of lakes, rivers, marshes, delta, estuaries and coastal zones*. Paris: UNESCO Man and the Biosphere Programme (MAB) report series No. **5**.
— 1973: *Ecology and rational use of island ecosystems*. Paris: UNESCO Man and the Biosphere Programme (MAB) report series No. **11**.
— 1973: *Conservation of natural areas and of the genetic material they contain*. Paris: UNESCO Man and the Biosphere Programme (MAB) report series no. **12**.
— 1974a: *Taskforce on criteria and guidelines for the choice and establishment of biosphere reserves. Final report*. Paris: UNESCO Man and the Biosphere Programme (MAB)

report series No. **22**.

— 1974b: *Impact of human activities on mountain and tundra ecosystems*. Paris: UNESCO Man and the Biosphere Programme (MAB) report series No. **14**.

UNITED KINGDOM 1970: 'The Countryside in 1970', 3rd Conference, Report 4: *Refuse disposal*. London: HMSO.

UNITED NATIONS (annually): *Demographic yearbook*. New York: UNO.

— (annually): *Statistical yearbook*. New York: UNO.

— 1977: Water supply and management. Proc. UN Water Conference, Mar del Plata, Argentina, 4 vols.

DEPARTMENT OF SOCIAL AFFAIRS 1953: *The determinants and consequences of population trends*. New York: UN Population Study **17**.

DEPARTMENT OF ECONOMIC AND SOCIAL AFFAIRS 1958: *The future growth of world population*. New York: UN Population Study **28**.

— 1964: *World population prospects as assessed in 1963*: New York: UN Population Study **41**.

US CONGRESS 1970: *Phosphates in detergents and the eutrophication of America's waters*. Washington, DC: House of Representatives Report 91–1004.

US DEPARTMENT OF AGRICULTURE 1955: *Water*. Yearbook of agriculture 1955. Washington, DC: US Government Printing Office.

US DEPARTMENT OF THE INTERIOR 1967: *Surface mining and our environment*. Report of the Strip and Surface Mine Study Policy Committee. Washington, DC: US Government Printing Office.

US FEDERAL COUNCIL FOR SCIENCE AND TECHNOLOGY, COMMITTEE ON ENVIRONMENTAL QUALITY 1968: *Noise-sound without value*. Washington, DC: US Government Printing Office.

US GEOLOGICAL SURVEY 1970: *Mercury in the environment*. Washington, DC: USGS Professional Paper 713.

US NATIONAL COMMISSION ON MATERIALS POLICY 1973: *Material needs and the environment today and tomorrow*. Washington DC: US Government Printing Office.

US PRESIDENT'S COUNCIL ON ENVIRONMENTAL QUALITY 1971: *Environmental quality 1971*. Washington, DC: US Government Printing Office.

UTTON, A. E. and TECLAFF, L. (eds.) 1978: *Water in a developing world: the management of a critical resource*. Boulder, Colo. and London: Westview Press.

VALE, T. R. 1975: Ecology and environmental issues of the Sierra Redwood (*Sequoiadendron giganteum*) now restricted to California. *Environmental Conservation* **2**, 179–88.

VAN DER LEEDEN, F. 1975: *Water resources of the world. Selected statistics*. Port Washington, N.Y.: Water Information Center.

VAN DER MEIDEN, H. A. 1974: Forests and raw material shortage. *Agriculture and Environment* **1**, 139–52.

VAN DER SCHALIE, H. 1972: WHO project Egypt 10: a case history of a schistosomiasis control project. In M. Taghi Farvar and J.P. Milton (eds.), 116–36.

VAN DYNE, G. M. (ed.) 1969: *The ecosystem concept in resource management*. London and New York: Academic Press.

VAN HYLCKAMA, T. E. A. 1975: Water resources. In W. W. Murdoch (ed.), *Environment*. Sunderland, Mass.: Sinauer, 2nd edn., 147–65.

VANN, A. and ROGERS, P. (eds.) 1972: *Human ecology and world development*. London and

New York: Plenum Press.

VAN OSTEN, R. (ed.) 1972: *World national parks: progress and opportunities*. Brussels: Hayez.

VAN RENSBURG, H. J. 1969: *Management and utilization of pastures in east Africa (Kenya, Tanzania, Uganda)*. FAO Pasture and Fodder Crop Series **3**. Rome: FAO.

VAN TASSEL, A. J. (ed.) 1970: *Environmental side—effects of rising industrial output*. Lexington, Mass.: Heath Lexington.

VARLEY, M. E. 1974: *Whole ecosystems*. Open University Course s323 Block A, Unit 5.

VAYDA, A. P. (ed.) 1969: *Environment and cultural behaviour*. New York: Natural History Press.

VERNEY, R. B. 1972: *Sinews for survival: a report on the management of natural resources*. London: HMSO for the Department of the Environment.

VITOUSEK, P., GOSZ, J. R., GRIER, C. C., MELILLO, J. M., REINERS, W. A., and TODD, R. L. 1979: Nitrate losses from disturbed ecosystems. *Science* **204**, 469–74.

VOKES. F. M. 1976: The abundance and availability of mineral resources. In G. J. S. Govett and M. H. Govett (eds.): *World mineral supplies. Assessment and perspective*. Amsterdam and Oxford: Elsevier Developments in Economic Geology, **3**, 65–97.

WAGAR, J. A. 1970: Growth versus the quality of life. *Science* 168, 1179–84.

— 1974: Recreational carrying capacity reconsidered. *Journal of Forestry* **72**, 274–8.

WALFORD, L. A. 1958: *Living resources of the sea*. New York: Ronald Press.

WALKER, C. 1971: *Environmental pollution by chemicals*. London: Hutchinson.

WALL, G. and WRIGHT, C. 1977: *The environmental impact of outdoor recreation*. Waterloo Ont.: University Department of Geography Publication Series No. **11**.

WALTER, H. 1978: Impact of human activity on wildlife. In K. A. Hammond, G. Macinko and W. B. Fairchild (eds.): *Sourcebook on the Environment. A Guide to the Literature*. Chicago and London: University of Chicago Press, 241–62.

WANG, F. F. H. and MCKELVEY, V. E. 1976: Marine mineral resources. In G. J. S. Govett and M. H. Govett (eds.): *World mineral supplies: assessment and prospect*. Amsterdam, Oxford and New York: Elsevier Developments in Economic Geology **3**, 221–86.

WARD, B. and DUBOS, R. 1972: *Only one earth*: Harmondsworth: Penguin. New York: Norton (*as the care and maintenance of a small planet*).

WARD, B. 1979: *Progress for a small planet*. Harmondsworth: Penguin (Pelican books) New York: Norton.

WARD, M. A. (ed.) 1970: *Man and his environment*. Part 1. Oxford and New York: Pergamon Press.

WAREING, P. F. and COOPER, J. P. 1971: *Potential crop production: a case study* London and New York: Heinemann Education.

WARREN, A. and GOLDSMITH, F. B. (eds.) 1974: *Conservation in practice*. Chichester and New York: Wiley.

WATER RESOURCES BOARD 1966: *Water supplies in SE England*. London: HMSO.

— 1969: *Planning our future water supply*. London: HMSO.

— 1970: *Water resources in the north*. London: HMSO.

WATT, K. E. F. 1965: Community stability and the strategy of biological control. *Canad. Entomol.* **97**, 887–95.

— 1968: *Ecology and resource management: a quantitative approach*. New York and Maidenhead: McGraw-Hill.

WATT, K. E. F., MOLLOY, L. F., VARSHNEY, C. K., WEEKS, D. and WIROSARDJONO, S., 1977: *The unsteady state. Environmental problems, growth and culture.* Honolulu: University of Hawaii Press for the East-West Center.

WATTERS, R. F. 1960: The nature of shifting cultivation: a review of recent research. *Pacific Viewpoint* **1**, 59–99.

WATTS, D. 1971: *Principles of biogeography*. London and New York: McGraw-Hill.

WEBB, M. 1979: *The chemistry, biochemistry and biology of cadmium.* Amsterdam, London and New York: Elsevier—North Holland. Topics in Environmental Health, vol. 2.

WECK, J. and WIEBECKE, C. 1961: *Weltwirtschaft und Deutschlands Forst-und Holzwirtschaft.* Munich: Bayerischer Landwirtschaftsverlag.

WEINBERG, A. M. 1968: Raw materials unlimited. *Texas Quarterly* **11**, 92–102.

WEINER, N. 1950: *The human use of human beings*. Boston: Houghton Mifflin.

WEISZ, J. A. 1970: The environmental effects of surface mining and mineral waste generation. In A. J. Van Tassel (ed.), 291–312.

WESLEY, J. P. 1974: *Ecophysics.* Springfield, Ill.: C. C. Thomolac.

WESTING, A. H. 1976: *Ecological consequences of the second Indochina war.* Stockholm: SIPRI/Almqvist and Wiksell.

— 1977: *Weapons of mass destruction and the environment.* London: Taylor and Francis for SIPRI.

WESTMAN, W. E. 1977: How much are nature's services worth? *Science* **197**, 960–64.

WESTOBY, J. C. 1963: The role of forest industries in the attack on economic underdevelopment. *Unasylva* **16**, 168–201.

WHITE, G. F. (ed.) 1964: *Choice of adjustment to floods.* Chicago: University of Chicago Department of Geography Research Paper **93**.

— 1966: *Alternatives in water management.* Washington, DC: NAS/NRC Publication 1408.

— (ed.) 1974: *Natural hazards. Local, national, global.* New York, Toronto and London: OUP.

— (ed.) 1977: *Environmental effects of complex river development.* Boulder, Colo.: Westview Press. London: Benn/Zwemmer.

— 1980: Environment. *Science* **209**, 183–90.

WHITMORE, T. C. 1975: *Tropical rain forests of the Far East.* Oxford: Clarendon Press.

WHITTAKER, R. H. 1975: *Communities and ecosystems.* New York: Macmillan, 2nd edn.

WHITTAKER, R. H. and LIKENS, G. E. 1975: The biosphere and man. In H. Lieth and R. H. Whittaker (eds.) *Primary productivity of the biosphere.* Berlin, Heidelberg and New York: Springer-Verlag Ecological Studies, Vol. 14, 305–28.

WIEGERT, R. G. and EVANS, F. C. 1967: Investigations of secondary productivity in grasslands. In K. Petrusewicz (ed.) *Secondary productivity of terrestrial ecosystems.* Warsaw and Cracow: PWN, 499–518.

WIENER, A. 1972: The development of Israel's water resources. *American Scientist* **60**, 466–73.

WIENS, J. A. (ed.) 1971: *Ecosystem structure and function.* Corvallis: Oregon State UP.

WIGLEY, T. M. L., JONES, P. D. and KELLEY, P. M. 1980: Scenarios for a warm high-CO_2 world. *Nature, Lond.* **283**, 17–21.

WILBER, C. K. 1978: Population in western economic theory. *Populi* **5** (3), 14–29.

WILLARD, B. E. and MARR, J. W. 1970: Effects of human activities on alpine tundra ecosystems in Rocky Mountain National Park, Colorado. *Biol. Cons.* **2**, 257–65.

WILLIAMS, J. E. and ZELINSKY, W. 1970: On some patterns of international tourist flows. *Econ. Geogr.* 549–67.

WILSON, E. O. and WILLIS, E. O. 1975: Applied biogeography. In M. L. Cody and J. M. Diamond (eds.) *Ecology and evolution of communities.* Cambridge, Mass. and London: Belknap Press/Harvard Univ. Press, 522–34.

WINDHORST, H.-W. 1978: *Geografie der Wald- und Forstwirtschaft.* Stuttgart: B. G. Tenbuer.

WINDSOR, M. and COOPER, M. 1977: Farmed fish, cows and pigs. *New Sci.* **75**, 740–2.

WINSLOW, J. H. (ed.) 1977: *The Melanesian environment.* Canberra: ANU Press.

de WINTER, F. and COX, M. (eds.) 1979: *Sun—mankind's future source of energy.* Oxford and New York: Pergamon Press, 3 vols.

WITTWER, S. H. 1974: Maximum production capacity of food crops. *BioScience* **24**, 216–24.

— 1975: Food production: technology and the resource base. *Science* **188**, 579–84.

WOLFE, R. I. 1966: Recreational travel: the new migration. *Canadian Geographer*, **10**, 1–14.

WOLFERS, D. 1971: The case against zero growth. *In. J. Env. Studs.* **1**, 227–32.

WOLLMAN, N. 1960: *A preliminary report on the supply of and demand for water in the US as estimated for 1980 and 2000.* Washington, DC: US Senate Select Committee on Water Resources 86th Congress 2nd Session, Committee Report 32.

WOLMAN, A. 1965: The metabolism of cities. *Sci. Amer.* **218** (9), 179–90.

WOLMAN, M. G. 1971: The nation's rivers. *Science* **174**, 905–18.

WOODIS, J. 1971: An introduction to neo-colonialism. In R. Revelle *et al.* (eds.), 303–12.

WOODLEY, W. L., SIMPSON, J., BIONDINI, R., and BERKELEY, J. 1977: Rainfall results 1970–1975: Florida area cumulus experiment. *Science* **195**, 735–74.

WOODWELL, G. M. 1963: The ecological effects of radiation. *Sci. Amer.* **208** (6), 2–11.

— 1967a: Toxic substances and ecological cycles. *Sci. Amer.* **216** (3), 24–31.

— 1967b: Radiation and the pattern of nature. *Science* **156**, 461–70.

— 1970a: The energy cycle of the biosphere. In *Scientific American* (ed.), 25–35.

— 1970b: Effects of pollution on the structure and physiology of ecosystems. *Science* **168**, 429–33.

— 1974: Success, succession and Adam Smith. *BioScience* **24**, 81–87.

WOODWELL, G. M., CRAIG, P. P. and JOHNSON, H. A. 1971: DDT in the biosphere: where does it go? *Science* **174**: 1101–7.

WOODWELL, G. M. and SMITH, H. H. (eds.) 1969: *Diversity and stability in ecological systems.* Upton, NY: Brookhaven Symp. Biol. **22**.

WOODWELL, G. M., WURSTER, C. F. and ISAACSON, P. A. 1967: DDT residues in an east coast estuary. *Science* **156**, 821–4.

WORTMAN, S. 1980: World food and nutrition: the scientific and technological base. *Science* **209**, 157–64.

WURSTER, C. F. 1968: DDT reduces photosynthesis by marine phytoplankton. *Science* **159**, 1474–5.

— 1969: Chlorinated hydrocarbon insecticides and the world ecosystem. *Biol. Cons.* **1**, 123–9.

WYATT, T. 1976: Food chains in the sea. In D. H. Cushing and J. J. Walsh (eds.) *The ecology of the seas.* Oxford: Blackwell, 341–59.

WYNNE-EDWARDS, V. C. 1962: *Animal dispersal in relation to social behavior.* London and

Edinburgh: Oliver and Boyd. New York: Hafner.
YOUNG, E. 1975: Technological and economic aspects of game management and utilization in Africa. In R. L. Reid (ed.) *Proc. III World Conference on Animal Production*. Sydney: University Press, 132–41.
YOUNG, G. 1973: *Tourism: blessing or blight?* Harmondsworth: Penguin.
ZELINSKY, W. 1966: *A prologue to population geography*. New York: Prentice-Hall.
— 1970: Beyond the exponentials: the role of geography in the great transition. *Econ. Geogr*. **46**, 498–535.
ZELINSKY, W., KOSINSKI, L. A. and PROTHERO, R. M. (eds.) 1970: *Geography and a crowding world*. New York: Oxford University Press.
ZEUNER, F. E. 1964: *A history of domesticated animals*. London: Methuen.

Index